Urological Cancer 2021

Urological Cancer 2021

Editors
José I. López
Claudia Manini

MDPI • Basel • Beijing • Wuhan • Barcelona • Belgrade • Manchester • Tokyo • Cluj • Tianjin

Editors
José I. López
Cruces University Hospital
Barakaldo, Spain

Claudia Manini
San Giovanni Bosco Hospital
Turin, Italy

Editorial Office
MDPI
St. Alban-Anlage 66
4052 Basel, Switzerland

This is a reprint of articles from the Special Issue published online in the open access journal *Cancers* (ISSN 2072-6694) (available at: https://www.mdpi.com/journal/cancers/special_issues/_UC).

For citation purposes, cite each article independently as indicated on the article page online and as indicated below:

LastName, A.A.; LastName, B.B.; LastName, C.C. Article Title. *Journal Name* **Year**, *Volume Number*, Page Range.

ISBN 978-3-0365-3048-2 (Hbk)
ISBN 978-3-0365-3049-9 (PDF)

Cover image courtesy of José I. López

© 2022 by the authors. Articles in this book are Open Access and distributed under the Creative Commons Attribution (CC BY) license, which allows users to download, copy and build upon published articles, as long as the author and publisher are properly credited, which ensures maximum dissemination and a wider impact of our publications.

The book as a whole is distributed by MDPI under the terms and conditions of the Creative Commons license CC BY-NC-ND.

Contents

About the Editors ix

Estibaliz López-Fernández, Javier C. Angulo, José I. López and Claudia Manini
Urological Cancer Panorama in the Second Year of the COVID-19 Pandemic
Reprinted from: *Cancers* **2022**, *14*, 493, doi:10.3390/cancers14030493 1

Yusuke Sugino, Takeshi Sasaki, Manabu Kato, Satoru Masui, Kouhei Nishikawa, Takashi Okamoto, Shinya Kajiwara, Takuji Shibahara, Takehisa Onishi, Shiori Tanaka, Hideki Kanda, Hiroshi Matsuura and Takahiro Inoue
Prognostic Effect of Preoperative Psoas Muscle Hounsfield Unit at Radical Cystectomy for Bladder Cancer
Reprinted from: *Cancers* **2021**, *13*, 5629, doi:10.3390/cancers13225629 7

Javier González, Jeffrey J. Gaynor and Gaetano Ciancio
Renal Cell Carcinoma with or without Tumor Thrombus Invading the Liver, Pancreas and Duodenum
Reprinted from: *Cancers* **2021**, *13*, 1695, doi:10.3390/cancers13071695 17

Laure Grelier, Michael Baboudjian, Bastien Gondran-Tellier, Anne-Laure Couderc, Robin McManus, Jean-Laurent Deville, Ana Carballeira, Raphaelle Delonca, Veronique Delaporte, Laetitia Padovani, Romain Boissier, Eric Lechevallier and Xavier Muracciole
Stereotactic Body Radiotherapy for Frail Patients with Primary Renal Cell Carcinoma: Preliminary Results after 4 Years of Experience
Reprinted from: *Cancers* **2021**, *13*, 3129, doi:10.3390/cancers13133129 31

Eileen Shiuan, Anupama Reddy, Stephanie O. Dudzinski, Aaron R. Lim, Ayaka Sugiura, Rachel Hongo, Kirsten Young, Xian-De Liu, Christof C. Smith, Jamye O'Neal, Kimberly B. Dahlman, Renee McAlister, Beiru Chen, Kristen Ruma, Nathan Roscoe, Jehovana Bender, Joolz Ward, Ju Young Kim, Christine Vaupel, Jennifer Bordeaux, Shridar Ganesan, Tina M. Mayer, Gregory M. Riedlinger, Benjamin G. Vincent, Nancy B. Davis, Scott M. Haake, Jeffrey C. Rathmell, Eric Jonasch, Brian I. Rini, W. Kimryn Rathmell and Kathryn E. Beckermann
Clinical Features and Multiplatform Molecular Analysis Assist in Understanding Patient Response to Anti-PD-1/PD-L1 in Renal Cell Carcinoma
Reprinted from: *Cancers* **2021**, *13*, 1475, doi:10.3390/cancers13061475 39

Mathilda Jing Chow, Yan Gu, Lizhi He, Xiaozeng Lin, Ying Dong, Wenjuan Mei, Anil Kapoor and Damu Tang
Prognostic and Therapeutic Potential of the OIP5 Network in Papillary Renal Cell Carcinoma
Reprinted from: *Cancers* **2021**, *13*, 4483, doi:10.3390/cancers13174483 55

Annick Laruelle, Claudia Manini, Elena Iñarra and José I. López
Metastasis, an Example of Evolvability
Reprinted from: *Cancers* **2021**, *13*, 3653, doi:10.3390/cancers13153653 81

Aloÿse Fourquet, Lucien Lahmi, Timofei Rusu, Yazid Belkacemi, Gilles Créhange, Alexandre de la Taille, Georges Fournier, Olivier Cussenot and Mathieu Gauthé
Restaging the Biochemical Recurrence of Prostate Cancer with [^{68}Ga]Ga-PSMA-11 PET/CT: Diagnostic Performance and Impact on Patient Disease Management
Reprinted from: *Cancers* **2021**, *13*, 1594, doi:10.3390/cancers13071594 87

Umang Swami, Jennifer Anne Sinnott, Benjamin Haaland, Nicolas Sayegh, Taylor Ryan McFarland, Nishita Tripathi, Benjamin L. Maughan, Nityam Rathi, Deepika Sirohi, Roberto Nussenzveig, Manish Kohli, Sumanta K. Pal and Neeraj Agarwal
Treatment Pattern and Outcomes with Systemic Therapy in Men with Metastatic Prostate Cancer in the Real-World Patients in the United States
Reprinted from: *Cancers* **2021**, *13*, 4951, doi:10.3390/cancers13194951 . . . . **101**

Sazan Rasul, Tim Wollenweber, Lucia Zisser, Elisabeth Kretschmer-Chott, Bernhard Grubmüller, Gero Kramer, Shahrokh F. Shariat, Harald Eidherr, Markus Mitterhauser, Chrysoula Vraka, Werner Langsteger, Marcus Hacker and Alexander R. Haug
Response and Toxicity to the Second Course of 3 Cycles of ^{177}Lu-PSMA Therapy Every 4 Weeks in Patients with Metastatic Castration-Resistant Prostate Cancer
Reprinted from: *Cancers* **2021**, *13*, 2489, doi:10.3390/cancers13102489 . . . . **119**

Andrea Fulco, Francesco Chiaradia, Luigi Ascalone, Vincenzo Andracchio, Antonio Greco, Manlio Cappa, Marcello Scarcia, Giuseppe Mario Ludovico, Vincenzo Pagliarulo, Camillo Palmieri and Stefano Alba
Multiparametric Magnetic Resonance Imaging-Ultrasound Fusion Transperineal Prostate Biopsy: Diagnostic Accuracy from a Single Center Retrospective Study
Reprinted from: *Cancers* **2021**, *13*, 4833, doi:10.3390/cancers13194833 . . . . **131**

Zin W. Myint, Ramon C. Sun, Patrick J. Hensley, Andrew C. James, Peng Wang, Stephen E. Strup, Robert J. McDonald, Donglin Yan, William H. St. Clair and Derek B. Allison
Evaluation of Glutaminase Expression in Prostate Adenocarcinoma and Correlation with Clinicopathologic Parameters
Reprinted from: *Cancers* **2021**, *13*, 2157, doi:10.3390/cancers13092157 . . . . **139**

Shea P. Connell, Robert Mills, Hardev Pandha, Richard Morgan, Colin S. Cooper, Jeremy Clark, Daniel S. Brewer and The Movember GAP1 Urine Biomarker Consortium
Integration of Urinary EN2 Protein & Cell-Free RNA Data in the Development of a Multivariable Risk Model for the Detection of Prostate Cancer Prior to Biopsy
Reprinted from: *Cancers* **2021**, *13*, 2102, doi:10.3390/cancers13092102 . . . . **151**

Oscar Selvaggio, Ugo Giovanni Falagario, Salvatore Mariano Bruno, Marco Recchia, Maria Chiara Sighinolfi, Francesca Sanguedolce, Paola Milillo, Luca Macarini, Ardeshir R. Rastinehad, Rafael Sanchez-Salas, Eric Barret, Franco Lugnani, Bernardo Rocco, Luigi Cormio and Giuseppe Carrieri
Intraoperative Digital Analysis of Ablation Margins (DAAM) by Fluorescent Confocal Microscopy to Improve Partial Prostate Gland Cryoablation Outcomes
Reprinted from: *Cancers* **2021**, *13*, 4382, doi:10.3390/cancers13174382 . . . . **167**

Subas Neupane, Jaakko Nevalainen, Jani Raitanen, Kirsi Talala, Paula Kujala, Kimmo Taari, Teuvo L.J. Tammela, Ewout W. Steyerberg and Anssi Auvinen
Prognostic Index for Predicting Prostate Cancer Survival in a Randomized Screening Trial: Development and Validation
Reprinted from: *Cancers* **2021**, *13*, 435, doi:10.3390/cancers13030435 . . . . **175**

Takahiro Kimura, Shun Sato, Hiroyuki Takahashi and Shin Egawa
Global Trends of Latent Prostate Cancer in Autopsy Studies
Reprinted from: *Cancers* **2021**, *13*, 359, doi:10.3390/cancers13020359 . . . . **189**

Giovana de Godoy Fernandes, Bruna Pedrina, Patrícia de Faria Lainetti, Priscila Emiko Kobayashi, Verônica Mollica Govoni, Chiara Palmieri, Veridiana Maria Brianezi Dignani de Moura, Renée Laufer-Amorim and Carlos Eduardo Fonseca-Alves
Morphological and Molecular Characterization of Proliferative Inflammatory Atrophy in Canine Prostatic Samples
Reprinted from: *Cancers* **2021**, *13*, 1887, doi:10.3390/cancers13081887 . . . . **201**

Sara Elena Rebuzzi, Giuseppe Luigi Banna, Veronica Murianni, Alessandra Damassi, Emilio Francesco Giunta, Filippo Fraggetta, Ugo De Giorgi, Richard Cathomas, Pasquale Rescigno, Matteo Brunelli and Giuseppe Fornarini
Prognostic and Predictive Factors in Advanced Urothelial Carcinoma Treated with Immune Checkpoint Inhibitors: A Review of the Current Evidence
Reprinted from: *Cancers* **2021**, *13*, 5517, doi:10.3390/cancers13215517 . . . . **219**

Linda Silina, Fatlinda Maksut, Isabelle Bernard-Pierrot, François Radvanyi, Gilles Créhange, Frédérique Mégnin-Chanet and Pierre Verrelle
Review of Experimental Studies to Improve Radiotherapy Response in Bladder Cancer: Comments and Perspectives
Reprinted from: *Cancers* **2021**, *13*, 87, doi:10.3390/cancers13010087 . . . . **237**

Makito Miyake, Nobutaka Nishimura, Kota Iida, Tomomi Fujii, Ryoma Nishikawa, Shogo Teraoka, Atsushi Takenaka, Hiroshi Kikuchi, Takashige Abe, Nobuo Shinohara, Eijiro Okajima, Takuto Shimizu, Shunta Hori, Norihiko Tsuchiya, Takuya Owari, Yasukiyo Murakami, Rikiya Taoka, Takashi Kobayashi, Takahiro Kojima, Naotaka Nishiyama, Hiroshi Kitamura, Hiroyuki Nishiyama and Kiyohide Fujimoto
Intravesical Bacillus Calmette–Guérin Treatment for T1 High-Grade Non-Muscle Invasive Bladder Cancer with Divergent Differentiation or Variant Morphologies
Reprinted from: *Cancers* **2021**, *13*, 2615, doi:10.3390/cancers13112615 . . . . **259**

Dai Koguchi, Kazumasa Matsumoto, Yuriko Shimizu, Momoko Kobayashi, Shuhei Hirano, Masaomi Ikeda, Yuichi Sato and Masatsugu Iwamura
Prognostic Impact of AHNAK2 Expression in Patients Treated with Radical Cystectomy
Reprinted from: *Cancers* **2021**, *13*, 1748, doi:10.3390/cancers13081748 . . . . **273**

Chin-Li Chen, En Meng, Sheng-Tang Wu, Hsing-Fan Lai, Yi-Shan Lu, Ming-Hsin Yang, Chih-Wei Tsao, Chien-Chang Kao and Yi-Lin Chiu
Targeting *S1PR1* May Result in Enhanced Migration of Cancer Cells in Bladder Carcinoma
Reprinted from: *Cancers* **2021**, *13*, 4474, doi:10.3390/cancers13174474 . . . . **287**

Soichiro Shimura, Kazumasa Matsumoto, Yuriko Shimizu, Kohei Mochizuki, Yutaka Shiono, Shuhei Hirano, Dai Koguchi, Masaomi Ikeda, Yuichi Sato and Masatsugu Iwamura
Serum Epiplakin Might Be a Potential Serodiagnostic Biomarker for Bladder Cancer
Reprinted from: *Cancers* **2021**, *13*, 5150, doi:10.3390/cancers13205150 . . . . **307**

About the Editors

José I. López

José I. López is head of Department of Pathology at the Cruces University Hospital and principal investigator of the Biomarker in Cancer Unit at the Biocruces-Bizkaia Health Research Institute. He graduated at the Faculty of Medicine, University of the Basque Country, Leioa, Spain, and trained in Pathology at the Hospital Universitario 12 de Octubre, Madrid, Spain. He received his PhD degree at the Universidad Complutense of Madrid, Spain. Dr. López has served as pathologist for more than 30 years in several hospitals in Spain, and is subspecialized in Uropathology, where he has published more than 190 peer-reviewed articles and reviews. Dr. López is interested in translational uropathology in general and in renal cancer in particular, and collaborates with several international research groups unveiling the genomic landscape of urological cancer. Intratumor heterogeneity, tumor sampling, tumor microenvironment, tumor ecology, immunotherapy, and basic mechanisms of carcinogenesis are his main topics of interest.

Claudia Manini

Claudia Manini is head of the Department of Pathology at the San Giovanni Bosco Hospital in Turin, Italy. She graduated from the Faculty of Medicine and Surgery and post-graduated in Surgical Pathology at the University of Turin, Turin, Italy. Dr. Manini has served as pathologist for more than 25 years in several hospitals in Italy developing an expertise in diagnostic uropathology, neuropathology and gynecopathology. She is also affiliated to the Department of Public Health and Pediatric Sciences at the University of Turin. Her main interest is translational pathology.

Editorial

Urological Cancer Panorama in the Second Year of the COVID-19 Pandemic

Estibaliz López-Fernández [1,2], Javier C. Angulo [3,4], José I. López [5,6,*] and Claudia Manini [7,8,*]

1. FISABIO Foundation, 46020 Valencia, Spain; estibaliz.lopez@universidadeuropea.es
2. Faculty of Health Sciences, European University of Valencia, 46023 Valencia, Spain
3. Clinical Department, Faculty of Medical Sciences, European University of Madrid, 28005 Madrid, Spain; javier.angulo@universidadeuropea.es
4. Department of Urology, University Hospital of Getafe, 28907 Madrid, Spain
5. Department of Pathology, Cruces University Hospital, 48903 Barakaldo, Spain
6. Biocruces-Bizkaia Health Research Institute, 48903 Barakaldo, Spain
7. Department of Pathology, San Giovanni Bosco Hospital, 10154 Turin, Italy
8. Department of Public and Pediatric Health Sciences, University of Turin, 10124 Turin, Italy
* Correspondence: joseignacio.lopez@osakidetza.eus (J.I.L.); claudia.manini@aslcittaditorino.it (C.M.); Tel.: +34-946006336 (J.I.L.); +39-0112402737 (C.M.)

A total of 22 contributions conforms this Special Issue that covers a wide spectrum of contemporary issues in urological cancer, a group of neoplasms with high incidence, prevalence, and mortality rates, especially in the male population of Western countries [1]. Renal cancer, with five contributions (three articles, one communication, and one perspective), prostate cancer and allied conditions, with ten (eight articles, one communication, and one review), and urinary bladder, with seven (five articles and two reviews), provide a comprehensive panorama of what has occurred in Urologic Oncology during 2021, the second year of the COVID-19 pandemic. Most papers in this collection deal with different aspects of cancer therapy, but diagnostic and prognostic subjects, migration-related topics, reviews, and preneoplastic lesions are also considered. Overall, this translational compilation of viewpoints around urological cancer will enrich multidisciplinary interactions for patient benefit.

1. Renal Cancer

Three papers analyze three different approaches to face against renal cancer, that is, surgery, radiotherapy, and immune checkpoint inhibition, thus reflecting the broad diversity of therapeutic possibilities available in this complex disease. In a context of different treatments, a precise definition of strict criteria for a rationale patient selection would be advisable.

Gonzalez et al. [2] analyze the post-surgical complications and survival benefit of radical surgery in a series of 18 locally advanced clear cell renal cell carcinomas (CCRCC) invading the inferior vena cava, pancreas, duodenum, and liver, establishing the technical feasibility of a radical resection that includes complex procedures derived from transplant surgery. A deep surgical experience seems, however, mandatory for a successful implementation of this therapeutic option. The usefulness of the stereotactic body radiotherapy as an alternative treatment to surgery has been explored by Grelier et al. [3]. Twenty-three patients have been treated with this option during a 4-year period. All the cases had a biopsy for pathological confirmation or renal cell carcinoma (RCC) prior to the procedure. All were organ-confined tumors (T1/2) and, as expected, CCRCC was the most frequently found neoplasm (73.9%). The dose administered oscillated between 24 Gy and 35 Gy and fractions from 3 and 5. The obtained results indicate that this technique could be seriously considered as a promising therapeutic alternative for patients with poor physical health. However, further studies are necessary to support the exact role of this treatment in RCC.

PD-1/PD-L1 axis blockade drugs, alone or in combination with other drugs, are being introduced in the therapeutic armamentarium in several neoplasms, advanced CCRCC included. However, its benefit to patients is far from being generalized and some controversies still exist in patient selection [4]. Shiuan et al. [5] have analyzed the patient characteristics, clinical correlates, and molecular parameters (multiplexed immunofluorescence, whole exome sequencing, T-cell receptor sequencing, RNA sequencing) in a series of RCC (clear cell, papillary, sarcomatoid, chromophobe, and undifferentiated) treated with nivolumab (84%) and atezolizumab (16%) in first (8.5%), second (29.8%), third (34%) and fourth (27.7%) lines. The authors conclude that PD-L1 immunostaining alone does not provide enough information to predict response in these patients.

Chow et al. [6] have analyzed the role of OPA interacting protein 5 (OIP5) in 20 papillary renal cell carcinomas (PRCC) and 37 (CCRCC). OIP5 up-regulation has been associated with biological aggressiveness in a broad spectrum of malignant neoplasms, PRCC included. For such a purpose, the authors have constructed a 66-gene multigene panel (Overlap66), including PLK1 gene, which effectively stratifies high-risk PRCC thus allowing to treat them with PLK1 inhibitors. The authors conclude that Overlap66 analysis and PLK1 inhibitors should be added to the list of personalized therapies in PRCC.

In a perspective, Laruelle et al. [7] analyze the metastatic process of CCRCC as an example of tumor evolvability, where a sociological perspective to the problem provided by Game Theory may shed some additional light to our knowledge of cancer evolutionary mechanisms. The authors hypothesize that the development of metastases in malignant tumors respond to the necessity of a subset of tumor cells to achieve Nash equilibria and friendlier environments far away from the primary tumor.

2. Prostate Cancer

Three out of 10 papers in this section deal with advanced prostate cancer. Fourquet et al. [8] evaluate the role of [68Ga]Ga-PSMA-11 PET/CT in predicting therapy efficacy, diagnostic usefulness, and patient management in 294 patients with biochemical recurrence. The authors conclude that this technique shows a high performance in locating prostate cancer recurrence sites. Additionally, [68Ga]Ga-PSMA-11 PET/CT impacts therapeutic management in two out of three patients studied. The use of PSMA radioligands with PET/CT should be considered as a first line imaging in patients with biochemical recurrence. Patient outcomes and hormonal therapy patterns have been retrospectively studied in a large observational real-world database of patients with metastatic prostate cancer in the USA by Swami et al. [9]. The authors conclude that novel hormonal therapies and doxetacel were underutilized in the series analyzed. In the third contribution within this subheading, Rasul et al. [10] analyze the response and toxicity of three cycles of 177Lu-PSMA in patients with castration-resistant prostate cancers. The authors conclude on their study that an intensive PSMA-radioligand therapy is well tolerated by patients with metastatic castration-resistant prostate cancers and is associated with promising overall survivals.

Four contributions analyze different aspects of the diagnostic process in prostate cancer. For example, Fulco et al. [11] evaluate the diagnostic accuracy of multiparametric magnetic resonance imaging-ultrasound fusion trans-perineal prostate biopsy. This single center retrospective study of 272 patients shows that a total of 16.7% of clinically significant tumors would have been undetected with standard biopsy methods. The authors conclude that the combined targeted and standard biopsy methodology is advisable to minimize the risk of missing clinically significant prostate cancer. Myint et al. [12] analyze the immunohistochemical expression of glutaminase in a series of 154 cancer and 41 benign prostate samples. RNA-Seq data of 246 prostate cancer samples are also obtained from The Cancer Genome Atlas. The authors conclude that although glutaminase expression is higher in prostate cancer than in benign prostate tissue this difference does not seem to be statistically significant. Connell et al. [13] develop a multivariable risk model termed ExoGrail for the non-invasive detection of prostate cancer prior to biopsy. They have observed that this model is able to reduce the number of unnecessary biopsies by 35%

when compared with current standards of care. Selvaggio et al. [14] use a fluorescent confocal microscope to intraoperatively analyze the ablation margins in 10 patients to improve partial prostate gland cryoablation outcomes. They conclude that this technique is feasible and reliable to reduce disease recurrence and functional complications of focal therapy in prostate cancer.

Neupane et al. [15] have developed a full prognostic index for predicting the survival of localized prostate cancer up to 15 to 20 years using a multivariable complementary log-log regression model. Age at diagnosis, trial arm, PSA at diagnosis, European Association Urology (EAU) risk group, treatment modality, mode of detection, and biochemical recurrence are taken into account in the study but not co-morbidity as it has not got a great impact on prostate cancer-specific survival. A simplified risk score tool has also been developed for early diagnosis and to predict survival at 10 years considering the three parameters commonly used in daily clinical practice (age, PSA, and EAU risk group). Both the full and simplified prognostic index have shown a superior performance than the EAU risk group, being the latest also more accurate in risk estimation than D'Amico risk classification and the cancer of prostate risk assessment (CAPRA) risk score. The authors conclude that further validations of both prognostic indexes are needed.

Kimura et al. [16] review the global trends of latent prostate cancer in Western and Asian countries considering variables such as step-sectioning versus random and/or single-sectioning, thickness, age, and race. The increased prevalence observed over time in Asia, as compared to the stable numbers among Western countries, could be explained due to PSA screening strategies and changes to Westernized lifestyles. The authors conclude that it is mandatory to agree in the diagnostic methods and molecular analyses to obtain homogeneous data that could even reconsider the nowadays definition of latent prostate cancer.

De Godoy Fernandes et al. [17] analyze the proliferative inflammatory atrophy in canine prostatic samples as a preneoplastic lesion with potential progression to prostate cancer. Proliferative atrophy shows a high proliferative rate in concordance with the overexpression observed in Ki67, CK5, high molecular weight cytokeratins, and p63. On the other hand, p53 and MDM2 are not deregulated as they are in advanced stages of prostatic carcinogenesis. Androgen receptor and PTEN are both downregulated activating the anti-apoptotic pathway. The authors conclude that high proliferative indexes and low levels of androgen receptor and PTEN are useful biomarkers to predict potential preneoplastic lesions even though more studies should be conducted.

3. Bladder Cancer

Two out of seven papers in this section are reviews. Rebuzzi et al. [18] evaluate the prognostic and predictive factors of advanced urothelial carcinoma treated with immune checkpoints inhibitors (ICIs), with the intention to identify biomarkers to select the patients at higher chances of responding to ICI treatment that may merit further prospective investigation. Their main conclusions are that current evidence on the predictive and prognostic value of PD-L1 expression is limited by the different assays used for each anti-PD1 or anti-PD-L1 agent in the clinical trials evaluated, and thereof their clinical value remains inconclusive. On the other hand, the authors position towards the value of sequencing of the tumor fraction of the cell-free DNA (ctDNA) is an interesting method to detect residual disease and anticipate disease relapse after treatment using different biomarkers that include FGFR3, XPD, HER2, and TMB. What is more, the serial monitoring of ctDNA as a tumor tissue surrogate could be of value to stratify metastatic urothelial carcinoma and could act as a treatment response marker [18,19].

Silina et al. [20] reviews bladder sparing strategies with the use of radiotherapy and chemotherapy combined, a very interesting therapeutic approach for invasive bladder cancer, that is emerging also as a therapeutic possibility in the context of medical system collapse for scheduled surgeries suffered by the COVID-19 health crisis. Based on the issue that strategies to radio-sensitize tumors and spare normal bladder tissue to improve treatment safety, a systematic search and literature review is performed. The conclusion is that a long way remains ahead before experimental research with cell lines and animal models regarding the combination of radiation with different agents can be optimized.

Five additional articles present original investigations regarding invasive bladder cancer. Miyake et al. [21] present a retrospective study of a very large series of patients with T1 high-grade urothelial carcinoma and evaluate the prognostic role of divergent differentiation and variant morphologies when Bacillus Calmette-Guérin (BCG) endovesical treatment is used. This topic presents the clinical implications of urothelial cancer heterogeneity and is of particular interest in this era of BCG shortage, where new treat modalities such as chemo-hyperthermia to optimize adjuvant treatment after transurethral resection are being investigated [22,23] or even early cystectomy [24] are controversial. With the limitations of the study design using inverse probability of treatment weighting analysis, the authors found that not only recurrence-free or progression-free rates are affected, but also cancer-specific mortality was affected by variant morphologies (nested, microcystic, micropapillary, lympho-epithelioma-like, plasmacytoid, giant cell, sarcomatoid, lipid-rich and clear cell variants), but not for urothelial cancer with divergent (squamous, glandular and trophoblastic) differentiation. This analysis is very interesting to optimize BCG therapy and promote a radical treatment once variant morphologies are identified.

The remining four original articles on invasive bladder cancer in the Issue deal the eternal dilemma to find the optimal prognostic marker in patients with invasive bladder cancer treated with radical cystectomy. It is still surprising that despite decades of intensive research the optimal tissue markers for urothelial cancer remains undetermined. Recent developments regarding basal/luminal phenotype markers are consolidating to predict disease prognosis, but other new immunohistochemical evaluations such as fibroblast activating protein (FAP) or pro-renin receptor (PRR) are also newly discovered [25,26].

Two very interesting novel pathways and their prognostic implications are investigated in the Special Issue. Koguchi et al. 2021 [27] evaluate the expression of AHNAK2 (AHNAK Nucleoprotein 2), a protein coding gene, that has been identified as a possible tumor marker and therapeutic target in different malignancies including clear cell renal cell carcinoma, papillary thyroid carcinoma, lung and pancreatic duct adenocarcinoma, and others. Their study of 120 patients with bladder cancer undergoing radical cystectomy finds an association between high AHNAK2 expression and classical aggressive pathologic findings, but also as an independent prognostic marker for recurrence-free survival and disease-specific mortality in this set of patients [27]. Chen et al. [28] investigate S1PR1 expression, a G protein-coupled receptor in vascular endothelial and immune cells. Their experimental cell-line and human tissue approach using transurethral resection specimens, and also the search in gene expression database collection NCBI-GEO, has evidenced that high S1PR1 expression in neoplasia inversely correlates with cell motility; thus, targeting *S1PR1* may result in the enhanced migration of bladder cancer cells. The implications of this finding merit further research to inhibit the progression of metastases.

The development of bladder cancer diagnostic markers is an unmet need in current Urologic Oncology. Shimura et al. [29] use a micro-dot blot array to evaluate EPPK1 (epiplakin) expression in patients with bladder neoplasia and controls. Their results are very stimulating because this cytoskeletal linker protein, that connects to intermediate filaments and controls their reorganization, is a promising molecule to act as diagnostic biomarkers for patients with urothelial carcinoma of the bladder. As revealed, the area under the curve is 78%. Immunohistochemical evaluation in a series of 127 patients with a bladder tumor treated with radical cystectomy did not confirm the role of this marker with cancer prognosis. Of course, validation and large-scale studies are needed to confirm these very promising results.

Finally, Sugino et al. [30] have evaluated another impressive prognosticator for bladder cancer, based on imaging studies alone. The preoperative non-contrast CT-scan evaluation axial image at the third lumbar vertebral level of psoas muscle Hounsfield unit before radical cystectomy correlates with the clinical prognosis. This factor, assessed in 177 consecutive surgically treated patients, behaves as a major predictor of overall survival, together with patient age, patient sex, and clinical staging of the disease. Of course, this is not a specific tumor marker as it probably reflects patient frailty before surgery. Tentatively, this prognostic factor can be used to predict overall survival, regardless of the condition of bladder cancer itself. Therefore, its use in Urologic Oncology can be deemed spurious. However, based on the fact that imaging modalities for preoperative assessment of patients undergoing cystectomy are universally used, its potential to be incorporated in preoperative assessment tools merits validation.

4. Conclusions

Patients with urological cancer are at a higher risk to be more severely affected by the infection than the general population due to their inherent immunosuppression status either related directly to the disease or secondarily induced by chemotherapy, radiotherapy, immune checkpoint inhibition, or other oncologic treatments. The global shut-down provoked by the SARS-CoV-2 pandemic in 2020 forced the urologists to reorganize their routine activity prioritizing the surgical activity and deferring non-urgent pathologies in the wake of the limited operating room capacities and hospital beds occupied by, or reserved for, the avalanche of patients with severe respiratory symptoms needing urgent medical care. Unfortunately, the high mutational capacity of the virus makes the resolution of this global problem a pending issue.

Author Contributions: E.L.-F., J.C.A., J.I.L. and C.M. conceived, designed, and wrote the manuscript. All authors have read and agreed to the published version of the manuscript.

Funding: The authors received no external funding.

Conflicts of Interest: The authors declare no conflict of interest.

References

1. Siegel, R.L.; Miller, K.D.; Fuchs, H.E.; Jemal, A. Cancer statistics, 2021. *CA Cancer J. Clin.* **2021**, *71*, 7–33. [CrossRef] [PubMed]
2. González, J.; Gaynor, J.; Ciancio, G. Renal cell carcinoma with or without tumor thrombus invading the liver, pancreas and duodenum. *Cancers* **2021**, *13*, 1695. [CrossRef]
3. Grelier, L.; Baboudjian, M.; Gondran-Tellier, B.; Couderc, A.; McManus, R.; Deville, J.; Carballeira, A.; Delonca, R.; Delaporte, V.; Padovani, L.; et al. Stereotactic body radiotherapy for frail patients with primary renal cell carcinoma: Preliminary results after 4 years of experience. *Cancers* **2021**, *13*, 3129. [CrossRef] [PubMed]
4. Nunes-Xavier, C.E.; Angulo, J.C.; Pulido, R.; López, J.I. A critical insight into the clinical translation of PD-1/PD-L1 blockade therapy in clear cell renal cell carcinoma. *Curr. Urol. Rep.* **2019**, *20*, 1. [CrossRef] [PubMed]
5. Shiuan, E.; Reddy, A.; Dudzinski, S.; Lim, A.; Sugiura, A.; Hongo, R.; Young, K.; Liu, X.; Smith, C.; O'Neal, J.; et al. Clinical features and multiplatform molecular analysis assist in understanding patient response to anti-PD-1/PD-L1 in renal cell carcinoma. *Cancers* **2021**, *13*, 1475. [CrossRef] [PubMed]
6. Chow, M.; Gu, Y.; He, L.; Lin, X.; Dong, Y.; Mei, W.; Kapoor, A.; Tang, D. Prognostic and therapeutic potential of the OIP5 network in papillary renal cell carcinoma. *Cancers* **2021**, *13*, 4483. [CrossRef] [PubMed]
7. Laruelle, A.; Manini, C.; Iñarra, E.; López, J. Metastasis, an example of evolvability. *Cancers* **2021**, *13*, 3653. [CrossRef] [PubMed]
8. Fourquet, A.; Lahmi, L.; Rusu, T.; Belkacemi, Y.; Créhange, G.; de la Taille, A.; Fournier, G.; Cussenot, O.; Gauthé, M. Restaging the biochemical recurrence of prostate cancer with [68Ga]Ga-PSMA-11 PET/CT: Diagnostic performance and impact on patient disease management. *Cancers* **2021**, *13*, 1594. [CrossRef]
9. Swami, U.; Sinnott, J.; Haaland, B.; Sayegh, N.; McFarland, T.; Tripathi, N.; Maughan, B.; Rathi, N.; Sirohi, D.; Nussenzveig, R.; et al. Treatment pattern and outcomes with systemic therapy in men with metastatic prostate cancer in the real-world patients in the United States. *Cancers* **2021**, *13*, 4951. [CrossRef] [PubMed]
10. Rasul, S.; Wollenweber, T.; Zisser, L.; Kretschmer-Chott, E.; Grubmüller, B.; Kramer, G.; Shariat, S.; Eidherr, H.; Mitterhauser, M.; Vraka, C.; et al. Response and toxicity to the second course of 3 cycles of 177Lu-PSMA therapy every 4 weeks in patients with metastatic castration-resistant prostate cancer. *Cancers* **2021**, *13*, 2489. [CrossRef]
11. Fulco, A.; Chiaradia, F.; Ascalone, L.; Andracchio, V.; Greco, A.; Cappa, M.; Scarcia, M.; Ludovico, G.; Pagliarulo, V.; Palmieri, C.; et al. Multiparametric magnetic resonance imaging-ultrasound fusion transperineal prostate biopsy: Diagnostic accuracy from a single center retrospective study. *Cancers* **2021**, *13*, 4833. [CrossRef]
12. Myint, Z.; Sun, R.; Hensley, P.; James, A.; Wang, P.; Strup, S.; McDonald, R.; Yan, D.; Clair, W.H.S.; Allison, D. Evaluation of glutaminase expression in prostate adenocarcinoma and correlation with clinicopathologic parameters. *Cancers* **2021**, *13*, 2157. [CrossRef] [PubMed]

13. Connell, S.; Mills, R.; Pandha, H.; Morgan, R.; Cooper, C.; Clark, J.; Brewer, D. Integration of urinary EN2 protein & cell-free RNA data in the development of a multivariable risk model for the detection of prostate cancer prior to biopsy. *Cancers* **2021**, *13*, 2102. [CrossRef]
14. Selvaggio, O.; Falagario, U.; Bruno, S.; Recchia, M.; Sighinolfi, M.; Sanguedolce, F.; Milillo, P.; Macarini, L.; Rastinehad, A.; Sanchez-Salas, R.; et al. Intraoperative digital analysis of ablation margins (DAAM) by fluorescent confocal microscopy to improve partial prostate gland cryoablation outcomes. *Cancers* **2021**, *13*, 4382. [CrossRef]
15. Neupane, S.; Nevalainen, J.; Raitanen, J.; Talala, K.; Kujala, P.; Taari, K.; Tammela, T.; Steyerberg, E.; Auvinen, A. Prognostic index for predicting prostate cancer survival in a randomized screening trial: Development and validation. *Cancers* **2021**, *13*, 435. [CrossRef]
16. Kimura, T.; Sato, S.; Takahashi, H.; Egawa, S. Global trends of latent prostate cancer in autopsy studies. *Cancers* **2021**, *13*, 359. [CrossRef]
17. de Godoy Fernandes, G.; Pedrina, B.; de Faria Lainetti, P.; Kobayashi, P.; Govoni, V.; Palmieri, C.; de Moura, V.; Laufer-Amorim, R.; Fonseca-Alves, C. Morphological and molecular characterization of proliferative inflammatory atrophy in canine prostatic samples. *Cancers* **2021**, *13*, 1887. [CrossRef]
18. Rebuzzi, S.; Banna, G.; Murianni, V.; Damassi, A.; Giunta, E.; Fraggetta, F.; De Giorgi, U.; Cathomas, R.; Rescigno, P.; Brunelli, M.; et al. Prognostic and predictive factors in advanced urothelial carcinoma treated with immune checkpoint inhibitors: A review of the current evidence. *Cancers* **2021**, *13*, 5517. [CrossRef]
19. Vandekerkhove, G.; Lavoie, J.-M.; Annala, M.; Murtha, A.J.; Sundahl, N.; Walz, S.; Sano, T.; Taavitsainen, S.; Ritch, E.; Fazli, L.; et al. Plasma ctDNA is a tumor tissue surrogate and enables clinical-genomic stratification of metastatic bladder cancer. *Nat. Commun.* **2021**, *12*, 184. [CrossRef] [PubMed]
20. Silina, L.; Maksut, F.; Bernard-Pierrot, I.; Radvanyi, F.; Créhange, G.; Mégnin-Chanet, F.; Verrelle, P. Review of experimental studies to improve radiotherapy response in bladder cancer: Comments and perspectives. *Cancers* **2021**, *13*, 87. [CrossRef] [PubMed]
21. Miyake, M.; Nishimura, N.; Iida, K.; Fujii, T.; Nishikawa, R.; Teraoka, S.; Takenaka, A.; Kikuchi, H.; Abe, T.; Shinohara, N.; et al. Intravesical bacillus Calmette–Guérin treatment for T1 high-grade non-muscle invasive bladder cancer with divergent differentiation or variant morphologies. *Cancers* **2021**, *13*, 2615. [CrossRef]
22. Zhao, H.; Chan, V.W.; Castellani, D.; Chan, E.O.; Ong, W.L.K.; Peng, Q.; Moschini, M.; Krajewski, W.; Pradere, B.; Ng, C.F.; et al. Intravesical chemohyperthermia vs. Bacillus Calmette-Guerin instillation for intermediate- and high-risk non-muscle invasive bladder cancer: A systematic review and meta-analysis. *Front. Surg.* **2021**, *8*, 775527. [CrossRef]
23. Plata, A.; Guerrero-Ramos, F.; Garcia, C.; González-Díaz, A.; Gonzalez-Valcárcel, I.; de la Morena, J.M.; Díaz-Goizueta, F.J.; Del Álamo, J.F.; Gonzalo, V.; Montero, J.; et al. Long-term experience with hyperthermic chemotherapy (HIVEC) using Mitomycin-C in patients with non-muscle invasive bladder cancer in Spain. *J. Clin. Med.* **2021**, *10*, 5105. [CrossRef] [PubMed]
24. Nishimura, N.; Miyake, M.; Iida, K.; Miyamoto, T.; Tomida, R.; Numakura, K.; Inokuchi, J.; Yoneyama, T.; Matsumura, Y.; Yajima, S.; et al. Prognostication in Japanese patients with Bacillus Calmette-Guerin-unresponsive non-muscle-invasive bladder cancer undergoing early radical cystectomy. *Int. J. Urol.* **2021**. [CrossRef]
25. Calvete-Candenas, J.; Larrinaga, G.; Errarte, P.; Martín, A.M.; Dotor, A.; Esquinas, C.; Nunes-Xavier, C.E.; Pulido, R.; López, J.I.; Angulo, J.C. The coexpression of fibroblast activation protein (FAP) and basal-type markers (CK 5/6 and CD44) predicts prognosis in high-grade invasive urothelial carcinoma of the bladder. *Hum. Pathol.* **2019**, *91*, 61–68. [CrossRef]
26. Larrinaga, G.; Calvete-Candenas, J.; Solano-Iturri, J.D.; Martín, A.M.; Pueyo, A.; Nunes-Xavier, C.E.; Pulido, R.; Dorado, J.F.; López, J.I.; Angulo, J.C. (Pro)renin Receptor is a novel independent prognostic marker in invasive urothelial carcinoma of the bladder. *Cancers* **2021**, *13*, 5642. [CrossRef] [PubMed]
27. Koguchi, D.; Matsumoto, K.; Shimizu, Y.; Kobayashi, M.; Hirano, S.; Ikeda, M.; Sato, Y.; Iwamura, M. Prognostic impact of AHNAK2 expression in patients treated with radical cystectomy. *Cancers* **2021**, *13*, 1748. [CrossRef]
28. Chen, C.; Meng, E.; Wu, S.; Lai, H.; Lu, Y.; Yang, M.; Tsao, C.; Kao, C.; Chiu, Y. Targeting S1PR1 may result in enhanced migration of cancer cells in bladder carcinoma. *Cancers* **2021**, *13*, 4474. [CrossRef] [PubMed]
29. Shimura, S.; Matsumoto, K.; Shimizu, Y.; Mochizuki, K.; Shiono, Y.; Hirano, S.; Koguchi, D.; Ikeda, M.; Sato, Y.; Iwamura, M. Serum epiplakin might be a potential serodiagnostic biomarker for bladder cancer. *Cancers* **2021**, *13*, 5150. [CrossRef]
30. Sugino, Y.; Sasaki, T.; Kato, M.; Masui, S.; Nishikawa, K.; Okamoto, T.; Kajiwara, S.; Shibahara, T.; Onishi, T.; Tanaka, S.; et al. Prognostic effect of preoperative psoas muscle Hounsfield Unit at radical cystectomy for bladder cancer. *Cancers* **2021**, *13*, 5629. [CrossRef]

Article

Prognostic Effect of Preoperative Psoas Muscle Hounsfield Unit at Radical Cystectomy for Bladder Cancer

Yusuke Sugino [1], Takeshi Sasaki [1], Manabu Kato [1], Satoru Masui [1], Kouhei Nishikawa [1], Takashi Okamoto [2], Shinya Kajiwara [2], Takuji Shibahara [2], Takehisa Onishi [2], Shiori Tanaka [3], Hideki Kanda [3], Hiroshi Matsuura [3] and Takahiro Inoue [1,*]

[1] Department of Nephro-Urologic Surgery and Andrology, Mie University Graduate School of Medicine, 2-174 Edobashi, Tsu 514-8507, Mie, Japan; y-sugino@med.mie-u.ac.jp (Y.S.); t-sasaki@med.mie-u.ac.jp (T.S.); katouro@med.mie-u.ac.jp (M.K.); s-masui@med.mie-u.ac.jp (S.M.); kouheini@med.mie-u.ac.jp (K.N.)
[2] Department of Urology, Ise Red Cross Hospital, 1-471-2 Funae, Ise 516-8512, Mie, Japan; t.okamoto555.medic@gmail.com (T.O.); shinya6226@gmail.com (S.K.); shibaharauro@chive.ocn.ne.jp (T.S.); takehisa@ise.jrc.or.jp (T.O.)
[3] Mie Prefectural General Medical Center, Department of Urology, 5450-132 Hinaga, Yokkaichi 510-8561, Mie, Japan; t-shiori@med.mie-u.ac.jp (S.T.); hideki-kanda@mie-gmc.jp (H.K.); hiroshi-matsuura@mie-gmc.jp (H.M.)
* Correspondence: tinoue28@med.mie-u.ac.jp; Tel.: +81-5-9231-5026

Simple Summary: Radical cystectomy (RC) is the standard treatment for patients with advanced bladder cancer. Since RC is a highly invasive procedure, it is necessary to carefully predict the prognosis before surgery and to determine the surgical indication. According to the results of the retrospective analysis of our 177 RC cases, we found the Hounsfield units of the psoas muscle at the third lumbar vertebral level to be a prognostic factor. Univariate and multivariate analyses revealed that age, sex, clinical T stage, and psoas muscle Hounsfield units were significant preoperative factors for overall survival. Furthermore, risk classification using these four factors was useful for predicting the prognosis of patients with RC.

Abstract: Radical cystectomy (RC) is the standard treatment for patients with advanced bladder cancer. Since RC is a highly invasive procedure, the surgical indications in an aging society must be carefully judged. In recent years, the concept of "frailty" has been attracting attention as a term used to describe fragility due to aging. We focused on the psoas muscle Hounsfield unit (PMHU) and analyzed its appropriateness as a prognostic factor together with other clinical factors in patients after RC. We retrospectively analyzed the preoperative prognostic factors in 177 patients with bladder cancer who underwent RC between 2008 and 2020. Preoperative non-contrast computed tomography axial image at the third lumbar vertebral level was used to measure the mean Hounsfield unit (HU) and cross-sectional area (mm^2) of the psoas muscle. Univariate analysis showed significant differences in age, sex, clinical T stage, and PMHU. In multivariate analysis using the Cox proportional hazards model, age (hazard ratio (HR) = 1.734), sex (HR = 2.116), cT stage (HR = 1.665), and PMHU (HR = 1.758) were significant predictors for overall survival. Furthermore, using these four predictors, it was possible to stratify the prognosis of patients after RC. Finally, PMHU was useful as a simple and significant preoperative factor that correlated with prognosis after RC.

Keywords: bladder cancer; urothelial carcinoma; radical cystectomy; frailty; prognostic factor; psoas muscle; Hounsfield units

1. Introduction

Radical cystectomy (RC) is the gold standard treatment for patients with muscle invasive bladder cancer, patients with selected T1 high-grade non-muscle invasive bladder cancer, and patients with carcinoma in situ resistant to Bacillus Calmette–Guérin treatment [1]. RC remains one of the most invasive urological procedures, and its surgical indication needs to be carefully assessed in an aging society. According to the "Annual Report on the Aging Society 2020" from the Cabinet Office of Japan, the total population of Japan is 126.17 million as of 2019, of which 35.89 million are aged 65 years or older. Japan is facing an aging society ahead of the rest of the world. A systematic review has reported that perioperative mortality within 90 days after RC significantly increases in the elderly, and overall survival (OS) and cancer-specific survival (CSS) also decrease with age [2]. Aging is clearly a risk factor for RC. In recent years, the concept of "frailty" has been attracting attention as a term that expresses fragility due to aging [3]. If the prognosis can be predicted more accurately by assessing not only the clinical stage and age of the patient but also malnutrition and muscle weakness associated with decreased physical activity, it can help in deciding whether to perform surgery. Although few established definitions with regard to the elderly or frailty have been reported, there have been some attempts to define them with various assessments [3–5]. Several reports linking frailty and sarcopenia to predict prognosis in patients with bladder cancer have been reported [3,6–10]. As an indicator that may objectively represent frailty, we focused on the psoas muscle Hounsfield unit (PMHU), which is defined as the mean computed tomographic attenuation value of the psoas muscle. The main aim of this study is to assess the utility of PMHU as a preoperative prognostic marker in patients receiving RC for bladder cancer.

2. Materials and Methods

2.1. Study Design and Patients

We retrospectively reviewed the records of consecutive patients who underwent open radical cystectomy (ORC), laparoscopic radical cystectomy (LRC), or robot-assisted laparoscopic radical cystectomy (RARC) for bladder cancer at Mie University Hospital, Ise Red Cross Hospital, and Mie Prefectural General Medical Center. A total of 177 patients (113 patients at Mie University Hospital, 42 patients at Ise Red Cross Hospital, and 23 patients at Mie Prefectural General Medical Center) were enrolled.

2.2. Image Analyses

Abdominal non-contrast computed tomography (NCCT) images (1–5 mm-thick slices) taken within 3 months before RC were used to measure the imaging factors related to the psoas muscle. Four urologists (Y.S., S.K., T.O., and S.T.) freehand outlined each psoas muscle at the third lumbar vertebral level in the axial NCCT image (Figure 1).

The right and left total areas were used for the psoas muscle area (PMA) (mm^2). The mean value of the mean computed tomographic attenuation value of the right and left psoas muscle was used for the PMHU (HU). The psoas mass index (PMI) (cm^2/m^2) was calculated by normalizing the cross-sectional area by height [11].

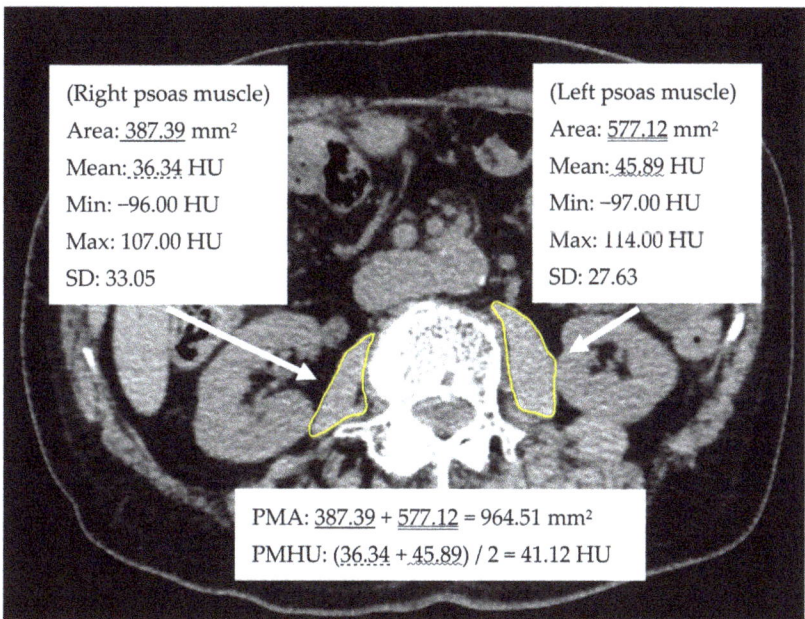

Figure 1. Measurement of the psoas muscle area (PMA) and the mean psoas muscle Hounsfield unit (PMHU) on the preoperative axial non-contrast computed tomography image at the third lumbar vertebral level.

2.3. Statistical Analyses

All statistical analyses were performed using EZR version 1.33 [12]. Student's *t*-test or Mann–Whitney U test was performed for comparisons between groups of continuous variables. Categorical variables were analyzed using the chi-squared test or Fisher's exact test. The survival curve was estimated using the Kaplan–Meier method and analyzed using the log-rank test. Cox proportional hazards analysis was used to calculate the hazard ratio (HR) and 95% confidence interval (CI) in univariate and multivariate analyses. In all tests, $p < 0.05$ was considered statistically significant.

3. Results

Of the 177 patients, 26 (14.7%) were women and 151 (85.3%) were men. The median age was 70 (quartile: 66–76) years, and the median follow-up period was 1002 (quartile: 358–1989) days. The 5-year OS, CSS, and recurrence-free survival (RFS) rates were 59.7%, 71.3%, and 48.7%, respectively. The median OS, CSS, and RFS were 8.86 years (95% CI, 5.00 years to not reached), not reached (95% CI, not reached to not reached), and 4.70 years (95% CI, 2.66 to 9.24 years), respectively. Platinum-based neoadjuvant and adjuvant chemotherapy were performed in 75 (42.4%) and 22 (12.4%) patients, respectively. The surgical procedures were ORC, LRC, and RARC in 119 (67.2%), 38 (21.5%), and 20 (11.3%) patients, respectively. There were 114 (64.4%) and 29 (16.4%) perioperative complications of Clavien–Dindo grade ≥ 2 and ≥ 3, respectively. The median length of hospital stay after RC was 26 (quartile: 22–41) days. The histopathological diagnosis after RC was urothelial carcinoma in 138 (77.8%) patients, and some histological variants were found in 39 (22.0%) patients. There were 71 (40.1%) patients with pT ≥ 3, 33 (18.6%) patients with pN positivity, and 89 (50.3%) patients with LVI positivity, all of which were significant prognostic factors for OS ($p < 0.01$, log-rank test). Body mass index, PMI, PMA, and PMHU were significantly different between men and women (Table 1).

Table 1. Patient characteristics.

Variables		Total Cases (n = 177)	Men (n = 151)	Women (n = 26)	p-Value
Median age at RC (IQR)		70 (66–76)	70 (66–76)	72.5 (65–76)	0.687
Median BMI (IQR)		22.92 (20.94–24.95)	23.11 (21.20–25.07)	21.22 (19.03–22.68)	<0.01
ASA-PS	1	21 (12%)	17 (11%)	4 (15%)	0.755
	2	120 (68%)	103 (68%)	17 (65%)	
	3	36 (20%)	31 (21%)	5 (19%)	
Clinical T stage, n (%)	NMIBC	51 (29%)	45 (30%)	6 (23%)	0.163
	2	70 (40%)	61 (40%)	9 (35%)	
	3	27 (15%)	19 (13%)	8 (31%)	
	4	29 (16%)	26 (17%)	3 (12%)	
Clinical N stage, n (%)	0	154 (87%)	132 (87%)	22 (85%)	0.752
	≥1	23 (13%)	19 (13%)	4 (15%)	
Neoadjuvant chemotherapy, n (%)	No	102 (58%)	90 (60%)	12 (46%)	0.207
	Yes	75 (42%)	61 (40%)	14 (54%)	
Median PMA (mm^2) (IQR)		1129 (894–1455)	1242 (983–1510)	679 (521–887)	<0.01
Median PMI (cm^2/m^2) (IQR)		4.31 (3.41–5.41)	4.66 (3.66–5.60)	2.88 (2.34–3.89)	<0.01
Median PMHU (HU) (IQR)		43.14 (39.26–47.54)	43.40 (39.56–47.75)	40.93 (34.56–45.16)	0.043

IQR = interquartile range, BMI = body mass index, ASA-PS; American Society of Anesthesiologists physical status, NMIBC = non-muscle invasive bladder cancer, PMA = psoas muscle area, PMI = psoas mass index, PMHU = psoas muscle Hounsfield unit.

For variables related to the psoas muscle, it was considered inappropriate to apply the same cutoff value for men and women; thus, the lower limit of the interquartile range for each sex (25 percentile) was used as the cutoff value.

The results of Cox proportional hazards analysis for OS were shown in Table 2.

Univariate analysis showed significant differences in age, sex, cT stage, and PMHU ($p < 0.05$). The median OS stratified by PMHU alone in the not-low and low PMHU groups were 9.24 years (95% CI, 6.40 years to not reached) and 2.78 years (2.06 years to not reached), respectively, and there was a significant difference among them ($p = 0.014$). Multivariate analysis using the four factors that were significantly different in the univariate analysis showed significant differences in all factors ($p < 0.05$).

We focused on these four factors to develop a risk classification for predicting OS in patients with bladder cancer after RC (Table 3). The Kaplan–Meier curve for OS according to the number of risks was shown in Figure 2.

Table 2. Cox proportional hazards analysis of overall survival in patients undergoing radical cystectomy.

Variables	Category	Univariate			Multivariate		
		HR	95% CI	p Value	HR	95% CI	p Value
Age	<70	Reference			Reference		
	≥70	2.093	1.239–3.533	0.006	1.734	1.010–2.977	0.046
Sex	Men	Reference			Reference		
	Women	2.210	1.189–4.109	0.012	2.116	1.132–3.954	0.019

Table 2. Cont.

Variables	Category	Univariate			Multivariate		
		HR	95% CI	p Value	HR	95% CI	p Value
ASA-PS	1,2	Reference					
	3	1.624	0.926–2.849	0.091			
Clinical T stage	<3	Reference			Reference		
	≥3	1.782	1.705–2.956	0.025	1.665	1.001–2.769	0.049
Clinical N stage	0	Reference					
	≥1	0.659	0.283–1.530	0.33			
Neoadjuvant chemotherapy	No	Reference					
	Yes	0.877	0.512–1.502	0.633			
BMI	Not low	Reference					
	Low	0.909	0.507–1.631	0.749			
PMA	Not low	Reference					
	Low	1.332	0.692–2.564	0.391			
PMI	Not low	Reference					
	Low	1.019	0.530–1.959	0.954			
PMHU	Not low	Reference			Reference		
	Low	1.924	1.132–3.270	0.016	1.758	1.014–3.048	0.044

HR = hazard ratio, CI = confidence interval, ASA-PS = American Society of Anesthesiologists physical status, BMI = body mass index, PMA = psoas muscle area, PMI = psoas mass index, PMHU = psoas muscle Hounsfield unit.

Table 3. Risk factors and risk category.

Risk Factors	0	1
1. Age	<70	≥70
2. Sex	Men	Women
3. Clinical T stage	<3	≥3
4. PMHU	Men: ≥39.56 HU Women: ≥34.56 HU	Men: <39.56 HU Women: <34.56 HU
Risk Category		
Low-risk	If <2 risk factors present	
Intermediate-risk	If 2 risk factors present	
High-risk	If ≥3 risk factors present	

PMHU = psoas muscle Hounsfield unit.

Based on this result, we defined a group with one or fewer risk factors as a low-risk group, a group with two risk factors as an intermediate-risk group, and a group with three or more risk factors as a high-risk group (Table 3). The Kaplan–Meier curve and HR for each risk category were shown in Figure 3 and Table 4, respectively.

The median OS by our risk category in the low-risk, intermediate-risk, and high-risk groups (Table 3) were not reached (95% CI, 8.86 years to not reached), 6.40 years (95% CI, 2.67 years to not reached), and 2.06 years (95% CI, 0.94 to 2.78 years), respectively ($p < 0.01$). There were no significant differences among the risk groups in terms of postoperative hospital stay and the incidence of complications.

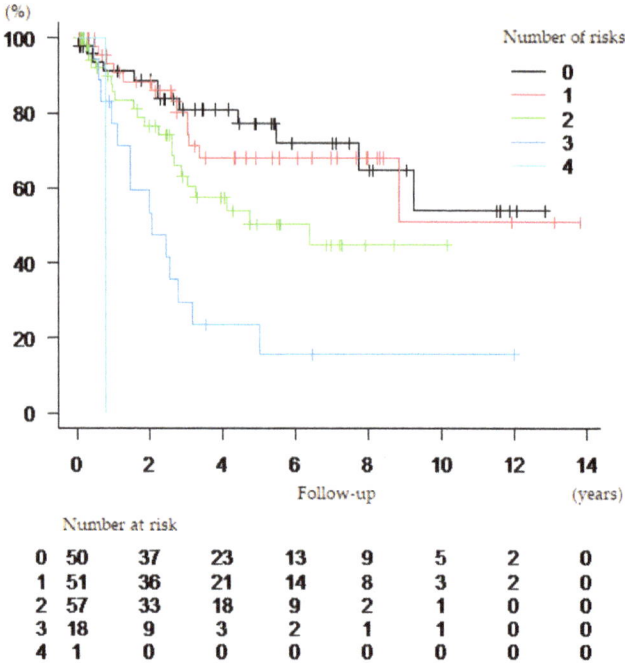

Figure 2. Kaplan–Meier curves for overall survival according to the number of risks.

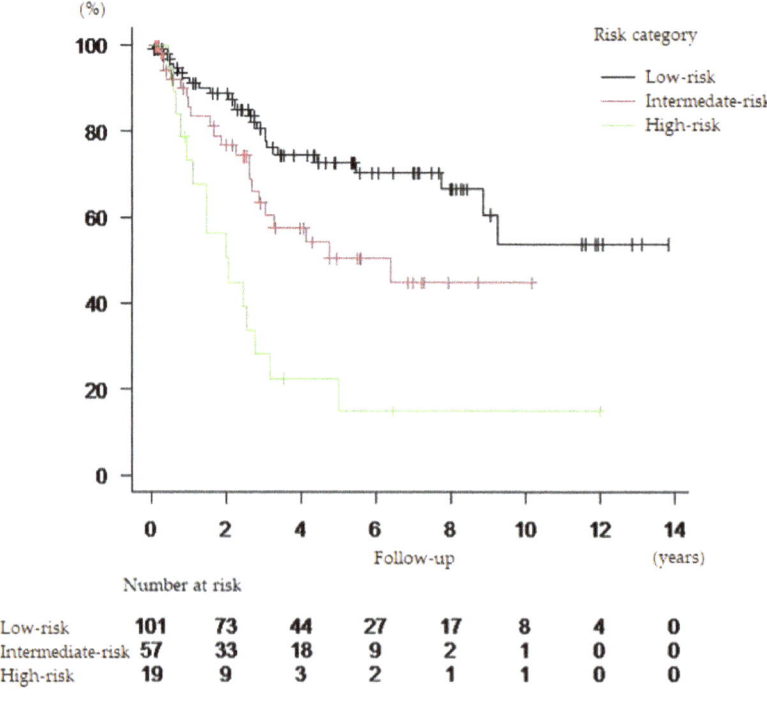

Figure 3. Kaplan–Meier curves for overall survival according to risk category.

Table 4. Hazard ratio for each risk category.

Risk Category	n	HR	95% CI	p Value
Low-risk	101		Reference	
Intermediate-risk	57	1.902	1.061–3.411	0.031
High-risk	19	4.597	2.408–8.775	<0.01

HR = hazard ratio, CI = confidence interval.

4. Discussion

In the present study, we investigated the significance of preoperative PMHU on the prognosis after RC in patients with bladder cancer. We showed that PMHU is a new prognostic marker along with sex, age, and cT stage, and that these predictors could be used to preoperatively stratify the prognosis of patients with bladder cancer after RC.

In recent years, medical technology innovations have led to minimally invasive surgery, but RC remains highly invasive because of the long operation time and high incidence of perioperative complication rates [2]. Elderly people had a higher 90-day mortality rate and more early complications after RC than younger people [2], and it is often discussed whether RC should be performed, especially for elderly patients.

In order to judge the indication for RC, it is required not only to evaluate the surgical tolerance such as cardiac function and respiratory function, but also to predict the postoperative prognosis to some extent. LVI and pN positivity and pT3 or higher grade have already been reported as postoperative factors related to prognosis [13], and similar results were obtained with our cases. However, pathological factors are the information that can only be known after RC. There are few reports on prognostic markers that are useful in deciding whether or not to perform RC itself.

In an aging society, the concept of frailty is drawing attention. Especially in the elderly, comprehensive assessment of patients' frailty as well as their age may be useful for treating diseases and improving their quality of life [14,15]. Diagnosis of frailty requires the measurement of grip strength and walking speed [4,5], but there are few facilities that can be incorporated into the daily practice of treating patients with urological malignancies. In particular, it is not desirable to impose a heavy preoperative evaluation on patients with their malignancies.

Regarding the assessment of frailty, NCCT images are easy to acquire and non-invasive. PMA [16], psoas muscle volume (PMV) [9,17,18], PMI [9,11], mean PMHU [19], skeletal muscle index (SMI) [6,8,20], intramuscular adipose tissue content (IMAC) [21], and so on, have been reported as representative imaging factors for frailty. According to a systematic review by Cao et al., SMI, PMI, muscle attenuation, and IMAC were useful for assessing the risk of postoperative complications as NCCT-assessed sarcopenia indices [22]. Analysis of our data did not reveal significant results for PMA and PMI, but PMHU correlated with the prognosis of patients with bladder cancer after RC.

PMHU can be measured very simply and easily. No special training or software is required. NCCT images are always acquired in patients before RC, without the burden of adding new special tests to the patients. Low PMHU reflects skeletal muscle fat infiltration and may indirectly be used to assess frailty [23]. Increase in fat infiltration within skeletal muscle might precede loss of skeletal muscle volume during the progression of cancer cachexia [24].

PMHU is a factor that reflects muscle quality, while PMA and PMI are factors that reflect muscle mass [20,22]. In our data, this may be one of the reasons why only PMHU, not PMA and PMI, showed a significant correlation with OS. Assessing psoas muscle mass with an NCCT image at the L3 level is very simple but may not necessarily reflect systemic skeletal muscle mass [25]. PMV and SMI may reflect systemic muscle mass better than PMA and PMI, but at the expense of ease of measurement [18]. We consider PMHU to be a more practical predictor because of its ease of measurement and its accuracy for prognosis.

Frailty was also evaluated as a risk factor for perioperative complications in patients with bladder cancer [9,26]. Prediction of patients' prognosis preoperatively may avoid surgical invasive procedures which would cause more harm than benefit in patients. According to our results, the median OS of high-risk patients who have three or more preoperative poor prognosis factors defined by age, sex, cT stage, and PMHU was only 2.06 years. Although our data would not give definitive prediction of perioperative complication, high-risk patients with poor prognosis in our risk classification might not recommend RC. Bladder sparing therapy combined with transurethral resection of the bladder tumor, chemotherapy, and radiation therapy are controversial, but they may be a good treatment option for patients with an apparently poor prognosis [9,27].

By preoperatively diagnosing frailty, it may be possible to improve the prognosis of cancer patients if interventions such as exercise therapy and nutritional guidance can be performed earlier [17,28]. Exercise therapy and essential amino acid supplement drinks have been shown to be useful in recovery from serious illnesses [29] and vitamin D supplementation was useful in improving sarcopenia in the elderly [30]. It goes without saying that treatment of the bladder cancer itself is important for improving the OS. However, in an aging society, we must understand the physical function and nutritional status of each patient before intervening with diverse treatments. To that end, the role of the rehabilitation team, which includes registered dietitians and physiotherapists, is also important, and more than ever it is necessary to deepen cooperation within the team.

There are several limitations of our study. The present study is a small retrospective study. Our results need to be prospectively validated in a larger cohort. In addition, the subjects of this study were limited to the Japanese population, and there is room for consideration of differences between races, especially regarding the cutoff value. Furthermore, the present study did not directly investigate the relationship between PMHU and frailty. In the future, the direct relationship between the already reported diagnostic factors of frailty and PMHU must be investigated [4,5]. In addition we believe that it should be examined in more detail whether perioperative nutrition therapy and physiotherapy for patients with low PMHU can improve the bladder cancer patients' survival and quality of life.

5. Conclusions

PMHU was a preoperative predictor of prognosis in patients with bladder cancer who were about to undergo RC. The prognosis of patients could be stratified before RC using age, sex, cT stage, and PMHU. Not only can PMHU be measured without burdening patients and clinicians, but also this risk classification helps determine whether to perform RC in patients with bladder cancer before surgery.

Author Contributions: Conceptualization, Y.S. and T.I.; data curation, Y.S., T.O. (Takashi Okamoto), S.K. and S.T.; formal analysis, Y.S.; funding acquisition, T.I.; investigation, Y.S.; methodology, Y.S. and T.I.; project administration, Y.S. and T.I.; supervision, T.S. (Takuji Shibahara), M.K., S.M., K.N., T.S. (Takeshi Sasaki), T.O. (Takehisa Onishi), H.K., H.M. and T.I.; writing—original draft, Y.S. and T.I.; writing—review and editing, Y.S., T.S. (Takeshi Sasaki), K.N. and T.I. All authors have read and agreed to the published version of the manuscript.

Funding: This work was financially supported by Grants-in-Aid from the Ministry of Education for Science and Culture of Japan (21H03066).

Institutional Review Board Statement: The study was conducted according to the guidelines of the Declaration of Helsinki, and approved by the Institutional Review Board of Mie university (approval code H2020-110).

Informed Consent Statement: Informed consent was obtained in the form of opt-out on the web-site. Those who opted out were excluded.

Data Availability Statement: The data are available to other researchers on written request to the corresponding author.

Conflicts of Interest: The authors declare no conflict of interest.

References

1. Stein, J.P.; Lieskovsky, G.; Cote, R.; Groshen, S.; Feng, A.-C.; Boyd, S.; Skinner, E.; Bochner, B.; Thangathurai, D.; Mikhail, M.; et al. Radical Cystectomy in the Treatment of Invasive Bladder Cancer: Long-Term Results in 1,054 Patients. *J. Clin. Oncol.* **2001**, *19*, 666–675. [CrossRef] [PubMed]
2. Fonteyne, V.; Ost, P.; Bellmunt, J.; Droz, J.P.; Mongiat-Artus, P.; Inman, B.; Paillaud, E.; Saad, F.; Ploussard, G. Curative Treatment for Muscle Invasive Bladder Cancer in Elderly Patients: A Systematic Review. *Eur. Urol.* **2018**, *73*, 40–50. [CrossRef] [PubMed]
3. Chappidi, M.R.; Kates, M.; Patel, H.D.; Tosoian, J.J.; Kaye, D.R.; Sopko, N.A.; Lascano, D.; Liu, J.-J.; McKiernan, J.; Bivalacqua, T.J. Frailty as a marker of adverse outcomes in patients with bladder cancer undergoing radical cystectomy. *Urol. Oncol. Semin. Orig. Investig.* **2016**, *34*, 256.e1–256.e6. [CrossRef] [PubMed]
4. Cruz-Jentoft, A.J.; Baeyens, J.P.; Bauer, J.M.; Boirie, Y.; Cederholm, T.; Landi, F.; Martin, F.C.; Michel, J.-P.; Rolland, Y.; Schneider, S.M.; et al. Sarcopenia: European consensus on definition and diagnosis: Report of the European Working Group on Sarcopenia in Older People. *Age Ageing* **2010**, *39*, 412–423. [CrossRef]
5. Chen, L.-K.; Woo, J.; Assantachai, P.; Auyeung, T.-W.; Chou, M.-Y.; Iijima, K.; Jang, H.C.; Kang, L.; Kim, M.; Kim, S.; et al. Asian Working Group for Sarcopenia: 2019 Consensus Update on Sarcopenia Diagnosis and Treatment. *J. Am. Med. Dir. Assoc.* **2020**, *21*, 300–307.e2. [CrossRef]
6. Mayr, R.; Gierth, M.; Zeman, F.; Reiffen, M.; Seeger, P.; Wezel, F.; Pycha, A.; Comploj, E.; Bonatti, M.; Ritter, M.; et al. Sarcopenia as a comorbidity-independent predictor of survival following radical cystectomy for bladder cancer. *J. Cachexia Sarcopenia Muscle* **2018**, *9*, 505–513. [CrossRef]
7. Fukushima, H.; Takemura, K.; Suzuki, H.; Koga, F. Impact of Sarcopenia as a Prognostic Biomarker of Bladder Cancer. *Int. J. Mol. Sci.* **2018**, *19*, 2999. [CrossRef]
8. Fukushima, H.; Yokoyama, M.; Nakanishi, Y.; Tobisu, K.-I.; Koga, F. Sarcopenia as a Prognostic Biomarker of Advanced Urothelial Carcinoma. *PLoS ONE* **2015**, *10*, e0115895. [CrossRef]
9. Saitoh-Maeda, Y.; Kawahara, T.; Miyoshi, Y.; Tsutsumi, S.; Takamoto, D.; Shimokihara, K.; Hayashi, Y.; Mochizuki, T.; Ohtaka, M.; Nakamura, M.; et al. A low psoas muscle volume correlates with a longer hospitalization after radical cystectomy. *BMC Urol.* **2017**, *17*, 87. [CrossRef]
10. Hu, X.; Dou, W.-C.; Shao, Y.-X.; Liu, J.-B.; Xiong, S.-C.; Yang, W.-X.; Li, X. The prognostic value of sarcopenia in patients with surgically treated urothelial carcinoma: A systematic review and meta-analysis. *Eur. J. Surg. Oncol. (EJSO)* **2019**, *45*, 747–754. [CrossRef]
11. Hamaguchi, Y.; Kaido, T.; Okumura, S.; Kobayashi, A.; Hammad, A.; Tamai, Y.; Inagaki, N.; Uemoto, S. Proposal for new diagnostic criteria for low skeletal muscle mass based on computed tomography imaging in Asian adults. *Nutrition* **2016**, *32*, 1200–1205. [CrossRef]
12. Kanda, Y. Investigation of the freely available easy-to-use software 'EZR' for medical statistics. *Bone Marrow Transplant.* **2013**, *48*, 452–458. [CrossRef]
13. Mari, A.; Kimura, S.; Foerster, B.; Abufaraj, M.; D'Andrea, D.; Gust, K.M.; Shariat, S.F. A systematic review and meta-analysis of lymphovascular invasion in patients treated with radical cystectomy for bladder cancer. *Urol. Oncol. Semin. Orig. Investig.* **2018**, *36*, 293–305. [CrossRef]
14. Robinson, S.M.; Denison, H.J.; Cooper, C.; Sayer, A.A. Prevention and optimal management of sarcopenia: A review of combined exercise and nutrition interventions to improve muscle outcomes in older people. *Clin. Interv. Aging* **2015**, *10*, 859–869. [CrossRef]
15. Billot, M.; Calvani, R.; Urtamo, A.; Sánchez-Sánchez, J.L.; Ciccolari-Micaldi, C.; Chang, M.; Roller-Wirnsberger, R.; Wirnsberger, G.; Sinclair, A.; Vaquero-Pinto, M.N.; et al. Preserving Mobility in Older Adults with Physical Frailty and Sarcopenia: Opportunities, Challenges, and Recommendations for Physical Activity Interventions. *Clin. Interv. Aging* **2020**, *15*, 1675–1690. [CrossRef]
16. Smith, A.B.; Deal, A.M.; Yu, H.; Boyd, B.; Matthews, J.; Wallen, E.M.; Pruthi, R.S.; Woods, M.E.; Muss, H.; Nielsen, M.E. Sarcopenia as a Predictor of Complications and Survival Following Radical Cystectomy. *J. Urol.* **2014**, *191*, 1714–1720. [CrossRef]
17. Miyake, M.; Morizawa, Y.; Hori, S.; Marugami, N.; Shimada, K.; Gotoh, D.; Tatsumi, Y.; Nakai, Y.; Inoue, T.; Anai, S.; et al. Clinical impact of postoperative loss in psoas major muscle and nutrition index after radical cystectomy for patients with urothelial carcinoma of the bladder. *BMC Cancer* **2017**, *17*, 237. [CrossRef]
18. Matsubara, Y.; Nakamura, K.; Matsuoka, H.; Ogawa, C.; Masuyama, H. Pre-treatment psoas major volume is a predictor of poor prognosis for patients with epithelial ovarian cancer. *Mol. Clin. Oncol.* **2019**, *11*, 376–382. [CrossRef]
19. Zhuang, C.-L.; Shen, X.; Huang, Y.-Y.; Zhang, F.-M.; Chen, X.-Y.; Ma, L.-L.; Chen, X.-L.; Yu, Z.; Wang, S.-L. Myosteatosis predicts prognosis after radical gastrectomy for gastric cancer: A propensity score–matched analysis from a large-scale cohort. *Surgery* **2019**, *166*, 297–304. [CrossRef]
20. Van Rijssen, L.B.; van Huijgevoort, N.C.; Coelen, R.J.; Tol, J.A.; Haverkort, E.B.; Nio, C.Y.; Busch, O.R.; Besselink, M.G. Skeletal Muscle Quality is Associated with Worse Survival After Pancreatoduodenectomy for Periampullary, Nonpancreatic Cancer. *Ann. Surg. Oncol.* **2017**, *24*, 272–280. [CrossRef]

21. Kitajima, Y.; Hyogo, H.; Sumida, Y.; Eguchi, Y.; Ono, N.; Kuwashiro, T.; Tanaka, K.; Takahashi, H.; Mizuta, T.; Ozaki, I.; et al. Severity of non-alcoholic steatohepatitis is associated with substitution of adipose tissue in skeletal muscle. *J. Gastroenterol. Hepatol.* **2013**, *28*, 1507–1514. [CrossRef]
22. Cao, Q.; Xiong, Y.; Zhong, Z.; Ye, Q. Computed Tomography-Assessed Sarcopenia Indexes Predict Major Complications following Surgery for Hepatopancreatobiliary Malignancy: A Meta-Analysis. *Ann. Nutr. Metab.* **2019**, *74*, 24–34. [CrossRef]
23. Aubrey, J.; Esfandiari, N.; Baracos, V.E.; Buteau, F.A.; Frenette, J.; Putman, C.T.; Mazurak, V.C. Measurement of skeletal muscle radiation attenuation and basis of its biological variation. *Acta Physiol.* **2014**, *210*, 489–497. [CrossRef]
24. Kliewer, K.L.; Ke, J.-Y.; Tian, M.; Cole, R.M.; Andridge, R.R.; Belury, M.A. Adipose tissue lipolysis and energy metabolism in early cancer cachexia in mice. *Cancer Biol. Ther.* **2014**, *16*, 886–897. [CrossRef]
25. Rutten, I.J.; Ubachs, J.; Kruitwagen, R.F.; Beets-Tan, R.G.; Damink, S.O.; Van Gorp, T. Psoas muscle area is not representative of total skeletal muscle area in the assessment of sarcopenia in ovarian cancer. *J. Cachexia Sarcopenia Muscle* **2017**, *8*, 630–638. [CrossRef]
26. Palumbo, C.; Knipper, S.; Pecoraro, A.; Rosiello, G.; Luzzago, S.; Deuker, M.; Tian, Z.; Shariat, S.F.; Simeone, C.; Briganti, A.; et al. Patient frailty predicts worse perioperative outcomes and higher cost after radical cystectomy. *Surg. Oncol.* **2020**, *32*, 8–13. [CrossRef]
27. Ploussard, G.; Daneshmand, S.; Efstathiou, J.A.; Herr, H.W.; James, N.D.; Rödel, C.M.; Shariat, S.F.; Shipley, W.U.; Sternberg, C.N.; Thalmann, G.N.; et al. Critical Analysis of Bladder Sparing with Trimodal Therapy in Muscle-invasive Bladder Cancer: A Systematic Review. *Eur. Urol.* **2014**, *66*, 120–137. [CrossRef]
28. PERIOP OG Working Group; Tully, R.; Loughney, L.; Bolger, J.; Sorensen, J.; McAnena, O.; Collins, C.G.; Carroll, P.A.; Arumugasamy, M.; Murphy, T.J.; et al. The effect of a pre- and post-operative exercise programme versus standard care on physical fitness of patients with oesophageal and gastric cancer undergoing neoadjuvant treatment prior to surgery (The PERIOP-OG Trial): Study protocol for a randomised controlled trial. *Trials* **2020**, *21*, 638. [CrossRef]
29. Jones, C.; Eddleston, J.; McCairn, A.; Dowling, S.; McWilliams, D.; Coughlan, E.; Griffiths, R. Improving rehabilitation after critical illness through outpatient physiotherapy classes and essential amino acid supplement: A randomized controlled trial. *J. Crit. Care* **2015**, *30*, 901–907. [CrossRef]
30. Muir, S.W.; Montero-Odasso, M. Effect of Vitamin D Supplementation on Muscle Strength, Gait and Balance in Older Adults: A Systematic Review and Meta-Analysis. *J. Am. Geriatr. Soc.* **2011**, *59*, 2291–2300. [CrossRef]

Article

Renal Cell Carcinoma with or without Tumor Thrombus Invading the Liver, Pancreas and Duodenum

Javier González [1], Jeffrey J. Gaynor [2] and Gaetano Ciancio [2,3,*]

[1] Department of Urology, Hospital General Universitario Gregorio Marañón, 28007 Madrid, Spain; fcojavier.gonzalez@salud.madrid.org
[2] Department of Surgery, Miami Transplant Institute, University of Miami Miller school of Medicine, Miami, FL 33136, USA; jgaynor@med.miami.edu
[3] Department of Surgery and Urology, University of Miami Miller School of Medicine, Jackson Memorial Hospital, Miami, FL 33136, USA
* Correspondence: gciancio@med.miami.edu

Simple Summary: Renal cell carcinoma rarely invades the surrounding visceral structures. While surgical extirpation has been the mainstay of treatment for the localized disease, the role of surgery in cases of venous involvement, adjacent invasion or distant metastasis remains controversial. Furthermore, the surgical option may represent a challenge. A large series of locally advanced renal cancer with involvement of the liver, pancreas, and/or duodenum, sometimes in conjunction with tumor thrombus extending inside the inferior vena cava is herein reported. Our series establishes the technical feasibility of this complex surgical procedure with acceptable complication rates, no perioperative death, and potential for durable response. With the use of new systemic therapy schedules, these patients will probably have a better opportunity of survival extension.

Abstract: Background: The purpose of this study is to report the outcomes of a series of patients with locally advanced renal cell carcinoma (RCC) who underwent radical nephrectomy, tumor thrombectomy, and visceral resection. Patients and methods: 18 consecutive patients who underwent surgical treatment in the period 2003-2019 were included. Neoplastic extension was found extending into the pancreas, duodenum, and liver in 9(50%), 2(11.1%), and 7(38.8%) patients, respectively. Seven patients (38.8%) presented also inferior vena cava tumor thrombus level I ($n = 3$), II ($n = 2$), or III ($n = 2$). The resection was tailored according to the degree of invasiveness. Demographics, clinical presentation, disease characteristics, surgical details, 30-day postoperative complications, and overall survival (OS) were analyzed. Results: Median age was 56 years (range: 40–76). Median tumor size was 14.5 cm (range, 8.8–22), and 10 cm (range: 4–15) for those cases with pancreatico-duodenal and liver involvement, respectively. Median estimated blood loss (EBL) was 475 mL (range: 100–4000) and resulted higher for those cases requiring thrombectomy (300 mL vs. 750 mL). Nine patients (50%) required transfusions with a median requirement of 4 units (range: 2–8). No perioperative deaths were registered in the first 30 days. Overall complication rate was 44.4%. Major complications were detected in 6/18 patients (33.3%). Overall median follow-up was 24 months (range: 0–108). Five-year OS (actuarial) rate was 89.9% and 75%, for 9/11 patients with pancreatico-duodenal involvement and 6/7 patients with liver invasion, respectively. Conclusion: Our series establishes the technical feasibility of this procedure with acceptable complication rates, no deaths, and potential for durable response.

Keywords: renal cell carcinoma; tumor thrombus; metastasectomy; postoperative complications; oncological outcomes

1. Introduction

Renal cell carcinoma (RCC) is the most common malignant tumor of the kidney. Over 73,750 new cases will be diagnosed in 2020 [1–3]. RCC has a myriad of presentations, infrequently extends into the inferior vena cava (IVC) [4,5] and approximately one third of RCC are diagnosed in advanced stages of the disease [6]. Most of the 14,830 estimated deaths from RCC for 2020, will be associated with these stages [1]. While surgical extirpation has been the mainstay of treatment for the localized disease [3], the role of surgery in cases of venous involvement, adjacent invasion or distant metastasis remains controversial and the surgical option may represent a challenge [4,5,7–9].

There are to date no clear guidelines for the surgical management of large renal masses with contiguous extension and/or synchronous metastases to surrounding organs such as the pancreas, duodenum or liver at the time of initial nephrectomy due in part to still insufficient data and the lack of prospective randomized trials. However, resection for metastatic RCC disease has increasingly been performed with acceptable morbidity and mortality rates in the last few years [8,9]. The use of surgical strategy as a first line of treatment in this context would serve a multiple purpose: (i) it provides a psychological benefit to the patient who may feel "treated", (ii) it provides symptomatic relief in the event of the presence of symptoms (rather frequent in this subgroup) (iii) it provides enough tissue sample to be used for assessment on the heterogeneity of the disease (usually present) and on the morphological, immunohistochemical and molecular variants involved, serving commonly as a guide for subsequent systemic treatment (type of agent/s and treatment sequence among the many currently available), (iv) reduces the burden of disease, thus avoiding the immunological sink generated by the overwhelming amount of malignancy and enabling an effective immune response in the patient against it, and (v) such an aggressive resection has revealed a modest survival benefit in favor of this approach in opposition to observation or systemic treatment alone [10].

We present a contemporary series of 18 patients with locally advanced renal masses invading adjacent visceral structures in which complete removal of gross disease was achieved by means of radical nephrectomy in conjunction with thrombectomy, resection of pancreas, duodenum and/or liver with low morbidity and mortality rates by using the surgical techniques derived from the field of transplantation applied to these complex cases. With this approach in the front line, we provide for symptomatic relief in most our patients, obtain enough sample of tissue to guide further systemic treatment, and observe durable responses.

2. Materials and Methods

After obtaining Institutional Review Board approval and following the ethical principles of the Helsinki Declaration (as revised in 2013), a retrospective chart review was performed on 18 consecutive patients who underwent surgical treatment for large, aggressive, locally advanced RCC cases between June 2003 and November 2019 at our institution. All relevant data on demographics, clinical presentation, disease characteristics, surgical details, and postoperative complications were collected and analyzed. Postoperative complications were assessed using the Clavien-Dindo Classification System [11], and defined as occurring within a 30-days period of the intervention date. Overall survival (OS) was ascertained by the review of medical records or using the Social Security Death Index (SSDI) database [12] when necessary. Abdominal computed tomography and/or magnetic resonance imaging were used to diagnose the renal tumor, delineate the tumor thrombus inside the inferior vena cava, and depict the extent of the invasion to adjacent organs (Figure 1).

Surgical Technique

Our transplant techniques for piggy-back liver mobilization and en-bloc mobilization of spleen and pancreas in order to facilitate the resection of large renal tumors have been previously described [4,5,7,13]. Briefly, the renal artery was ligated early during the surgery by medially mobilizing the involved kidney [14]. The goal of early renal artery ligation was to decompress collateral circulation and decrease blood loss during the procedure. The use of the maneuvers derived from transplantation surgery for either side, allowed wide exposure of the retroperitoneal space thus facilitating the removal of the renal mass and the resection of variable segments of other organs involved. All visceral resections were performed by a single surgeon and with curative intention, and tailored according to the particular degree of tumor invasiveness. Intracaval involvement required IVC exploration, thrombus withdrawal, and IVC resection/reconstruction as previously described [13–15].

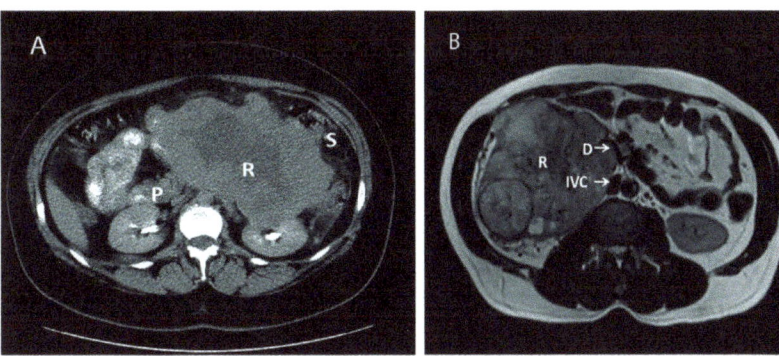

Figure 1. Computed tomography (CT) and magnetic resonance imaging (MRI) of two different patients with large renal mass. (**A**): CT of a large left renal mass with lack of soft tissue planes between pancreas, spleen, and splenic flexure; (**B**): MRI of a right large solid and cystic enhancing mass replacing the right renal parenchyma with mass effect and compression of proximal duodenum. P: Pancreas; R: Renal Tumor; S: Splenic flexure of colon; IVC: Inferior Vena Cava: D: Duodenum.

At the time of resection, neoplastic extension was found extending outside the perirenal tissue into the pancreas in 9/18 patients (50%), mostly coming from the left side. The second portion of duodenum was affected in two right-sided tumors (18%), a left side tumor invaded the third and four portions of the duodenum in one patient (9%), and the remaining case (9%) showed tumor invasion extended into both pancreas and duodenum.

These 9 patients underwent variable pancreatic resections that included pancreaticoduodenectomy (i.e., Whipple procedure) (9%), partial or subtotal pancreatectomy (63%), and combined body and tail pancreatectomy (9%). One patient underwent spleen preservation along with distal pancreatectomy. However, the spleen and the ipsilateral adrenal were also excised en-bloc in 7 (63%) and 9 (100%) cases, respectively. The duodenum was preserved whenever possible, although partial resection was considered necessary in cases of exclusive serosal involvement (27%). Intracaval extension was additionally detected in 3 of the patients (27%); one case of level II tumor thrombus(TT) according to Neves and Zincke [16] (showing concomitant invasion of the left colon, thus requiring additional left hemicolectomy), and two cases of level III TT (i.e., IIIa and IIIb tumor thrombi according to our classification) [17].

Partially occluding (level I or II) thrombi were managed by tangential cavectomy and primary closure. Completely occluding (level II and III) tumor thrombi required circumferential cavectomy (i.e., segment located between the caval bifurcation and the inferior border of the major hepatic veins entrance) without vascular reconstruction. The level IIIb tumor thrombus was "milked" and controlled below the major hepatic veins before the cavectomy was attempted (Figure 2) [18].

RCC was infiltrating the right lobe in three patients. Partial right lobectomy was performed en-bloc with the renal tumor and adrenal gland in three patients, resection of segments two and three in one patient, and wedge-resection of the liver mass in other three patients. Four of these patients harbored also an IVC tumor thrombus (level I in $n = 3$, and level II in $n = 1$) that required additional thrombectomy, tangential cavectomy, and primary closure. Of note, one of the patients that underwent a partial right lobectomy had two previous unsuccessful ipsilateral renal artery embolization attempts complicated by massive bleeding before referred to our institution (Figure 3).

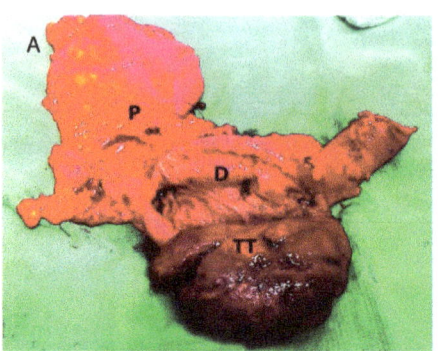
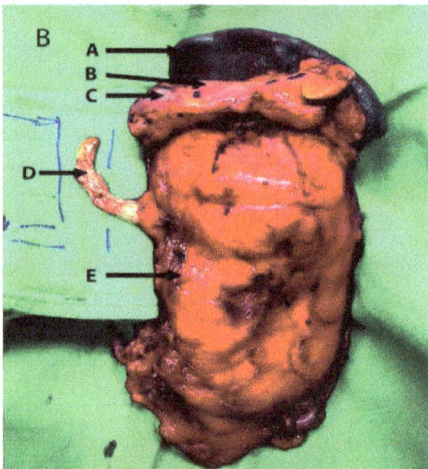
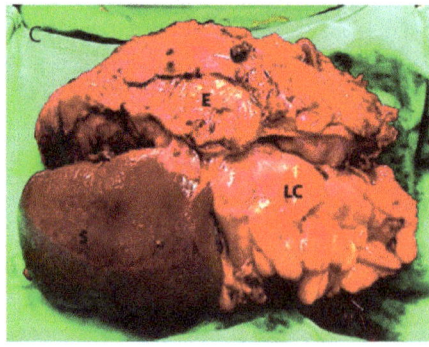

Figure 2. Renal tumor involving adjacent organs. (**A**): Head of the pancreas resected along with a portion of duodenum; (**B**): Left radical nephrectomy specimen with resection of IVC tumor thrombus, pancreas and spleen; (**C**): Left kidney tumor resected with adjacent organs, the distal pancreas was not visualized. P: Head of pancreas; D: 2nd portion of duodenum; TT: Tumor thrombus inside the second portion of the duodenum; A: Spleen; B: Pancreas; C: Pancreatic metastatic mass; D: Tumor Thrombus going inside the inferior vena cava; E: Kidney Tumor; S: Spleen; LC: Left Colon.

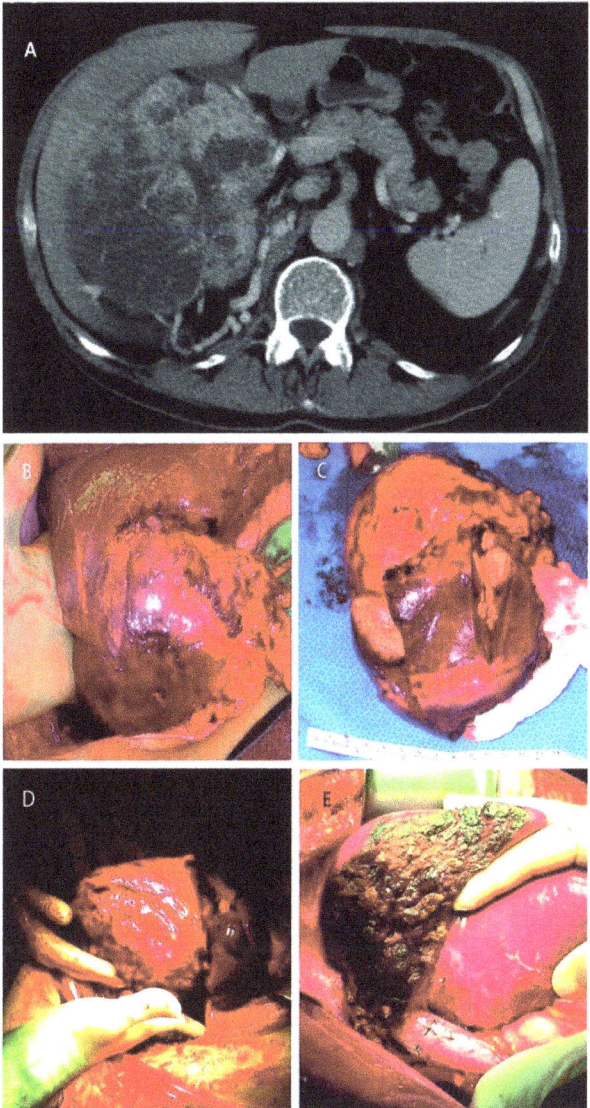

Figure 3. (**A**): Computed tomography of large right renal mass with collaterals from the liver; (**B**): Right renal mass with renal artery and vein ligated but still with blood flow from the liver; (**C**): The resected specimens of right renal mass with right lobe of the liver; (**D**): Right renal mass infiltrating the right lobe of the liver; (**E**): Remaining liver after almost complete right lobectomy.

Extended lymph node dissection (including all the potentially affected palpable lymph nodes, and paracaval, inter aortocaval and paraaortic templates depending on the particular features of the case) completed the procedure in all cases.

3. Results

3.1. Patient Demographics

Over a period of 16 years, a total of 18 patients underwent radical nephrectomy, tumor thrombectomy (when present), and resection of visceral structures invaded by disease.

Table 1 lists pertinent demographic information. The median age at the time of resection was 56 years (range: 40–76 years). There were seven males (38.9%) and 11 females (61.1%). Except for four cases (22%) in which an incidental mass was found on imaging, all the remaining cases were symptomatic (78%). Complaints in symptomatic patients ranged from abdominal pain (44%) to gastrointestinal (GI) bleeding (5.5%) due to tumor erosion into duodenum. One patient presented with collapsed lung due to endobronchial RCC metastases and required left upper lobe resection prior to the abdominal intervention [15]. None of the patients had obstructive jaundice. Median tumor size was 14.5 cm (range, 8.8–22 cm), and 10 cm (range: 4–15 cm) for those cases with pancreatico-duodenal and liver involvement, respectively.

Table 1. Demographic information.

Parameter	Value
No. of patients	18
Male, n (%)	7 (38.9%)
Female, n (%)	11 (61.1%)
Age, median (years)	56 (40–76)
Follow up, median (months)	24 months (0–108); 3 deaths at 13, 24, and 108 months post-surgery, respectively
Asymptomatic, n (%)	4 (22%)
Symptomatic, n (%)	14 (78%)
Abdominal pain	8 (42%)
Hematuria	3 (16.6%)
Fatigue	3 (16.6%)
GI bleeding	1 (5.5%)
Collapsed lung	1 (5.5%)

GI: gastrointestinal.

3.2. Intra-Operative Outcomes

Overall median estimated blood loss (EBL) was 475 mL (range: 100–4000 mL). Median EBL for those requiring pancreatic ($n = 11$) and liver ($n = 7$) resections was 300 mL (range, 100–2500 mL) and 500 mL (range 200–4000), respectively; while median EBL for those having ($n = 7$) vs. not having ($n = 11$) IVC exploration (along with either liver resection, duodenal resection or pancreatectomy) was 750 mL (range, 200–4000 mL) and 300 mL (range 100–600), respectively. Nine of the 18 patients required blood transfusions during surgery or in the immediate postoperative period; the median transfusion requirement for the nine patients who required transfusions was four units of packed red blood cells (range: 2–8 units).

3.3. Surgical Outcomes

The intervention provided complete or partial symptomatic relief in 12/14 cases (85%) and 2/14 cases (15%), respectively. Fatigue improved considerably in all patients presenting it, but remained after the intervention in two of them (66%). No perioperative deaths were registered in this series during the first 30 days after the intervention. Overall complication rate was 44.4% (8/18 patients). Major complications (Clavien-Dindo grade $\geq$ III) were detected in 6 of the 18 patients (33.3%); 45% (5/11) of the patients with pancreatico-duodenal involvement and 14% (1/7) of the patients showing liver invasion. Postoperative complications included isolated intra-abdominal collection (2), wound infection (1), pancreatitis (1), pancreatic leak (1), and atrial fibrillation with pulmonary embolism that required prolonged intubation (1) for those patients with pancreatico-duodenal invasion, and postoperative intra-abdominal collection associated to pulmonary embolism that required prolonged intubation in one of the patients with liver involvement. Tables 2 and 3 list operative outcomes and classified complications, respectively.

Table 2. Operative details and outcomes.

Parameter	Value
Side of lesion, n (%)	
Left	8 (38.8%)
Right	11 (61.2%)
IVC involvement [1], n (%)	
Level I	3 (16.6%)
Level II	2 (11.1%)
Level III [2]	2 (11.1%)
Tumor size (cm)	
Pancreatico-duodenal involvement	14.5 (8.8–22)
Liver involvement	10 (4–15)
EBL, median (mL)	475
Pancreatico-duodenal involvement	300 (100–2500)
Liver involvement	500 (200–4000)
IVC exploration	750 (200–4000)
No IVC exploration	300 (100–600)
Transfusion, median (PRBC units)	1U for the total group of 18 patients; 4U among the 9 patients who required transfusions
Liver resection [3], n (%)	7 (38.8%)
Pancreaticoduodenectomy, n (%)	1 (5.5%)
Partial/subtotal pancreatectomy, n (%)	7 (38.8%)
Distal pancreatectomy, n (%)	1 (5.5%)
Other organs removed, n (%)	2 (11%)
Partial duodenum	3 (16.6%)
Spleen	7 (38.8%)
Adrenal gland	18 (100%)
Left colon	1 (5.5%)

IVC: inferior vena cava; PRBC: packed red blood cells; U: units; [1] Acoording to the Neves-Zincke Classification System; [2] One of the patients presented a level IIIa tumor thrombus (below the major hepatic veins), while the other presented a level IIIb tumor thrombus (at the level of the major hepatic veins); [3] Partial right lobectomy in 3 patients, resection of segments 2 and 3 in 1 patient, and wedge resection in 3 patients.

Table 3. Postoperative complications according to the Clavien-Dindo Classification System.

Complication	n (%)	Grade
Collection	3 (16.6%)	IIIa
Pulmonary [1]	2 (11.1%)	Iva
Atrial fibrillation	1 (5.5%)	II
Wound infection	1 (5.5%)	II
Deep venous thrombosis	1 (5.5%)	IIIa
Pancreatitis	1 (5.5%)	II
Pancreatic leak	1 (5.5%)	IIIa

[1] Pulmonary embolization with prolonged intubation.

All tumors were classified as T4 according to the American Joint Committee on Cancer (AJCC) 2009 TNM staging system [19]. Table 4 lists pathological characteristics for all specimens. Among the specimens showing pancreatico-duodenal involvement, histological examination revealed no positive margins, and RCC conventional type in 9 (81%), chromophobe type in 1 (9%), and metastatic poorly differentiated carcinoma from the adrenal in 1 (9%). Specimens presenting liver involvement showed RCC conventional type in 5 patients (57%), chromophobe type in 1 (14%) and poorly differentiated squamous cell carcinoma in 1 (14%). Median tumor grade was higher (IV vs. III) for those tumors invading the pancreas or duodenum than for those presenting liver infiltration. However, neither categorization of Fuhrman grade nor identification of the source was possible in the case showing the squamous cell carcinoma invading the liver.

Table 4. Pathological characteristics and survival outcomes.

Variable	Value
AJCC pT4, n (%)	18 (100%)
RCC conventional type, n (%)	14 (77.7%)
Fuhrman Grade, median	
Pancreatico-duodenal involvement	IV
Liver involvement	III
Lymph node metastases, n (%)	
Pacreatico-duodenal involvement	1 (9%)
Liver involvement	2 (28.5%)
Median survival from time of resection (months)	
Pancreatico-duodenal involvement	36 (13–108)
Liver involvement	24 (12–96)
Actuarial 5-yr OS (%)	
Pancreatico-duodenal involvement	84.6%
Liver involvement	75%

RCC: Renal cell carcinoma; yr: years; OS: overall survival.

A total of 50 lymph nodes were removed from the 11 patients showing pancreatico-duodenal involvement. Only one of the patients showed invasion by RCC (conventional type). Twenty lymph nodes were removed from the seven patients showing liver involvement. Pathologic examination of these specimens resulted positive in two of these patients (RCC clear cell type and poorly differentiated squamous cell carcinoma, respectively).

3.4. Oncological Outcomes

Overall median follow-up was 24 months (range: 0–108 months). Two of the 11 patients with pancreatico-duodenal involvement returned to their local community physicians and were lost for further follow-up. Overall survival (OS) was ascertained in the remaining nine patients showing a five-year (actuarial) rate of 89.9%. Excluding the two patients with no post-surgical follow-up, the median survival time from the date of resection was 36 months (range: 13–108 months). One patient (9%) lived for nine years but developed multiple metastases during the follow-up (87 months) requiring additional resection and adjuvant systemic therapy (sunitinib); another patient (9%) remained alive after seven years but need systemic chemotherapy for the treatment of lymphoma at the time of last follow-up. The remaining six patients (66.7%) had no evidence of disease as of last visit.

One of the seven patients with liver involvement returned to his local community physician for follow-up. OS was ascertained in the remaining six patients showing a five-year (actuarial) OS rate of 75%. Excluding the single patient with no post-surgical follow-up, the median survival time from the date of resection was 24 months (range: 12–96 months). One patient died after two years of multiple metastases.

4. Discussion

The main purpose of this study was to review our experience in the removal of renal tumors in conjunction or not with tumor thrombus that required additional liver, pancreas and/or duodenum. We described the surgical outcomes of a contemporary series including a total of 18 consecutive patients. Eleven of these patients, underwent radical nephrectomy and resection of the pancreas and/or duodenum. Three of them presented a TT inside the IVC, making their surgical management even more complex. We also included seven patients with renal tumors invading the liver (four of them with TT involving the IVC) in whom simultaneous radical nephrectomy, removal of TT if necessary, and liver resection was performed. Our series of multiorgan resection was associated with a major complication rate of 33.3% (n = 6/18) in the absence of post-operative mortality. These outcomes confirm again the technical feasibility already established by other series [20,21]. More importantly, this study reconfirms that in experience hands and properly performed, this approach is worth the effort since shows curative intent and provides better quality of life by means of symptomatic relief.

In this regard, Yezhelyev et al. reported their experience of 25 patients that underwent simultaneous radical nephrectomy and major hepatectomy for RCC [20]. Eight of those patients presented a TT in the IVC. Direct liver invasion by contiguity was the most common indication for hepatectomy. Ten patients in these series (40%) presented postoperative complications including one death in the perioperative period. Karellas et al. also described their experience of 38 patients with pT4 RCC invading adjacent organs. The liver was the most commonly visceral structure resected en-bloc with RCC (n = 10), but they also reported pancreatic resection in six patients. They recorded two perioperative deaths, but no report of post-operative complications was provided. Only 1 patient was alive and free of disease at five years. The median time from surgical resection to death was 11.7 months. Their conclusion was that once advanced RCC involved the adjacent structures the prognosis was poor [21].

Eleven patients in our series showed either pancreatico-duodenal (n = 9) or isolated duodenal involvement (n = 2) at the time of the radical nephrectomy. Although the involvement of the pancreas from RCC is rare, is still more frequent that isolated duodenal involvement. In this regard, Margulis et al. reported their experience in the management of these cases [22], which resulted quite similar to the Karellas et al. [21] in terms of morbidity and perioperative mortality. They reported 12 out of 30 patients (40%) presenting pT4 disease at debut with direct invasion into adjacent organs demonstrated by preoperative imaging. The pancreas was involved in only three of these patients, none of them invading the duodenum. Despite the aggressive surgical approach, the disease recurred in 10 of those 12 patients (83.3%) at a median time of 2.3 months, and only five (41.6%) were alive at the time of the report [22].

Resection of RCC metastasizing into the duodenum has been rarely described in a number of case reports. Most of these metastases were metachronous and frequently coursed with upper GI bleeding, in the same way that one of the patients included in our series [23–25]. Conversely, duodenal resection at the time of the radical nephrectomy has been only reported once [26]. In our series, four of the patients had the duodenum resected at the time of the radical nephrectomy; including a Whipple procedure, two partial duodenal resections, and an excision of the third and fourth portions of the duodenum in conjunction with distal pancreatectomy and splenectomy.

RCC represents an entity with high potential for metastatic spreading. Recent epidemiological datasets report that approximately 17% of patients debut with metastases, and almost one third of the remaining patients will develop local recurrences or distant metastases during the follow-up [1]. Although metastasectomy in locally advanced or metastatic RCC has been used since the 1930s, to date no prospective randomized controlled trials have evaluated the clinical benefit of this approach. Conversely, surgical tumor debulking is currently supported by retrospective experiences showing rather favorable overall and progression free survival rates [27]. Actually, two recent systematic reviews suggested that complete metastasectomy is associated with better survival (range, 36.5–142 months vs. 8.4–27 months) and/or symptomatic control when compared to no or incomplete metastasectomy [28]. Comparable outcomes have been reported in the context of the resection of multiple metastatic sites, local/distant lymph nodes, and local recurrences [29]. In fact, disease-free intervals after complete resection may be notable, thus avoiding systemic treatment requirements at least in the first instance [30]. These observations are in the line of our experience with actuarial five-year OS rates reaching 89.9% and 75% for those patients showing pancreatico-duodenal and liver involvement, respectively.

The difference in terms of survival between these two subgroups seems notable in favor of the subgroup with exclusive pancreatico-duodenal involvement. Although in cases of pancreatico-duodenal involvement postoperative complications are more frequent, the actuarial five-year OS is approximately 15% higher. A possible explanation for these results would rely on differences in the pattern of disease dissemination between the subgroups.

While pancreatic-duodenal involvement occurs mainly by direct contiguity, liver involvement may result from the coexistence of direct invasion or hematogenous dissemination. Probably, the latter would impact negatively the prognosis. Conversely, in cases in which pancreatic resection affected the cephalic region, surgical reconstruction after removal is technically more demanding and, with no doubt, more subject to complications than the resection of a portion of the liver parenchyma. Nevertheless, intraoperative bleeding in these cases is commonly more exuberant and difficult to manage intraoperatively.

Although there are no current clear guidelines regarding patient selection in order to determine who will benefit most from surgery, the decision-making regarding the resection of synchronous neighboring visceral metastases seems to rely on different factors such as comorbid patient conditions, performance status, a number of different prognostic risk factors, as well as number and location of the sites involved. According to the National Comprehensive Cancer Network guidelines, metastasectomy should be considered in those patients with clear or non-clear cell variants who initially present with primary RCC in conjunction with oligometastatic involvement or develop metastases after a prolonged free-disease interval from initial nephrectomy [31]. In addition, the European Society for Medical Oncology recommends that the decision-making process should be upon a multidisciplinary team trying to identify patients with adequate performance status, resectable oligometastatic disease, or low/intermediate risk in which complete resection is achievable [32], given that the best outcomes after metastasectomy have been reported in such cases. Only one of our patients showed a solitary lung metastasis that was resected before attempting the abdominal intervention.

All of them were considered suitable for surgery in terms of performance status and comorbidity, and the decision-making regarding the extent of the local resection was commonly made upon the direct vision of the operative field. Nevertheless, considering this possibility, extensive counseling was made in a particular basis upon the findings provided by preoperative cross-sectional imaging.

Doubts regarding complete resection predicting short-term failure for disease control has led to concepts combining surgery either with presurgical or adjuvant systemic therapy. The neoadjuvant approach followed by complete surgical excision has been evaluated in small retrospective series during the cytokine era [33]. The greater effect in terms of efficacy in response or downsizing the tumor burden (reaching a median reduction of 9–14% in primary tumor diameter) obtained with targeted therapies opened the gates for further multimodality treatment schedules in this context. Surgery following tyrosine kinase inhibitors is overall safe [34], making candidates initially unsuitable for complete resection amenable for surgical approach reconsideration if downsizing or significant response is confirmed after at least one cycle of targeted therapy [35,36]. Although not prospectively studied, acceptable outcomes have been reported with this approach in small retrospective series [34]. The largest experience is based on a 22-patient series from three different institutions in which primary excision associated with concomitant or deferred metastasectomy was performed after at least one cycle of tyrosine kinase inhibitors [37]. However, most of the patients included continued systemic treatment once recovered from surgery, making the evaluation of the long-term free-relapse survival periods observed difficult to attribute to either surgery or systemic treatment alone, or the combination of both treatment modalities acting synergically.

Currently, no role for systemic adjuvant therapy after metastasectomy has been established. A randomized, double-blinded, placebo-controlled multicenter phase-III trial conducted by the Eastern Cooperative Oncology Group-American College of Radiology Imaging Network group assessed the role of adjuvant pazopanib vs. placebo in patients with no evidence of disease after metastasectomy [38]. The patients received treatment for a year and were stratified according to the number of sites of resected disease as well as disease-free interval. The study was unable to meet a primary end-point for disease-free survival (HR 0.85; CI95% 0.55–1.31; $p = 0.47$), actually showing a trend to worse overall survival with pazopanib. Comparable results have been reported with sorafenib in the postmetastasectomy setting [39]. Currently, different ongoing trials are evaluating the use of sunitinib and a number of immune checkpoint inhibitors in this context, but there is not enough evidence to date to support their use over observation.

It remains unclear what role metastasectomy or surgical consolidation will play in the next future, given the morphing landscape of systemic immunotherapy combined with targeted therapy, and new data on different tumor molecular features acting as potential biomarkers for response to treatment. So far, the clinical experience should guide the recommendations for metastasectomy in these patients. Therefore, our surgical outcomes allow us to advocate in favor of a surgical approach if considered feasible and indicated. However, this study is not without limitations. The series herein reported is retrospective in nature and reflects the experience of a single high-volume surgeon from a referral center. Hence, these outcomes should be evaluated accordingly, given that may be not reproduced by teams of lower experience or working in different clinical settings.

5. Conclusions

We reported a large series of locally advanced RCC with involvement of the liver, pancreas, and/or duodenum, sometimes in conjunction with TT extending inside the IVC. Our series establishes the technical feasibility of this complex surgical procedure by means of the application of the techniques derived from transplant surgery with acceptable complication rates, no perioperative death, and potential for durable response. In addition, by using this strategy in the front line we obtain symptomatic relief in most of the patients and provide for an optimal tissue sample that may be used to guide the further systemic therapeutic approach required. With the use of new systemic therapy schedules, these patients will probably have a better opportunity of survival extension. In our opinion, the management of these rare cases should be safely entrusted to an experienced surgeon/surgical team, but exclusively in high volume referral centers, where complex procedures, such as the ones reported, can be successfully performed.

Author Contributions: Conceptualization, J.G., J.J.G., and G.C.; methodology, J.G., J.J.G., and G.C.; investigation, J.G., J.J.G., and G.C.; data curation, J.J.G. and G.C.; writing—original draft preparation, J.G., J.J.G., and G.C.; writing—review and editing, J.G., J.J.G., G.C. All authors have read and agreed to the published version of the manuscript.

Funding: This research received no external funding.

Institutional Review Board Statement: The study was conducted according to the guidelines of the Declaration of Helsinki, and approved by the Institutional Review Board of THE MIAMI TRANSPLANT INSTITUTE (protocol code 20200791 and date of approval 17 August 2020).

Informed Consent Statement: Informed consent was obtained from all subjects involved in the study.

Data Availability Statement: Data supporting reported results may be obtained upon request to the corresponding author.

Conflicts of Interest: The authors declare no conflict of interest.

References

1. Siegel, R.L.; Miller, K.D.; Jemal, A. Cancer statistics, 2020. *CA Cancer J. Clin.* **2020**, *70*, 7–30. [CrossRef] [PubMed]
2. Sun, M.; Thuret, R.; Abdollah, F.; Lughezzani, G.; Schmitges, J.; Tian, Z.; Shariat, S.F.; Montorsi, F.; Patard, J.J.; Perrotte, P.; et al. Age-adjusted incidence, mortality, and survival rates of stage-specific renal cell carcinoma in North America: A trend analysis. *Eur. Urol.* **2011**, *59*, 135–141. [CrossRef] [PubMed]
3. Frees, S.K.; Kamal, M.M.; Nestler, S.; Levien, P.M.; Bidnur, S.; Brenner, W.; Thomas, C.; Jaeger, W.; Thüroff, J.W.; Roos, F.C. Risk-adjusted proposal for >60 months follow up after surgical treatment of organ-confined renal cell carcinoma according to life expectancy. *Int. J. Urol.* **2019**, *26*, 385–390. [CrossRef] [PubMed]
4. Ciancio, G.; Gonzalez, J.; Shirodkar, S.P.; Angulo, J.C.; Soloway, M.S. Liver transplantation techniques for the surgical management of renal cell carcinoma with tumor thrombus in the inferior vena cava: Step-by-step description. *Eur. Urol.* **2011**, *59*, 401–406. [CrossRef] [PubMed]

5. Ciancio, G.; Livingstone, A.S.; Soloway, M. Surgical management of renal cell carcinoma with tumor thrombus in the renal and inferior vena cava: The University of Miami experience in using liver transplantation techniques. *Eur. Urol.* **2007**, *51*, 988–994. [CrossRef]
6. Gupta, K.; Miller, J.D.; Li, J.Z.; Russell, M.W.; Charbonneau, C. Epidemiologic and socioeconomic burden of metastatic renal cell carcinoma (mRCC): A literature review. *Cancer Treat. Rev.* **2008**, *34*, 193–205. [CrossRef]
7. Ciancio, G.; Hawke, C.; Soloway, M. The use of liver transplant techniques to aid in the surgical management of urological tumors. *J. Urol.* **2000**, *164*, 665–672. [CrossRef]
8. Pikoulis, E.; Margonis, G.A.; Antoniou, E. Surgical Management of Renal Cell Cancer Liver Metastases. *Scand J. Surg.* **2016**, *105*, 263–268. [CrossRef]
9. Benhaim, R.; Oussoultzoglou, E.; Saeedi, Y.; Mouracade, P.; Bachellier, P.; Lang, H. Pancreatic metastasis from clear cell renal cell carcinoma: Outcome of an aggressive approach. *Urology* **2015**, *85*, 135–140. [CrossRef]
10. Chatzizacharias, N.A.; Rosich-Medina, A.; Dajani, K.; Harper, S.; Huguet, E.; Liau, S.S.; Praseedom, R.K.; Jah, A. Surgical management of hepato-pancreatic metastasis from renal cell carcinoma. *World J. Gastrointest. Oncol.* **2017**, *9*, 70–77. [CrossRef]
11. Clavien, P.A.; Barkun, J.; de Oliveira, M.L.; Vauthey, J.N.; Dindo, D.; Schulick, R.D.; de Santibañes, E.; Pekolj, J.; Slankamenac, K.; Bassi, C.; et al. The Clavien-Dindo classification of surgical complications: Five-year experience. *Ann. Surg.* **2009**, *250*, 187–196. [CrossRef] [PubMed]
12. Huntington, J.T.; Butterfield, M.; Fisher, J.; Torrent, D.; Bloomston, M. The Social Security Death Index (SSDI) most accurately reflects true survival for older oncology patients. *Am. J. Cancer Res.* **2013**, *3*, 518–522. [PubMed]
13. Ciancio, G.; Vaidya, A.; Shirodkar, S.; Manoharan, M.; Hakky, T.; Soloway, M. En bloc mobilization of the pancreas and spleen to facilitate resection of large tumors, primarily renal and adrenal, in the left upper quadrant of the abdomen: Techniques derived from multivisceral transplantation. *Eur. Urol.* **2009**, *55*, 1106–1111. [CrossRef]
14. Ciancio, G.; Vaidya, A.; Soloway, M. Early ligation of the renal artery using the posterior approach: A basic surgical concept reinforced during resection of large hypervascular renal cell carcinoma with or without inferior vena cava thrombus. *BJU Int.* **2003**, *92*, 488–489. [CrossRef] [PubMed]
15. Manoharan, A.; Lugo-Baruqui, A.; Ciancio, G. Radical Nephrectomy and Pulmonary Lobectomy for Renal Cell Carcinoma with Tumor Thrombus Extension into the Inferior Vena Cava and Pulmonary Arteries. *Anticancer Res.* **2020**, *40*, 5837–5844. [CrossRef] [PubMed]
16. Neves, R.J.; Zincke, H. Surgical treatment of renal cancer with vena cava extension. *Br. J. Urol.* **1987**, *59*, 390–395. [CrossRef] [PubMed]
17. Ciancio, G.; Vaidya, A.; Savoie, M.; Soloway, M. Management of renal cell carcinoma with level III thrombus in the inferior vena cava. *J. Urol.* **2002**, *168*, 1374–1377. [CrossRef]
18. González, J.; Gorin, M.A.; Garcia-Roig, M.; Ciancio, G. Inferior vena cava resection and reconstruction: Technical considerations in the surgical management of renal cell carcinoma with tumor thrombus. *Urol. Oncol.* **2014**, *32*, 34–e19. [CrossRef] [PubMed]
19. Swami, U.; Nussenzveig, R.H.; Haaland, B.; Agarwal, N. Revisiting AJCC TNM staging for renal cell carcinoma: Quest for improvement. *Ann. Transl. Med.* **2019**, *7*, S18. [CrossRef] [PubMed]
20. Yezhelyev, M.; Master, V.; Egnatashvili, V.; Kooby, D.A. Combined nephrectomy and major hepatectomy: Indications, outcomes, and recommendations. *J. Am. Coll. Surg.* **2009**, *208*, 410–418. [CrossRef]
21. Karellas, M.E.; Jang, T.L.; Kagiwada, M.A.; Kinnaman, M.D.; Jarnagin, W.R.; Russo, P. Advanced-stage renal cell carcinoma treated by radical nephrectomy and adjacent organ or structure resection. *BJU Int.* **2009**, *103*, 160–164. [CrossRef] [PubMed]
22. Margulis, V.; Sanchez-Ortiz, R.F.; Tamboli, P.; Cohen, D.D.; Swanson, D.A.; Wood, C.G. Renal cell carcinoma clinically involving adjacent organs: Experience with aggressive surgical management. *Cancer* **2007**, *109*, 2025–2030. [CrossRef] [PubMed]
23. Peters, N.; Lightner, C.; McCaffrey, J. An Unusual Case of Gastrointestinal Bleeding in Metastatic Renal Cell Carcinoma. *Case Rep. Oncol.* **2020**, *13*, 738–741. [CrossRef] [PubMed]
24. Munir, A.; Khan, A.M.; McCarthy, L.; Mehdi, S. An Unusual Case of Renal Cell Carcinoma Metastasis to Duodenum Presenting as Gastrointestinal Bleeding. *JCO Oncol. Pract.* **2020**, *16*, 49–50. [CrossRef] [PubMed]
25. Mohamed, M.O.; Al-Rubaye, S.; Reilly, I.W.; McGoldrick, S. Renal cell carcinoma presenting as an upper gastrointestinal bleeding. *BMJ Case Rep.* **2015**, *2015*. [CrossRef] [PubMed]
26. Schlussel, A.T.; Fowler, A.B.; Chinn, H.K.; Wong, L.L. Management of locally advanced renal cell carcinoma with invasion of the duodenum. *Case Rep. Surg.* **2013**, *2013*, 596362. [CrossRef] [PubMed]
27. Gonzalez, J.; Gaynor, J.J.; Alameddine, M.; Esteban, M.; Ciancio, G. Indications, complications, and outcomes following surgical management of locally advanced and metastatic renal cell carcinoma. *Expert Rev. Anticancer. Therapy* **2018**, *18*, 237–250. [CrossRef]
28. Dabestani, S.; Marconi, L.; Hofmann, F.; Stewart, F.; Lam, T.B.L.; E Canfield, S.; Staehler, M.; Powles, T.; Ljungberg, B.; Bex, A. Local treatments for metastases of renal cell carcinoma: A systematic review. *Lancet Oncol.* **2014**, *15*, e549–e561. [CrossRef]
29. Achkar, T.; Maranchie, J.K.; Appleman, L.J. Metastasectomy in advanced renal cell carcinoma: A systematic review. *Kidney Cancer* **2019**, *3*, 31–40. [CrossRef]
30. Alt, A.L.; Boorjian, S.A.; Ms, C.M.L.; Costello, B.A.; Leibovich, B.C.; Blute, M.L. Survival after complete surgical resection of multiple metastases from renal cell carcinoma. *Cancer* **2011**, *117*, 2873–2882. [CrossRef] [PubMed]
31. National Comprehensive Cancer Network. Kidney Cancer (Version 2.2020). Available online: Https://www.nccn.org/professionals/physician_gls/pdf/kidney.pdf (accessed on 2 October 2020).

32. Escudier, B.; Porta, C.; Schmidinger, M.; Rioux-Leclercq, N.; Bex, A.; Khoo, V.; Grünwald, V.; Gillessen, S.; Horwich, A. ESMO Guidelines Committee. Renal cell carcinoma: ESMO clinical practice guidelines for diagnosis, treatment, and follow-up. *Ann. Oncol.* **2019**, *30*, 706–720. [CrossRef]
33. Krishnamurthi, V.; Novick, A.C.; Bukowski, R.M. Efficacy of multimodality therapy in advanced renal cell carcinoma. *Urology* **1998**, *51*, 933–937. [CrossRef]
34. Margulis, V.; Matin, S.F.; Tannir, N.; Tamboli, P.; Swanson, D.A.; Jonasch, E.; Wood, C.G. Surgical morbidity associated with administration of targeted molecular therapies before cytoreductive nephrectomy or resection of locally recurrent renal cell carcinoma. *J. Urol.* **2008**, *180*, 94–98. [CrossRef] [PubMed]
35. Karam, J.A.; Rini, B.I.; Varella, L.; Garcia, J.A.; Dreicer, R.; Choueiri, T.K.; Jonasch, E.; Matin, S.F.; Campbell, S.C.; Wood, C.G.; et al. Metastasectomy after targeted therapy in patients with advanced renal cell carcinoma. *J Urol.* **2011**, *185*, 439–444. [CrossRef] [PubMed]
36. Shuch, B.; Riggs, S.B.; LaRochelle, J.C.; Kabbinavar, F.F.; Avakian, R.; Pantuck, A.J.; Belldegrun, A.S.; Patard, J.-J. Neoadjuvant targeted therapy and advanced kidney cancer: Observations and implications for a new treatment paradigm. *BJU Int.* **2008**, *102*, 692–696. [CrossRef] [PubMed]
37. Patard, J.-J.; Thuret, R.; Raffi, A.; Laguerre, B.; Bensalah, K.; Culine, S. Treatment with sunitinib enabled complete resection of massive lymphadenopathy not previously amenable to excision in a patient with renal cell carcinoma. *Eur. Urol.* **2009**, *55*, 237–239. [CrossRef] [PubMed]
38. Sternberg, C.N.; Hawkins, R.E.; Wagstaff, J.; Salman, P.; Mardiak, J.; Barrios, C.H.; Zarba, J.J.; Gladkov, O.A.; Lee, E.; Szczylik, C.; et al. Randomized, double-blind phase III study of pazopanib versus placebo in patients with metastatic renal cell carcinoma who have no evidence of disease following metastasectomy: A trial of the ECOG-ACRIN cancer research group (E2810). *J. Clin. Oncol.* **2019**, *37*, 4502. [CrossRef]
39. Procopio, G.; Apollonio, G.; Cognetti, F.; Miceli, R.; Milella, M.; Mosca, A.; Chiuri, V.E.; Bearz, A.; Morelli, F.; Ortega, C.; et al. Sorafenib versus observation following radical metastasectomy for clear-cell renal cell carcinoma: Results from the phase 2 randomized open label RESORT study. *Eur. Urol. Oncol.* **2019**, *2*, 699–707. [CrossRef]

Communication

Stereotactic Body Radiotherapy for Frail Patients with Primary Renal Cell Carcinoma: Preliminary Results after 4 Years of Experience

Laure Grelier [1], Michael Baboudjian [2,*], Bastien Gondran-Tellier [2], Anne-Laure Couderc [3], Robin McManus [2], Jean-Laurent Deville [4], Ana Carballeira [5], Raphaelle Delonca [2], Veronique Delaporte [2], Laetitia Padovani [1], Romain Boissier [2], Eric Lechevallier [2] and Xavier Muracciole [1]

[1] Department of Urology and Kidney Transplantation, Aix-Marseille University, Assistance Publique–Hôpitaux de Marseille (AP-HM), Conception Academic Hospital, 13005 Marseille, France; Laure.GRELIER@ap-hm.fr (L.G.); Laetitia.Padovani@ap-hm.fr (L.P.); Xavier.MURACCIOLE@ap-hm.fr (X.M.)
[2] Department of Radiotherapy, La Timone Hospital, Assistance Publique-Hopitaux de Marseille (AP-HM), 13005 Marseille, France; bastien.gondran@ap-hm.fr (B.G.-T.); robinmcmanus@rcsi.ie (R.M.); raphaelle.delonca@ap-hm.fr (R.D.); Veronique.DELAPORTE@ap-hm.fr (V.D.); romain.boissier@ap-hm.fr (R.B.); eric.lechevallier@ap-hm.fr (E.L.)
[3] Internal Medicine, Geriatrics and Therapeutic University, Assistance Publique–Hopitaux de Marseille (AP-HM), 13005 Marseille, France; Anne-Laure.COUDERC@ap-hm.fr
[4] Department of Oncology, La Timone Hospital, Aix-Marseille University, Assistance Publique–Hopitaux de Marseille (AP-HM), 13005 Marseille, France; Jean-laurent.DEVILLE@ap-hm.fr
[5] Department of Radiology, Aix-Marseille University, Assistance Publique–Hopitaux de Marseille (AP-HM), Conception Academic Hospital, 13005 Marseille, France; ana.carballeira-alvarez@ap-hm.fr
* Correspondence: Michael.BABOUDJIAN@ap-hm.fr

Citation: Grelier, L.; Baboudjian, M.; Gondran-Tellier, B.; Couderc, A.-L.; McManus, R.; Deville, J.-L.; Carballeira, A.; Delonca, R.; Delaporte, V.; Padovani, L.; et al. Stereotactic Body Radiotherapy for Frail Patients with Primary Renal Cell Carcinoma: Preliminary Results after 4 Years of Experience. *Cancers* **2021**, *13*, 3129. https://doi.org/10.3390/cancers13133129

Academic Editors: José I. López and Jean-Yves Blay

Received: 26 May 2021
Accepted: 21 June 2021
Published: 23 June 2021

Publisher's Note: MDPI stays neutral with regard to jurisdictional claims in published maps and institutional affiliations.

Copyright: © 2021 by the authors. Licensee MDPI, Basel, Switzerland. This article is an open access article distributed under the terms and conditions of the Creative Commons Attribution (CC BY) license (https://creativecommons.org/licenses/by/4.0/).

Simple Summary: Surgical therapy is currently the standard of care for the treatment of primary renal cell carcinoma (RCC). Alternative strategies such as stereotactic body radiotherapy (SBRT) have emerged as potentially curative treatment approaches. In this study, we show a promising short-term local control effect of SBRT in the management of primary RCC. The treatment was well tolerated with no high-grade side effects. The main advantages are the outpatient management without anesthesia and the non-invasive approach. Thus, SBRT appears to be a promising alternative to surgery, or ablative therapy, to treat primary RCC in patients with poor physical health. Future studies are needed to definitively assess the place of SBRT in the RCC treatment portfolio.

Abstract: Introduction: The aim of this study was to report the oncological outcomes and toxicity of stereotactic body radiotherapy (SBRT) to treat primary renal cell carcinoma (RCC) in frail patients unfit for surgery or standard alternative ablative therapies. Methods: We retrospectively enrolled 23 patients who had SBRT for primary, biopsy-proven RCC at our tertiary center between October 2016 and March 2020. Treatment-related toxicities were defined using CTCAE, version 4.0. The primary outcome was local control which was defined using the Response Evaluation Criteria in Solid Tumors. Results: The median age, Charlson score and tumor size were 81 (IQR 79–85) years, 7 (IQR 5–8) and 40 (IQR 28–48) mm, respectively. The most used dose fractionation schedule was 35 Gy (78.3%) in five or seven fractions. The median duration of follow-up for all living patients was 22 (IQR 10–39) months. Local recurrence-free survival, event-free survival, cancer-specific survival and overall survival were 96 (22/23), 74 (18/23), 96 (22/23) and 83% (19/23), respectively. There were no grade 3–4 side effects. No patients required dialysis during the study period. No treatment-related deaths or late complications were reported. Conclusion: SBRT appears to be a promising alternative to surgery or ablative therapy to treat primary RCC in frail patients.

Keywords: stereotactic body radiotherapy; renal cell carcinoma; frail patients; oncological outcomes

1. Introduction

Renal cell carcinoma (RCC) represents around 3% of all cancers, with the highest incidence occurring in Western countries [1]. Surgical therapy is currently the standard of care for the treatment of primary RCC in fit patients with adequate renal function [2]. Given the demographics of patients with RCC, many older patients have comorbidities which may preclude them from major surgery. Alternative strategies such as cryotherapy and radiofrequency ablation have emerged as potentially curative treatment approaches for patients who refuse or are unsuitable for surgery [3–5]. However, these minimally invasive therapies are limited to small renal masses, distant from vascular structures and the upper urinary tract [6,7]. By contrast, stereotactic body radiotherapy (SBRT) is an emerging noninvasive treatment and does not necessitate inpatient hospital treatment. SBRT is delivered in single or multiple treatment sessions, and is typically associated with low toxicity and excellent local control rates in a variety of malignancies [8].

RCC is usually considered resistant to radiation delivered using conventional fractionation schedules (1.8–2 Gy per fraction). The current literature reports encouraging results of SBRT on primary RCC in terms of local control and acceptable toxicity [9]. However, there is still insufficient evidence to recommend this therapy.

The aim of our study was to report the oncological outcomes and toxicity of SBRT for frail patients with primary RCC unfit for surgery or standard alternative ablative therapies.

2. Materials and Methods

2.1. Population Study

The study was approved by the Ethics Committee of the French Urological Association (CERU_2020/014). We retrospectively reviewed the charts of all patients who had SBRT for primary RCC at our tertiary center between October 2016 and March 2020. Data were extracted from medical files and collected in a pseudo-anonymized database in accordance with GDPR regulations. Eligible patients were included who were medically unfit for surgery and were poor candidates for cryotherapy and radiofrequency ablation (tumor size > 4 cm, near from vascular pedicle or upper urinary tract). Patients who could not tolerate the prolonged supine positioning necessary for successful treatment, or with a history of abdominal or pelvic radiotherapy were excluded. No restrictions were applied regarding the histological subtype, size or stage of the tumor. All patients had a proven histological diagnosis of RCC by percutaneous tumor biopsy under computed tomography (CT) guidance and all indications for radiotherapy were confirmed by our multidisciplinary team (MDT).

2.2. Stereotactic Body Radiotherapy

Patients were positioned with a TomoTherapy™ (Accuray, Madison, WI, USA) device, which delivered arc x-ray therapy of 6 MV. Prior to CT, patients were immobilized with the BlueBAG™ (Elekta, Stockholm, Sweden) BodyFIX Vacuum Cushions system and the abdominal compression plate (ACP) to minimize breathing motion. The gross tumor volume (GTV) was contoured on different reconstructions of the simulation CT with breath-phased 3D CT scan to encompass the motion and to create an internal target volume (ITV). For the last patients included, the GTV was contoured on the different respiratory phases of a 4D simulation scan. A 5 mm expansion was given to derive the planning target volume (PTV). The dosimetry was evaluated on Accuray's VOLO™ (Accuray, Madison, WI, USA) software. The objective prescription isodose was 90%. Sessions were spaced 48 h apart and were delivered on non-consecutive days (one day interval). SBRT procedures were adapted from consensus statements from the International Radiosurgery Oncology Consortium for primary renal cell carcinoma [10,11]. The total dose administered was in accordance with De Meerleer's guidelines [12].

2.3. Follow-up and Endpoints

Patients were followed up with every 3 months for the first two years, and twice annually thereafter. The follow-up included physical examination, glomerular filtration rate (eGFR) which was estimated by the Chronic Kidney Disease Epidemiology Collaboration (CKD-EPI) equation, and radiological examination with CT-scan, MRI or ultrasound. Treatment-related toxicities were defined using Common Terminology Criteria for Adverse Events, version 4.0. The primary outcome was local control, which was defined using Response Evaluation Criteria in Solid Tumors, version 1.1 [13]. Secondary outcomes included treatment-related toxicities, evolution of renal function, event-free survival, cancer-specific survival and overall survival.

2.4. Statistical Analysis

Descriptive statistics were delineated for the available variables. Quantitative variables were reported as median and interquartile range (IQR) and analyzed by Mann–Whitney U Test. Categorical variables were described as numbers and percentages and were analyzed by Chi-squared test. Kaplan–Meier curves were generated for all time-to-event endpoints. Statistical analyses were performed using R Version 4.0.2 (Foundation for Statistical Computing, Vienna, Austria). A p-value of ≤ 0.05 was considered statistically significant.

3. Results

Between October 2016 and March 2020, 24 patients underwent SBRT for primary RCC in our center and were assessed for eligibility. One patient declined to continue radiation therapy after initial sessions and was excluded. Twenty-three patients were included in the final analysis.

3.1. Baseline Characteristics

Patient and tumor characteristics are summarized in Table 1. The median age in the study cohort was 81 (IQR 79–85) years. Eight female patients (34.8%) were included and the median Charlson Comorbidity score was 7 (IQR 5–8). Pathologic confirmation before treatment was achieved in all cases. Overall, 73.9% (17/23) of patients harbored clear cell RCC. Stages T1a, T1b and T2 were recorded in 56.5, 39.1 and 4.4% of cases, respectively. A high Fuhrman grade (G3–4) was recorded in 30.4% of all patients. Four patients (17.4%) had M1 disease. The median maximal tumor size was 40 (IQR 28–48) mm.

Table 2 shows dose fractionation schedules. The most used dose schedule was 35 Gy (78.3%) in five fractions (43.5%, 10/23) or seven fractions (34.8%, 8/23). No complications or technical difficulties were recorded during sessions.

3.2. Recurrence and Survival

The median duration of follow-up for all living patients was 22 (IQR 10–39) months. At the time of most recent follow-up, we did not reach 50% recurrence, events or death in the study group. One patient experienced local recurrence after 36 months of follow-up. We observed a significant correlation between time of follow-up and decreased size of primary tumor (Figure 1).

Lymph node and metastatic recurrences were recorded in one and four cases, respectively, including two cases in patients who were already M1 before SBRT. A total of four deaths were observed: one case related to disease progression and three deaths from another cause. Thus, after a median follow-up of 22 months, local recurrence-free survival, event-free survival, cancer-specific survival and overall survival were 96, 74, 96 and 83%, respectively (Figure 2).

Table 1. Baseline characteristics.

Variables	Overall Cohort (n = 23)
Gender, n (%)	
Male	15 (65.2)
Female	8 (34.8)
Median (IQR) age, years	81 (79–85)
ECOG performance score, n (%)	
0	10 (43.5)
1	11 (47.8)
≥2	2 (8.7)
Median (IQR) Charlson score	7 (5–8)
Comorbidities, n (%)	
Diabetes mellitus	8 (34.8)
Arterial hypertension	14 (60.9)
Median (IQR) creatinine clearance, mL/mn/1.73 m^2	57 (34–75)
Tumor stage, n (%)	
T1a	13 (56.5)
T1b	9 (39.1)
T2	1 (4.4)
Metastatic stage, n (%)	
M0	19 (82.6)
M1	4 (17.4)
Tumor side, n (%)	
Right	13 (56.5)
Left	10 (43.5)
Fuhrman grade, n (%)	
Low grade (1–2)	14 (60.9)
High grade (3–4)	7 (30.4)
Unknown	2 (8.7)
RCC type, n (%)	
Clear cell	17 (73.9)
Papillary	2 (8.7)
Other	4 (17.4)
Median (IQR) tumor size, mm	40 (28–48)

Legend: ECOG: Eastern Cooperative Oncology Group; RCC: renal cell carcinoma.

Table 2. Dose fractionation schedules.

Dose (Gy)/Fraction (Fr)	n (%)
35 Gy/5 Fr	10 (43.5)
35 Gy/7 Fr	8 (34.8)
36 Gy/3 Fr	4 (17.4)
24 Gy/3 Fr	1 (4.3)

Legend: Gy: gray; Fr: fraction.

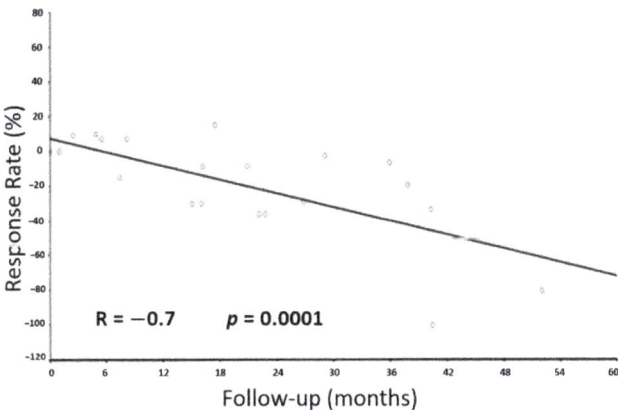

Figure 1. Correlation between tumor size and follow-up.

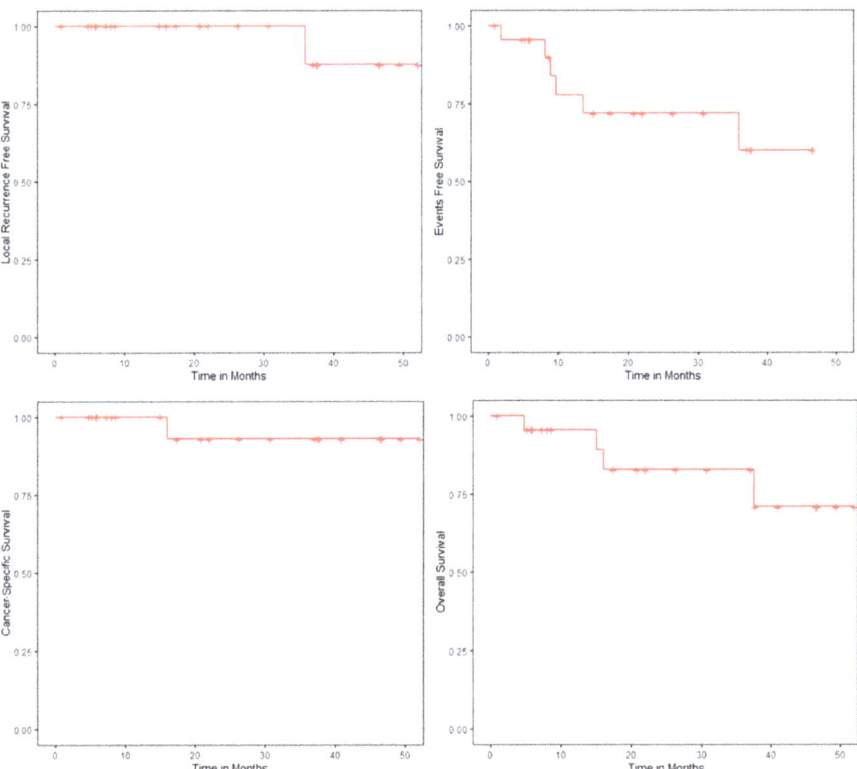

Figure 2. Kaplan–Meier estimates of local recurrence-free survival; event-free survival; cancer-specific survival and overall survival.

3.3. Evolution of Renal Function

The median baseline eGFR was 57 (IQR 34–75) mL per minute. The median change in eGFR at the most recent follow-up was −7 (IQR −17; 0) mL per minute ($p = 0.15$). No patients required dialysis during the study period.

3.4. Toxicity

A total of five side effects (21.7%) were recorded. Grade 2 asthenia was observed in two cases. One grade 2 epigastralgia and two grade 1 episodes of nausea were also recorded. There were no grade 3–4 side effects recorded. No treatment-related deaths or late complications were recorded.

4. Discussion

Over the past decade, SBRT has emerged as a potential treatment option for primary RCC [4]. Defining a potential target population for SBRT is a major challenge. The main advantages are the outpatient management without anesthesia and the non-invasive approach. The majority of patients were either old patients with multiple comorbidities, or patients with tumors inaccessible to standard focal therapies. SBRT has been implemented in our practices from 2016 to treat primary RCC in patients with poor physical health who were medically inoperable and unfit for cryotherapy and radiofrequency ablation. To our knowledge and following a review of the literature, we report the first series of SBRT in this patient population.

Our study reports the promising short-term local control effects of SBRT with a TomoTherapy™ (Accuray, Madison, WI, USA) device. After a median follow-up of 22 months, only one patient (4%) experienced local recurrence. In this case, a percutaneous biopsy of the tumor was performed because intratumoral hypermetabolic foci increased and was positive without an increase in tumor size. Our results correlate with previous data from Chang et al. who reported no local recurrence after a median follow-up of 19 months in their series of 16 patients treated with 30 to 40 Gy in five fractions [14]. In a prospective cohort of 37 patients receiving either a single fraction of 26 Gy or three fractions of 14 Gy for inoperable renal cell carcinoma, Siva et al. reported a freedom from local progression at 2 years of 100% [15]. Additionally, our results support those of a meta-analysis which enrolled 126 patients from 10 studies [9]. The most commonly employed fractionation schedule was 40 Gy, delivered over five fractions, and a local control was reported of 93.91% (range: 84–100%) [9]. A pooled multi-institutional analysis of IROCK focused on a cohort of 223 patients treated for primary RCC by stereotaxic ablative radiotherapy [11]. The patients were divided into two groups: those receiving a single fraction with an average dose of 25 Gy (14–26 Gy), and those receiving multiple fractions with an average total dose of 40 Gy (2 to 10 fractions). The local control rate at five years was 97.8% and was similar between both groups. The authors reported in the group of patients treated with a multifractional regimen a significantly lower rate of nausea but a poorer progression-free survival and cancer-specific survival. The authors raised the possibility of an abscopal effect in the monofraction group. The immunosensitive nature of CCR was also mentioned in the study by Siva et al. [15]. The possibility of SBRT-mediated immunomodulation is based on preclinical in vivo data and then on reported cases of patients with an abscopal effect [16,17]. Our team modeled the synergy between SBRT and checkpoint inhibitor therapies, emphasizing the advantage of delivering these two treatments concomitantly [18]. In addition, we developed a new concept of immunologically effective dose (IED) varying with the radiotherapy regimen used, complementary to the classic linear–quadratic formula [19]. In our series, some patients presented with an increase in tumor size 12–15 months after SBRT sessions no greater than 20%, but decrease after that time, as reported by Ponski et al. [20] and Chang J.H. [14]. This initial, and temporary, increase in tumor size could be explained by an immunological stimulating effect of SBRT.

The safety of SBRT was confirmed in the present study; 21.7% grade 1–2 side effects with no grade 3 or higher event occurring. Our results are very similar to previous data from

Siva et al. [15] and Correa et al. [21], which reported 3 and 1.5% occurrence, respectively, of grade ≥3 side effects. Furthermore, SBRT had a modest impact on renal function, with a mean reduction in eGFR at last follow-up of −7 mL/min. These outcomes are in agreement with previous data [22] and comparable to those of nephron-sparing approaches such as PN or ablative therapies [23–25].

The present study has several limitations that should be acknowledged. The main limitations concern its retrospective design, the small cohort included and the short follow-up. Conversely, the main strength of this preliminary study was to report the outcomes of SBRT in a cohort of frail patients which has not previously been reported in the literature. All patients presented with a progressive tumor after active surveillance and/or a tumor with high histological grade. Future studies are essential to determine the place of SBRT in the therapeutic arsenal of RCC. An ongoing clinical trial will provide additional important findings on the subject [26]. To date, there is no prospective randomized study comparing SBRT to standard ablative therapies. The meta-analysis by Kunkle et al. [5] that included 1375 primary RCC patients treated with radiofrequency ablation or cryoablation reported a local control rate of 87.1 and 94.8%, respectively, with a median follow-up of 18.7 months. In addition, the average tumor diameter was 26 mm, against 40mm in our study.

5. Conclusions

SBRT appears to be a promising alternative to surgery or ablative therapy to treat primary RCC in patients with poor physical health. Future studies are needed to definitively assess the place of SBRT in the RCC treatment portfolio.

Author Contributions: Study concept and design: L.G., M.B. and X.M.; Acquisition of data: R.D., L.G., M.B. and B.G.-T.; Analysis and interpretation of data: A.C., L.G., M.B., B.G.-T., E.L. and X.M.; Drafting of the manuscript: L.G., R.M., M.B. and X.M.; Critical revision of the manuscript for important intellectual content: A.-L.C., J.-L.D., L.P., V.D. and R.B.; Statistical analysis: B.G.-T. and X.M. All authors have read and agreed to the published version of the manuscript.

Funding: None.

Institutional Review Board Statement: The study was conducted according to the guidelines of the Declaration of Helsinki and approved by the Institutional Review Board of the French Urological Association (CERU_2020/014).

Informed Consent Statement: Informed consent was not required due to the retrospective design.

Data Availability Statement: The data presented in this study are available on request from the corresponding author.

Conflicts of Interest: The authors declare no conflict of interest.

References

1. Ferlay, J.; Colombet, M.; Soerjomataram, I.; Dyba, T.; Randi, G.; Bettio, M.; Gavin, A.; Visser, O.; Bray, F. Cancer incidence and mortality patterns in Europe: Estimates for 40 countries and 25 major cancers in 2018. *Eur. J. Cancer* **2018**, *103*, 356–387. [CrossRef] [PubMed]
2. Ljungberg, B.; Albiges, L.; Abu-Ghanem, Y.; Bensalah, K.; Dabestani, S.; Fernández-Pello, S.; Giles, R.H.; Hofmann, F.; Hora, M.; Kuczyk, M.A.; et al. European Association of Urology Guidelines on Renal Cell Carcinoma: The 2019 Update. *Eur. Urol.* **2019**, *75*, 799–810. [CrossRef]
3. Zargar, H.; Atwell, T.D.; Cadeddu, J.A.; de la Rosette, J.J.; Janetschek, G.; Kaouk, J.H.; Matin, S.F.; Polascik, T.J.; Zargar-Shoshtari, K.; Thompson, R.H. Cryoablation for Small Renal Masses: Selection Criteria, Complications, and Functional and Oncologic Results. *Eur. Urol.* **2016**, *69*, 116–128. [CrossRef]
4. Trudeau, V.; Larcher, A.; Boehm, K.; Dell'Oglio, P.; Sun, M.; Tian, Z.; Briganti, A.; Shariat, S.F.; Jeldres, C.; Karakiewicz, P.I. Comparison of Postoperative Complications and Mortality Between Laparoscopic and Percutaneous Local Tumor Ablation for T1a Renal Cell Carcinoma: A Population-based Study. *Urology* **2016**, *89*, 63–67. [CrossRef]
5. Kunkle, D.A.; Uzzo, R.G. Cryoablation or radiofrequency ablation of the small renal mass: A meta-analysis. *Cancer* **2008**, *113*, 2671–2680. [CrossRef]

6. Salagierski, M.; Wojciechowska, A.; Zając, K.; Klatte, T.; Thompson, R.H.; Cadeddu, J.A.; Kaouk, J.; Autorino, R.; Ahrar, K.; Capitanio, U.; et al. The Role of Ablation and Minimally Invasive Techniques in the Management of Small Renal Masses. *Eur. Urol. Oncol.* **2018**, *1*, 395–402. [CrossRef]
7. Kunkle, D.A.; Egleston, B.L.; Uzzo, R.G. Excise, ablate or observe: The small renal mass dilemma–a meta-analysis and review. *J. Urol.* **2008**, *179*, 1227–1233. [CrossRef]
8. Lo, S.S.; Fakiris, A.J.; Chang, E.L.; Mayr, N.A.; Wang, J.Z.; Papiez, L.; Teh, B.S.; McGarry, R.C.; Cardenes, H.R.; Timmerman, R.D. Stereotactic body radiation therapy: A novel treatment modality. *Nat. Rev. Clin. Oncol.* **2010**, *7*, 44–54. [CrossRef]
9. Siva, S.; Pham, D.; Gill, S.; Corcoran, N.M.; Foroudi, F. A systematic review of stereotactic radiotherapy ablation for primary renal cell carcinoma. *BJU Int.* **2012**, *110*, 737–743. [CrossRef]
10. Siva, S.; Daniels, C.P.; Ellis, R.J.; Ponsky, L.; Lo, S.S. Stereotactic ablative body radiotherapy for primary kidney cancer: What have we learned from prospective trials and what does the future hold? *Future Oncol.* **2016**, *12*, 601–606. [CrossRef]
11. Siva, S.; Louie, A.V.; Warner, A.; Muacevic, A.; Gandhidasan, S.; Ponsky, L.; Ellis, R.; Kaplan, I.; Mahadevan, A.; Chu, W.; et al. Pooled analysis of stereotactic ablative radiotherapy for primary renal cell carcinoma: A report from the International Radiosurgery Oncology Consortium for Kidney (IROCK). *Cancer* **2018**, *124*, 934–942. [CrossRef]
12. De Meerleer, G.; Khoo, V.; Escudier, B.; Joniau, S.; Bossi, A.; Ost, P.; Briganti, A.; Fonteyne, V.; Van Vulpen, M.; Lumen, N.; et al. Radiotherapy for renal-cell carcinoma. *Lancet Oncol.* **2014**, *15*, 170–177. [CrossRef]
13. Eisenhauer, E.A.; Therasse, P.; Bogaerts, J.; Schwartz, L.H.; Sargent, D.; Ford, R.; Dancey, J.; Arbuck, S.; Gwyther, S.; Mooney, M.; et al. New response evaluation criteria in solid tumours: Revised RECIST guideline (version 1.1). *Eur. J. Cancer* **2009**, *45*, 228–247. [CrossRef]
14. Chang, J.H.; Cheung, P.; Erler, D.; Sonier, M.; Korol, R.; Chu, W. Stereotactic Ablative Body Radiotherapy for Primary Renal Cell Carcinoma in Non-surgical Candidates: Initial Clinical Experience. *Clin. Oncol. (R. Coll. Radiol.)* **2016**, *28*, 109–114. [CrossRef]
15. Siva, S.; Pham, D.; Kron, T.; Bressel, M.; Lam, J.; Tan, T.H.; Chesson, B.; Shaw, M.; Chander, S.; Gill, S.; et al. Stereotactic ablative body radiotherapy for inoperable primary kidney cancer: A prospective clinical trial. *BJU Int.* **2017**, *120*, 623–630. [CrossRef]
16. Rodríguez-Ruiz, M.E.; Vanpouille-Box, C.; Melero, I.; Formenti, S.C.; Demaria, S. Immunological Mechanisms Responsible for Radiation-Induced Abscopal Effect. *Trends Immunol.* **2018**, *39*, 644–655. [CrossRef]
17. Weichselbaum, R.R.; Liang, H.; Deng, L.; Fu, Y.X. Radiotherapy and immunotherapy: A beneficial liaison? *Nat. Rev. Clin. Oncol.* **2017**, *14*, 365–379. [CrossRef]
18. Serre, R.; Benzekry, S.; Padovani, L.; Meille, C.; André, N.; Ciccolini, J.; Barlesi, F.; Muracciole, X.; Barbolosi, D. Mathematical Modeling of Cancer Immunotherapy and Its Synergy with Radiotherapy. *Cancer Res.* **2016**, *76*, 4931–4940. [CrossRef]
19. Serre, R.; Barlesi, F.; Muracciole, X.; Barbolosi, D. Immunologically effective dose: A practical model for immuno-radiotherapy. *Oncotarget* **2018**, *9*, 31812–31819. [CrossRef] [PubMed]
20. Ponsky, L.; Lo, S.S.; Zhang, Y.; Schluchter, M.; Liu, Y.; Patel, R.; Abouassaly, R.; Welford, S.; Gulani, V.; Haaga, J.R.; et al. Phase I dose-escalation study of stereotactic body radiotherapy (SBRT) for poor surgical candidates with localized renal cell carcinoma. *Radiother. Oncol.* **2015**, *117*, 183–187. [CrossRef] [PubMed]
21. Correa, R.J.M.; Louie, A.V.; Zaorsky, N.G.; Lehrer, E.J.; Ellis, R.; Ponsky, L.; Kaplan, I.; Mahadevan, A.; Chu, W.; Swaminath, A.; et al. The Emerging Role of Stereotactic Ablative Radiotherapy for Primary Renal Cell Carcinoma: A Systematic Review and Meta-Analysis. *Eur. Urol. Focus* **2019**, *5*, 958–969. [CrossRef]
22. Siva, S.; Correa, R.J.M.; Warner, A.; Staehler, M.; Ellis, R.J.; Ponsky, L.; Kaplan, I.D.; Mahadevan, A.; Chu, W.; Gandhidasan, S.; et al. Stereotactic Ablative Radiotherapy for $\geq$T1b Primary Renal Cell Carcinoma: A Report From the International Radiosurgery Oncology Consortium for Kidney (IROCK). *Int. J. Radiat. Oncol. Biol. Phys.* **2020**, *108*, 941–949. [CrossRef]
23. Scosyrev, E.; Messing, E.M.; Sylvester, R.; Campbell, S.; Van Poppel, H. Renal function after nephron-sparing surgery versus radical nephrectomy: Results from EORTC randomized trial 30904. *Eur. Urol.* **2014**, *65*, 372–377. [CrossRef]
24. Guillotreau, J.; Haber, G.P.; Autorino, R.; Miocinovic, R.; Hillyer, S.; Hernandez, A.; Laydner, H.; Yakoubi, R.; Isac, W.; Long, J.A.; et al. Robotic partial nephrectomy versus laparoscopic cryoablation for the small renal mass. *Eur. Urol.* **2012**, *61*, 899–904. [CrossRef] [PubMed]
25. Pan, X.W.; Cui, X.M.; Huang, H.; Huang, Y.; Li, L.; Wang, Z.J.; Qu, F.J.; Gao, Y.; Cui, X.G.; Xu, D.F. Radiofrequency ablation versus partial nephrectomy for treatment of renal masses: A systematic review and meta-analysis. *Kaohsiung J. Med. Sci.* **2015**, *31*, 649–658. [CrossRef] [PubMed]
26. Siva, S.; Chesson, B.; Bressel, M.; Pryor, D.; Higgs, B.; Reynolds, H.M.; Hardcastle, N.; Montgomery, R.; Vanneste, B.; Khoo, V.; et al. TROG 15.03 phase II clinical trial of Focal Ablative STereotactic Radiosurgery for Cancers of the Kidney—FASTRACK II. *BMC Cancer* **2018**, *18*, 1030. [CrossRef] [PubMed]

Article

Clinical Features and Multiplatform Molecular Analysis Assist in Understanding Patient Response to Anti-PD-1/PD-L1 in Renal Cell Carcinoma

Eileen Shiuan [1,2], Anupama Reddy [3], Stephanie O. Dudzinski [1,4], Aaron R. Lim [1,2], Ayaka Sugiura [1,5], Rachel Hongo [6], Kirsten Young [7], Xian-De Liu [8], Christof C. Smith [9,10], Jamye O'Neal [6], Kimberly B. Dahlman [6], Renee McAlister [6], Beiru Chen [11], Kristen Ruma [11], Nathan Roscoe [11], Jehovana Bender [11], Joolz Ward [12], Ju Young Kim [11], Christine Vaupel [11], Jennifer Bordeaux [11], Shridar Ganesan [13,14], Tina M. Mayer [13,14], Gregory M. Riedlinger [15], Benjamin G. Vincent [9,10,16], Nancy B. Davis [6], Scott M. Haake [6,17], Jeffrey C. Rathmell [5], Eric Jonasch [8], Brian I. Rini [6], W. Kimryn Rathmell [6] and Kathryn E. Beckermann [6,*]

1. Medical Scientist Training Program, Vanderbilt University, Nashville, TN 37232, USA; eileen.f.shiuan@vanderbilt.edu (E.S.); stephanie.o.dudzinski@vanderbilt.edu (S.O.D.); aaron.lim@vanderbilt.edu (A.R.L.); ayaka.sugiura@vanderbilt.edu (A.S.)
2. Program in Cancer Biology, Vanderbilt University, Nashville, TN 37232, USA
3. Prism Bioanalytics, North Carolina Biotechnology Center, Morrisville, NC 27560, USA; areddy@prismbioanalytics.com
4. Department of Biomedical Engineering, Vanderbilt University, Nashville, TN 37232, USA
5. Department of Pathology, Microbiology, and Immunology, Vanderbilt University, Nashville, TN 37232, USA; jeff.rathmell@vumc.org
6. Division of Hematology/Oncology, Department of Medicine, Vanderbilt University Medical Center, Nashville, TN 37232, USA; rachel.hongo@vumc.org (R.H.); jamye.oneal@vumc.org (J.O.); kim.dahlman@vumc.org (K.B.D.); renee.k.mcalister@vumc.org (R.M.); nancy.davis@vumc.org (N.B.D.); scott.haake@vumc.org (S.M.H.); brian.rini@vumc.org (B.I.R.); kimryn.rathmell@vumc.org (W.K.R.)
7. Health Science Center, College of Medicine, The University of Tennessee, Memphis, TN 38163, USA; kyoung50@uthsc.edu
8. Department of Genitourinary Medical Oncology, The University of Texas MD Anderson Cancer Center, Houston, TX 77030, USA; XLiu10@mdanderson.org (X.-D.L.); EJonasch@mdanderson.org (E.J.)
9. Lineberger Comprehensive Cancer Center, University of North Carolina at Chapel Hill, Chapel Hill, NC 27514, USA; chrsmit@email.unc.edu (C.C.S.); benjamin_vincent@med.unc.edu (B.G.V.)
10. Department of Microbiology and Immunology, University of North Carolina at Chapel Hill, Chapel Hill, NC 27514, USA
11. Navigate Biopharma Services, Inc., A Novartis Subsidiary, Carlsbad, CA 92008, USA; beiru.chen@navigatebp.com (B.C.); kristen.ruma@navigatebp.com (K.R.); nathan.roscoe@navigatebp.com (N.R.); jo.bender@navigatebp.com (J.B.); ju_young.kim@navigatebp.com (J.Y.K.); christine.vaupel@navigatebp.com (C.V.); jennifer.bordeaux@navigatebp.com (J.B.)
12. Thermo Fisher Scientific, Sacramento, CA 95605, USA; Joolz.Ward@ThermoFisher.com
13. Rutgers Cancer Institute of New Jersey, New Brunswick, NJ 08901, USA; ganesash@cinj.rutgers.edu (S.G.); mayertm@cinj.rutgers.edu (T.M.M.)
14. Department of Medicine, Rutgers Robert Wood Johnson Medical School, Rutgers University, New Brunswick, NJ 08901, USA
15. Department of Pathology, Rutgers Robert Wood Johnson Medical School, Rutgers University, New Brunswick, NJ 08901, USA; gr338@cinj.rutgers.edu
16. Curriculum in Bioinformatics and Computational Biology, Computational Medicine Program, University of North Carolina at Chapel Hill, Chapel Hill, NC 27514, USA
17. Nashville Veterans Affairs Medical Center, Nashville, TN 37212, USA
* Correspondence: katy.beckermann@vumc.org; Tel.: +1-615-343-6653; Fax: +1-615-343-7602

Citation: Shiuan, E.; Reddy, A.; Dudzinski, S.O.; Lim, A.R.; Sugiura, A.; Hongo, R.; Young, K.; Liu, X.-D.; Smith, C.C.; O'Neal, J.; et al. Clinical Features and Multiplatform Molecular Analysis Assist in Understanding Patient Response to Anti-PD-1/PD-L1 in Renal Cell Carcinoma. *Cancers* **2021**, *13*, 1475. https://doi.org/10.3390/cancers13061475

Academic Editor: José I. López

Received: 19 February 2021
Accepted: 16 March 2021
Published: 23 March 2021

Publisher's Note: MDPI stays neutral with regard to jurisdictional claims in published maps and institutional affiliations.

Copyright: © 2021 by the authors. Licensee MDPI, Basel, Switzerland. This article is an open access article distributed under the terms and conditions of the Creative Commons Attribution (CC BY) license (https://creativecommons.org/licenses/by/4.0/).

Simple Summary: Immune checkpoint inhibitor (ICI) therapy has proven effective for many cancer patients, but predicting which patients with renal cell carcinoma (RCC) will respond has been challenging. We analyzed clinical characteristics and molecular parameters of a cohort of patients with RCC treated with anti-programmed death 1 (PD-1)/PD-L1 therapy to determine factors that correlate with patient outcome. We found that the composition of circulating immune cells in the blood, development of immune-related toxicities, and gene expression patterns within the tumor correlate with patient response. In addition, we see that high expression of PD-L1 and lower numbers of unique T cell clones in RCC tumors are associated with improved survival. In summary, our findings corroborate previously published work and introduce new potential factors impacting response to ICI therapy that deserve further investigation.

Abstract: Predicting response to ICI therapy among patients with renal cell carcinoma (RCC) has been uniquely challenging. We analyzed patient characteristics and clinical correlates from a retrospective single-site cohort of advanced RCC patients receiving anti-PD-1/PD-L1 monotherapy (N = 97), as well as molecular parameters in a subset of patients, including multiplexed immunofluorescence (mIF), whole exome sequencing (WES), T cell receptor (TCR) sequencing, and RNA sequencing (RNA-seq). Clinical factors such as the development of immune-related adverse events (odds ratio (OR) = 2.50, 95% confidence interval (CI) = 1.05–5.91) and immunological prognostic parameters, including a higher percentage of circulating lymphocytes (23.4% vs. 17.4%, p = 0.0015) and a lower percentage of circulating neutrophils (61.8% vs. 68.5%, p = 0.0045), correlated with response. Previously identified gene expression signatures representing pathways of angiogenesis, myeloid inflammation, T effector presence, and clear cell signatures also correlated with response. High PD-L1 expression (>10% cells) as well as low TCR diversity ($\leq$644 clonotypes) were associated with improved progression-free survival (PFS). We corroborate previously published findings and provide preliminary evidence of T cell clonality impacting the outcome of RCC patients. To further biomarker development in RCC, future studies will benefit from integrated analysis of multiple molecular platforms and prospective validation.

Keywords: renal cell carcinoma; PD-1; PD-L1; biomarkers; immune checkpoint inhibitors

1. Introduction

Over the past decade, immune checkpoint inhibitors (ICIs), including antibodies against the programmed death 1 (PD-1) receptor, its ligand (PD-L1), and cytotoxic T-lymphocyte-associated protein 4 (CTLA-4), have become a mainstay of treatment against cancer. Patients with metastatic renal cell carcinoma (RCC) have overall response rates (ORRs) to single-agent PD-1/PD-L1 blockade in the first- and second-line setting of approximately 16–34% [1–4]. The current standard of care in the frontline setting is combination therapy using anti-PD-1/PD-L1 with either anti-CTLA-4 or a vascular endothelial growth factor (VEGF)-targeting agent, which yields ORRs of 40–60% [4–9].

Predicting response to ICI therapy in patients with RCC has proven to be difficult. The predictive value of tumor PD-L1 expression and mutational burden (TMB), which are used as companion diagnostic biomarkers for other tumor types, remains equivocal in RCC, with a number of studies demonstrating no correlation with response [2,3,6,7,9–11]. Results of standard clinical tests on peripheral blood, including elevated absolute lymphocyte count (ALC) [12], or lower absolute neutrophil count (ANC) [12,13], neutrophil-to-lymphocyte ratio (NLR) [12–15], and monocyte-to-lymphocyte ratio (MLR) [14], have been associated with better response in solid tumors but not prospectively validated. In addition, the relationship between body mass index (BMI) and response is still disputable, with studies demonstrating improved response and survival in RCC patients with higher BMI [16,17] and others showing the opposite [18].

Molecular studies have shed light on the biological response behind anti-PD-1 monotherapy, such as the presence of endogenous retroviruses [19,20] and differential expression of gene signatures, including T cell effector function [10], interferon (IFN) or tumor necrosis factor (TNFα) signaling [21], and metabolic gene signatures [22]. In the randomized trials IMmotion150 and IMmotion151, molecular signatures of response were assessed in treatment-naïve patients who received sunitinib, the combination of atezolizumab (PD-L1 antibody) and bevacizumab (vascular endothelial growth factor (VEGF) antibody), or, in the case of IMmotion150, atezolizumab monotherapy [10,23]. A T effector signature correlated with response in both ICI monotherapy and combination therapy arms, while the angiogenic signature correlated with response in anti-VEGF monotherapy and combination therapy. Furthermore, responders to atezolizumab monotherapy generally had lower inflammatory myeloid signatures and higher T effector signatures, while patients with high suppressive myeloid signatures were more likely to achieve response if VEGF inhibition was combined with checkpoint inhibition [10]. More in-depth analysis from IMmotion150 further categorized RCC patients based on integration of various molecular parameters into seven molecular subsets, which appeared to have differential clinical outcomes to sunitinib versus atezolizumab plus bevacizumab [23].

Other investigations have attempted to correlate genetic drivers of RCC with response to ICI. The role of *PBRM1*, the second most commonly mutated gene in clear cell RCC (ccRCC) and a component of the chromatin remodeling complex, is heavily contested, with several studies yielding mixed results [11,24–27]. Recently, Braun et al. analyzed 592 ccRCC samples from patients in prospective trials of PD-1 blockade using whole exome sequencing (WES), RNA-sequencing (RNA-seq), and immunofluorescence (IF) analysis [11]. The authors found that although TMB and CD8+ T cell infiltration do not correlate with response, additional chromosomal abnormalities are specifically associated with response or resistance to anti-PD-1 monotherapy. For example, chromosomal loss of 9p21.3 was associated with decreased response among tumors with high CD8+ T cell infiltration. In addition, truncating mutations in *PBRM1* were associated with higher angiogenesis gene expression and lower IL6-JAK-STAT3 signaling, as well as improved survival with anti-PD-1 monotherapy [11,24,25].

In this study, we evaluated an institutional cohort of patients with RCC who received single-agent anti-PD-1/PD-L1 to study both clinical and molecular correlates of response. In our cohort of 94 patients, we analyzed clinical characteristics and laboratory data during the course of ICI treatment. Our study shows the feasibility of performing multiple molecular analyses, including mIF, WES, TCR sequencing, and RNA-seq, on a biomarker cohort created from available archived tumor samples.

2. Materials and Methods

2.1. Patient Population and Data Collection

Patients with RCC who had been treated with single-agent anti-PD-1/PD-L1 at Vanderbilt University Medical Center (VUMC) between 2007 and 2017 were identified under an investigator review board (IRB)-approved protocol and verified to have sufficient documentation to assess response to therapy, defined by a minimum of a baseline and three-month computed tomography (CT) scan after initiating ICI therapy (Figure 1A). Objective response was evaluated by investigators. Of the 94 patients, 18 had available archived formalin-fixed, paraffin-embedded (FFPE) tumor tissue specimens at VUMC. These specimens and their matched normal samples, when available, combined with eight external tumor specimens from Rutgers University, were used for molecular studies in the biomarker cohort. A total of 26 patients were represented by 21 primary and 16 metastatic tumor samples.

2.2. DNA and RNA Extraction

DNA and RNA were extracted from FFPE RCC and normal tissue using the Maxwell 16 FFPE Plus LEV DNA Purification Kit (Promega (Madison, WI, USA) #AS1135) and RNA Purification Kit (Promega #AS1260) respectively, on the Promega Maxwell 16 instrument.

2.3. mIF Immunohistochemistry

2.3.1. Immunofluorescence Staining

Antibody validation for the PD-1, PD-L1, and cytokeratin panel was performed as previously indicated in Johnson et al. [28] and can be found in Table S1. The CD4, CD8, CD25, FoxP3, Ki67, cytokeratin panel was performed using a fully automated staining protocol on the Bond Rx (Leica). Slides were dewaxed using the Bond Rx followed by antigen retrieval in Epitope Retrieval Solution 2 (Leica) buffer at 95 °C for 20 min. Primary antibodies were incubated for 1 h, detected with either EnVision+ HRP Mouse or EnVision+ HRP Rabbit for 30 min and slides were heat-cycled in ER1 (Leica) buffer for 20 min at 95 °C after each round of primary, secondary, and Opal fluorophore staining. Slides were stained with 1:100 dilution of mouse anti-CD4 (4B12, Agilent, Santa Clara, CA, USA), detected with Opal 520 (Akoya Biosciences, Marlborough, MA, USA), 1:400 dilution of mouse anti-CD8 (C8/144B, Agilent), detected with Opal 620 (Akoya Biosciences), 1:100 dilution of rabbit anti-FoxP3 (D2W8E, Cell Signaling Technology, Danvers, MA, USA), detected with Opal 540 (Akoya Biosciences), 1:400 dilution of rabbit anti-CD25 (SP176, Sigma Aldrich, St. Louis, MO, USA), detected with Opal 570 (Akoya Biosciences), 1:1000 dilution of mouse anti-Ki67 (MIB-1, Agilent), detected with Opal 650 (Akoya Biosciences), and 1:400 dilution of mouse anti-cytokeratin (AE1/AE3, Agilent), detected with Opal 690, and finally, incubated for 10 min with spectral 4′,6-diamidino-2-phenylindole (DAPI) (Akoya Biosciences).

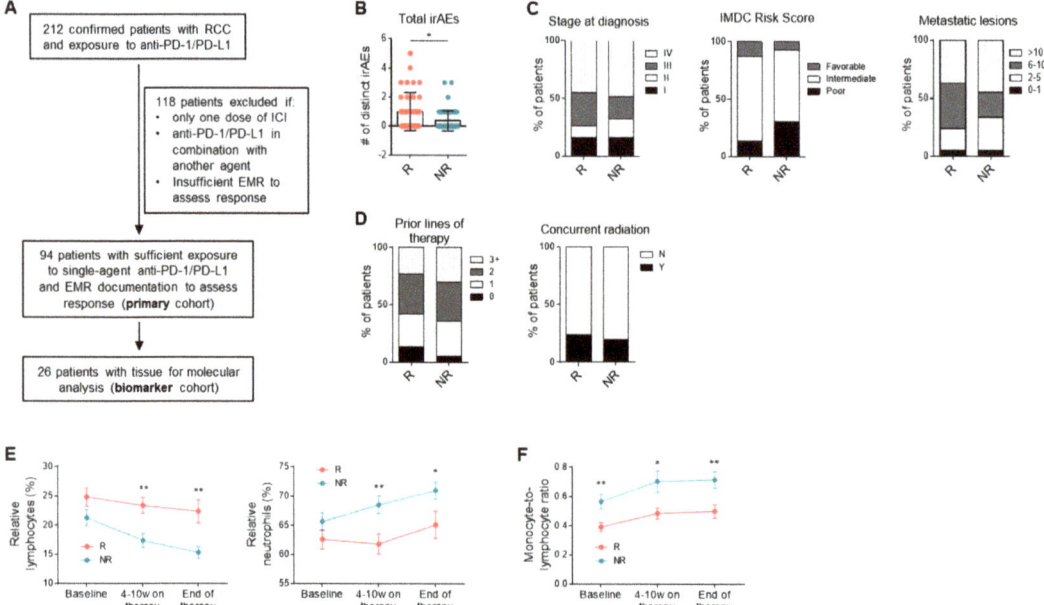

Figure 1. Clinical correlates and response to single-agent anti-PD-1/PD-L1. (**A**) Flow diagram depicting subdivision of primary and biomarker patient cohorts. (**B**) Number of immune-related adverse events (irAEs) experienced by responders and non-responders. Data shown are averages ± standard deviation (SD). * $p < 0.05$, unpaired Mann–Whitney U-test ($n = 38, 56$). (**C**) Stage at diagnosis, International Metastatic RCC Database Consortium (IMDC) risk score, and number of metastatic lesions of at initiation of ICI therapy. (**D**) Prior lines of therapy before ICI therapy, as well as concurrent radiation during ICI therapy. (**E**) Percentage of lymphocytes (two-way analysis of variance (ANOVA): response effect, $p < 0.0001$; time effect, $p = 0.012$; interaction, $p = 0.44$) and neutrophils (response effect, $p = 0.0002$; time effect, $p = 0.063$; interaction, $p = 0.55$) in peripheral blood of responders compared to non-responders at baseline, 4 to 10 weeks of therapy, and end of therapy ($n = 37, 53$). Data shown are averages ± SEM. * $p < 0.05$, ** $p < 0.01$, post-hoc two-tailed unpaired Welch's t test, uncorrected for multiple comparisons. (**F**) Monocyte-to-lymphocyte ratio (MLR) (response effect, $p < 0.0001$; time effect, $p = 0.047$; interaction, $p = 0.91$) in responders compared to non-responders ($n = 36, 52$).

2.3.2. Sample Imaging

Fluorescence imaging was obtained as indicated in Johnson et al. [28]. The CD4/CD8/ CD25/FoxP3/Ki67/CK assay was imaged using Vectra 3 software (Akoya Biosciences), where the whole slide was scanned at 4× for DAPI, fluorescein isothiocyanate (FITC), CY3, Texas Red, and Cy5, and an automated algorithm was used to enrich for areas with CD25 and FoxP3 staining. The images were then reviewed by a pathologist and adjusted to ensure tumor areas were included in the imaging before 20× multispectral images were acquired of up to 40 fields of view. Accepted images were processed with AQUA import tool (Navigate) to generate spectrally unmixed images for analysis.

2.4. WES and Analysis

Raw sequencing data, paired fastq files, were processed using Genome Analysis Toolkit (GATK) [29]. Briefly, reads were aligned using Burrows-Wheeler Aligner Maximum Exact Match (BWA-MEM) [30] against the hg38 human reference and duplicate reads were marked. Joint variant calling was performed using Haplotype Caller [31]. Analysis was performed using Terra [32] on Google Cloud Platform. Variants were annotated using Annotate Variation (ANNOVAR) [33] and filtered using standard filtering criteria to obtain high quality, likely somatic variants, and variant classification. The following filters were used: (1) quality filters: quality by depth, read position bias, (2) exclusion of non-exonic variants, (3) exclusion of variants with missing genotypes found in >25% of the samples, (4) exclusion of variants found in matched normal samples, (5) exclusion of synonymous variants or variants of unknown function, and (6) max population frequency < 0.001 using the following population datasets: ExAC [34], gnomAD [35], 1000 G [36], and ESP6500 [37].

Driver gene-filtering was performed as follows. Genes implicated in prior studies [38] and implicated in OncoKB [39] were retained. For the remaining genes, we applied the following filters for downstream analysis: (1) non-synonymous/synonymous ratio >2.5 or frequency of frameshifts >33%, (2) transcript size <15,000 amino acids, and (3) variability ≤ 0.06 (number of variants as a function of variant size in the ESP6500 database [37]). For the mutation heatmap, genes were sorted based on frequency, and variants were color-coded based on their effect.

2.5. TCR Sequencing and Analysis

Analysis of TCR sequencing data was performed using MIGEC and VDJtools [40]. Briefly, unique molecular barcodes (UMIs) were used to create consensus reads associated with each molecule, reducing polymerase chain reaction (PCR) and sequencing errors. Hierarchical clustering was performed on TCR genes to verify that samples from the same patient clustered together and to detect cross-contamination. TCR diversity was correlated with clinical variables, and median diversity was used as a cutoff for PFS curves.

2.6. RNA-seq and Analysis

RNA-seq libraries were generated with the Illumina TruSeq RNA Access kit, according to the manufacturer's protocol. Sequencing was performed using 2 × 100 read chemistry on an Illumina HiSeq 2500 system. Alignment of FASTQ files was performed using STAR v2.4.2a with default arguments, with downstream quantification performed using Salmon v0.8.2 in quantification mode. Quality control of BAM files was performed using Picard Tools v1.86 "CollectRnaSeqMetrics". Gene count matrices were analyzed using DESeq2 [41], and normalized gene expression counts were obtained. We performed principal component analysis (PCA) to identify groupings or outliers from the data and compared them to clinical covariates. Differential analysis was performed to identify differential genes between responders and non-responders (adjusted p-value (Benjamini–Hochberg) < 0.1). Angiogenic, T effector, inflammatory myeloid signatures, and the expanded immune and antigen-presenting gene panel were defined as previously published [10]. The contribution of immune cell types in each of the samples was identified using deconvolution methods [42]. Clear cell subtypes ccA and ccB were defined by the ClearCode34 gene [43].

2.7. Statistical Analysis

Analyses of clinical characteristics were performed using GraphPad Prism v6. Comparisons of clinical characteristics between responders and non-responders were carried out using two-way analysis of variance (ANOVA) with post-hoc unpaired Welch's t test at individual time points for laboratory data, Mann–Whitney U-tests for continuous and ordinal variables with non-normal distributions (i.e., number of irAEs), and Pearson χ^2 or Fisher's exact tests for categorical variables (i.e., previous nephrectomy). Survival analysis was performed using Kaplan–Meier estimation of survival functions followed by log-rank testing. All tests were two-sided, and a p-value less than 0.05 was statistically significant. Statistical analysis did not correct for multiple comparisons.

Analyses using molecular platforms and clinical variables were performed using R v3.6.1 (https://www.r-project.org/, accessed on 13 May 2020) and GraphPad Prism v6 (San Diego, CA, USA). For mIF, only primary clear cell RCC samples were included in comparative analyses to prevent confounding due to tissue or tumor type, and comparison of mIF data was evaluated by Mann–Whitney U-tests. Gene-level mutations obtained from WES and gene expression data were used to identify associations with clinical covariates. Survival analysis was used to identify associations with PFS, and differential analysis was used to identify associations with clinical response.

3. Results

3.1. Study Population

Of the 212 patients who were verified to have RCC and treated with any ICI therapy, 94 patients had sufficient electronic medical record (EMR) documentation and treatment with single-agent anti-PD-1 or PD-L1 therapy to assess response (Figure 1A). The demographic and treatment information for these patients are summarized in Table 1 (Table S6). Clinical characteristics were similar to those reported in trials investigating single-agent anti-PD-1/PD-L1, though our cohort had a considerably lower proportion of patients in the favorable International Metastatic RCC Database Consortium (IMDC) risk group [2,15]. Thirty-eight patients (40.4%) were considered as responders (complete response (CR), partial response (PR), mixed response), and 56 (59.6%) as non-responders (stable disease (SD) and progressive disease (PD)). Median PFS and overall survival (OS) of all patients was 6.6 months (95% confidence interval (CI): 4.4–8.7) and 23.5 months (20.4–34.1), respectively. The clinical characteristics for the biomarker cohort are detailed in Table S2 and were enriched for responders (50%) compared to the primary cohort.

3.2. Clinical Correlates and Response to Anti-PD-1/PD-L1

Consistent with previous reports, the number of irAEs experienced while on ICI therapy was higher in responders than non-responders (n = 38, 56; p = 0.012) (Figure 1B), and the odds of response were increased in patients who experienced at least one irAE compared to those who experienced none (odds ratio (OR) = 2.50, 95% CI = 1.057–5.911) (Table S3). Among patients who experienced an irAE, the type of irAE that occurred, the highest grade of irAE experienced, and the percentage of patients requiring oral and/or intravenous steroid administration did not differ significantly between responders and non-responders (Table S3). Stage at diagnosis, IMDC risk score, number of metastatic lesions at initiation of ICI therapy, number of previous lines of systemic therapy, and concurrent radiation with ICI therapy were not associated with the probability of objective response (Figure 1C, D). Although the number of metastatic sites did not differ between responders and non-responders, the presence of pancreatic metastasis correlated with decreased likelihood of response (OR = 0.257, 95% CI = 0.0683–0.968) (Table S4). Other demographic (age, gender, BMI), and treatment characteristics (previous nephrectomy, radiation, antiangiogenic agent, mammalian target of rapamycin (mTOR) inhibitor, or IL-2 therapy) were not significantly different between groups (Figure S1A,B).

Evaluation of patient laboratory values showed that while no difference in percentage of lymphocytes (n = 37, 53; 24.8% vs. 21.3%, p = 0.10) or neutrophils (n = 37, 53; 62.5% vs. 65.7%, p = 0.18) of total leukocytes was seen at time of ICI initiation, there was a significantly higher percentage of lymphocytes (23.4% vs. 17.4, p = 0.0015) and lower percentage of neutrophils (61.8% vs. 68.5%, p = 0.0045) in responders compared to non-responders early during the course of ICI therapy that was sustained until the end of therapy ($p = 0.0030$, $p = 0.038$) (Figure 1E). Additionally, MLR was lower in responders compared to non-responders throughout all measured timepoints ($n = 36, 52$; two-way ANOVA; response effect $p < 0.0001$, time effect $p = 0.047$, interaction $p = 0.91$) (Figure 1F).

Table 1. Clinical characteristics of primary cohort of patients with RCC.

Clinical Characteristic	Primary Cohort, $n = 94$	Responders (CR, PR, Mixed), $n = 38$	Non-Responders (PD, SD), $n = 56$
Best response to ICI therapy (%)			
CR	2 (2.1)	2 (5.3)	0 (0.0)
PR	23 (24.5)	23 (60.5)	0 (0.0)
SD	18 (19.1)	0 (0.0)	18 (32.1)
PD	38 (40.4)	0 (0.0)	38 (67.9)
Mixed	13 (13.8)	13 (34.2)	0 (0.0)
Median age at initiation of ICI (range), year	63 (27–82)	62 (27–79)	63 (31–82)
Sex (%)			
Male	71 (75.5)	30 (78.9)	41 (73.2)
Female	23 (24.5)	8 (21.1)	15 (26.8)
Stage at diagnosis (%)			
I	15 (16.0)	6 (15.8)	9 (16.1)
II	13 (13.8)	4 (10.5)	9 (16.1)
III	22 (23.4)	11 (28.9)	11 (19.6)
IV	44 (46.8)	17 (44.7)	27 (48.2)
Histology			
Clear cell	79 (84.0)	32 (84.2)	47 (83.9)
Papillary	4 (4.3)	1 (2.6)	3 (5.4)
Sarcomatoid	2 (2.1)	1 (2.6)	1 (1.8)
Chromophobe	2 (2.1)	0 (0.0)	2 (3.6)
Undifferentiated	7 (7.4)	4 (10.5)	3 (5.4)
IMDC risk group (%)			
Favorable	9 (9.6)	5 (13.2)	4 (7.1)
Intermediate	63 (67.0)	28 (73.7)	35 (62.5)
Poor	22 (23.4)	5 (13.2)	17 (30.4)
Previous therapies (%)			
Nephrectomy	90 (95.7)	35 (92.1)	55 (98.2)
Radiation	32 (34.0)	13 (34.2)	19 (33.9)
Anti-angiogenic agent	81 (86.2)	30 (78.9)	51 (91.1)
mTOR inhibitor	25 (26.6)	10 (26.3)	15 (26.8)
High-dose IL-2	22 (23.4)	11 (28.9)	11 (19.6)
ICI agent (%)			
Nivolumab	79 (84.0)	28 (73.7)	51 (91.1)
Atezolizumab	15 (16.0)	10 (26.3)	5 (8.9)
ICI line of therapy (%)			
First-line	8 (8.5)	5 (13.2)	3 (5.4)
Second-line	28 (29.8)	11 (28.9)	17 (30.4)
Third-line	32 (34.0)	13 (34.2)	19 (33.9)
Fourth-line+	26 (27.7)	9 (23.7)	17 (30.4)
Median duration of ICI therapy (range), days	189 (12–1637)	329 (28–1637) ****	98 (12–769) ****
Median survival (95% CI), months			
PFS	6.6 (4.4–8.7)	11.1 (9.0–23.6) ####	3.1 (2.7–5.7) ####
OS	23.5 (20.4–34.1)	43.6 (29.4–not reached) ####	16.4 (10.6–23.0) ####

**** $p < 0.0001$, two-tailed Mann-Whitney U test. #### $p < 0.0001$, log-rank test. Abbreviations: CR, complete response; PR, partial response; SD, stable disease; PD, progression of disease; ICI, immune checkpoint inhibitor; IMDC, International Metastatic RCC Database Consortium; irAE, immune-related adverse event; CI, confidence interval; PFS, progression-free survival; OS, overall survival.

3.3. PD-L1 Expression and Immune Milieu

To further our understanding of the biology underlying response to PD-1/PD-L1 blockade in patients with RCC, we performed multiplatform molecular profiling in the biomarker cohort (Table S7). PD-L1 expression was assessed on tumor, non-tumor, and all cells. The percentage of PD-L1-expressing cells was increased in responders compared to non-responders, most significantly among non-tumor cells ($n = 7, 11; p = 0.0058$) (Figure 2A,B). Prior work in melanoma suggests that the density of both PD-1 and PD-L1 expression, quantified by the interaction score, is a stronger predictor than PD-L1 alone [28]. Although PD-1 expression in non-tumor cells was not different between responders and non-responders (Figure S2A), the PD-1/PD-L1 interaction score was non-significantly higher in responders ($n = 7, 11; p = 0.055$) (Figure 2C) and a score over 200 correlated with improved PFS ($n = 6, 12$; hazard ratio (HR) = 0.38, 95% CI = 0.11–0.70) (Figure 2D). Regardless of cell type measured, PD-L1 expression correlated with PFS when using a threshold of >5% when measured on either tumor cells ($n = 10, 8$; HR = 0.31, 95% CI = 0.047–0.48) or non-tumor cells ($n = 7, 11$; HR = 0.36, 95% CI = 0.098–0.67) (Figure 2E). When looking across all cells, higher overall PD-L1 expression correlated with improved survival, which was significant when using a cutoff of 10% of all cells ($n = 6, 12$; HR = 0.30, 95% CI = 0.079–0.53) (Figure S2B).

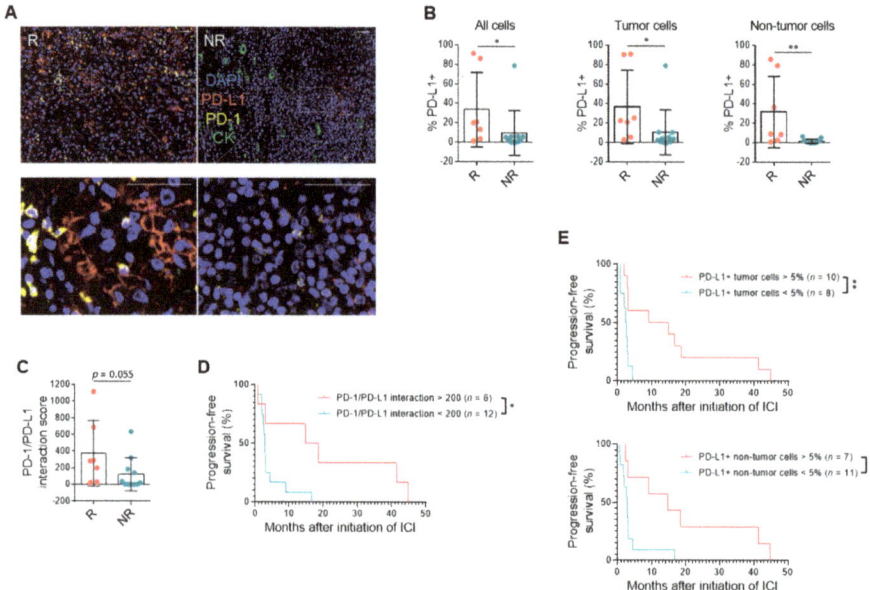

Figure 2. PD-L1 expression is associated with response to anti-PD-1/PD-L1 therapy. (**A**) Representative mIF images of primary tumors stained for DAPI (blue), PD-L1 (red), PD-1 (yellow), and cytokeratin (CK) (green) from a responder and non-responder. Scale bar: 50 µm. (**B**) Quantification of PD-L1 expression on all cells, tumor cells, and non-tumor cells in primary tumors from responders compared to non-responders. Data shown are averages ± SD. * $p < 0.05$, ** $p < 0.01$, unpaired Mann–Whitney U-test ($n = 7, 10$). (**C**) PD-1/PD-L1 interaction scores in responders versus non-responders. Data shown are averages ± SD. Unpaired Mann–Whitney U-test ($n = 7, 10$). (**D**) PFS of patients based on PD-1/PD-L1 interaction score threshold of 200. * $p < 0.05$, log-rank test ($n = 6, 12$). (**E**) PFS of patients based on PD-L1 expression of tumor ($n = 10, 8$) or non-tumor cells ($n = 7, 11$) with a threshold of 5%. * $p < 0.05$, ** $p < 0.01$, log-rank test.

To better understand the presence and impact of other cells in the tumor immune microenvironment in RCC and response to immunotherapy, mIF measured both immune stimulatory and suppressive components. The amount of total, CD4+, and CD8+ T cell infiltration did not correlate with response (Figure S2C). Immune-suppressive components in the RCC microenvironment such as T regulatory cells (Tregs), macrophages, myeloid-derived suppressor cells (MDSCs), and cells expressing indoleamine 2,3-dioxygenase (IDO1) were lower in responders, though not significantly ($n = 7, 10; p = 0.16, 0.29, 0.39, 0.41$) (Figure S2D).

3.4. TMB and Driver Mutations Do Not Correlate with ICI Response

WES was performed in patients who had matched primary tumor, metastasis, and/or adjacent normal tissue. Matched samples from the same patient showed a high degree of similarity quantified by the Jaccard index and clustered together as expected (Table S8). Consistent with previous reports, the classical driver mutations associated with ccRCC, including alterations in VHL, PBRM1, BAP1, and SETD2, were identified in tumor samples. Truncating mutations were the most common, followed by missense and synonymous mutations (Figure 3A). However, responders and non-responders did not cluster based on WES analysis, and no single gene or mutation significantly correlated with response. Non-synonymous PBRM1 mutations trended towards response (OR = 15.00) but were not statistically significant ($p = 0.10$). Additionally, TMB calculated based on all unfiltered variants did not correlate with response to therapy (Figure 3B). Thus, tumor mutational profile did not predict likelihood of response to ICI.

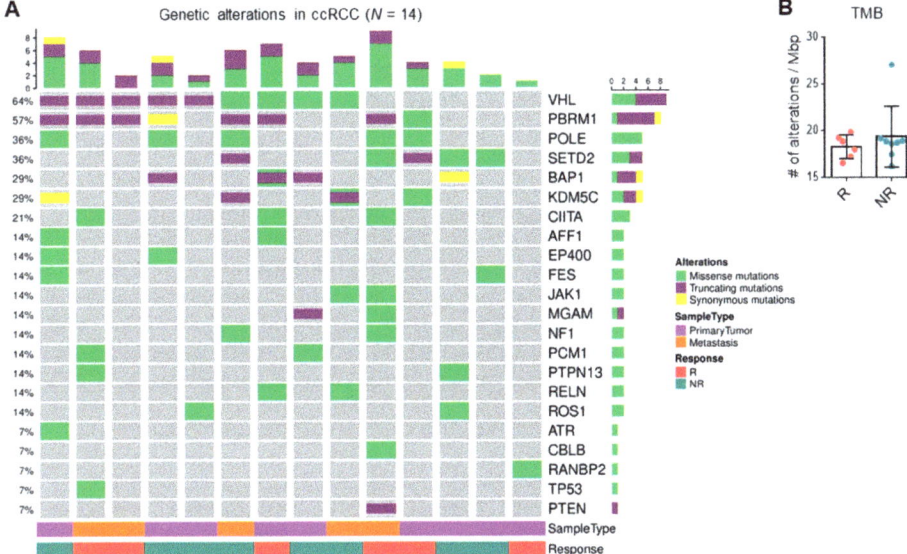

Figure 3. Tumor mutational burden (TMB) and driver mutations do not correlate with response. (**A**) Quantification of missense, truncating, and synonymous mutations in top 22 altered genes found in clear cell RCC samples. (**B**) TMB of responders ($n = 6$) compared to non-responders ($n = 8$) calculated based on all unfiltered variants. Data shown are averages ± SD.

3.5. TCR Clonal Diversity Does Not Correlate with Response but May Impact Survival

Lower TCR diversity may represent purposeful expansion of specific tumor antigen-driven TCR clones, indicating an existing anti-tumor T cell response that can be further enhanced with immunotherapy. Previous studies in ccRCC have demonstrated that a polyclonal infiltrating T cell population is indicative of an exhausted, poorly cytotoxic T cell phenotype compared to an oligoclonal population [44]. As expected, unsupervised clustering of TCR clones in our samples showed that TCRs were not generally shared across patients, and multiple samples from a single individual clustered together (Figure 4A, Table S9). Although TCR diversity was not significantly different in objective response ($n = 5, 7$) (Figure 4B), a lower TCR diversity (<644 clonotypes) among tumor-infiltrating lymphocytes suggested improved survival ($n = 6, 6$; PFS HR = 0.49, 95% CI = 0.13–1.6; OS HR = 0.32, 95% CI = 0.055–0.98) (Figure 4C).

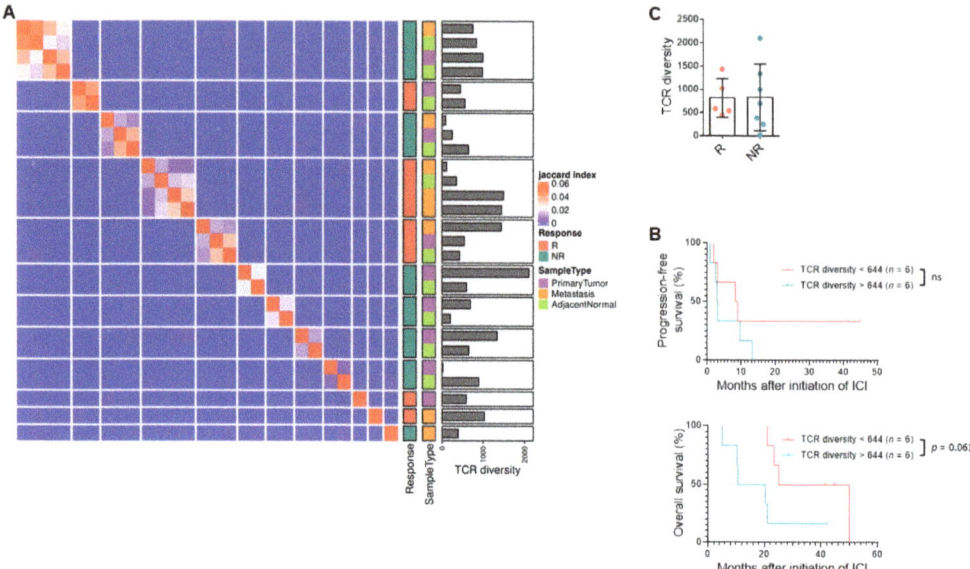

Figure 4. TCR clonal diversity does not correlate with response but may impact survival. (**A**) Heatmap depicting unsupervised clustering analysis of Jaccard index representing similarity among primary tumor, metastatic tumor, and adjacent normal samples from 12 patients in biomarker cohort based on TCR sequencing. (**B**) PFS and OS of patients with low versus high intra-tumoral TCR diversity based on median cutoff of 644 clonotypes. Log-rank test ($n = 6, 6$). (**C**) Quantification of TCR diversity in primary tumors from responders compared to non-responders. Data shown are averages ± SD.

3.6. Gene Expression Patterns in RCC Suggest Response to Single-Agent Immunotherapy

Prior work has suggested that gene expression signatures correlate with response to anti-PD-1/PD-L1 therapy [10,21,22]. Differential gene expression analysis in our study demonstrated genes that were significantly differentially expressed between responders and non-responders ($n = 8, 7$) (Figure 5A, Table S10) and gene set variation analysis revealed differentially expressed pathways, including immune and metabolic pathways, between responders and non-responders (Table S5). Furthermore, deconvolution analysis revealed a non-significantly higher proportion of tumor-infiltrating immune cells ($p = 0.34$), including M1 macrophages ($p = 0.094$), among responders compared to non-responders (Figure 5B). Consistent with mIF results, proportions of other immune cell types, including CD8+ T cells, were not different between responders and non-responders (Figure S3A). Evaluation using the previously published ClearCode34 gene set [43] showed that responders to PD-1/PD-L1 blockade tended to cluster together with a ccB profile, while non-responders were largely grouped under the ccA profile (Figure S3B).

Unsupervised clustering analysis showed that responders and non-responders tended to cluster separately based on key gene expression pathways, including angiogenesis, myeloid, and T effector signature scores previously defined by McDermott et al. [10] and shown in Figure 5C. Responders in this cohort tended to have a lower angiogenic and higher T effector signature. Expression of the expanded immune and antigen-presenting gene panel [10], which includes other stimulatory cytokines and immune checkpoint proteins in addition to the T effector signature, was upregulated in responders compared to non-responders (Figure 5D). In summary, previously identified gene expression signatures including angiogenic, T effector, expanded immune, and clear cell subtype (ClearCode34) gene signatures correlated with response to anti-PD-1/PD-L1 monotherapy.

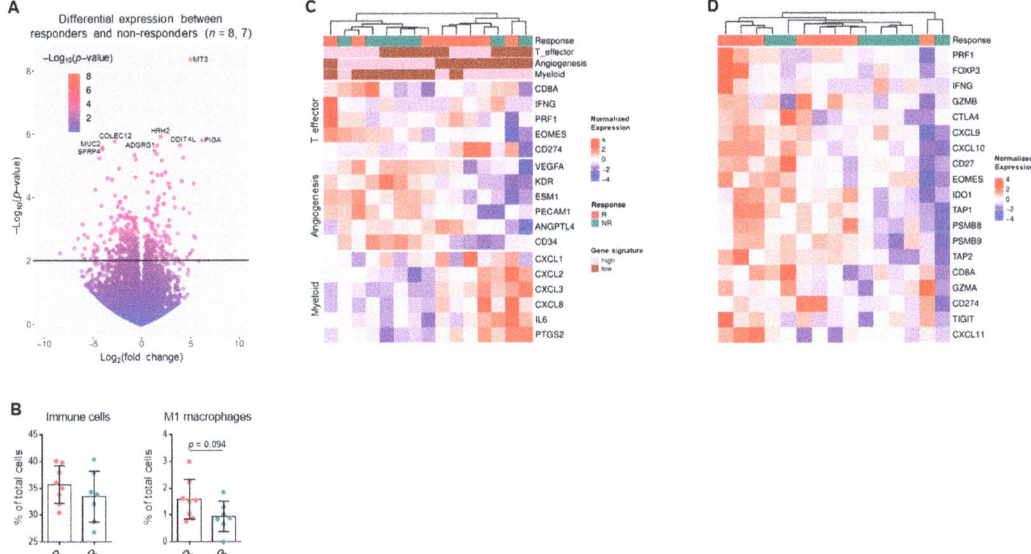

Figure 5. Gene expression patterns in RCC suggest response to single-agent immunotherapy. (**A**) Volcano plot depicting statistical significance against fold-change expression of differentially expressed genes from responders (*n* = 8) and non-responders (*n* = 7) in biomarker cohort based on RNA-seq data. Top eight differentially expressed genes labeled. (**B**) Percentages of all immune cells and M1 macrophages in bulk tumor samples from responders compared to non-responders based on deconvolution analysis. Data shown are averages ± SD, unpaired Mann–Whitney U-test (*n* = 8, 7). (**C**) Heatmap depicting clustering of tumor samples by T effector, angiogenesis, and myeloid signature scores. (**D**) Heatmap depicting clustering of tumor samples by expanded immune and antigen-presenting gene expression panel.

4. Discussion

The results of this retrospective study identify clinical and laboratory characteristics associated with response to ICI in patients with RCC and explore how these characteristics relate to novel biomarker platforms. Patients who experienced at least one irAE were more likely to respond to ICI. These results replicate prior associations between irAE and patient response to ICI [16,17] and suggest that immune reactivity occurs not only at the site of the tumor but also at non-tumor tissue sites. While such association requires large-scale validation, irAE in patients with mixed response on imaging or concerns for pseudoprogression may aid in decision-making for clinicians. Patients with pancreatic metastases were less likely to respond to anti-PD-1/PD-L1. This is consistent with results reported by Singla et al., who have shown that pancreatic metastases of RCC are typically VEGF-driven and refractory to ICI treatment [45].

Patients with response to ICI commonly had higher lymphocyte and lower neutrophil percentages, corroborating work that has shown NLR at baseline and after several weeks on treatment as a predictor of response in RCC. Patients with higher NLR have lower ORR, PFS, and OS on ICI therapy [12]. These findings suggest that higher levels of suppressive myeloid cells induced by the tumor and lower levels of activated circulating lymphocytes contribute to the correlation between high NLR and poor response to ICI therapy. The clinical application of NLR awaits further prospective study.

PD-L1 testing has been fraught with challenges, including tumor heterogeneity, differing antibodies, varying percent expression cutoff, and differing target cell populations of the analysis. PD-L1 in this mIF assay showed correlation with response across cell types assessed, but, similar to prior studies [2,3,6,7,10], not all patients with response had expression of PD-L1 at the protein level. Unlike our findings, prior studies generally did not find an association between overall PD-L1 expression and response to ICI therapy.

Potential reasons for these discrepant results include the fact that the PD-L1 antibody in our assay does not share the same clonality or manufacturer as those used in clinical trials and our staining and detection methods differ from traditional immunohistochemistry (IHC). The novelty of this study was the ability to look not only at PD-L1 but at other cells in the tumor immune microenvironment. In non-responders, there were higher immune suppressive components such as Tregs and macrophages, while in responders, higher total T cell infiltration was found. These data suggest that, in addition to PD-L1, the presence and function of cells in the microenvironment are associated with response to ICI. Larger validation studies incorporating additional cell types covering both immune effector as well as suppressive functions are likely to add to the understanding of response to ICI in RCC.

Braun et al. recently used WES, RNA-seq, IF, and copy number analysis to study response in a similar clinical context as this study for patients with ccRCC receiving ICI therapy [11]. They utilized clinical correlates from patients enrolled on Checkmate010 and Checkmate025, with the majority of this group receiving nivolumab following VEGF-targeted therapy. Our biomarker cohort differs in that the majority of patients received anti-PD-1/PD-L1 monotherapy in the third or later line of treatment and that responses were durable in the responders (median PFS, 13.4 months) compared to non-responders (median PFS, 2.76 months). Similar to Braun et al., we found that TMB and CD8+ T cell infiltration were not associated with clinical response. We did not observe a significant correlation between PBRM1 mutations and response as Braun et al. did, and this may in part be due to differences in cohorts and inclusion of non-truncating PBRM1 mutations in correlative analyses. Additionally, Braun et al. performed copy number variation analysis that we did not, which showed that additional chromosomal aberrations can help discriminate response.

Based on data from studies using ClearCode34 [43] and IMmotion150 [10], we reviewed gene expression patterns to determine if specific pathways or previously established signatures could be differentially enriched in this study. In our small cohort, patients with response demonstrated differentially expressed genes involved in inflammatory signaling and metabolic pathways. Response tended to correlate more with the ccB subtype, as well as lower angiogenic, higher T effector, and higher expanded immune signatures. This is the first study to date suggesting an association between clear cell subtype and response to ICI therapy. Previous studies utilizing patient cohorts not treated with checkpoint inhibitors demonstrated that ccB tumors tend to have a worse prognosis compared to ccA tumors [43]. Thus, our work suggests that ICI therapy may provide greater clinical benefit, specifically in patients with ccB tumors.

Compared to IMotion150, we observed similar patterns in responders with high T effector signature and high expanded immune infiltrate signature. In IMotion150, treatment-naïve patients with high T effector combined with high myeloid-suppressive signatures tended not to do as well with single-agent PD-L1 blockade [10]. In this cohort, low angiogenesis and high T effector signatures were more likely to respond, consistent with previously published studies [10,23]. While results between this study and IMotion150 were discrepant regarding ability to respond with a high myeloid signature, this may be due to differences in patient population and line of therapy. While the patients in IMotion150 were treatment-naïve, patients in this study had several lines of therapy between sample collection and ICI treatment. Standard of care treatments in RCC such as VEGF inhibition have been shown to alter the tumor immune microenvironment and enhance response to anti-PD-1/PD-L1 monotherapy [46]. Samples were from archival FFPE tissue, including prior nephrectomy samples when the biology of a patient's response may have been very different. Sequential biopsy while on therapy and at time of progression would improve our understanding of the changes in the tumor microenvironment and immune milieu that we hypothesize may occur after treatment with VEGF inhibition or other therapies.

New data from this study are preliminary results suggesting that the presence of lower TCR clonality should be investigated further for its potential to distinguish patient outcome

This analysis suggests that having a lower TCR diversity may correlate with improved PFS and OS and specific tumor clonotypes may drive the response to anti-PD-1/PD-L1 monotherapy. Future directions include identifying the tumor antigens associated with these clonotypes. In our cohort, we did not find overall TMB and alterations in commonly mutated genes like PBRM1 to be associated with response. Thus, neoantigens derived from these common mutations may not be the primary drivers of the adaptive T cell response in RCC; instead, the role of other factors such as endogenous retroviral genes [19,20] or larger chromosomal abnormalities [11] should be studied in shaping the TCR repertoire.

We note that this study has limitations, including being conducted at a single site in a retrospective manner. The use of archival FFPE specimens may skew analyses due to suboptimal molecular quality, though many others have similarly utilized this preservation method for their studies. RCC is also characterized by intratumoral heterogeneity and thus a single tissue sample analyses likely underestimated the complex biology of response. Similarly, future studies would benefit from focused analyses of varying histologies including non-ccRCC such as papillary. Although we demonstrate patterns and further our understanding of the biology that correlate with response to ICI in RCC, we are unable to definitively classify responders and non-responders based on these data. The clinical and biomarker cohorts were both limited by number of patients and samples included in each. In the future, more rigorous analyses will benefit from larger cohorts and accounting for multiplicity of testing. While limited, ultimately, this study independently confirms gene expression data from previously published work while also showing the feasibility to study more than one biomarker in order to improve understanding behind the complex biology underlying response to ICI treatment in RCC.

5. Conclusions

Response to ICI therapy remains challenging to predict in RCC, but these data build upon previous work and suggest that PD-L1 staining alone does not give sufficient information to predict response [11,24]. Checkpoint inhibitors elicit a complex biology that will require a combination of biomarkers to predict response. Platforms analyzing TCR diversity, gene expression, multiplex IHC or IF, and chromosomal alterations or endogenous retroviruses will need to be assessed in large prospective clinical trials moving forward with the goals of developing sensitive and specific biomarkers that can be used in the clinic.

Supplementary Materials: The following are available online at https://www.mdpi.com/2072-6694/13/6/1475/s1, Figure S1: Patient characteristics and prior therapies, Figure S2: PD-L1 expression is associated with survival, Figure S3: Gene expression patterns in RCC suggest response to single-agent immunotherapy, Table S1: Antibody panels used in multiplexed immunofluorescence assay, Table S2: Clinical characteristics of biomarker cohort of patients with RCC, Table S3: irAEs during ICI therapy in primary patient cohort, Table S4: Sites of metastasis prior to initiation of ICI in primary patient cohort, Table S5: Selected differentially expressed pathways from gene set variation analysis. Table S6: Clinical laboratory data from primary cohort, Table S7: mIF data from biomarker cohort, Table S8: Filtered variant data from biomarker cohort, Table S9: TCR-sequencing data from biomarker cohort, Table S10: Normalized gene expression counts from biomarker cohort.

Author Contributions: E.S. reviewed clinical information, analyzed data, created figures and wrote the manuscript; A.R. performed bioinformatic analysis and created figures; A.R.L., A.S., S.O.D. and R.M. reviewed clinical information; R.H., K.Y., J.O., K.B.D., T.M.M. and J.C.R. helped in obtaining samples and performing biomarker assays; B.C., K.R., N.R., J.B. (Jehovana Bender), J.W., J.Y.K., C.V. and J.B. (Jennifer Bordeaux) performed mIF staining and analysis; S.G. provided patient samples; C.C.S. and B.G.V. performed RNA-seq and bioinformatic analysis; S.M.H., J.C.R., E.J., B.I.R. and W.K.R. contributed to the design and interpretation of studies and writing of the manuscript; K.E.B. wrote the IRB, created the clinical database, reviewed clinical information, isolated RNA, coordinated biomarker platform assays, analyzed data, and wrote the manuscript; Conceptualization, E.S., S.M.H., JCR, E.J., B.I.R., W.K.R. and K.E.B.; Methodology, E.S., A.R. and K.E.B.; Software, A.R., C.C.S. and B.G.V.; Validation, E.S., A.R. and K.E.B.; Formal Analysis, E.S., A.R., C.C.S. and B.G.V.; Investigation, E.S., S.O.D., A.R.L., A.S., R.H., K.Y., J.O., K.B.D., R.M., B.C., K.R., N.R., J.B. (Jehovana Bender), J.W.,

J.Y.K., C.V., J.B. (Jennifer Bordeaux), T.M.M., J.C.R. and K.E.B.; Resources, E.S., R.H., K.Y., X.-D.L., J.O., K.B.D., S.G., T.M.M., G.M.R., J.C.R. and K.E.B.; Data Curation, E.S., A.R., S.O.D., A.R.L., A.S., R.M. and K.E.B.; Writing—Original Draft Preparation, E.S., A.R. and K.E.B.; Writing—Review and Editing, E.S., A.R., S.O.D., A.R.L., A.S., R.H., K.Y., XDL, C.C.S., J.O., K.B.D., RC, B.C., K.R., N.R., J.B. (Jehovana Bender), J.W., J.Y.K., C.V., J.B. (Jennifer Bordeaux), S.G., T.M.M., G.M.R., B.G.V., N.B.D., S.M.H., J.C.R., E.J., B.I.R., W.K.R. and K.E.B.; Visualization, E.S. and A.R.; Supervision, B.I.R., W.K.R. and K.E.B.; Project Administration, B.I.R., W.K.R. and K.E.B.; Funding Acquisition, W.K.R. and K.E.B. All authors have read and agreed to the published version of the manuscript.

Funding: This study was supported in part by NIH grants: T32 GM007347 (E.S., S.O.D., A.R.L., A.S.), F30 CA216891-01 (E.S.), F30 CA224559 (S.O.D.), F30 CA225136-01 (C.C.S.), K12 CA090625 (K.E.B.), 5P30 CA068485-21 (K.B.D.), 2P50 CA098131-11 (K.B.D.). The AACR (W.K.R., A.R.L.), the DoD W81XWH-18-2-0052 (S.M.H., E.J., W.K.R., K.E.B., S.G.), the UNC University Cancer Research Fund (B.G.V.), and the TJ Martell Foundation (K.B.D.).

Institutional Review Board Statement: This study was conducted according to the guidelines of the Declaration of Helsinki and approved by the IRB of Vanderbilt University Medical Center (protocol #160979).

Informed Consent Statement: Waiver of consent was approved by the IRB of Vanderbilt University Medical Center as the study involved minimal risk to the subjects.

Data Availability Statement: The data presented in this study are available in the Supplementary Materials.

Acknowledgments: The authors thank all the patients who made this study possible.

Conflicts of Interest: W.K.R. reports research support from Incyte and clinical trial support from Novartis, Pfizer, Roche, Bristol Myers Squibb, Merck, Tracon, Calithera, and the NIH/CTEP. K.E.B. reports research support from Bristol Myers Squibb and has served as a consultant for Aravive and Exelexis. S.G. has served as a consultant for Merck, Novartis, Roche, Foundation Medicine, Foghorn Therapeutics, Silagene and Inspirata Inc, and has received research funding from M2Gen. K.B.D. and J.O. report research support from Kadmon, Inc. B.G.V. reports equity from GeneCentric Therapeutics. T.M. reports research funding from Pfizer, Sotio, and Merck and has consulted for AstraZeneca.

References

1. De Velasco, G.; Miao, D.; Voss, M.H.; Hakimi, A.A.; Hsieh, J.J.; Tannir, N.M.; Tamboli, P.; Appleman, L.J.; Rathmell, W.K.; Van Allen, E.M.; et al. Tumor Mutational Load and Immune Parameters across Metastatic Renal Cell Carcinoma Risk Groups. *Cancer Immunol. Res.* **2016**, *4*, 820–822. [CrossRef]
2. Motzer, R.J.; Escudier, B.; McDermott, D.F.; George, S.; Hammers, H.J.; Srinivas, S.; Tykodi, S.S.; Sosman, J.A.; Procopio, G.; Plimack, E.R.; et al. Nivolumab versus Everolimus in Advanced Renal-Cell Carcinoma. *N. Engl. J. Med.* **2015**, *373*, 1803–1813. [CrossRef] [PubMed]
3. McDermott, D.F.; Lee, J.-L.; Szczylik, C.; Donskov, F.; Malik, J.; Alekseev, B.Y.; Larkin, J.M.G.; Matveev, V.B.; Gafanov, R.A.; Tomczak, P.; et al. Pembrolizumab monotherapy as first-line therapy in advanced clear cell renal cell carcinoma (accRCC): Results from cohort A of KEYNOTE-427. *J. Clin. Oncol.* **2018**, *36*, 4500. [CrossRef]
4. Atkins, M.B.; Jegede, O.; Haas, N.B.; McDermott, D.F.; Bilen, M.A.; Drake, C.G.; Sosman, J.A.; Alter, R.S.; Plimack, E.R.; Rini, B.I.; et al. Phase II study of nivolumab and salvage nivolumab + ipilimumab in treatment-naïve patients (pts) with advanced renal cell carcinoma (RCC) (HCRN GU16-260). *J. Clin. Oncol.* **2020**, *38*, 5006. [CrossRef]
5. Motzer, R.J.; Tannir, N.M.; McDermott, D.F.; Frontera, O.A.; Melichar, B.; Choueiri, T.K.; Plimack, E.R.; Barthélémy, P.; Porta, C.; George, S.; et al. Nivolumab plus Ipilimumab versus Sunitinib in Advanced Renal-Cell Carcinoma. *N. Engl. J. Med.* **2018**, *378*, 1277–1290. [CrossRef]
6. Rini, B.I.; Powles, T.; Atkins, M.B.; Escudier, B.; McDermott, D.F.; Suarez, C.; Bracarda, S.; Stadler, W.M.; Donskov, F.; Lee, J.L.; et al. Atezolizumab plus bevacizumab versus sunitinib in patients with previously untreated metastatic renal cell carcinoma (IMmotion151): A multicentre, open-label, phase 3, randomised controlled trial. *Lancet* **2019**, *393*, 2404–2415. [CrossRef]
7. Rini, B.I.; Plimack, E.R.; Stus, V.; Gafanov, R.; Hawkins, R.; Nosov, D.; Pouliot, F.; Alekseev, B.; Soulières, D.; Melichar, B. Pembrolizumab plus Axitinib versus Sunitinib for Advanced Renal-Cell Carcinoma. *N. Engl. J. Med.* **2019**, *380*, 1116–1127. [CrossRef] [PubMed]
8. Motzer, R.J.; Penkov, K.; Haanen, J.; Rini, B.; Albiges, L.; Campbell, M.T.; Venugopal, B.; Kollmannsberger, C.; Negrier, S.; Uemura, M. Avelumab plus Axitinib versus Sunitinib for Advanced Renal-Cell Carcinoma. *N. Engl. J. Med.* **2019**, *380*, 1103–1115. [CrossRef]

9. Motzer, R.J.; Robbins, P.B.; Powles, T.; Albiges, L.; Haanen, J.B.; Larkin, J.; Mu, X.J.; Ching, K.A.; Uemura, M.; Pal, S.K.; et al. Avelumab plus axitinib versus sunitinib in advanced renal cell carcinoma: Biomarker analysis of the phase 3 JAVELIN Renal 101 trial. *Nat. Med.* **2020**, *26*, 1733–1741. [CrossRef]
10. McDermott, D.F.; Huseni, M.A.; Atkins, M.B.; Motzer, R.J.; Rini, B.I.; Escudier, B.; Fong, L.; Joseph, R.W.; Pal, S.K.; Reeves, J.A.; et al. Clinical activity and molecular correlates of response to atezolizumab alone or in combination with bevacizumab versus sunitinib in renal cell carcinoma. *Nat. Med.* **2018**, *24*, 749–757. [CrossRef]
11. Braun, D.A.; Hou, Y.; Bakouny, Z.; Ficial, M.; Angelo, M.S.; Forman, J.; Ross-Macdonald, P.; Berger, A.C.; Jegede, O.A.; Elagina, L.; et al. Interplay of somatic alterations and immune infiltration modulates response to PD-1 blockade in advanced clear cell renal cell carcinoma. *Nat. Med.* **2020**, *26*, 909–918. [CrossRef]
12. Huszno, J.; Kolosza, Z.; Mrochem-Kwarciak, J.; Rutkowski, T.; Skladowski, K. The Role of Neutrophil-Lymphocyte Ratio, Platelet-Lymphocyte Ratio, and Platelets in the Prognosis of Metastatic Renal Cell Carcinoma. *Oncology* **2019**, *97*, 7–17. [CrossRef]
13. Capone, M.; Giannarelli, D.; Mallardo, D.; Madonna, G.; Festino, L.; Grimaldi, A.M.; Vanella, V.; Simeone, E.; Paone, M.; Palmieri, G.; et al. Baseline neutrophil-to-lymphocyte ratio (NLR) and derived NLR could predict overall survival in patients with advanced melanoma treated with nivolumab. *J. Immunother. Cancer* **2018**, *6*, 74. [CrossRef] [PubMed]
14. De Giorgi, U.; Procopio, G.; Giannarelli, D.; Sabbatini, R.; Bearz, A.; Buti, S.; Basso, U.; Mitterer, M.; Ortega, C.; Bidoli, P.; et al. Association of Systemic Inflammation Index and Body Mass Index with Survival in Patients with Renal Cell Cancer Treated with Nivolumab. *Clin. Cancer Res.* **2019**, *25*, 3839–3846. [CrossRef] [PubMed]
15. Martini, D.J.; Liu, Y.; Shabto, J.M.; Carthon, B.C.; Hitron, E.E.; Russler, G.A.; Caulfield, S.; Kissick, H.T.; Harris, W.B.; Kucuk, O.; et al. Novel Risk Scoring System for Patients with Metastatic Renal Cell Carcinoma Treated with Immune Checkpoint Inhibitors. *Oncology* **2019**, *25*, e484–e491. [CrossRef]
16. Cortellini, A.; Bersanelli, M.; Buti, S.; Cannita, K.; Santini, D.; Perrone, F.; Giusti, R.; Tiseo, M.; Michiara, M.; Di Marino, P.; et al. A multicenter study of body mass index in cancer patients treated with anti-PD-1/PD-L1 immune checkpoint inhibitors: When overweight becomes favorable. *J. Immunother. Cancer* **2019**, *7*, 57. [CrossRef] [PubMed]
17. Labadie, B.W.; Liu, P.; Bao, R.; Crist, M.; Fernandes, R.; Ferreira, L.; Graupner, S.; Poklepovic, A.S.; Duran, I.; Vareki, S.M.; et al. BMI, irAE, and gene expression signatures associate with resistance to immune-checkpoint inhibition and outcomes in renal cell carcinoma. *J. Transl. Med.* **2019**, *17*, 1–12. [CrossRef]
18. Boi, S.K.; Orlandella, R.M.; Gibson, J.T.; Turbitt, W.J.; Wald, G.; Thomas, L.; Rosean, C.B.; Norris, K.E.; Bing, M.; Bertrand, L.; et al. Obesity diminishes response to PD-1-based immunotherapies in renal cancer. *J. Immunother. Cancer* **2020**, *8*, e000725. [CrossRef]
19. Panda, A.; De Cubas, A.A.; Stein, M.; Riedlinger, G.; Kra, J.; Mayer, T.; Smith, C.C.; Vincent, B.G.; Serody, J.S.; Beckermann, K.E.; et al. Endogenous retrovirus expression is associated with response to immune checkpoint pathway in clear cell renal cell carcinoma. *JCI Insight* **2018**, *3*. [CrossRef] [PubMed]
20. Smith, C.C.; Beckermann, K.E.; Bortone, D.S.; De Cubas, A.A.; Bixby, L.M.; Lee, S.J.; Panda, A.; Ganesan, S.; Bhanot, G.; Wallen, E.M.; et al. Endogenous retroviral signatures predict immunotherapy response in clear cell renal cell carcinoma. *J. Clin. Investig.* **2018**, *128*, 4804–4820. [CrossRef]
21. Ayers, M.; Lunceford, J.; Nebozhyn, M.; Murphy, E.; Loboda, A.; Kaufman, D.R.; Albright, A.; Cheng, J.D.; Kang, S.P.; Shankaran, V.; et al. IFN-γ–related mRNA profile predicts clinical response to PD-1 blockade. *J. Clin. Investig.* **2017**, *127*, 2930–2940. [CrossRef]
22. Ascierto, M.L.; McMiller, T.L.; Berger, A.E.; Danilova, L.; Anders, R.A.; Netto, G.J.; Xu, H.; Pritchard, T.S.; Fan, J.; Cheadle, C.; et al. The Intratumoral Balance between Metabolic and Immunologic Gene Expression Is Associated with Anti–PD-1 Response in Patients with Renal Cell Carcinoma. *Cancer Immunol. Res.* **2016**, *4*, 726–733. [CrossRef]
23. Motzer, R.J.; Banchereau, R.; Hamidi, H.; Powles, T.; McDermott, D.; Atkins, M.B.; Escudier, B.; Liu, L.-F.; Leng, N.; Abbas, A.R.; et al. Molecular Subsets in Renal Cancer Determine Outcome to Checkpoint and Angiogenesis Blockade. *Cancer Cell* **2020**, *38*, 803–817.e4. [CrossRef] [PubMed]
24. Miao, D.; Margolis, C.A.; Gao, W.; Voss, M.H.; Li, W.; Martini, D.J.; Norton, C.; Bossé, D.; Wankowicz, S.M.; Cullen, D.; et al. Genomic correlates of response to immune checkpoint therapies in clear cell renal cell carcinoma. *Science* **2018**, *359*, 801–806. [CrossRef] [PubMed]
25. Braun, D.A.; Ishii, Y.; Walsh, A.M.; Van Allen, E.M.; Wu, C.J.; Shukla, S.A.; Choueiri, T.K. Clinical Validation of PBRM1 Alterations as a Marker of Immune Checkpoint Inhibitor Response in Renal Cell Carcinoma. *JAMA Oncol.* **2019**, *5*, 1631–1633. [CrossRef] [PubMed]
26. Turajlic, S.; Xu, H.; Litchfield, K.; Rowan, A.; Chambers, T.; Lopez, J.I.; Nicol, D.; O'Brien, T.; Larkin, J.; Horswell, S.; et al. Tracking Cancer Evolution Reveals Constrained Routes to Metastases: TRACERx Renal. *Cell* **2018**, *173*, 581–594.e12. [CrossRef]
27. Liu, X.-D.; Kong, W.; Peterson, C.B.; McGrail, D.J.; Hoang, A.; Zhang, X.; Lam, T.; Pilie, P.G.; Zhu, H.; Beckermann, K.E.; et al. PBRM1 loss defines a nonimmunogenic tumor phenotype associated with checkpoint inhibitor resistance in renal carcinoma. *Nat. Commun.* **2020**, *11*, 1–14. [CrossRef]
28. Johnson, D.B.; Bordeaux, J.M.; Kim, J.-Y.; Vaupel, C.A.; Rimm, D.L.; Ho, T.H.; Joseph, R.W.; Daud, A.I.; Conry, R.M.; Gaughan, E.M.; et al. Quantitative Spatial Profiling of PD-1/PD-L1 Interaction and HLA-DR/IDO-1 Predicts Improved Outcomes of anti-PD-1 Therapies in Metastatic Melanoma. *Clin. Cancer Res.* **2018**, *24*, 5250–5260. [CrossRef]
29. Depristo, M.A.; Banks, E.; Poplin, R.; Garimella, K.V.; Maguire, J.R.; Hartl, C.; Philippakis, A.A.; Del Angel, G.; Rivas, M.A.; Hanna, M.; et al. A framework for variation discovery and genotyping using next-generation DNA sequencing data. *Nat. Genet.* **2011**, *43*, 491–498. [CrossRef]

30. Li, H.; Durbin, R. Fast and accurate long-read alignment with Burrows–Wheeler transform. *Bioinformatics* **2010**, *26*, 589–595. [CrossRef]
31. Poplin, R.; Ruano-Rubio, V.; DePristo, M.A.; Fennell, T.J.; Carneiro, M.O.; Van der Auwera, G.A.; Kling, D.E.; Gauthier, L.D.; Levy-Moonshine, A.; Roazen, D.; et al. Scaling accurate genetic variant discovery to tens of thousands of samples. *BioRxiv* **2017**. [CrossRef]
32. Voss, K.; Van der Auwera, G.; Gentry, J. Full-stack genomics pipelining with GATK4 + WDL + Cromwell. *F1000Research* **2017**, *6*. [CrossRef]
33. Yang, H.; Wang, K. Genomic variant annotation and prioritization with ANNOVAR and wANNOVAR. *Nat. Protoc.* **2015**, *10*, 1556–1566. [CrossRef]
34. Lek, M.; Karczewski, K.J.; Minikel, E.V.; Samocha, K.E.; Banks, E.; Fennell, T.; O'Donnell-Luria, A.H.; Ware, J.S.; Hill, A.J.; Cummings, B.B.; et al. Analysis of protein-coding genetic variation in 60,706 humans. *Nature* **2016**, *536*, 285–291. [CrossRef]
35. Minikel, E.V.; Karczewski, K.J.; Martin, H.C.; Cummings, B.B.; Whiffin, N.; Alföldi, J.; MacArthur, D.G.; Genome Aggregation Database (gnomAD) Production Team; Genome Aggregation Database (gnomAD) Consortium; Schreiber, S.L.; et al. Evaluating potential drug targets through human loss-of-function genetic variation. *BioRxiv* **2019**. [CrossRef]
36. The 1000 Genomes Project Consortium. An integrated map of genetic variation from 1092 human genomes. *Nature* **2012**, *491*, 56–65. [CrossRef]
37. Fu, W.; O'Connor, T.D.; Jun, G.; Kang, H.M.; Abecasis, G.; Leal, S.M.; Gabriel, S.; Rieder, M.J.; Altshuler, D.; Shendure, J.; et al. Analysis of 6515 exomes reveals the recent origin of most human protein-coding variants. *Nature* **2013**, *493*, 216–220. [CrossRef]
38. Ricketts, C.J.; De Cubas, A.A.; Fan, H.; Smith, C.C.; Lang, M.; Reznik, E.; Bowlby, R.; Gibb, E.A.; Akbani, R.; Beroukhim, R.; et al. The Cancer Genome Atlas Comprehensive Molecular Characterization of Renal Cell Carcinoma. *Cell Rep.* **2018**, *23*, 313–326.e5. [CrossRef]
39. Chakravarty, D.; Gao, J.; Phillips, S.M.; Kundra, R.; Zhang, H.; Wang, J.; Rudolph, J.E.; Yaeger, R.; Soumerai, T.; Nissan, M.H.; et al. OncoKB: A Precision Oncology Knowledge Base. *JCO Precis. Oncol.* **2017**, *2017*, 1–16. [CrossRef] [PubMed]
40. Shugay, M.; Bagaev, D.V.; Turchaninova, M.A.; Bolotin, D.A.; Britanova, O.V.; Putintseva, E.V.; Pogorelyy, M.V.; Nazarov, V.I.; Zvyagin, I.V.; Kirgizova, V.I.; et al. VDJtools: Unifying Post-analysis of T Cell Receptor Repertoires. *PLoS Comput. Biol.* **2015**, *11*, e1004503. [CrossRef] [PubMed]
41. Love, M.I.; Huber, W.; Anders, S. Moderated estimation of fold change and dispersion for RNA-seq data with DESeq2. *Genome Biol.* **2014**, *15*, 550. [CrossRef]
42. Finotello, F.; Mayer, C.; Plattner, C.; Laschober, G.; Rieder, D.; Hackl, H.; Krogsdam, A.; Loncova, Z.; Posch, W.; Wilflingseder, D.; et al. Molecular and pharmacological modulators of the tumor immune contexture revealed by deconvolution of RNA-seq data. *Genome Med.* **2019**, *11*, 34. [CrossRef] [PubMed]
43. Brannon, A.R.; Reddy, A.; Seiler, M.; Arreola, A.; Moore, D.T.; Pruthi, R.S.; Wallen, E.M.; Nielsen, M.E.; Liu, H.; Nathanson, K.L.; et al. Molecular stratification of clear cell renal cell carci-noma by consensus clustering reveals distinct subtypes and survival patterns. *Genes Cancer* **2010**, *1*, 152–163. [CrossRef] [PubMed]
44. Giraldo, N.A.; Becht, E.; Vano, Y.; Petitprez, F.; Lacroix, L.; Validire, P.; Sanchez-Salas, R.; Ingels, A.; Oudard, S.; Moatti, A.; et al. Tumor-Infiltrating and Peripheral Blood T-cell Immunophenotypes Predict Early Relapse in Localized Clear Cell Renal Cell Carcinoma. *Clin. Cancer Res.* **2017**, *23*, 4416–4428. [CrossRef] [PubMed]
45. Singla, N.; Xie, Z.; Zhang, Z.; Gao, M.; Yousuf, Q.; Onabolu, O.; McKenzie, T.; Tcheuyap, V.T.; Ma, Y.; Choi, J.; et al. Pancreatic tropism of metastatic renal cell carcinoma. *JCI Insight* **2020**, *5*. [CrossRef]
46. Yang, J.; Yan, J.; Liu, B. Targeting VEGF/VEGFR to Modulate Antitumor Immunity. *Front. Immunol.* **2018**, *9*, 978. [CrossRef]

Article

Prognostic and Therapeutic Potential of the OIP5 Network in Papillary Renal Cell Carcinoma

Mathilda Jing Chow [1,2,3], Yan Gu [1,2,3], Lizhi He [4], Xiaozeng Lin [1,2,3], Ying Dong [1,2,3], Wenjuan Mei [1,2,3], Anil Kapoor [1,2,3,*] and Damu Tang [1,2,3,*]

[1] Department of Surgery, McMaster University, Hamilton, ON L8S 4K1, Canada; zhouj32@mcmaster.ca (M.J.C.); guy3@mcmaster.ca (Y.G.); linx36@mcmaster.ca (X.L.); dongy87@mcmaster.ca (Y.D.); wenjuanmei1986@gmail.com (W.M.)
[2] Urological Cancer Center for Research and Innovation (UCCRI), St Joseph's Hospital, Hamilton, ON L8N 4A6, Canada
[3] The Research Institute of St Joe's Hamilton, St Joseph's Hospital, Hamilton, ON L8N 4A6, Canada
[4] Cutaneous Biology Research Center, Harvard Medical School and Massachusetts General Hospital, Boston, MA 02115, USA; LHE3@mgh.harvard.edu
* Correspondence: akapoor@mcmaster.ca (A.K.); damut@mcmaster.ca (D.T.); Tel.: +1-905-522-1155 (ext. 35218) (A.K.); +1-905-522-1155 (ext. 35168) (D.T.)

Simple Summary: Papillary renal cell carcinoma (pRCC) is an aggressive kidney cancer. Currently, there are no effective prognostic biomarkers and lack of efficacious therapies in treating pRCC. We report a novel and critical pRCC oncogenic factor OIP5. Its expression is increased in pRCC and the upregulation is associated with adverse features. High levels of OIP5 effectively predict pRCC recurrence and fatality. OIP5 promotes pRCC cell proliferation and tumor formation through complex processes. A 66-gene multigene panel (Overlap66) was constructed. Overlap66 is novel and robustly predicts pRCC recurrence and fatality. High risk pRCCs stratified by Overlap66 are associated with immune suppression. Furthermore, PLK1 is a component gene of Overlap66; PLK1 inhibitor significantly reduced OIP5-promoted pRCC cell proliferation in vitro and tumor growth in vivo. Collectively, Overlap66 can effectively stratifies high-risk pRCCs and these tumors can be treated with PLK1 inhibitors. Our findings can be explored for personalized therapy in pRCC patients.

Abstract: Papillary renal cell carcinoma (pRCC) is an aggressive but minor type of RCC. The current understanding and management of pRCC remain poor. We report here OIP5 being a novel oncogenic factor and possessing robust prognostic values and therapeutic potential. OIP5 upregulation is observed in pRCC. The upregulation is associated with pRCC adverse features (T1P < T2P < CIMP, Stage1 + 2 < Stage 3 < Stage 4, and N0 < N1) and effectively stratifies the fatality risk. OIP5 promotes ACHN pRCC cell proliferation and xenograft formation; the latter is correlated with network alterations related to immune regulation, metabolism, and hypoxia. A set of differentially expressed genes (DEFs) was derived from ACHN OIP5 xenografts and primary pRCCs ($n = 282$) contingent to OIP5 upregulation; both DEG sets share 66 overlap genes. Overlap66 effectively predicts overall survival ($p < 2 \times 10^{-16}$) and relapse ($p < 2 \times 10^{-16}$) possibilities. High-risk tumors stratified by Overlap66 risk score possess an immune suppressive environment, evident by elevations in Treg cells and PD1 in CD8 T cells. Upregulation of PLK1 occurs in both xenografts and primary pRCC tumors with OIP5 elevations. PLK1 displays a synthetic lethality relationship with OIP5. PLK1 inhibitor BI2356 inhibits the growth of xenografts formed by ACHN OIP5 cells. Collectively, the OIP5 network can be explored for personalized therapies in management of pRCC patients.

Keywords: OIP5; papillary renal cell carcinoma; PLK1; tumorigenesis; therapy; biomarkers

1. Introduction

Renal cell carcinoma (RCC) accounts for approximately 85% of kidney cancer cases. RCC can be classified as clear cell RCC (ccRCC, 75%) and non-ccRCC (nccRCC, 25%) [1].

In the latter group, papillary RCC (pRCC) constitutes 50–64% of total incidence [2,3]. Morphologically, pRCC consists of two subtypes: type 1 pRCC (T1P) and type 2 pRCC (T2P) [4]. T1P and T2P are often associated with low nuclear grade (Fuhrman 1–2) and high nuclear grade (Fuhrman 3–4) tumors respectively [4,5], providing a clinical basis for T2P tumors having poor prognosis [6–9]. Genetically, while T1P tumors typically have alterations in the *MET* gene leading to abnormal MET activation [10], T2P tumors are heterogenous and contains: (1) mutations in the *FH* (fumarate hydratase) [11], *CDKN2A*, *SETD2*, *BAP1*, and *PBRM1* genes [12], (2) CpG island methylator phenotype (CIMP), and (3) activation of the NFR2-ARE (antioxidant response element) pathway [12]. Among the T2P tumors, CIMP subtype show a particularly low possibility of overall survival [12].

While these morphological and molecular subtyping offers a primary prognostic assessment of pRCC, significant improvement is needed to enhance patient counselling and management. Effective prediction of the risk of pRCC relapse is essential in offering personalized treatments; this risk assessment is particularly important in the light that surgery remains the primary treatment for localized pRCC with a relapse rate of nearly 40% [13]. Furthermore, therapeutic options for recurrent and metastatic pRCCs are limited and non-effective, which was partly a result of treatments being extrapolated from ccRCC studies. For instance, sunitinib is a standard of care for patients with metastatic pRCC [14], despite the therapeutic benefits being low and not as effective as for ccRCC [15]. The lack of effective prognostic biomarkers and therapeutic options highlight the unmet need for a more thorough investigation of the critical factors regulating pRCC progression.

Opa interacting protein 5 (OIP5) was discovered as an Opa (*Neisseria gonorrhoeae* opacity- associated) interacting protein [16]. The protein is highly enriched in human testis (https://www.proteinatlas.org/ENSG00000104147-OIP5/tissue, accessed on 1 August 2021) [17]; its upregulation is associated with adverse clinical features in multiple cancer types, including leukemia [18], ccRCC [19], glioma [20], and the cancers of the liver [21,22], lung [23], breast [24], gastric [25,26], and bladder [27–30]. Functionally, knockdown of OIP5 was reported to attenuate the proliferation of bladder cancer cells [29], as well as colorectal and gastric cells in vitro [25]. Building on these limited studies (n = 17 articles in PubMed on 5 April 2021) reporting a relevance of OIP5 in oncogenic events, much more remains unanswered for OIP5-facilitated oncogenesis, particular in the context of pRCC, as OIP5 has yet to be reported in studies related to pRCC.

We provide the first comprehensive analysis of OIP5's oncogenic contributions in pRCC. OIP5 expression is significantly upregulated in pRCC; high levels of OIP5 correlate with adverse clinical characteristics of the disease, including stage, histological subtype (T2P), molecular subtype (CIMP), and lymph node metastasis. OIP5 expression robustly stratifies the risk of pRCC progression (progression-free survival) and fatality (overall survival and disease-specific survival). Functionally, OIP5 promotes pRCC cell proliferation in vitro and xenograft growth in vivo. Mechanistically, OIP5 facilitates pRCC progression along with network alterations; these changes show robust prognostic efficacies for rapid pRCC progression and fatality risk. Those of high-risk tumors display alterations in immune cell subsets including increases in the regulatory T (Treg) cell population. Treg cells are a major contributor to tumor-associated immune suppression [31]. Additionally, we identified polo-like kinase 1 (PLK1) as an OIP5-related gene in pRCC; the inhibition of PLK1 reduced OIP5-derived promotion of pRCC xenograft growth in vivo. Collectively, we report here (1) novel multigene sets derived from the OIP5 network that effectively predict the shortening of progression-free survival (PFS), overall survival (OS), and disease-specific survival (DSS) of pRCC, (2) an immune suppressive environment in pRCC tumors with OIP5 upregulation, and (3) inhibition of PLK1 as a potentially effective therapy in pRCC harboring OIP5 upregulation.

2. Materials and Methods

2.1. Cell Lines, Plasmid, and Retrovirus Infection

ACHN pRCC cell line and 786-O ccRCC cell line were purchased from ATCC and cultured in MEM and RPMI1640 respectively (Gibco, Carlsbad, CA, USA), both supplemented with 1% Penicillin-Streptomycin (Gibco, Carlsbad, CA, USA) and 10% fetal bovine serum (Life Technologies, Burlington, ON, USA). Cell lines were routinely checked for Mycoplasma contamination using a PCR kit (Abm, Cat#: G238). OIP5 cDNA plasmid was obtained from Origene (Cat: RG202255, Rockville, MD, USA) and subcloned into pBABE puro retroviral plasmid (from Dr. Tak Mak at University of Toronto). Packing of retrovirus and the subsequent transfection were performed following our published conditions [32].

2.2. Invasion and Soft Agar Assay

Insert chambers with a control or matrigel membrane (8 µM pore size) for 24-well plates (Life sciences Corning® BioCoat™, Glendale, AZ, USA) was used for invasion assay following manufacturer's instructions. Cells (10^4) were seeded into the top chamber; serum-free medium and 10% serum medium was added to the top and bottom chamber, respectively. Cells passing through the membrane were stained with crystal violet (0.5%). Soft agar assay was performed following our published conditions [32].

2.3. Colony Formation Assay and Proliferation Assay

Growth curves were generated by seeding 10^5 cells/per well into 6-well tissue culture plates. Cell numbers were counted every 2 days. Colony formation assay was conducted by seeding cells in six-well plates with 100, 500, 1000 cells for ACHN, and 100, 300, 500 for 786-O. Colonies were fixed by fixation buffer (2% formaldehyde) and stained by crystal violet (0.5%) after cultured for 2 weeks. Colony numbers were counted and analyzed.

2.4. Western Blot

Cell lysates were prepared, and western blot was carried out as we have previously published [32]. Antibodies used included Anti-Flag M2 (1:1500, Sigma-Aldrich, Oakville, ON, Canada) and Anti-OIP5 (1:500, Sigma-Aldrich, Oakville, ON, Canada).

2.5. Immunohistochemistry (IHC)

Kidney cancer TMA (KD29602) was purchased from US Biomax (Dervood, MD, USA). Slide was baked at 60 °C for 1 h, then de-paraffinized in 100% xylene and 70% EtOH series. Antigen retrieval buffer was prepared with sodium citrate buffer (PH = 6) in the steamer for 20 min. OIP5 (1:50, Sigma-Aldrich, Oakville, ON, Canada) antibodies were incubated at 4 °C overnight. Secondary anti-rabbit antibodies (Vector Laboratories, 1:200), VECTASTAIN ABC and DAB solution (Vector Laboratories) were subsequently added to the slides and incubated following our IHC protocol. Washes were performed by 1× PBS and distilled water. Slides were counterstained by haematoxylin (Sigma Aldrich, Oakville, ON, Canada) and image analysis was conducted with ImageScope software (Leica Microsystems Inc., Richmond Hill, ON, Canada). Staining intensity scores were calculated into HScore by the formula [HScore = (%Positive) × (Intensity) + 1]. Statistical analysis was performed by student t-test, and $p < 0.05$ was considered statistically significant.

Xenograft tumors were paraffin embedded and cut serially by microtome. OIP5 (1:50, Sigma-Aldrich), Anti-Phospho-Histone H3 (Ser 10) (1:200, Upstate Biotechnology Inc., Lake Placid, NY, USA), CDK2 (1:200, Santa Cruz, Dallas, TX, USA), and PLK1 (1:300, Novus Biologicals, Toronto, ON, Canada) antibodies were used in the analyses for the xenograft tumors.

2.6. Xenograft Tumor Formation and Treatment with PLK1 Inhibitor

ACHN OIP5 and ACHN EV were suspended in 0.1 mL MEM/Matrigel (BD) mixture with 1:1 volume and implanted subcutaneously into the left flank of 8-week-old nonobese diabetic/severe combined immunodeficiency (NOD/SCID) male mice (The Jackson Laboratory). The mice were monitored post-injection of cancer cells through observation and palpation. The size of the tumors was measured every two days by caliper. Tumor volume was calculated based on the formula V = L × W2 × 0.52. BI2536 PLK1 inhibitor (Selleckchem, Burlington, ON, Canada) was dissolved in 0.1 N HCl and diluted by 0.9% NaCl. Diluted BI2536 or 0.9% NaCl (negative control) was injected to mice intravenously via tail vein with a dosage of 50 mg/kg. The mice were euthanized when the tumor volume reached 1000 mm^3. The xenograft tumor, together with all the major organs, were photographed and collected. All tumors were cut in half, with one half fixed with 10% formalin (VWR, Mississauga, ON, Canada) and the other half stored in −80 °C. The formalin-fixed tissue was processed by department of Histology (St. Joseph's Health care, Hamilton, ON, Canada) and embedded in paraffin. All the animal works were performed according to the protocols approved by McMaster University Animal Research Ethics Board (16-06-23).

2.7. RNA Sequencing Analysis

RNA sequencing analysis was carried out following our established conditions [33]. RNA was extracted from ACHN EV (*n* = 3) and ACHN OIP5 (*n* = 3) xenografts using a miRNeasy Mini Kit (Qiagen, No. 217004) according to the manufacturer's instructions. RNA-seq libraries were generated with TruSeq Ribo Profile Mammalian Kit (Illumina, RPHMR12126) according to manufacturer's instruction. These libraries were sequenced in a paired end setting by Harvard Bauer Core Facility using Nextseq 500/550. RNA-seq reads were processed and analyzed using Galaxy (https://usegalaxy.org/, accessed on 31 May 2020). Specifically, low quality reads and adaptor sequences (AGATCGGAAGAG-CACACGTCTGAACTCCAGTCA: forward strand and AGATCGGAAGAGCGTCGTG-TAGGGAAAGAGTGT: reverse strand) were first removed. Alignment and read counts were performed using HISAT2 and Featurecounts respectively. Differential gene expression was determined using DESeq2. KEGG analysis and GSEA (Gene Set Enrichment Analysis) were also performed using Galaxy; the FGSEA (fast preranked GSEA) was used for GSEA analysis. Enrichment analyses were carried out using Metascape (https://metascape.org/gp/index.html#/main/step1, accessed on 1 September 2020) [34].

2.8. RNA Sequencing Analysis

Cox proportional hazards (Cox PH) regression analyses were performed using the R survival package. The PH assumption was tested. Cutoff points were estimated using Maximally Selected Rank Statistics (the Maxstat package, https://cran.r-project.org/web/packages/maxstat/maxstat.pdf, accessed on 8 August 2020). The TCGA PanCancer Atlas pRCC dataset available from cBioPortal [35,36] was used.

2.9. Examination of Gene Expression

Gene expressions were determined using the UALCAN platform (ualcan.path.uab.edu/home, accessed on 31 March 2021) [37].

2.10. Statistical Analysis

Kaplan-Meier survival analyses and logrank test were conducted by R *Survival* package and tools provided by cBioPortal. Cox regression analyses were performed using R *survival* package. Time-dependent receiver operating characteristic (tROC) analyses were carried out with R *timeROC* package. ROC and precision-recall (PR) profiles were constructed using the PRROC package in R. Two-tailed Student *t*-test, one-way ANOVA, and two-way ANOVA were performed for statistical analysis of two and more than two groups respectively, with $p < 0.05$ to be considered statistically significant. Tukey's test was performed for post-hoc analysis. Statistical analysis was conducted by GraphPad Prism 7 and data were presented as mean ± SEM/SD. A value of $p < 0.05$ was considered statistically significant.

3. Results

3.1. Association of OIP5 Upregulation with pRCC Tumorigenesis and Progression

OIP5 was reported to be a component gene in a multigene set predicting the risk of prostate cancer recurrence [38]; its upregulation associates with adverse features in ccRCC and bladder cancer [19,29], supporting a general involvement of OIP5 in urogenital cancers. To investigate this possibility, we examined OIP5 expression in pRCC using a tissue microarray (TMA) containing 40 pairs of pRCC and 74 pairs of ccRCC tumors with the adjacent non-tumor kidney (AJK) tissues from 20 and 37 patients, respectively. The pRCC patient population ($n = 20$) consists of 11 men and 9 women with most tumors being at T1 stage (Table 1). In comparison to AJK tissues, pRCC tumor tissues expressed a significant OIP5 upregulation (Figure 1A,B). OIP5 expression was further increased in advanced T stage tumors (Figure 1B). Consistent with a previous report, OIP5 upregulation occurred in Grade 2–3 ccRCC tumors compared to the AJK tissues; nonetheless, we could not demonstrate OIP5 upregulation in Grade 1 ccRCC compared to the AJK tissues (Figure S1), suggesting a role of OIP5 in ccRCC progression. By using the TCGA RNA-sequencing data organized by UALCAN (ualcan.path.uab.edu/home, 31 March 2021) [37], OIP5 upregulation at the mRNA level in pRCC tissues was observed (Figure 1c); the upregulations reflects the level of severity and the order of unfavorable outcome of pRCC with higher expression levels in T2P over T1P tumors, CIMP tumors over other subtypes (Figure 1D), stage 3 tumors over stages 1–2 tumors, stage 4 over stage 3 tumors (Figure 1E), and N1 (lymph node metastasis) over N0 tumors (Figure 1F). Consistent with its associations with adverse tumor features, OIP5 expression robustly stratifies pRCC tumors into a high- and low-risk group based on overall survival possibility (Figure 1G). Among the 10 patients in the OIP5-high group, seven died in a rapid time course (Figure 1G). Collectively, these observations support a strong association of OIP5 with pRCC tumorigenesis and progression.

Table 1. The clinical parameters of pRCC patients included in TMA.

Parameter	Age (Year)	Male (n)	Female (n)	T1 (n)	T2 (n)	T3 (n)
Details	49.5 (39.8–61)	11	9	13	6	1

Age: median (Q1/quartile 1–Q3) n: number of cases. All patients were without lymph node metastasis (N0) and distant metastasis (M0).

3.2. OIP5-Mediated Enhancement of pRCC Tumorigenesis along with Network Alterations

Attributed to the uncommon status of pRCC, there are only limited number of confirmed pRCC cell lines available. ACHN is the most widely used and confirmed metastatic pRCC cell line; the cells have the typical feature of c-MET polymorphism detected in pRCC [39,40]. ACHN is likely the only confirmed metastatic pRCC cell line [39]. To analyze the functional impact of OIP5 on pRCC tumorigenesis, we stably expressed OIP5 in ACHN cells (Figure 2A). In comparison to ACHN EV (empty vector) cells, ACHN OIP5 cells displayed elevated abilities for proliferation (Figure 2B), colony formation (Figure 2C; Figure S2A), invasion (Figure 2D; Figure S2B), and growth in soft agar (Figure 2E; Figure S2C). We have also established the EV and OIP5 stable lines in the commonly used 786-O ccRCC cells, and OIP5 overexpression did not affect all of the above oncogenic events observed in ACHN cells in vitro (data not shown), which suggests a certain level of specificity of OIP5 in promoting pRCC. In vivo, OIP5 enhanced the growth of ACHN cell-produced xenografts compared to tumors produced by ACHN EV cells (Figure 2F); mice bearing ACHN OIP5 tumors reached endpoint faster compared to animals with ACHN EV cell-produced tumors (Figure 2G). The overexpression of OIP5 in ACHN OIP5 tumors was confirmed (Figure S3A). The ACHN OIP5 tumors show a significant increase of CDK2 expression largely in the nuclei (Figure S3B); the functions of this are not clear as no upregulations of the relevant cyclins (cyclin A and cyclin E) was observed (data not shown).

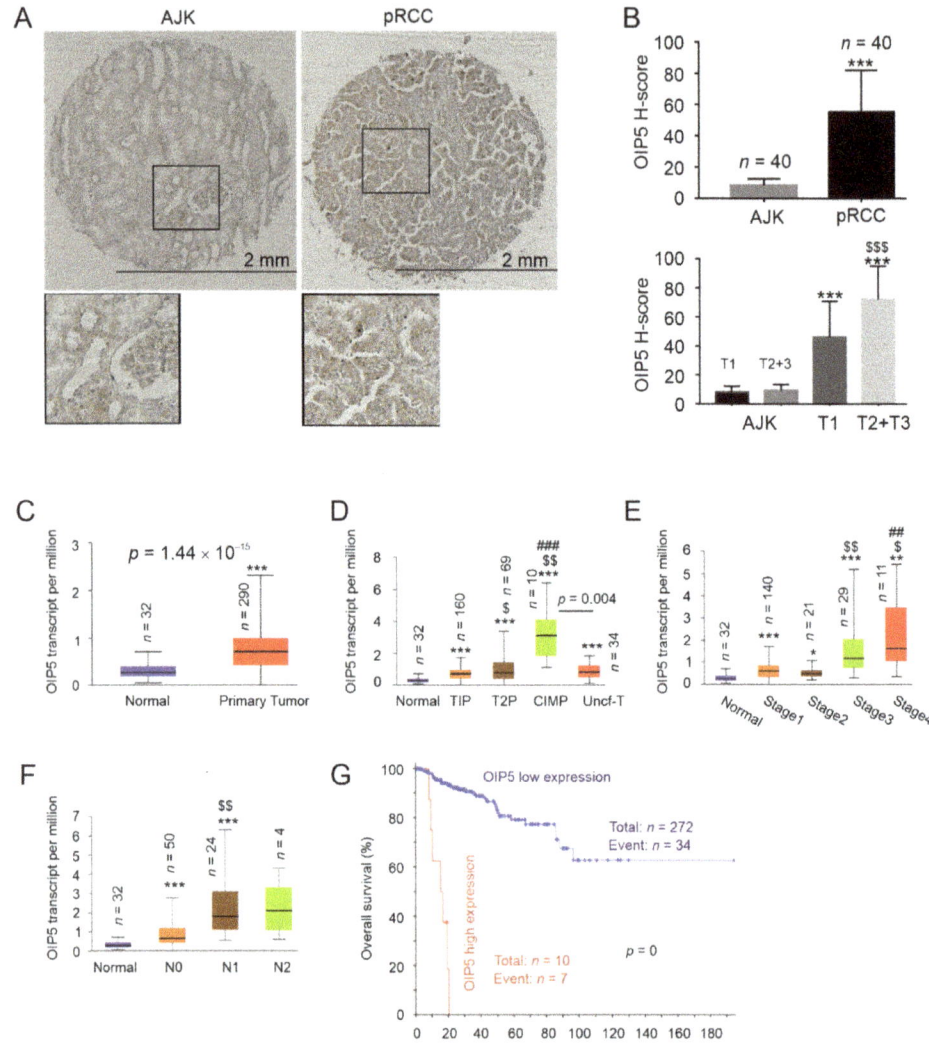

Figure 1. Upregulation of OIP5 associates with adverse features of pRCC and predicts poor overall survival. (**A**) IHC staining for OIP5 was performed using a RCC TMA; typical images of OIP5 staining in the adjacent kidney (AJK) and pRCC tumor tissues are presented. (**B**) Quantification of OIP5 IHC staining by H-score in the indicated tissues; means ± standard deviations (SDs) are graphed. Statistical analyses were performed using 2-tailed Student's t-test; ***: $p < 0.001$ compared to the respective AJK tissues, \$\$\$: $p < 0.001$ compared to T1 tumors. (**C**–**F**) OIP5 mRNA expressions in the indicated setting were analyzed using the TCGA dataset organized by UALCAN [37]. Student's t-test (**C**) and other indicated paired statistics were provided by UALCAN. *: $p < 0.05$, **: $p < 0.01$, ***: $p < 0.001$ compared to normal kidney tissues; \$: $p < 0.05$, \$\$: $p < 0.01$ compared to T1P (**D**), Stage 2 (**E**), and N0 (**F**); ##: $p < 0.01$, ###: $p < 0.001$ compared to T2P (**D**) and Stage 3 (**E**). (**G**) Survival analysis was performed using the TCGA Pancancer pRCC dataset within cBioPortal. Logrank test was performed. Cutoff point used to separate the high- and low-OIP5 expression groups was ≥ 2 z-score or 2SD. The graph was produced using tools provided by cBioPortal. The median months overall survival for patients in the high-OIP5 group was 15.48 months.

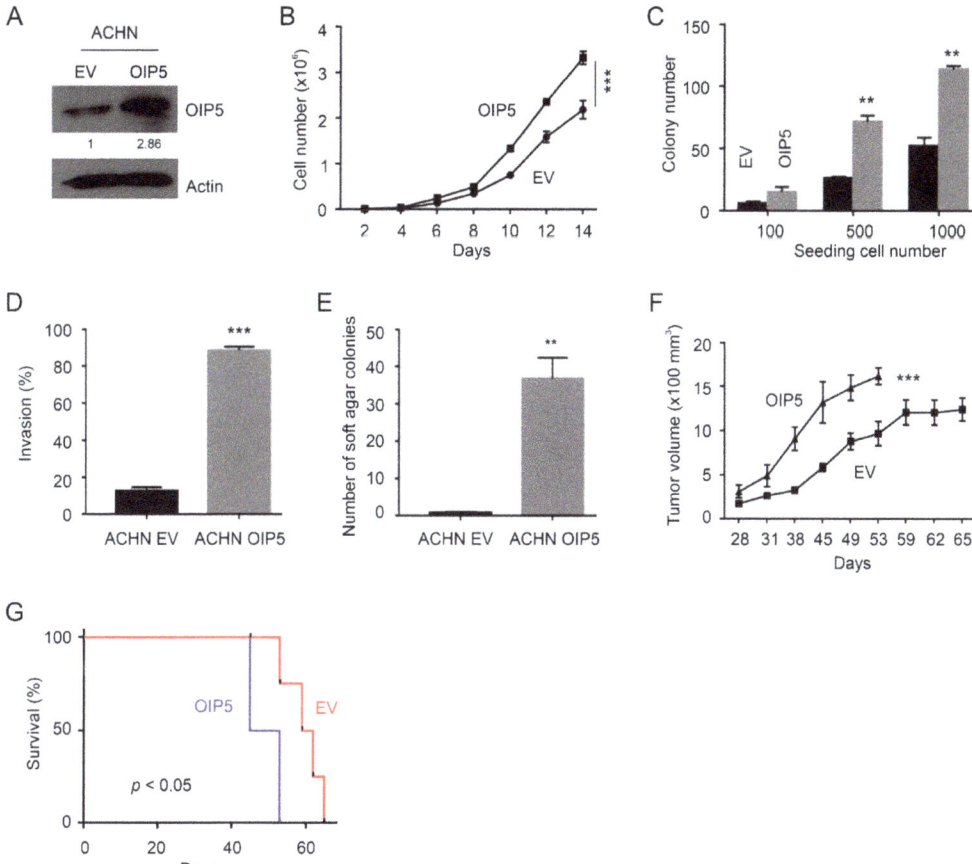

Figure 2. OIP5 promotes oncogenic processes of ACHN cells in vitro and in vivo. (**A**) ACHN empty vector (EV) and OIP5 stable lines. Western blot was carried out using anti-OIP5 and Actin antibodies. OIP5 expression was normalized to Actin and presented at fold changes to OIP5 expression in EV cells. (**B**) ACHN EV and ACHN OIP5 cells were seeded in 6-well plate at 10^5 cell/well; cell numbers were recorded at the indicated days. Experiments were repeated three times; means ± SDs are graphed. Statistical analysis was performed using 2-way ANOVA. ***: $p < 0.001$ between the two curves. (**C**) The indicated cells were seeded at the indicated number in 6-well plates. Colonies were formed following 2 weeks culture. Experiments were repeated three times; means ± SDs are graphed. **: $p < 0.01$ compared to the respective EV by Student's t-test (2-way). (**D,E**) Invasion and soft agar assays were repeated 3 times; means ± SDs are graphed. **: $p < 0.01$, ***: $p < 0.001$ compared to the respective EV control by Student's t-test (2-way). (**F,G**) Xenografts were produced in NOS/SCID mice (5 mice per group) using ACHN EV cells and ACHN OIP5 cells. Means ± SEM (standard error of the mean) are graphed; ***: $p < 0.001$ between the two curved by two-way ANOVA (**F**). Kaplan-Meier curve; statistical analysis was performed using logrank test (**G**).

To further analyze factors and networks utilized by OIP5 in enhancing ACHN cell-produced xenografts, RNA-sequencing (RNA-seq) was performed on ACHN EV and ACHN OIP5 tumors at three per group. Gene set enrichment analysis (GSEA) was conducted on differentially expressed genes obtained in the setting of OIP5 vs. EV. When enrichment in the oncogenic gene sets (C6) collection was analyzed using FGSEA (fast gene set enrichment analysis), we observed that genes downregulated (DN) in cells with activation (UP) of ERB2, MEK, and mTOR were also downregulated in ACHN OIP5 tumors compared to ACHN EV tumors (Figure 3A), suggesting OIP5 suppressing those genes that are downregulated by ERB2, MEK, and mTOR. Similarly, ACHN OIP5 tumors also display downregulation of EGFR-downregulated genes (Table S1). The serine/threonine kinase 33 (STK33) is a synthetic lethal interacting protein of KRAS mutant, i.e., cells expressing KRAS mutant rely on STK33 for survival [41]. Knockdown of STK33 in acute myeloid leukemia cells led to upregulation of a set of genes (STK33-UP) [41], suggesting a potential inhibition of these genes by STK33. These gene expressions were also reduced in ACHN OIP5 tumors (Figure 3A; Table S1). To test the reliability of the enrichment obtained by FGSEA, GSEA was further conducted using a more stringent platform: EGSEA. Ensemble gene set enrichment analysis produces a consensus gene set ranking (enrichment) with the combination of multiple (up to $n = 12$) algorithms [42]. With the maximal stringent condition using all 12 algorithms, EGSEA revealed within the top 12 ranks the downregulation of the ERB2- and MEK-suppressed gene sets in ACHN OIP5 tumors (Figure S4); downregulation of genes in cells with STK33 knockdown was observed in multiple setting (Figure S4) which is consistent with the enrichments derived from using FGSEA (Table S1). All top 12 ranked gene sets obtained by EGSA (Figure S4) are also included in those produced by FGSEA (Table S1). It is intriguing that VEGFA-suppressed genes in HUVEC (human umbilical vein endothelial cell) cells were also downregulated in ACHN OIP5 tumors (Figure S4; Table S1). Based on the overall gene set enrichment within the oncogenic gene set (C6, MSigDB) collection (Table S1), we can summarize that in ACHN OIP5 xenografts, the RB pathway is inhibited and the signaling processes of STK33, BMI1, EZH2, MYC, WNT, VEGFA, and EGFR/ERB2 are enhanced (Figure 3B).

We further examined gene set enrichment within the Hallmark gene set collection using FGSEA. The analyses revealed downregulations of inflammatory response, TNFα_via_NFκB signaling (NFκB-regulated genes in response to TNFα), and complement gene expression (Hallmark_Component, normalized enrichment score/NES: −1.48, padj 0.013) (Figure S5A; Table S2). Additionally, ACHN OIP5 xenografts exhibited upregulations in gene sets regulating fatty acid metabolism and cholesterol homeostasis (Figure 3C; Table S2). These enrichments were also produced by EGSEA (Figure S5B,C). Several processes are enhanced in ACHN OIP5 tumors, which include oxidative phosphorylation, the expression of E2F and MYC targets, EMT (epithelial mesenchymal transition), mTORC1 signaling, and adipogenesis (Table S2). Enrichment in glycolysis in ACHN OIP5 tumors was obtained by FGSEA (Table S2), which was also confirmed by KEGG pathway analysis using EGSEA (Figure S6). Evidence thus suggests a metabolic switch to Warburg metabolism in ACHN OIP5 tumors.

3.3. Association of OIP5-Related Differentially Expressed Genes with CIMP Subtype

In comparison to other pRCC subtypes, CIMP tumors have a Warburg metabolic shift [12], indicating an association between OIP5-affected genes and CIMP. This notion is supported by the elevation of OIP5 expression in CIPM pRCC tumors (Figure 1D). To investigate this possibility, we firstly defined the differentially expressed genes (DEGs) in ACHN OIP5 tumors vs. ACHN EV tumors as those with $p.adj < 0.05$ and fold change $> |1.5|$; a total of 1128 DEGs were derived (Table S3). In these DEGs, the top upregulated genes include WNT7A, FGF1, CNTN1 [43], SOX2, and others, which are known for their facilitative roles in tumorigenesis. The top 20 clusters enriched in these DEGs contain those that regulate urogenital system development, blood vessel morphogenesis, hippo pathway, cell surface receptor signaling, pathway in cancer, epithelial cell proliferation, and others (Figure 4A; Table S4). Individual terms in these enriched clusters form a network connection (Figure S7A). These pathways are clearly relevant to tumorigenesis. DEGs are clustered in ACHN OIP5 tumors vs. ACHN EV tumors (Figure 4B).

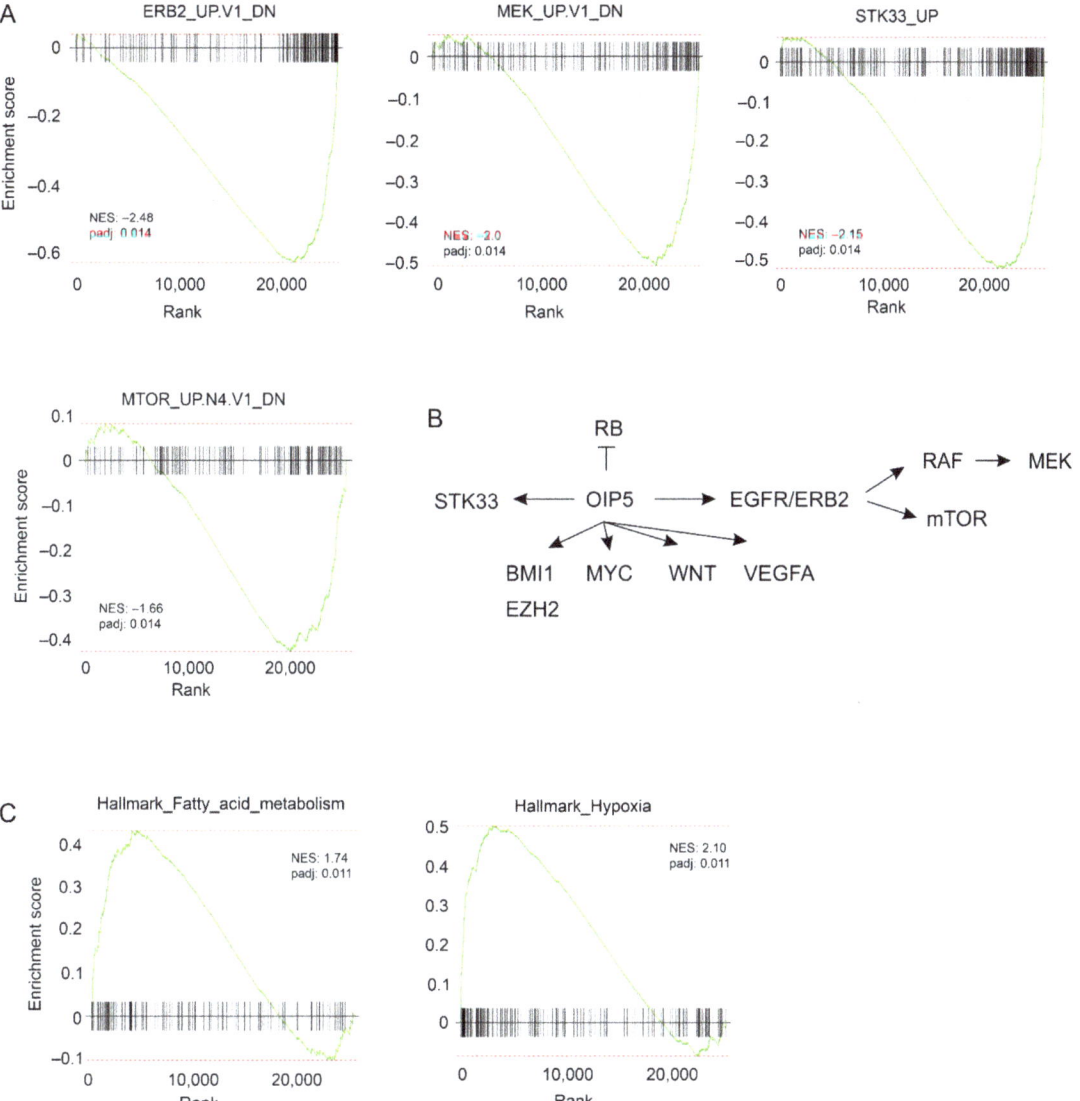

Figure 3. OIP5 induces network alterations during pRCC tumorigenesis. (**A**) GSEA (gene set enrichment analysis) on differentially expressed genes derived from the comparison of ACHN OIP5 tumors to ACHN EV tumors was performed with FGSEA within the Galaxy platform. The MSigDB oncogenic gene sets (C6) collection was used. (**B**) Summary of the major oncogenic gene sets affected in ACHN OIP5 tumors (see Table S1 for individual gene sets affected). (**C**) Enrichment of the indicated gene set within the MSigDB hallmark gene sets collection (see Table S2 for individual gene sets affected).

Table 2. Characterization of Overlap66 genes.

Gene	OS [1]	p Value	CIMP [2]	p Value	Tumor [3]	p Value
SLC7A11	+	0.0135 *	Up	8.4×10^{-5} ***	Up	1.68×10^{-5} ***
PCSK5 [4]	+	6.73×10^{-10} ***	Up	0.0085 **	Down	0.011 *
STC2 [4]	+	1.72×10^{-6} ***	Up	2.03×10^{-5} ***	None	NS
TEX15 [4]	+	0.00101 **	Up	0.074	Down	1.55×10^{-5} ***
ESCO2 [5]	+	6.89×10^{-10} ***	Up	0.0149 *	Up	1.67×10^{-12} ***
OIP5 [4]	+	7.41×10^{-12} ***	Up	8.07×10^{-5} ***	Up	1.44×10^{-15} ***
PLK1 [5]	+	5.58×10^{-15} ***	Up	0.0241 *	Up	$<1 \times 10^{-12}$ ***
ELOVL2 [5]	+	4.36×10^{-8} ***	Up	0.0197 *	Up	4.34×10^{-7} ***
LYPD6 [4]	N	NS	Up	0.0236 *	Down	4.82×10^{-8} ***
ATAD2 [5]	+	1.84×10^{-13} ***	Up	0.00446 **	Up	3.80×10^{-8} ***
ISM1 [4]	+	0.00237 **	N	NS	None	NS
TK1 [5]	+	1.51×10^{-10} ***	Up	0.0126 *	Up	5.97×10^{-8} ***
TRIB3 [5]	+	2.26×10^{-6} ***	Up	6.19×10^{-9} ***	Up	3.23×10^{-13} ***
KIAA1324L [4]	N	NS	Up	0.0346 *	None	NS
SLIT3 [5]	+	0.000677 ***	N	NS	Down	6.05×10^{-9} ***
COL14A1	+	0.00114 **	N	NS	Down	3.21×10^{-7} ***
FAM40B	N	NS	Up	0.0203 *	None	NS
STOX1	N	NS	N	NS	Down	1.89×10^{-14} ***
ABCA12 [5]	+	9.87×10^{-5} ***	Up	0.0182 *	Up	$<1 \times 10^{-12}$ ***
RGS20	N	NS	Up	0.0377 *	Up	1.63×10^{-12} ***
ACCN2	+	0.0105 *	Up	1.57×10^{-4} ***	None	NS
DPYSL3 [5]	+	2.18×10^{-6} ***	Up	3.37×10^{-4} ***	Down	7.92×10^{-5} ***
STAT4 [5]	+	0.024 *	N	NS	Up	7.34×10^{-8} ***
CALCRL [4]	+	0.0378 *	Up	0.0024 **	Down	7.29×10^{-11} ***
SRXN1	+	0.0258 *	Up	0.0102 *	Up	3.39×10^{-10} ***
FAR2	N	NS	Down[2a]	0.0049 **	Down	0.0165 *
TPD52 [5]	+	8.63×10^{-7} ***	Up	8.08×10^{-4} ***	Down	1.62×10^{-12} ***
ZNF239 [4]	+	8.96×10^{-6} ***	Up	0.00488 **	None	NS
C16orf75 [5]	+	1.3×10^{-10} ***	Up	2.56×10^{-12} ***	Up	6.13×10^{-10} ***
HEYL [5]	+	0.000855 ***	Up	0.0312 *	Down	2.52×10^{-6} ***
F2R [4]	+	7.34×10^{-7} ***	Up	1.89×10^{-5} ***	Down	2.81×10^{-9} ***
KCNJ8 [4]	+	0.000334 ***	N	NS	Down	4.02×10^{-5} ***
RAD54B [5]	+	0.000931 ***	Up	0.0197 *	Up	1.14×10^{-9} ***
KCNK1 [4]	+	4.35×10^{-6} ***	Up	0.00816 **	Down	0.00236 **
ZNF391 [5]	+	0.00543 **	N	NS	Down	4.24×10^{-4} ***
POLR3G [5]	+	0.000266 ***	N	NS	Up	1.74×10^{-5} ***
MEIS1	N	NS	Up	0.0046 **	Down	9.44×10^{-12} ***
MCM8 [5]	+	0.00612 **	N	NS	None	NS
SNX16 [5]	+	2.29×10^{-7} ***	Up	3.51×10^{-5} ***	None	NS

Table 2. Cont.

Gene	OS [1]	p Value	CIMP [2]	p Value	Tumor [3]	p Value
SPAG1 [5]	+	0.000246 ***	Up	5.15×10^{-4} ***	None	NS
CX3CL1 [5]	−	0.000679 ***	Down	1.98×10^{-6} ***	Up	4.47×10^{-8} ***
DYNC2LI1	−	0.00415 **	Down	2.01×10^{-4} ***	Up	1.62×10^{-12} ***
ACSS2 [4]	−	0.0214 *	Down	1.62×10^{-12} ***	Down	1.73×10^{-5} ***
HS3ST5 [4]	N	NS	Down	6.1×10^{-5} ***	None	NS
DPF3 [4]	N	NS	Down [2a]	0.0027 **	Down	3.98×10^{-5} ***
ZNF862	N	NS	Down	1.83×10^{-11} ***	Up	7.87×10^{-12} ***
LHPP [5]	−	0.00907 **	N	NS	Down	0.0496 *
PITPNM3	−	0.0391 *	N	NS	Down	7.08×10^{-11} ***
GNG7 [5]	−	0.000249 ***	N	NS	Down	3.30×10^{-9} ***
CHD5 [4]	N	NS	Down	6.26×10^{-5} ***	None	NS
CCDC106 [5]	−	0.000256 ***	Down	1.01×10^{-6} ***	None	NS
NBL1	-	0.0211 *	Down	4.47×10^{-5} ***	Up	$<1 \times 10^{-12}$ ***
LYNX1 [5]	−	0.00675 **	Down	2.29×10^{-5} ***	Down	2.29×10^{-8} ***
PHYHIP	N	NS	Down	4.79×10^{-4} ***	None	NS
NRXN3	N	NS	N	NS	Down	1.87×10^{-9} ***
TMEM130 [4]	N	NS	Down	2.25×10^{-12} ***	Down	4.59×10^{-12} ***
EREG [4]	N	NS	Down	0.00318 **	Up	1.70×10^{-12} ***
C2orf62	−	0.00479 **	Down	1.97×10^{-4} ***	Up	1.62×10^{-12} ***
CCDC135	−	0.0478 *	Down	$<1 \times 10^{-12}$ ***	Up	1.62×10^{-12} ***
SYCE1L	N	NS	Down	2.56×10^{-12} ***	Up	$<1 \times 10^{-12}$ ***
GAL3ST3	N	NS	Down	1.63×10^{-12} ***	Down	7.38×10^{-4} ***
SPATA18 [5]	−	1.82×10^{-7} ***	Down	1.62×10^{-12} ***	Up	1.62×10^{-12} ***
C6orf138 [4]	N	NS	Down	0.026 *	Up	1.62×10^{-12} ***
ABI3BP	N	NS	Down	5.83×10^{-11} ***	Up	$<1 \times 10^{-12}$ ***
CNTN6 [4]	N	NS	Down	5.07×10^{-11} ***	Up	$<1 \times 10^{-12}$ ***
SCEL [4]	−	0.0331 *	Down	1.66×10^{-12} ***	Up	$<1 \times 10^{-12}$ ***

1: prediction of overall survival determined by univariate Cox analysis; +, −, and N: gene expression positively, negatively, and not predicting OS, respectively. NS: not significant. 2: expression status in CIMP tumors, "Up": upregulation compared to T2P, "Down": downregulation compared to T1P, 2a: in comparison to T2P as the comparison to T1P being not significant, N: no changes. 3: tumor (n = 290) in comparison to normal tissues (n = 30). 4: these genes are in Overlap21. 5: these genes are in Overlap21plus. Expression analysis in "CIMP" and "Tumor" using the TCGA data (UALCAN). *: $p < 0.5$, **: $p < 0.01$, ***: $p < 0.001$.

To confirm the relevance of these DEGs derived from ACHN cell-produced xenografts in pRCC pathogenesis, we analyzed their relationship to DEGs derived from primary pRCCs relative to OIP5 expression. In the TCGA Pancancer pRCC dataset within cBioPortal, high OIP5 expression robustly separates pRCC tumors into a high and low risk group based on their overall survival (OS) possibilities (Figure 1G). From these two groups, we obtained 873 DEGs defined by $q < 0.05$ and fold change $\geq |2|$ (Table S5). These primary pRCC-derived DEGs share 66 overlap DEGs (Overlap66) with the xenograft-derived DEGs (Table 2; Figure 4C). The alterations in their expressions in normal kidney tissues (n = 30) and pRCC tumors (n = 290) at different stages are presented in Figure S8. The genes with further elevations in Stage 3–4 tumors include SLC7A11, PCSK5, STC2, PLK1, TK1, TRIB3, and SRXN1 (Figure S8).

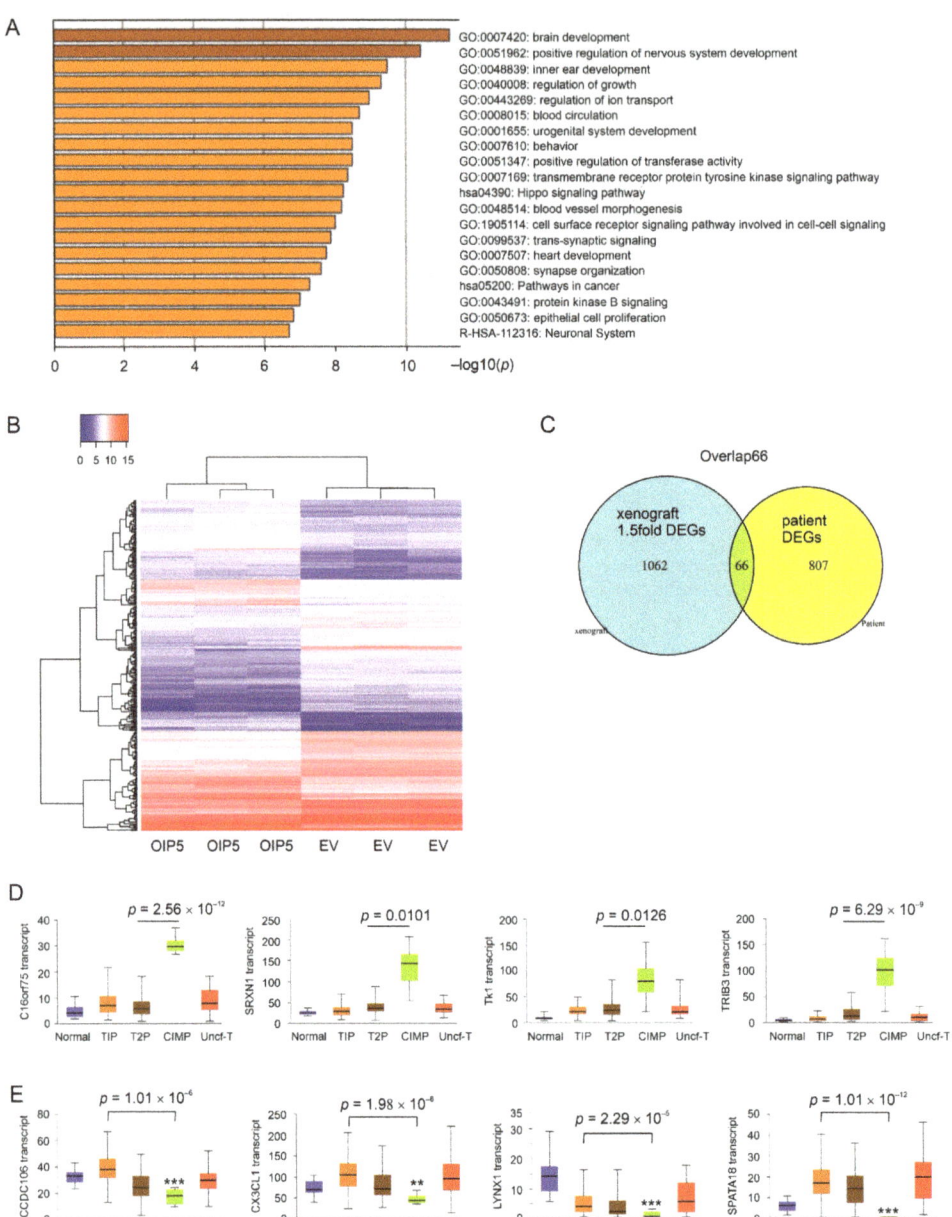

Figure 4. Pathway enrichment of OIP5 DEGs. DEGs were first defined as *p*.adj < 0.05 and fold changes > |1.5| in the comparison of ACHN OIP5 tumors (*n* = 3) vs. ACHN EV tumors (*n* = 3). (**A**) Pathway enrichment in these DEGs (Table S3) was then performed using the Metascape [34] platform. (**B**) Clustering of DEGs in ACHN OIP5 tumors and ACHN EV tumors. (**C**) The number of overlapping genes between primary (patient) pRCC-derived DEGs and DEGs obtained from xenografts at fold change > |1.5|. (**D**,**E**) The indicated DEGs were analyzed for expression in the histological subtypes of pRCC using the UALCAN platform [37]. DEGs positively (**D**) and negatively (**E**) predict shortening of OS (see Table 2 for details). **: $p < 0.01$, ***: $p < 0.001$ in comparison to normal kidney tissues.

Among these 66 DEGs, 8 and 41 genes are not known for associations with cancer and ccRCC respectively (Table S6A); only PLK1 was reported to be a component gene in a prognostic multigene of pRCC (Table S6A). Overlap66 is novel to pRCC. Forty-six out of 66 DEGs significantly predict overall survival (OS) possibility with some being individually efficacious based on their p values: 6.73×10^{-10} for PCSK5, 1.3×10^{-10} for C11orf75 (RMI2), 1.84×10^{-13} for ATAD2, and others (Table S6A). Furthermore, 33 DEGs retain their predictive significance after adjusting for age at diagnosis, sex, and T stage (Table S6A).

The potentials of the 33 DEGs as prognostic biomarkers are in accordance with their expression status in CIMP. C11orf75, SRXN1, TK1, and TRIB3 positively predict poor OS (Table 2; Table S6A); they are notably upregulated in CIMP tumors (Figure 4D). In reverse, CCDC106, CX3CL1, LYNX1, and SPATA18 are negatively associated with poor OS (Table 2; Table S6A); their expressions are particularly downregulated in CIMP tumors (Figure 4E). In all 46 genes with their expression associated with OS shortening, 11 show no alterations in gene expression in CIMP tumors (Table 2); for the remaining 35 genes, their positive and negative predictions of OS shortening correlate with their respective upregulation and downregulation in CIMP tumors (Table 2). This correlation of expression was not observed in tumors vs. non-tumor tissues (Table 2). In view of CIMP tumors having the poorest OS possibility [12], the association of these gene expression with CIMP tumors supports their potential as prognostic biomarkers.

3.4. Robust Prognostic Biomarker Potential of Overlap66 and Its Sub-Multigene Panels

Following the above analyses, we examined the OS-related prognostic potential of Overlap66 as a multigene panel. The expression data for these DEGs along with the relevant clinical data were retrieved from the Pancancer pRCC dataset within cBioPortal. Risk scores for individual tumors were calculated as $\sum(\text{coef}_i \times \text{Gene}_{i\text{exp}})n$ (coef_i: Cox coefficient of gene$_i$, Gene$_{i\text{exp}}$: expression of Gene$_i$, $n = 66$). Coefs were obtained using the multivariate Cox model. Overlap66 risk scores efficiently predict OS shortening using both univariate (UV) and multivariate (MV) Cox models (Figure S7B). The MV model consists of the risk scores, age at diagnosis, sex, and T stage (Figure S7B). With the cutoff point optimized using the Maximally Selected Rank Statistics (Figure S9), Overlap66 effectively stratifies the risk of fatality (possibility of OS) and relapse (progression-free survival/PFS) (Figure 5A,B). The discriminations of OS and PFS are with time-dependent area under the curve (tAUC) value of 94.6–91.3% in the time frame of 12.4 month (M) to 57.2 M (Figure 5A) and 93.7–86.7% for 10.8 M to 50.6 M (Figure 5B), respectively. Collective evidence supports Overlap66 being a novel and robust prognostic multigene panel for pRCC.

We further validated Overlap66 risk score in stratification of pRCC fatality risk using a recently developed R package: contpointr (https://github.com/thiele/cutpointr, accessed on 1 May 2021). An optimal cutoff point was obtained with Kernel smoothing model coupled with 1000 bootstrapping. This cutoff point classifies pRCC fatality risk at 0.78 sensitivity and 0.84 specificity or the sum of sensitivity and specificity (sum_sens_spec) value of 1.62 (Figure 6A). Risk stratifications of out-of-bag bootstrap samples ($n = 1000$) occurred most frequently at sum_sens_spec 1.6 (Figure 6B), which closely approximates sum_sens_spec 1.62 associated with the optimal cutoff point on the full cohort (Figure 6A). The fatality risk stratifications of the in-bag samples ($n = 1000$, average 63.2% of full samples) and the out-of-bag samples ($n = 1000$) were at the median sum_sens_spec values of 1.62 and 1.60 respectively. Taken together, these bootstrap analyses reveal a good out-of-sample performance of Overlap66 in classification of pRCC fatality risk, supporting Overlap66's application in real world. This potential is strengthened by the effectiveness of the risk classification with a range of cutoff points (Figure 6B,C).

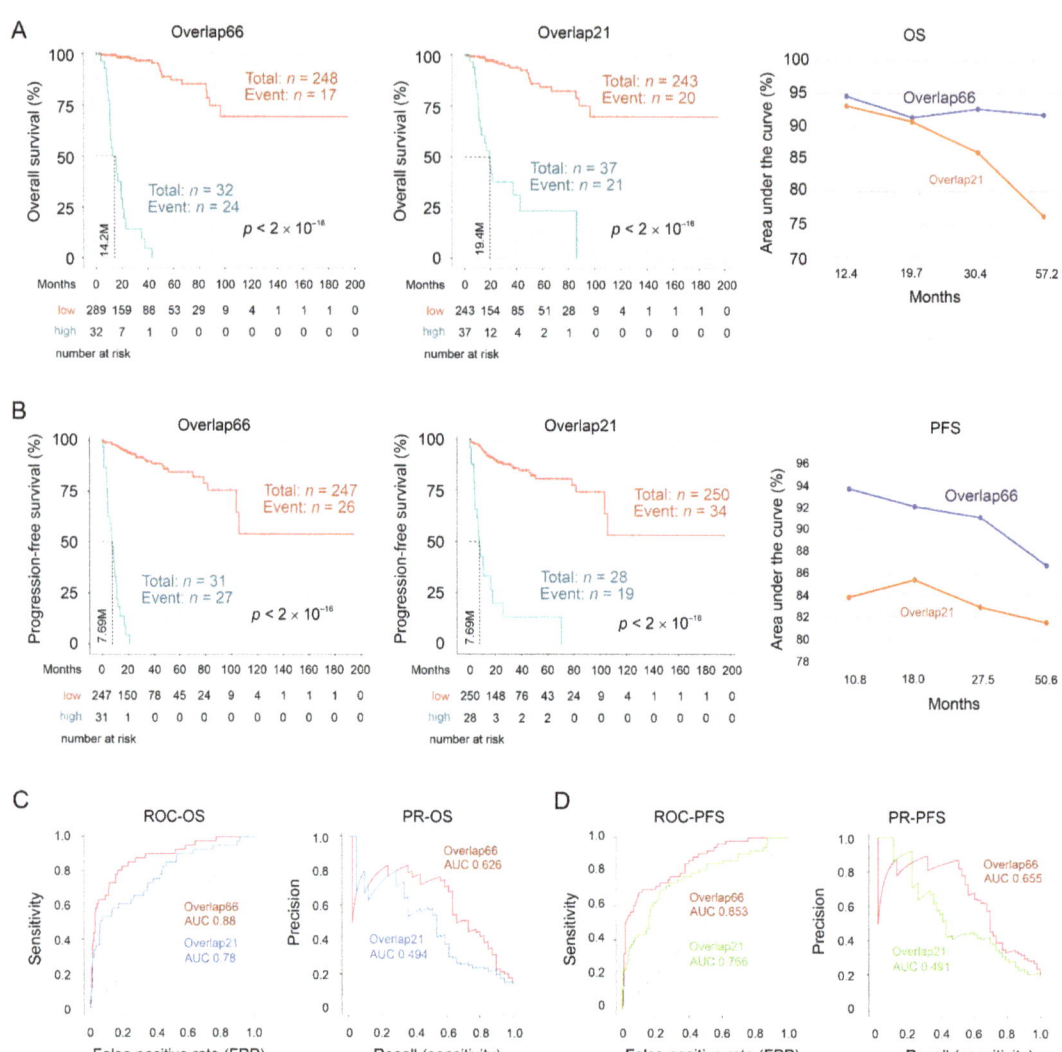

Figure 5. Stratification of the possibilities of overall survival (OS) and progression-free survival (PFS) by Overlap66 and Overlap21. (**A,B**) Cutoff points were determined by Maximally Selected Rank Statistics for the risk scores of Overlap66 (see Figure S9) and Overlap21. Kaplan Meier curves for OS (**A**) and PFS (**B**) are constructed, using the R survival package, with the populations at risk in the indicated follow-up period included. Statistical analyses were performed using logrank test. The median months of OS and PFS are indicated. Time-dependent ROC (receiver operating characteristic; tROC) curves were generated using the R *timeROC* package; time-dependent area under the curve (AUC) values for the indicated multigene sets are shown. (**C,D**) ROC and precision-recall (PR) curves for Overlap66 and Overlap21 in predicting OS and PFS possibilities were produced using the R PRROC.

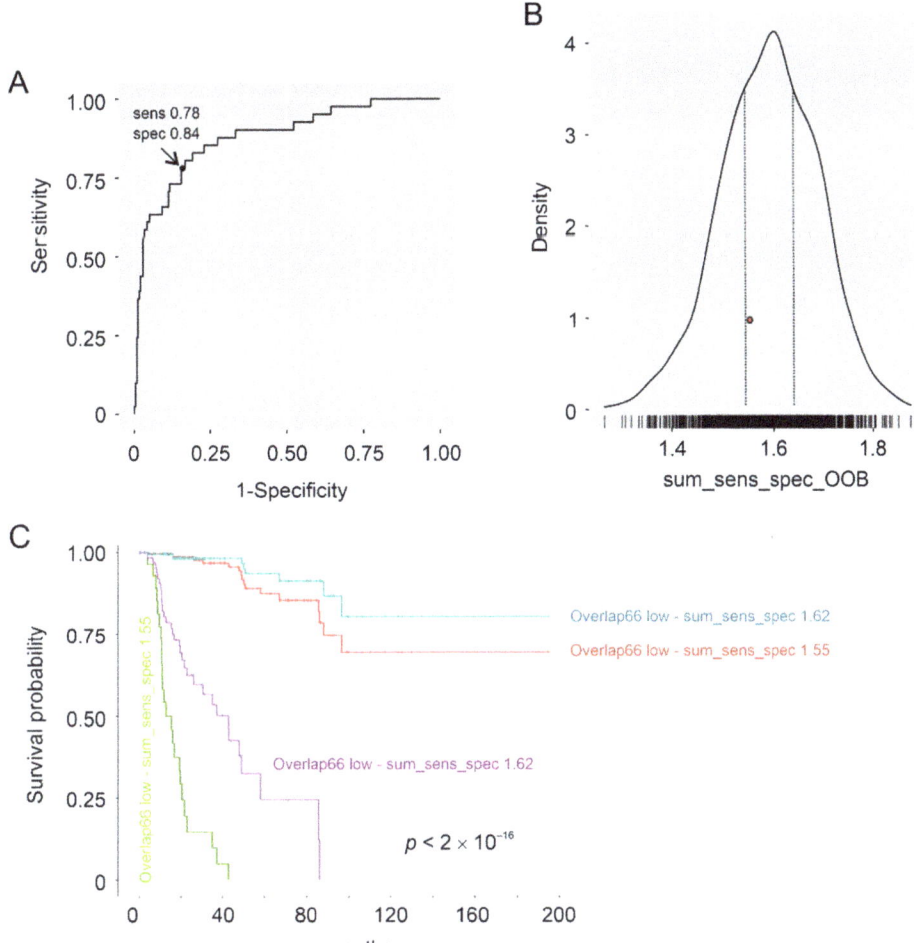

Figure 6. Validation of Overlap66 risk score in stratification of pRCC fatality risk. Cutoff points were estimated using Kernel smoothing method coupled with bootstrapping (n = 1000). The average in-bag and out-of-bag (OOB) bootstrap samples are 63.2% and 36.8% of the full sample size respectively. The analysis was performed using the cutpointr R package (https://github.com/thiele/cutpointr, accessed on 21 July 2021). (**A**) ROC curve with the optimal cutoff point indicated (arrow); sens: sensitivity, spec: specificity, and the sum_sens_spec: 1.62. (**B**) Distribution of out-of-bag (OOB) metric values. The most predictions occur in these OOB samples (n = 1000) at the sum_sens_sepc value 1.6. The region marked by the 2 dotted lines includes a range of sum_sens_sepc values that frequently stratify the fatality risk with high accuracy. The red dot represents a sum_sens_sepc value 1.55. (**C**) Classification of pRCC tumors into a high- and low-risk group using two indicated cutpoints; the sum_sens_sepc 1.62 cutoff point was obtained using Kernel method and the sum_sens_sepc 1.55 cutoff point (see the red dot in panel (**B**)) was derived using Maximally selected LogRank statistics (see Figure S9). The p value is for both separations.

We subsequently optimized Overlap66. As OIP5 expression was at 1.9 folds in ACHN OIP5 tumors compared to ACHN EV tumors (Table S3), we defined a subgroup of DEGs as those with $p.adj < 0.05$ and fold $\geq |1.9|$ in ACHN OIP5 tumors compared to ACHN EV tumors. These DEGs (n = 298) share 21 overlap genes (Overlap21) with primary pRCC-derived DEGs (Figure S7C, Table S6B). As expected, Overlap21 is a subgroup of Overlap66 (Table 2). Overlap21 risk scores predict OS possibility under both UV and MV Cox models with comparable efficiency as Overlap66, evident by Hazard ratio (HR) and 95% confident interval (CI) (Figure S7B). Similar prediction efficiencies for PFS between Overlap21 and Overlap66 were also observed (Figure S7B). Overlap21 effectively stratifies the risk of mortality and PFS; the discriminations possess high tAUC values (Figure 5A,B). In comparison, Overlap21 seems marginally less effective compared to Overlap66 in the discriminations of OS and PFS (Figure 5A,B). Nonetheless, the Overlap21-mediated predictions are clearly effective. Similar to Overlap66, Overlap 21 risk score is an independent predictor of poor OS after adjusting age at diagnosis, sex, and T stage (Table 3).

Table 3. Univariate and multivariate Cox analysis of Overlap66 and Overlap21 risk scores for pRCC OS.

Factors	Univariate Cox Analysis			Multivariate Cox Analysis		
	HR	95% CI	p-Value	HR	95% CI	p-Value
Overlap66	2.72	2.14–3.46	3.82×10^{-16} ***	3.03	2.29–4.01	1.15×10^{-14} ***
Overlap21	2.72	2.19–3.38	$<2 \times 10^{-16}$ ***	2.71	2.1–3.5	2.81×10^{-14} ***
Age	1.01	0.98–1.04	0.504	1.04 [i] 1.03 [ii]	1.01–1.08 1.003–1.064	0.0119 * 0.0333 *
Sex	0.67	0.34–1.36	0.268	0.80 [i] 1.45 [ii]	0.36–1.76 0.67–3.13	0.576 0.346
Tstage 1	5.13	2.73–9.62	3.53×10^{-7} ***	1.75 [i] 3.28 [ii]	0.81–3.76 1.61–6.65	0.154 0.001 **

Analyses were performed using the TCGA PanCancer pRCC dataset. Age: at diagnosis. Sex: male vs. female. Tstage 1: T stage 3 + 4 vs. Tstage 0: T stage 1 + 2. i and ii: in analysis with Overlap66 (i) and Overlap21 (ii). *, **, and ***: $p < 0.05$, $p < 0.01$, and $p < 0.001$ respectively

The utility of Overlap21 in assessing pRCC fatality risk is further illustrated by its impressive separation of disease-specific survival (DSS) risk (Figure 7A,C). DSS is more specific compared to OS in addressing factors contributing to cancer-caused deaths. Overlap66 did not perform well in DSS estimation (data not shown), which might be attributable to the small number of events (disease-specific death $n = 27$) in the context of the large number of variables ($n = 66$ in Overlap66). We thus generated Overlap21plus by using Overlap21 as the basis, and the rest of DEGs within Overlap66 were added if they remain risk factors for decreased OS after adjusting age at diagnosis, sex, and T stages (Table S6A). However, Overlap21plus was not superior to Overlap21 in the estimation of OS and PFS (data not shown). Nonetheless, the risk score of Overlap21plus predicts DSS risk in a comparable efficiency as Overlap21 (Figure S7B); its ability to classify DSS possibility was marginally superior to Overlap21 (Figure 7A–C).

Instead of using time-dependent ROC (receiver-operating characteristic) in evaluating the performance of Overlap66, Overlap21, and Overlap21plus for their prognostic prediction, we further examined their prediction performance using the intact population (i.e., without the time component) by both ROC-AUC and PR-AUC curves. The precision-recall (PR) curve is used to account for the imbalance nature of dataset; the event rates are 14.6% (41/280) for OS, 18.9% (53/280) for PFS, and 9.6% (27/280) for DSS, which are much less than 50%. PR-curve was suggested to evaluate biomarker's discriminative performance [44]. According to both ROC-AUC and PR-AUC curves, Overlap66 predicts OS and PFS possibilities better than Overlap21 (Figure 5C,D), while Overlap21plus holds a slight edge over Overlap21 in estimating DSS possibility (Figure 7D,E).

3.5. Alterations in Immune Cell Subsets in High-Risk pRCC Tumors

Tumor-associated immune cells play critical role in tumor initiation and progression [45,46], suggesting alterations of immune components in Overlap66-stratified high-risk pRCC tumors compared to those of low-risk. To examine this possibility, we profiled all 22 leukocyte subsets in 280 primary pRCC tumors within the TCGA Pancancer dataset using CIBERSORTx (https://cibersortx.stanford.edu/index.php, accessed on 21 July 2021) [47]. Significant alterations in several immune cell subsets between high-risk ($n = 32$) and low-risk tumors ($n = 248$) were detected (Figure 8). Increases in B naïve cells, T follicular helper cells (Tfh), CD4 T memory (activated) cells, and CD8 T ($p = 0.075$) cells were detected in high-risk local pRCC tumors (Figure 8A), indicating persistent immune reactions towards tumors; this scenario is not uncommon, evident by the co-existence of ATM-derived tumor surveillance (antioncogenic actions) with oncogenic actions during cancer initiation and progression [48]. However, CD8 T cells expressed an upregulation of programmed cell death protein 1 (PDCD1 or PD1) (Figure 8B), a major mechanism contributing to CD8 T cell exhaustion in cancer [49]. Additionally, T regulatory (Treg) cells suppress T cells activation via downregulation of CD80/86 in antigen-presenting dendritic cells [50] and a significant elevation of Treg cells was observed in high-risk pRCC tumors (Figure 8A). Alterations in M1 and M2 composition in high-risk pRCCs (Figure 8A) are consistent with the contributions of tumor-associated macrophages in cancer progression [51]. Decreases in macrophages M2 in high risk pRCC tumors is supported by a downregulation of β-2-adrenergic receptor (ADRB2) in these tumors (Figure 8C); the receptor was associated with M2 macrophages [52]. Reductions of activated mast cells in high-risk pRCC tumors (Figure 8A) suggest a downregulation of immune reactions in facilitating pRCC progression. While B naïve cells, CD8 T cells, M2 macrophages, and activated master cells are similarly clustered in both Overlap66 stratified high- and low-risk pRCCs (Figure S10), activated CD4 T memory cells, Tfh, Treg, and M1 macrophages in the high-risk tumors display different clustering patterns from their counterparts in the low-risk pRCCs (Figure 8D–G). Collectively, changes in immune components in high-risk pRCC tumors stratified by Overlap66 risk scores favor the development of an immune suppressive microenvironment, which might be a mechanism underpinning pRCC progression. This concept provides additional evidence supporting Overlap66 being a novel and effective prognostic biomarker for pRCC.

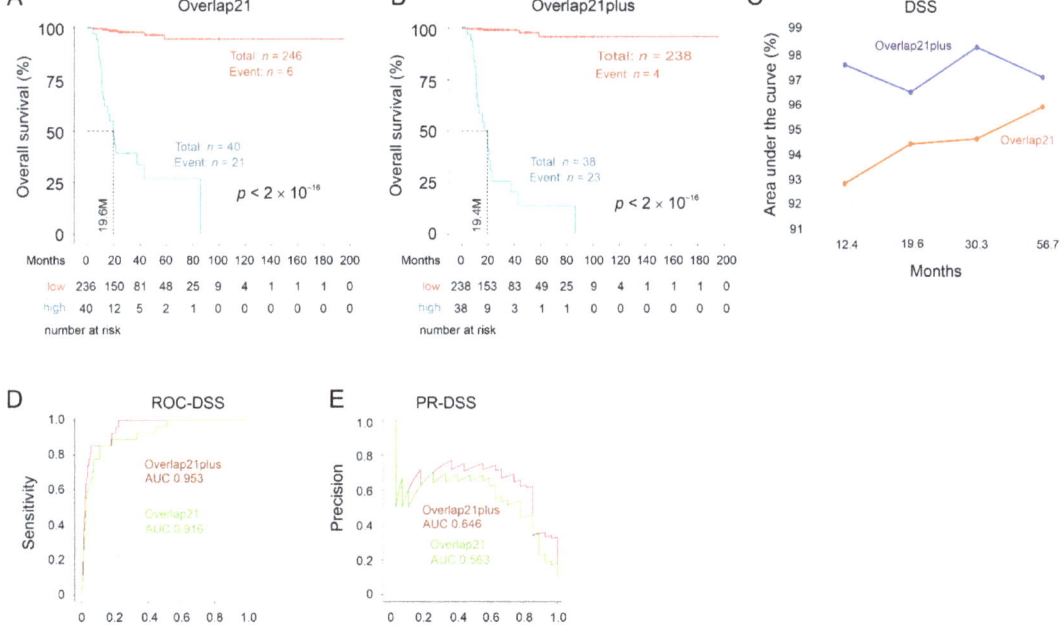

Figure 7. Stratification of the risk of disease-specific survival (DSS) by Overlap21 and Overlap21plus. (**A,B**) Separation of a high- and low-risk group based on DSS by Overlap21 and Overlap21plus risk scores. (**C**) tROC analysis. (**D,E**) ROC-AUC and PR-AUC curves for the indicated events.

3.6. Critical Contributions of PLK1 to OIP5-Promoted Growth of pRCC Tumors

PLK1 (Polo-like kinase 1) is one of the upregulated DEGs identified in relation to OIP5 upregulation in both xenograft tumors and primary pRCC, i.e., a component gene of Overlap66 (Table 2). In the same manner, both LYPD6 and PCSK5 were upregulated in primary pRCC tumors with elevated OIP5 expression and in ACHN OIP5 xenografts determined by RNA-seq (Table 2). By using real-time PCR, we confirmed LYPD6 (fold 2.32 ± 0.2/SD, $p < 0.5$) and PCSK5 (fold 2.6 ± 0.1, $p < 0.05$) upregulations in ACHN OIP5 tumors ($n = 4$) compared to ACHN EV tumors ($n = 6$). PLK1 upregulation in xenografts produced by ACHN OIP5 cells compared to those derived from ACHN EV cells was demonstrated by RNA-seq and real-time PCR (Figure 9A,B)). In primary pRCC tumors, OIP5 expression correlates with PLK1 expression with a Pearson correlation value of 0.7 (UALCAN, ualcan.path.uab.edu/home, 1 March, 2021). OIP5, which is also known as Mis18β, is an essential component of the Mis18 complex that is required to load a histone H3 variant CENP-A (centromere protein A) to centromere of newly synthesized DNA strand in early G1 phase [53,54]. PLK1 contributes to CENP-A loading via phosphorylation of M18BP1, a component of the Mis18 complex [55]. In line with this knowledge, we examined whether PLK1 kinase activity plays a role in OIP5-promoted pRCC growth.

PLK1 inhibitors have been developed and approved by FDA as Orpha Drug Designation for cancer therapy [56,57]. The PLK1 inhibitor BI2536 caused G2/M arrest with concurrent reduction in G1 phase in ACHN OIP5 cells without apparent effects on cell cycle distributions of ACHN EV cells at the conditions used (Figure 9C). We then treated mice bearing ACHN EV or ACHN OIP5 cell-produced xenograft tumors with BI2536 when tumors reached 100 mm^3. In the vehicle treatment group, the OIP5 tumors grew significantly faster compared to the EV tumor (Figure 9D). Administration of BI2536 had no effects on the growth of ACHN EV tumors but significantly inhibited the growth of ACHN OIP5 tumors (Figure 9E,F). In the presence of BI2536, ACHN OIP5 tumor showed marginally slower growth compared to ACHN EV tumors (Figure 9G). Inhibition of PLK1 significantly increases the survival of mice bearing ACHN OIP5 tumor (Figure 9H). As ACHN is a metastatic pRCC cell line [39], evidence supports inhibition of PLK1 being an option in treating metastatic pRCCs with OIP5 upregulation. Collectively, the above observations indicate synthetic lethality between OIP5 and PLK1 in metastatic pRCCs.

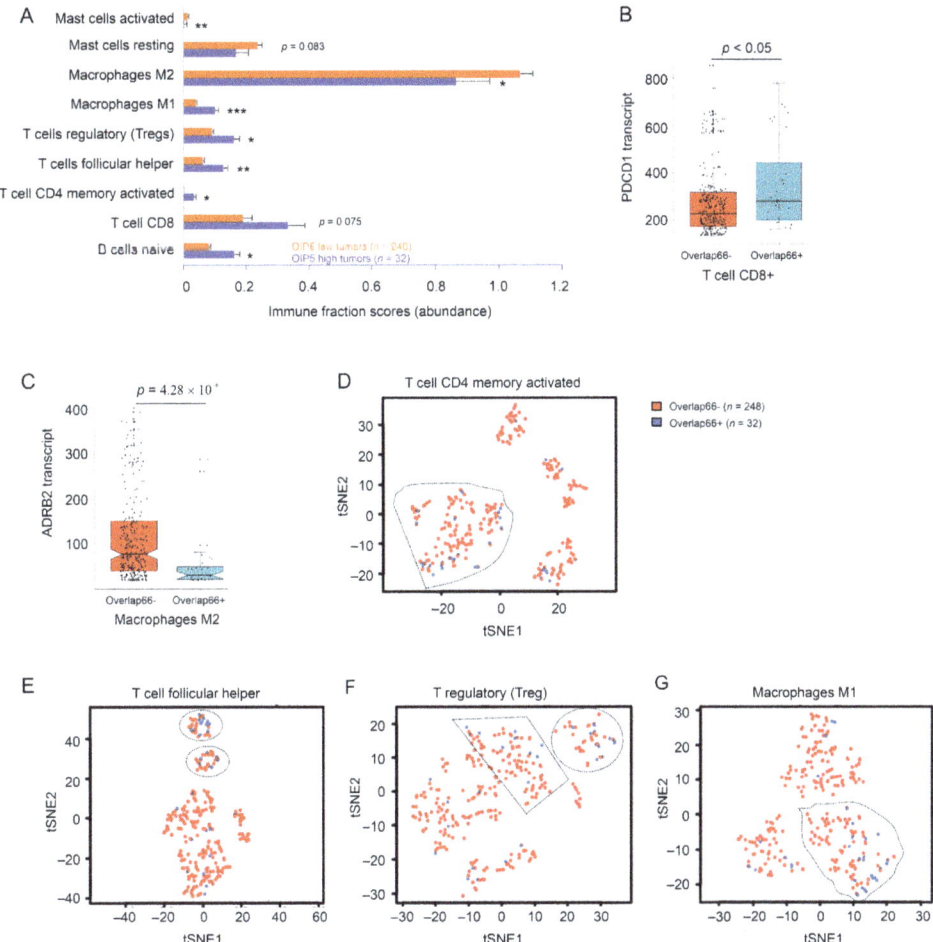

Figure 8. Changes in immune cells in pRCC tumors with high risk of fatality. RNA-seq profiles for 280 pRCC tumors were retrieved from cBioPortal and analyzed for immune cell profiles using the LM22 signature matrix and the CIBERSORTx program (https://cibersortx.stanford.edu/index.php, accessed on 21 July 2021) [47]. The analysis setting was with B-mode batch correction and 500 permutations (https://cibersortx.stanford.edu/index.php, accessed on 21 July 2021). (**A**) The abundance of the indicated immune cell subsets was determined by their immune fraction scores. Means ± SEMs in high-risk and low-risk tumors stratified by Overlap66 risk score are graphed; *: $p < 0.05$; **: $p < 0.01$; and ***: $p < 0.001$ in comparison to low-risk tumors by 2-tailed t-test. (**B**,**C**) Boxplots for the expression of PDCD1 and ADRB2 in low- and high-risk pRCC tumors; statistical analyses were conducted using Welch t-test with p-value adjusted with the Holm–Bonferroni (Holm) method. (**D**–**G**) Clustering of the indicated immune cell types associated with low risk (Overlap66−) and high risk (Overlap66+) tumors by tSNE (t-distributed stochastic neighbor embedding); the marked clusters are enriched with high-risk tumors. The graph was produced using CIBERSORTx (https://cibersortx.stanford.edu/index.php, accessed on 21 July 2021).

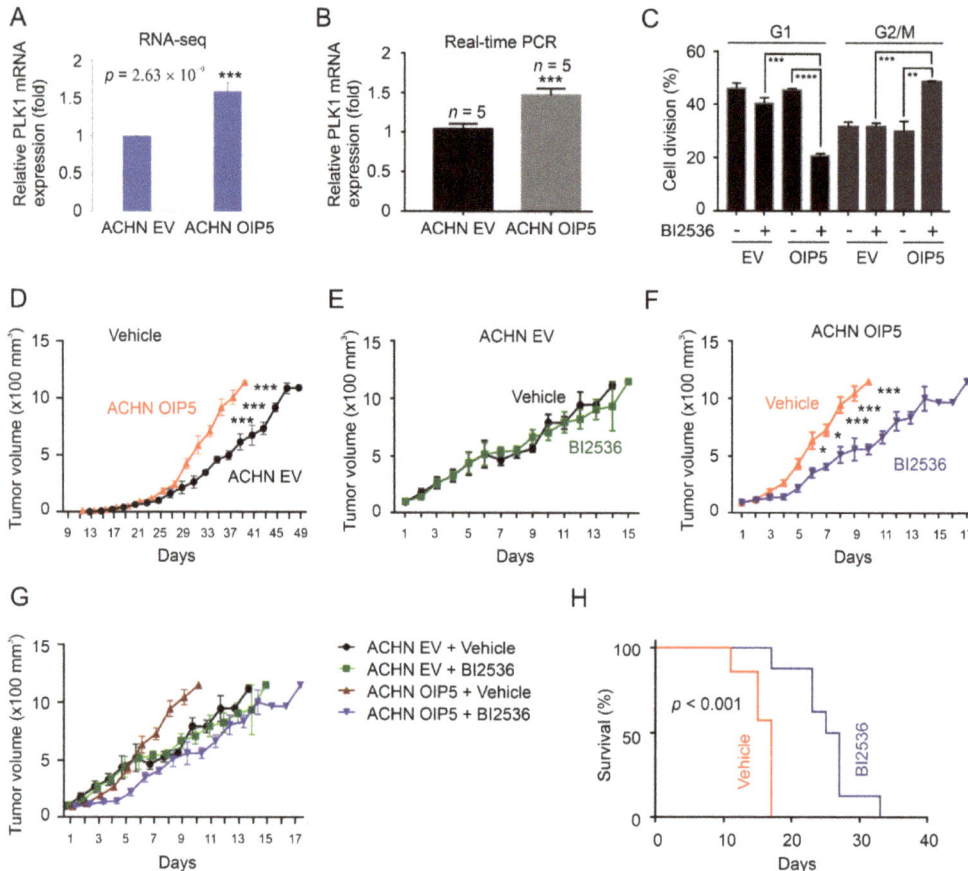

Figure 9. PLK1 inhibitor reduces OIP5-mediated pRCC tumorigenesis. (**A,B**) RNA-seq and real-time PCR analyses of PLK1 expression in ACHN EV and ACHN OIP5 tumors. RNA-seq was performed in 3 each from ACHN EV and ACHN OIP5 tumors. (**C**) ACHN EV and ACHN OIP5 cells were treated with DMSO (−) or the PLK1 inhibitor (BI2536) in 40 nM PLK1 inhibitor for 72 h, followed by quantification of cell cycle distributions. Experiments were repeated 3 times; means ± SEMs are graphed. **: $p < 0.01$, ***: $p < 0.001$, ****: $p < 0.0001$ in the indicated comparisons by 2-tailed Student's test. (**D–G**) Mice bearing ACHN EV or ACHN OIP5 tumors were treated with vehicle or BI2536 (50 mg/kg) intravenously. The overall profiles of tumor growth in the vehicle treated setting (**D**); tumor volumes were recorded following treatments (**E–G**). Statistical analyses were performed using 2-tailed Student's t-test; *: $p < 0.05$, ***: $p < 0.001$. (**H**) Kaplan Meier curve for the indicated mice reaching endpoints. Statistical analysis was performed using logrank test.

4. Discussion

Papillary RCC is a minor type of RCC compared to ccRCC which composes 75–80% of RCC cases. Nonetheless, pRCC can be as aggressive as ccRCC, particularly the T2P tumors which usually have more aggressive potential than ccRCC. As a minor RCC type, research on pRCC falls short compared to ccRCC. Therefore, the current understanding on pRCC remains limited, which presents a major concern particularly considering pRCC being associated with poor prognosis. The situation calls for improvement in risk assessment and personalized therapies in managing pRCC.

We provide here the first evidence for OIP5 being an important oncogenic factor of pRCC. This concept is supported by multiple pieces of evidence with respect to the impact of OIP5 upregulation on the tumorigenesis of pRCC cells in vitro and in vivo as well as the association of OIP5 upregulation with primary pRCC. Although we have made extensive efforts to knockdown OIP5 in ACHN cells, the attempts were not successful, suggesting OIP5 being essential for ACHN cell survival. This plausibility is in accordance with OIP5 initiating multiple processes critical for pRCC tumorigenesis, including those regulating urogenital system development, immune reaction, and others. Among these features is the expression status of OIP5 and its related DEGs within Overlap66 in CIMP. Although it remains to be determined whether OIP5 and these DEGs contribute to CIMP, this possibility seems likely. Among the pRCC subtypes, CIPM tumors are associated with a metabolic shift towards Warburg metabolism, which include enhancement of glycolysis, fatty acid and lipid metabolism, and hypoxia [12]. These are typical pathways enriched in ACHN OIP5 tumors (Figure 3; Figure S4–S6; Table S2).

OIP5 may also utilize other pathways in promoting pRCC. As an essential component (Mis18β) in the Mis18 complex, OIP5 is required for CENP-A loading and thus centromere formation [53,54]. This process is essential for genome stability, evident by the centromere-mediated chromosome segregation. In line with this concept, genes with function in maintaining genome stability are overrepresented in Overlap66; RMI2 (RecQmediated genome instability 2; C16ORF75) [58], RAD54B (RAD54 homolog B) [59], and PLK1 [60] all play roles in genome stability. Furthermore, pathway enrichment analysis of Overlap66 DEGs revealed the top pathways enriched being GO:0071168: protein localization to chromatin ($p < 0.0001$), GO:0140013: meiotic nuclear division ($p < 0.0001$), GO:0006790: sulfur compound metabolic process ($p < 0.001$), and GO:0000724: double-strand break repair via homologous recombination ($p < 0.001$).

One of the neighboring genes to OIP5 is OIP5-AS1 (OIP5 antisense RNA 1). According to GRCh38.p13 (Genome Reference Consortium Human Build 38 patch release 13) released in Feb 28, 2019, the OIP5 genes runs from 41,332,591 to 41,309,273 on chromosome 15 (https://www.ncbi.nlm.nih.gov/gene/11339, accessed on 29 August 2021), while the OIP5-AS1 gene runs from 41,282,697 to 41,313,338 on chromosome 15 (https://www.ncbi.nlm.nih.gov/gene/729082, accessed on 29 August 2021). While both genes have an overlap region of 4065 nucleotides, there is no evidence suggesting a regulatory relationship between OIP5 and OIP5-AS1 [61]. OIP5-AS1 encodes a long non-coding RNA (lnRNA) and possesses oncogenic activities via regulating a set of microRNAs [61]. For instance, OIP5-AS1 was reported to sponge miR-143-3P to enhance cervical cancer [62] and miR-186a-5p to facilitate hepatoblastoma [63]. However, the involvement of OIP5-AS1 in pRCC remains unknown. In view of both OIP5 and OIP5-AS1 being pro-oncogenesis and their adjacent genetic locations, potential functional connections between both in pRCC pathogenesis and progression is worthy of future investigation. In supporting this possibility, we noticed OIP5-AS1 being upregulated (1.37 folds, $p = 0.00464$ and $q = 0.0459$) in pRCC tumors expressing high levels of OIP5 compared to those with low levels of OIP5 expression (Figure 1G).

OIP5 is a tumor-associated antigen (TAA), owning to its largely restricted expression in human testis and its upregulation in multiple cancers [64,65]. We noticed that testis-associated proteins are also enriched in Overlap66, including OIP5, TEX15 (testis expressed 15, meiosis, and synapsis associated), SPAG1 (sperm associated antigen 1), and SPATA18 (spermatogenesis associated 18) (Table 2). It is thus tempting to propose an involvement of some testis events in pRCC tumorigenesis. OIP5 possesses a robust prognostic potential (Figure 1G). This predictive power is significantly strengthened in OIP5-derived multigene sets: Overlap66, Overlap21, and Overlap21plus. Because of the small number of component genes and its effectiveness in predicting OS, PFS, and DSS, Overlap21 may offer primary clinical application with the other two provide assisting roles. These multigene sets possess great potential to be implemented into clinical applications. This possibility is supported by a very good out-of-sample performance of Overlap66 in stratification of pRCC fatality risk (Figure 6) and these stratifications can be effective using a range of cutoff points (Figure 6B,C). Clinical applications of Overlap66 and its-related multigene panels may significantly improve our ability in predicting prognosis and potentially even the development of personalized therapies.

Although a recent phase 2 clinical trial suggests the MET inhibitor cabozantinib improves PFS and OS in patients with metastatic pRCC compared to a current standard of care with sunitinib [66], much more needs to be done to confirm its efficacy. The dependence on MET signaling is likely much less in T2P compared to T1P, which needs to be considered in using MET inhibitors in treating patients with T2P tumors. Our finding of PLK1 inhibitor being effective in inhibition of ACHN OIP5 tumor growth may have significant clinical applications in treating metastatic pRCC with OIP5 upregulation; this will offer a venue for potential utilization of personalized therapy in pRCC. This possibility can be readily explored as volasertib, a PLK1 inhibitor, has been granted Orpha Drug Designation status in treating AML (acute myeloid leukemia) in 2014 and rhabdomyosarcoma in 2020 (https://oncoheroes.com/press-releases-content/2020/10/14/volasertib-a-potential-new-treatment-for-rhabdomyosarcoma-receives-orphan-drug-designation-from-the-us-fda, accessed on 31 May 2021) by FDA. Even for BI2356 used in this study, its clinical safety was deemed acceptable based on multiple phase II clinical trials (NCT00701766, NCT00376623, and NCT00526149) on solid cancer. Intriguingly, we observed changes in immune cells in pRCC tumors stratified by Overlap66, an OIP5-derived multigene panel, including increases of Treg cells and PD1 upregulation in CD8 T cells (Figure 8). This suggests that these patients might benefit from rescuing of CD8 T cell exhaustion via PD1-based immune therapies. Treg action can be suppressed via CTLA-4 immune therapy. In this regard, combinations of PLK1 inhibitor and PD1 or CTLA-4 immune therapies might optimize personalized treatment. Collectively, this research enhances our understanding of pRCC and suggests novel means in predicting pRCC prognosis and in developing personalized therapy. Nonetheless, additional work is required to realize these potentials.

5. Conclusions

We report here a novel and thorough investigation of OIP5's contributions to pRCC. OIP5 upregulations robustly predict the survival possibility of pRCC patients. The multigene panel Overlap66, a portion of the OIP5 network, possesses an impressive prognostic potential in predicting pRCC progression, disease-specific survival, and overall survival; the predictions are associated with an excellent out-of-sample performance, indicating its potential clinical applications. Furthermore, PLK1 is among Overlap66 and displays synthetic lethality with OIP5; inhibition of PLK1 using BI2356 only suppresses the growth of xenograft tumors generated by ACHN OIP5 cells but not the growth of tumor produced by ACHN EV cells, supporting a targeted and personalized therapy for pRCCs with OIP5 elevations. Collectively, combinational use of Overlap66 and PLK1 inhibitors may open an era of personalized therapy in pRCC.

6. Patents

A US provisional patent (63/202,616) has been filed.

Supplementary Materials: The following are available online at https://www.mdpi.com/article/10.3390/cancers13174483/s1, Figure S1: OIP5 upregulation in ccRCC; Figure S2: OIP5 enhances ACHN cell proliferation, colony formation, and growth in soft agar; Figure S3: Increases in OIP5 and CDK2 expression in ACHN OIP5 tumors; Figure S4: Gene set enrichment in ACHN OIP5 tumors; Figure S5: Enrichment of h-Hallmarks Gene Set collection in ACHN OIP5 tumors; Figure S6: Enhancement of the glycolysis pathway in ACHN OIP5 tumors; Figure S7: Prediction of OS, PFS, and DSS possibilities by Overlaop66; Figure S8: Heatmap of Overlap66 gene expressions in the indicated tissues; Figure S9: Cutoff point of Overlap66 risk score in the estimation of OS shortening; Figure S10: Clustering of the indicated immune cell subsets with tSNE; Table S1: Enrichment of the C6 oncogenic gene sets collection within OIP5-related DEGs; Table S2: Enrichment of the Hallmark gene sets collection within OIP5-related DEGs; Table S3: Differentially expressed genes derived from the comparison between ACHN OIP5 xenografts and ACHN EV tumors; Table S4: Pathway enrichment in OIP5 related DEGs (p.adj < 0.05 and fold change > |1.5|) derived from the comparison between ACHN OIP5 xenografts and ACHN EV tumors; Table S5: DEGs derived from primary pRCC with respect to

OIP5 expression; Table S6A: Overlap66 genes' association with OS possibility and their oncogenic functions; Table S6B: Overlap21 and Overlap21plus gene lists.

Author Contributions: M.J.C. performed analyses of OIP5 expression in primary pRCC using TMA, constructed ACHN EV and OIP5 stable lines, examined the impact of OIP5 on a set of oncogenic events in ACHN cell in vitro and on the cell's ability of forming xenografts in vivo, and investigated the inhibition of PLK1 inhibitor on the growth of ACHN OIP5 cell-derived xenografts. Y.G. contributed to production of xenograft tumors. M.J.C. and L.H. performed RNA-seq analyses. X.L. and Y.D. examined OIP5-affected immune reactions. M.J.C., X.L. and W.M. contributed to the molecular manipulations of OIP5. A.K. contributed to clinical analysis of primary pRCC. D.T. performed prognostic analyses. All authors contributed to data interpretation. M.J.C., A.K. and D.T. designed this research. A.K. and D.T. supervised the research. M.J.C. and D.T. prepared the manuscript. M.J.C., Y.G., X.L. and A.K. revised the manuscript. All authors have read and agreed to the published version of the manuscript.

Funding: Y.G. is supported by Studentship provided by Ontario Graduate Student Fellowship and Research Institute of St Joe's Hamilton. Grant support was from Canadian Cancer Society (grant #: 319412) and CIHR to DT. This research was also supported by funds from Urological Cancer Centre for Research and Innovation (UCCRI), St Joseph's Hospital, Hamilton, ON L8N 4A6, Canada.

Institutional Review Board Statement: The study was conducted according to the guidelines of the Declaration of Helsinki and approved by the Institutional Review Board of McMaster University (16-06-23) and Hamilton Integrated Research Ethics Board (11-3472).

Informed Consent Statement: Not applicable.

Data Availability Statement: Data is present in this article and Supplementary Materials.

Acknowledgments: The results shown here are in part based on data generated by the TCGA Research Network (http://cancergenome.nih.gov/, accessed on 1 September 2020).

Conflicts of Interest: All authors declare no competing interests.

References

1. Cohen, H.T.; McGovern, F.J. Renal-cell carcinoma. *N. Engl. J. Med.* **2005**, *353*, 2477–2490. [CrossRef] [PubMed]
2. Twardowski, P.W.; Mack, P.C.; Lara, P.N., Jr. Papillary renal cell carcinoma: Current progress and future directions. *Clin. Genitourin. Cancer* **2014**, *12*, 74–79. [CrossRef] [PubMed]
3. Pai, A.; Brunson, A.; Brown, M.; Pan, C.X.; Lara, P.N., Jr. Evolving epidemiologic trends in nonclear cell renal cell cancer: An analysis of the california cancer registry. *Urology* **2013**, *82*, 840–845. [CrossRef] [PubMed]
4. Delahunt, B.; Eble, J.N. Papillary renal cell carcinoma: A clinicopathologic and immunohistochemical study of 105 tumors. *Mod. Pathol.* **1997**, *10*, 537–544.
5. Zbar, B.; Tory, K.; Merino, M.; Schmidt, L.; Glenn, G.; Choyke, P.; Walther, M.M.; Lerman, M.; Linehan, W.M. Hereditary papillary renal cell carcinoma. *J. Urol.* **1994**, *151*, 561–566. [CrossRef]
6. Delahunt, B.; Eble, J.N.; McCredie, M.R.; Bethwaite, P.B.; Stewart, J.H.; Bilous, A.M. Morphologic typing of papillary renal cell carcinoma: Comparison of growth kinetics and patient survival in 66 cases. *Hum. Pathol.* **2001**, *32*, 590–595. [CrossRef]
7. Kosaka, T.; Mikami, S.; Miyajima, A.; Kikuchi, E.; Nakagawa, K.; Ohigashi, T.; Nakashima, J.; Oya, M. Papillary renal cell carcinoma: Clinicopathological characteristics in 40 patients. *Clin. Exp. Nephrol.* **2008**, *12*, 195–199. [CrossRef]
8. Mejean, A.; Hopirtean, V.; Bazin, J.P.; Larousserie, F.; Benoit, H.; Chretien, Y.; Thiounn, N.; Dufour, B. Prognostic factors for the survival of patients with papillary renal cell carcinoma: Meaning of histological typing and multifocality. *J. Urol.* **2003**, *170*, 764–767. [CrossRef]
9. Wong, E.C.L.; Di Lena, R.; Breau, R.H.; Pouliot, F.; Finelli, A.; Lavallee, L.T.; So, A.; Tanguay, S.; Fairey, A.; Rendon, R.; et al. Morphologic subtyping as a prognostic predictor for survival in papillary renal cell carcinoma: Type 1 vs. Type 2. *Urol. Oncol.* **2019**, *37*, 721–726. [CrossRef]
10. Akhtar, M.; Al-Bozom, I.A.; Al Hussain, T. Papillary renal cell carcinoma (prcc): An update. *Adv. Anat. Pathol.* **2019**, *26*, 124–132. [CrossRef]
11. Cohen, D.; Zhou, M. Molecular genetics of familial renal cell carcinoma syndromes. *Clin. Lab. Med.* **2005**, *25*, 259–277. [CrossRef]
12. Cancer Genome Atlas Research Network; Linehan, W.M.; Spellman, P.T.; Ricketts, C.J.; Creighton, C.J.; Fei, S.S.; Davis, C.; Wheeler, D.A.; Murray, B.A.; Schmidt, L.; et al. Comprehensive molecular characterization of papillary renal-cell carcinoma. *N. Engl. J. Med.* **2016**, *374*, 135–145. [CrossRef] [PubMed]
13. Courthod, G.; Tucci, M.; Di Maio, M.; Scagliotti, G.V. Papillary renal cell carcinoma: A review of the current therapeutic landscape. *Crit. Rev. Oncol. Hematol.* **2015**, *96*, 100–112. [CrossRef] [PubMed]

14. Motzer, R.J.; Jonasch, E.; Boyle, S.; Carlo, M.I.; Manley, B.; Agarwal, N.; Alva, A.; Beckermann, K.; Choueiri, T.K.; Costello, B.A.; et al. Nccn guidelines insights: Kidney cancer, version 1.2021. *J. Natl. Compr. Canc. Netw.* **2020**, *18*, 1160–1170. [CrossRef]
15. Ravaud, A.; Oudard, S.; De Fromont, M.; Chevreau, C.; Gravis, G.; Zanetta, S.; Theodore, C.; Jimenez, M.; Sevin, E.; Laguerre, B.; et al. First-line treatment with sunitinib for type 1 and type 2 locally advanced or metastatic papillary renal cell carcinoma: A phase ii study (supap) by the french genitourinary group (getug)dagger. *Ann. Oncol. Off. J. Eur. Soc. Med. Oncol. ESMO* **2015**, *26*, 1123–1128. [CrossRef]
16. Williams, J.M.; Chen, G.C.; Zhu, L.; Rest, R.F. Using the yeast two-hybrid system to identify human epithelial cell proteins that bind gonococcal opa proteins: Intracellular gonococci bind pyruvate kinase via their opa proteins and require host pyruvate for growth. *Mol. Microbiol.* **1998**, *27*, 171–186. [CrossRef] [PubMed]
17. Pawlak, A.; Toussaint, C.; Levy, I.; Bulle, F.; Poyard, M.; Barouki, R.; Guellaen, G. Characterization of a large population of mrnas from human testis. *Genomics* **1995**, *26*, 151–158. [CrossRef]
18. Yazarloo, F.; Shirkoohi, R.; Mobasheri, M.B.; Emami, A.; Modarressi, M.H. Expression analysis of four testis-specific genes aurkc, oip5, piwil2 and taf7l in acute myeloid leukemia: A gender-dependent expression pattern. *Med. Oncol.* **2013**, *30*, 368. [CrossRef]
19. Gong, M.; Xu, Y.; Dong, W.; Guo, G.; Ni, W.; Wang, Y.; Wang, Y.; An, R. Expression of opa interacting protein 5 (oip5) is associated with tumor stage and prognosis of clear cell renal cell carcinoma. *Acta Histochem.* **2013**, *115*, 810–815. [CrossRef]
20. Chen, L.; Wang, R.; Gao, L.; Shi, W. Opa-interacting protein 5 expression in human glioma tissues is essential to the biological function of u251 human malignant glioma cells. *Cancer Control.* **2020**, *27*, 1073274820968914. [CrossRef]
21. Li, H.; Zhang, J.; Lee, M.J.; Yu, G.R.; Han, X.; Kim, D.G. Oip5, a target of mir-15b-5p, regulates hepatocellular carcinoma growth and metastasis through the akt/mtorc1 and beta-catenin signaling pathways. *Oncotarget* **2017**, *8*, 18129–18144. [CrossRef]
22. Li, Y.; Xiao, F.; Li, W.; Hu, P.; Xu, R.; Li, J.; Li, G.; Zhu, C. Overexpression of opa interacting protein 5 increases the progression of liver cancer via bmpr2/jun/chek1/rac1 dysregulation. *Oncol. Rep.* **2019**, *41*, 2075–2088. [CrossRef]
23. Li, S.; Xuan, Y.; Gao, B.; Sun, X.; Miao, S.; Lu, T.; Wang, Y.; Jiao, W. Identification of an eight-gene prognostic signature for lung adenocarcinoma. *Cancer Manag. Res.* **2018**, *10*, 3383–3392. [CrossRef] [PubMed]
24. Li, H.C.; Chen, Y.F.; Feng, W.; Cai, H.; Mei, Y.; Jiang, Y.M.; Chen, T.; Xu, K.; Feng, D.X. Loss of the opa interacting protein 5 inhibits breast cancer proliferation through mir-139-5p/notch1 pathway. *Gene* **2017**, *603*, 1–8. [CrossRef]
25. Chun, H.K.; Chung, K.S.; Kim, H.C.; Kang, J.E.; Kang, M.A.; Kim, J.T.; Choi, E.H.; Jung, K.E.; Kim, M.H.; Song, E.Y.; et al. Oip5 is a highly expressed potential therapeutic target for colorectal and gastric cancers. *BMB Rep.* **2010**, *43*, 349–354. [CrossRef] [PubMed]
26. Nakamura, Y.; Tanaka, F.; Nagahara, H.; Ieta, K.; Haraguchi, N.; Mimori, K.; Sasaki, A.; Inoue, H.; Yanaga, K.; Mori, M. Opa interacting protein 5 (oip5) is a novel cancer-testis specific gene in gastric cancer. *Ann. Surg. Oncol.* **2007**, *14*, 885–892. [CrossRef]
27. Afsharpad, M.; Nowroozi, M.R.; Mobasheri, M.B.; Ayati, M.; Nekoohesh, L.; Saffari, M.; Zendehdel, K.; Modarressi, M.H. Cancer-testis antigens as new candidate diagnostic biomarkers for transitional cell carcinoma of bladder. *Pathol. Oncol. Res. POR* **2019**, *25*, 191–199. [CrossRef] [PubMed]
28. Wang, D.; Chen, Z.; Lin, F.; Wang, Z.; Gao, Q.; Xie, H.; Xiao, H.; Zhou, Y.; Zhang, F.; Ma, Y.; et al. Oip5 promotes growth, metastasis and chemoresistance to cisplatin in bladder cancer cells. *J. Cancer* **2018**, *9*, 4684–4695. [CrossRef] [PubMed]
29. He, X.; Hou, J.; Ping, J.; Wen, D.; He, J. Opa interacting protein 5 acts as an oncogene in bladder cancer. *J. Cancer Res. Clin. Oncol.* **2017**, *143*, 2221–2233. [CrossRef]
30. He, X.; Ding, X.; Hou, J.; Ping, J.; He, J. Exploration of the pathways and interaction network involved in bladder cancer cell line with knockdown of opa interacting protein 5. *Pathol. Res. Pract.* **2017**, *213*, 1059–1066. [CrossRef]
31. Tanaka, A.; Sakaguchi, S. Targeting treg cells in cancer immunotherapy. *Eur. J. Immunol.* **2019**, *49*, 1140–1146. [CrossRef]
32. He, L.; Fan, C.; Kapoor, A.; Ingram, A.J.; Rybak, A.P.; Austin, R.C.; Dickhout, J.; Cutz, J.C.; Scholey, J.; Tang, D. Alpha-mannosidase 2c1 attenuates pten function in prostate cancer cells. *Nat. Commun.* **2011**, *2*, 307. [CrossRef] [PubMed]
33. Jiang, Y.; Lin, X.; Kapoor, A.; He, L.; Wei, F.; Gu, Y.; Mei, W.; Zhao, K.; Yang, H.; Tang, D. Fam84b promotes prostate tumorigenesis through a network alteration. *Ther. Adv. Med. Oncol.* **2019**, *11*, 1758835919846372. [CrossRef]
34. Zhou, Y.; Zhou, B.; Pache, L.; Chang, M.; Khodabakhshi, A.H.; Tanaseichuk, O.; Benner, C.; Chanda, S.K. Metascape provides a biologist-oriented resource for the analysis of systems-level datasets. *Nat. Commun.* **2019**, *10*, 1523. [CrossRef]
35. Cerami, E.; Gao, J.; Dogrusoz, U.; Gross, B.E.; Sumer, S.O.; Aksoy, B.A.; Jacobsen, A.; Byrne, C.J.; Heuer, M.L.; Larsson, E.; et al. The cbio cancer genomics portal: An open platform for exploring multidimensional cancer genomics data. *Cancer Discov.* **2012**, *2*, 401–404. [CrossRef]
36. Gao, J.; Aksoy, B.A.; Dogrusoz, U.; Dresdner, G.; Gross, B.; Sumer, S.O.; Sun, Y.; Jacobsen, A.; Sinha, R.; Larsson, E.; et al. Integrative analysis of complex cancer genomics and clinical profiles using the cbioportal. *Sci. Signal.* **2013**, *6*, pl1. [CrossRef]
37. Chandrashekar, D.S.; Bashel, B.; Balasubramanya, S.A.H.; Creighton, C.J.; Ponce-Rodriguez, I.; Chakravarthi, B.; Varambally, S. Ualcan: A portal for facilitating tumor subgroup gene expression and survival analyses. *Neoplasia* **2017**, *19*, 649–658. [CrossRef]
38. Jiang, Y.; Mei, W.; Gu, Y.; Lin, X.; He, L.; Zeng, H.; Wei, F.; Wan, X.; Yang, H.; Major, P.; et al. Construction of a set of novel and robust gene expression signatures predicting prostate cancer recurrence. *Mol. Oncol.* **2018**, *12*, 1559–1578. [CrossRef]
39. Brodaczewska, K.K.; Szczylik, C.; Fiedorowicz, M.; Porta, C.; Czarnecka, A.M. Choosing the right cell line for renal cell cancer research. *Mol. Cancer* **2016**, *15*, 83. [CrossRef] [PubMed]
40. Schmidt, L.; Junker, K.; Nakaigawa, N.; Kinjerski, T.; Weirich, G.; Miller, M.; Lubensky, I.; Neumann, H.P.; Brauch, H.; Decker, J.; et al. Novel mutations of the met proto-oncogene in papillary renal carcinomas. *Oncogene* **1999**, *18*, 2343–2350. [CrossRef] [PubMed]

41. Scholl, C.; Frohling, S.; Dunn, I.F.; Schinzel, A.C.; Barbie, D.A.; Kim, S.Y.; Silver, S.J.; Tamayo, P.; Wadlow, R.C.; Ramaswamy, S.; et al. Synthetic lethal interaction between oncogenic kras dependency and stk33 suppression in human cancer cells. *Cell* **2009**, *137*, 821–834. [CrossRef] [PubMed]
42. Alhamdoosh, M.; Law, C.W.; Tian, L.; Sheridan, J.M.; Ng, M.; Ritchie, M.E. Easy and efficient ensemble gene set testing with egsea. *F1000Research* **2017**, *6*, 2010. [CrossRef] [PubMed]
43. Gu, Y.; Li, T.; Kapoor, A.; Major, P.; Tang, D. Contactin 1: An important and emerging oncogenic protein promoting cancer progression and metastasis. *Genes* **2020**, *11*, 874. [CrossRef] [PubMed]
44. Brott, T.; Marler, J.R.; Olinger, C.P.; Adams, H.P., Jr.; Tomsick, T.; Barsan, W.G.; Biller, J.; Eberle, R.; Hertzberg, V.; Walker, M. Measurements of acute cerebral infarction: Lesion size by computed tomography. *Stroke* **1989**, *20*, 871–875. [CrossRef] [PubMed]
45. Angell, H.; Galon, J. From the immune contexture to the immunoscore: The role of prognostic and predictive immune markers in cancer. *Curr. Opin. Immunol.* **2013**, *25*, 261–267. [CrossRef] [PubMed]
46. Gentles, A.J.; Newman, A.M.; Liu, C.L.; Bratman, S.V.; Feng, W.; Kim, D.; Nair, V.S.; Xu, Y.; Khuong, A.; Hoang, C.D.; et al. The prognostic landscape of genes and infiltrating immune cells across human cancers. *Nat. Med.* **2015**, *21*, 938–945. [CrossRef]
47. Newman, A.M.; Steen, C.B.; Liu, C.L.; Gentles, A.J.; Chaudhuri, A.A.; Scherer, F.; Khodadoust, M.S.; Esfahani, M.S.; Luca, B.A.; Steiner, D.; et al. Determining cell type abundance and expression from bulk tissues with digital cytometry. *Nat. Biotechnol.* **2019**, *37*, 773–782. [CrossRef] [PubMed]
48. Reddy, J.P.; Peddibhotla, S.; Bu, W.; Zhao, J.; Haricharan, S.; Du, Y.C.; Podsypanina, K.; Rosen, J.M.; Donehower, L.A.; Li, Y. Defining the atm-mediated barrier to tumorigenesis in somatic mammary cells following erbb2 activation. *Proc. Natl. Acad. Sci. USA* **2010**, *107*, 3728–3733. [CrossRef]
49. Hashimoto, M.; Kamphorst, A.O.; Im, S.J.; Kissick, H.T.; Pillai, R.N.; Ramalingam, S.S.; Araki, K.; Ahmed, R. Cd8 t cell exhaustion in chronic infection and cancer: Opportunities for interventions. *Annu. Rev. Med.* **2018**, *69*, 301–318. [CrossRef]
50. Wing, K.; Onishi, Y.; Prieto-Martin, P.; Yamaguchi, T.; Miyara, M.; Fehervari, Z.; Nomura, T.; Sakaguchi, S. Ctla-4 control over foxp3+ regulatory t cell function. *Science* **2008**, *322*, 271–275. [CrossRef]
51. Ostuni, R.; Kratochvill, F.; Murray, P.J.; Natoli, G. Macrophages and cancer: From mechanisms to therapeutic implications. *Trends Immunol.* **2015**, *36*, 229–239. [CrossRef]
52. Lamkin, D.M.; Ho, H.Y.; Ong, T.H.; Kawanishi, C.K.; Stoffers, V.L.; Ahlawat, N.; Ma, J.C.Y.; Arevalo, J.M.G.; Cole, S.W.; Sloan, E.K. Beta-adrenergic-stimulated macrophages: Comprehensive localization in the m1-m2 spectrum. *Brain Behav. Immun.* **2016**, *57*, 338–346. [CrossRef]
53. Hayashi, T.; Fujita, Y.; Iwasaki, O.; Adachi, Y.; Takahashi, K.; Yanagida, M. Mis16 and mis18 are required for cenp-a loading and histone deacetylation at centromeres. *Cell* **2004**, *118*, 715–729. [CrossRef]
54. Fujita, Y.; Hayashi, T.; Kiyomitsu, T.; Toyoda, Y.; Kokubu, A.; Obuse, C.; Yanagida, M. Priming of centromere for cenp-a recruitment by human hmis18alpha, hmis18beta, and m18bp1. *Dev. Cell* **2007**, *12*, 17–30. [CrossRef]
55. McKinley, K.L.; Cheeseman, I.M. Polo-like kinase 1 licenses cenp-a deposition at centromeres. *Cell* **2014**, *158*, 397–411. [CrossRef]
56. Gutteridge, R.E.; Ndiaye, M.A.; Liu, X.; Ahmad, N. Plk1 inhibitors in cancer therapy: From laboratory to clinics. *Mol. Cancer* **2016**, *15*, 1427–1435. [CrossRef] [PubMed]
57. Raab, C.A.; Raab, M.; Becker, S.; Strebhardt, K. Non-mitotic functions of polo-like kinases in cancer cells. *Biochim. Biophys. Acta. Rev. Cancer* **2021**, *1875*, 188467. [CrossRef]
58. Hudson, D.F.; Amor, D.J.; Boys, A.; Butler, K.; Williams, L.; Zhang, T.; Kalitsis, P. Loss of rmi2 increases genome instability and causes a bloom-like syndrome. *PLoS Genet.* **2016**, *12*, e1006483. [CrossRef] [PubMed]
59. Yasuhara, T.; Suzuki, T.; Katsura, M.; Miyagawa, K. Rad54b serves as a scaffold in the DNA damage response that limits checkpoint strength. *Nat. Commun.* **2014**, *5*, 5426. [CrossRef] [PubMed]
60. Ehlen, A.; Martin, C.; Miron, S.; Julien, M.; Theillet, F.X.; Ropars, V.; Sessa, G.; Beaurepere, R.; Boucherit, V.; Duchambon, P.; et al. Proper chromosome alignment depends on brca2 phosphorylation by plk1. *Nat. Commun* **2020**, *11*, 1819. [CrossRef] [PubMed]
61. Ghafouri-Fard, S.; Dashti, S.; Farsi, M.; Hussen, B.M.; Taheri, M. A review on the role of oncogenic lncrna oip5-as1 in human malignancies. *Biomed. Pharmacother.* **2021**, *137*, 111366. [CrossRef]
62. Yang, J.; Jiang, B.; Hai, J.; Duan, S.; Dong, X.; Chen, C. Long noncoding rna opa-interacting protein 5 antisense transcript 1 promotes proliferation and invasion through elevating integrin alpha6 expression by sponging mir-143-3p in cervical cancer. *J. Cell Biochem.* **2019**, *120*, 907–916. [CrossRef] [PubMed]
63. Zhang, Z.; Liu, F.; Yang, F.; Liu, Y. Kockdown of oip5-as1 expression inhibits proliferation, metastasis and emt progress in hepatoblastoma cells through up-regulating mir-186a-5p and down-regulating zeb1. *Biomed. Pharmacother.* **2018**, *101*, 14–23. [CrossRef] [PubMed]
64. Salmaninejad, A.; Zamani, M.R.; Pourvahedi, M.; Golchehre, Z.; Hosseini Bereshneh, A.; Rezaei, N. Cancer/testis antigens: Expression, regulation, tumor invasion, and use in immunotherapy of cancers. *Immunol. Investig.* **2016**, *45*, 619–640. [CrossRef] [PubMed]
65. Tarnowski, M.; Czerewaty, M.; Deskur, A.; Safranow, K.; Marlicz, W.; Urasinska, E.; Ratajczak, M.Z.; Starzynska, T. Expression of cancer testis antigens in colorectal cancer: New prognostic and therapeutic implications. *Dis. Markers* **2016**, *2016*, 1987505. [CrossRef]
66. Pal, S.K.; Tangen, C.; Thompson, I.M., Jr.; Balzer-Haas, N.; George, D.J.; Heng, D.Y.C.; Shuch, B.; Stein, M.; Tretiakova, M.; Humphrey, P.; et al. A comparison of sunitinib with cabozantinib, crizotinib, and savolitinib for treatment of advanced papillary renal cell carcinoma: A randomised, open-label, phase 2 trial. *Lancet* **2021**, *397*, 695–703. [CrossRef]

Perspective

Metastasis, an Example of Evolvability

Annick Laruelle [1,2], Claudia Manini [3], Elena Iñarra [1,4] and José I. López [5,6,*]

[1] Department of Economic Analysis, University of the Basque Country (UPV/EHU), 48015 Bilbao, Spain; annick.laruelle@ehu.eus (A.L.); elena.inarra@ehu.eus (E.I.)
[2] IKERBASQUE, Basque Foundation of Science, 48011 Bilbao, Spain
[3] Department of Pathology, San Giovanni Bosco Hospital, 10154 Turin, Italy; claudia.manini@aslcittaditorino.it
[4] Public Economic Institute, University of the Basque Country (UPV/EHU), 48015 Bilbao, Spain
[5] Department of Pathology, Cruces University Hospital, 48903 Barakaldo, Spain
[6] Biocruces-Bizkaia Health Research Institute, 48903 Barakaldo, Spain
* Correspondence: joseignacio.lopez@osakidetza.eus

Simple Summary: Cancer is a complex disease. Modern molecular technologies are progressively unveiling its genetic and epigenetic complexity, but still many key issues remain unknown. Considering cancer as a social dysfunction in a community of individuals has provided new perspectives of analysis with promising results. This narrative considers both approaches with respect to the metastatic process, the final cause of death in most patients affected by this disease.

Abstract: This overview focuses on two different perspectives to analyze the metastatic process taking clear cell renal cell carcinoma as a model, molecular and ecological. On the one hand, genomic analyses have demonstrated up to seven different constrained routes of tumor evolution and two different metastatic patterns. On the other hand, game theory applied to cell encounters within a tumor provides a sociological perspective of the possible behaviors of individuals (cells) in a collectivity. This combined approach provides a more comprehensive understanding of the complex rules governing a neoplasm.

Keywords: cancer; metastasis; genomic analysis; microenvironment; tumor ecology; game theory

1. Introduction

Modern treatment modalities are obtaining longer survivals and even cure in a significant percentage of patients with cancer, transforming the disease into a chronic-like condition. However, cancer still remains today the leading cause of death in Western countries. Metastatic dissemination is responsible for more than 90% of tumor deaths and is a challenge for modern oncology. However, metastasis is an extremely inefficient process. It has been estimated that only 0.01% of circulating tumor cells succeed in developing it [1]. The acquisition of motility by tumor cells involves a complex self-organized and self-regulated systemic cellular organization emerging from Hopfield-like dynamics [2]. These dynamics tightly regulate cell migration and other critical functions like associative memory in the cell, as it has been demonstrated very recently in unicellular organisms like amoebae [3].

The initialization of the metastatic process is possible only under regulated cellular metabolic conditions. These particular conditions allow the occurrence of epithelial-to-mesenchymal transition processes that enable epithelial cells to acquire amoeboid motility through the development of a cascade of molecular events. Unveiling the intricate interactions between tumor cells themselves on the one hand, and between tumor and host cells on the other is not totally understood and remains one of the main next frontiers in oncology. Such an approach will benefit from an ecological perspective, a viewpoint that will be considered later in this narrative.

Evolvability is an ecological term that reflects the adaptive ability of a species to generate and maintain a heritable phenotypic diversification to prevent extinction [4]. At least theoretically, metastasis may be conceived as an example of tumor evolvability achieved by genetic and epigenetic modifications that confer migratory abilities to cells to escape to more "favorable" ecosystems. A significant amount of studies are being published focusing on the intricate genomic/epigenomic complexity of this issue, unveiling the multiple routes that enable malignant cells to acquire locomotion capacities. However, very few studies have analyzed the collective pressures and cell-to-cell interactions that may explain the reasons for which malignant cells decide to migrate far away. This new approach implies considering tumors as a sort of social dysfunction [5] and analyzing malignant cells and their microenvironment from an ecological viewpoint [6].

In this perspective, we approach the metastatic process from genomic/epigenomic, ecological, and sociological perspectives. On the one hand, we review some of the last salient genomic/epigenomic findings related to the acquisition of metastatic capacities focusing on a paradigmatic example of intratumor heterogeneity and metastatic selection: The clear cell renal cell carcinoma (CCRCC). Tumor/non-tumor cell interactions (endothelial cells, tumor-infiltrating lymphocytes, and cancer-associated fibroblasts) are also mentioned as a substantial part of the adaptive processes leading to tumor migration. Finally, we analyze cell-to-cell interactions using a game theory approach and hypothesize that metastasis may be simply a specific response of a subset of tumor cells. Such response would consist of searching for collective stability far away from the primary tumor to improve their collective wellness and prevent extinction.

2. Molecular Approach

The molecular characterization of CCRCC has been largely analyzed [7]. Clonal and sub-clonal evolutions generate in some of these carcinomas a high variability of metastatic patterns. While some tumors typically develop only one metastatic clone, others are able to develop several ones with specific capacities and topographic affinities [8]. CCRCC is a paradigmatic example of intratumor heterogeneity (ITH) [9]. Actually, some deterministic constraints with prognostic impact [10] and two different metastatic patterns [11] have been identified in this tumor. A rule-based classification system supported by unsupervised clustering comparing different genomic, histological, and clinical data has shown up to seven evolutionary subtypes [10]. Multiple clonal drivers, *BAP-1* driven, and *VHL* wild type displayed a punctuated evolution and accelerated clinical progression. On the other hand, CCRCC related to *PBRM-1* gene dysfunction (PBRM-1 → SETD2, PBRM-1 → PI3K, PBRM-1 → SCNA) pursued a less aggressive clinical progression and a branched evolution. The seventh subtype detected, the *VHL* mono-driver, showed a monoclonal structure without additional driver mutations.

Losses of 9p and 14q are molecular alterations linked with metastatic ability in CCRCC [11]. Tumors that follow a punctuated model of evolution display high chromosomal complexity but low ITH. Here, a single clone with high fitness fixes early in its evolution and occupies a significant percentage of the tumor mass following a Darwinian pattern. These cases develop multiple metastases early in their evolution (rapid progression). The branching pattern, however, is also a Darwinian example of tumor evolution, but in this case, there is high clonal and sub-clonal diversification. As a result, chromosomal complexity is low, but ITH is high. Branching-type tumors develop few metastases late in their evolution (attenuated progression).

Hypoxia promotes metastases, but this factor is not homogeneously distributed across the tumor. Actually, a tumor spatial specialization pattern has been very recently detected in CCRCC. Tumor areas with metastatic abilities are located in the tumor interior in a multi-regional analysis of 79 cases [12]. Tumor necrosis and higher Ki-67 index, histological grade, and chromosomal complexity were detected in the tumor center, where microenvironmental pressures and the struggle for survival are supposedly higher due to higher levels of hypoxia.

The metastatic capacity is associated with the acquisition of a stem cell-like phenotype because of the evolvability of a subset of tumor cells, the so-called metastasis-initiating cells [13]. Metastatic dissemination (local invasion, intravasation, blood and/or lymphatic circulation, extravasation and extravascular spread), dormancy, immune evasion, organ colonization, and local tropism development conform to the stepwise process followed by this special subset of traveling tumor cells. This concatenated sequence of events needs a dynamic epigenetic remodeling to adapt these cells to the ever-changing environments. Abnormal DNA methylation (methyl-binding proteins, post-translational histone modifications, miRNAs, lncRNAs, RNA methylation) is one of the epigenetic mechanisms most extensively studied and an actionable target for future treatments [14].

3. Microenvironmental Context

A question arises at this point: How many times can tumor cells migrate? In other words, can a metastasis metastasize? The answer is yes, as it has been demonstrated in genomic studies analyzing the clonality patterns of the widespread tumor seeding in renal, prostate, breast, pancreatic, and colo-rectal cancers [15,16]. Such complex polyclonal dissemination seems a cornerstone in the development of prostate cancer clones resistant to castration, as demonstrated by Gundem et al. [15].

The relationship between tumor cells and their microenvironment is crucial in cancer progression and metastatic process. Immune cells, neovascular endothelia, and fibroblasts are the best-studied tumor-associated cells. These stromal elements dynamically interact between and with tumor cells. In addition, this interplay is becoming a promising therapeutic target, with immune checkpoint blockade (PD-1, PD-L1, CTLA-4) and anti-angiogenesis as two of the most representative examples of precision therapies in CCRCC [17] and other neoplasms. In fact, the combination of both treatments in a subset of these patients has a positive effect based on the synergic actions of antiangiogenic and checkpoint blockers [17].

PD-1 and its ligand PD-L1 (B7-H1) blockade has been implemented in clinical practice as a promising therapeutic strategy in CCRCC and other tumors, but the immunohistochemical selection of candidates, with different antibodies and cut-offs, is still controversial [18]. By contrast, the soluble fraction of PD-L1 has been associated with prognosis and metastatic status in these tumors [19]. Immune checkpoint blockade seems especially useful in the so-called inflamed tumors, a group of aggressive CCRCC with abundant tumor-infiltrating lymphocytes linked to *BAP-1* gene inactivation and tendency to early metastases [20]. On the other hand, B7-H3, another immune regulator of the B7 family, has been directly implicated in the metastatic process [21], expanding the list of candidates for checkpoint blockade in the clinical practice.

CCRCC is a neoplasm associated with *VHL* gene malfunction that creates a pseudo-hypoxic status in neoplastic cells. This condition induces the initialization of the VEGF cascade [9], ending in neoangiogenesis. A new definition of microvessel density in CCRCC has been recently published [22]. These authors consider the sum of classical microvessel density and vasculogenic mimicry as the total microvessel density and correlate this sum with prognosis [22]. Since these neoplasms are highly vascularized, the antiangiogenic tyrosine-kinase inhibitors have had a relevant role in the therapeutic armamentarium, particularly in metastatic patients [23].

Not all CCRCC, however, display the same degree of angiogenesis. Predominantly angiogenic examples are typically associated with *PBRM-1* gene inactivation [20] and show a pancreatic tropism [24]. Interestingly, pancreatic metastases in angiogenic-type low-grade CCRCC have been detected significantly later in the tumor evolution compared with metastases in the rest of the sites [11].

Cancer-associated fibroblasts (CAF) are essential elements in the tumor microenvironment [25]. The interplay between tumor cells and CAF activates the epithelial-to-mesenchymal transition process providing a stem cell phenotype to epithelial tumor cells, thus enabling tumor cell migration [26]. In this interaction, CAF produce fibroblast activation protein-α (FAP), a protein associated with biological aggressiveness [27] and early metastases [28] in CCRCC. Other proteins produced by CAF, however, also contribute to tumor progression [29]. Coercive feedback signaling from cancer cells to CAF does exist, thus assuring the maintenance of FAP production and perpetuating the loop [26].

4. Sociological Approach

Game theory [30] is a branch of applied mathematics initially applied to economics that investigates interactions between individuals. It has been used to examine complex problems in biology [31] and, more recently, in oncology [32]. Considering cancer as a social dysfunction [5] in which cells interact following different strategies, game theory has opened up new perspectives to understand cancer dynamics.

The Hawk–Dove game (Figure 1) is a classical model in biology [31]. It assumes a population whose members have bilateral encounters to divide a resource (v), with some cost (c) associated with fight (c > v). In every encounter, each individual can behave as a "hawk" and escalate to a fight or as a "dove" and back down. Therefore, if one individual acts as a hawk while the other acts as a dove, the "hawk" gets the resource (v) and the dove gets nothing (0). If two "hawks" meet, there is a fight, the winner receives the resource (v) and the other faces the cost of the fight (−c). On average, the two hawks receive (v − c)/2. If two individuals act as dove, they share the resource. On average, the two doves get v/2.

	Hawk	Dove
Hawk	$\dfrac{v-c}{2}$	v
Dove	0	$\dfrac{v}{2}$

Figure 1. Hawk–Dove game matrix. Each of the two individuals chooses one strategy. The matrix summarizes the payoffs of the four possible results for the individual choosing the row strategy playing against an individual choosing the column strategy (v = resource, c = cost).

The concept of evolutionarily stable strategy (ESS) is used to solve the game. ESS in game theory captures the resilience of a strategy against another in the sense that it assumes that most members of the population play the ESS but a small proportion of members, here termed mutants, chooses a different strategy. In this context, each mutant's expected payoff is smaller than the expected payoff of an individual who plays the ESS. Consequently, mutants are driven out of the population. As a consequence, the Hawk–Dove game has a unique ESS, which consists of a fraction, v/c, of the population which plays hawk while the remaining population plays dove.

This description is appropriate for analyzing the interactions between cancer cells and cancer-associated fibroblasts in CCRCC [25] that we have previously mentioned in this narrative. The resource, in this case, is the diffusible factor fibroblast activation protein-α (FAP) that improves cancer fitness, while the cost is the effort of producing and excreting it to the medium.

It is thus theoretically possible to predict that a homogeneous neoplasm will achieve its ESS. However, there is no such thing as a perfectly homogeneous tumor in real life, so a heterogeneous population must be considered. In the heterogeneous Hawk–Dove game [33], the population is divided into two types. In consequence, each individual conditions his/her action depending on his/her opponent's type. In the heterogeneous game, the previously described ESS is not stable any longer. Therefore, the game will have two ESSs. Each ESS is characterized by one of the two types of individuals being discriminated. In such a context, one type of individual receives a higher payoff than the other. Furthermore, the payoff of the discriminated individuals in the heterogeneous game is smaller than the payoff obtained in the homogeneous one.

However, there is an unanswered question in oncology: Why does a malignant tumor metastasize? From an ecological viewpoint, the goal of cell migration is tumor survival. From a sociological perspective, however, it can be hypothesized that the ultimate common endpoint of neoplastic cells that migrate far away is to achieve a higher payoff elsewhere. Here, the set of discriminated cells in the heterogeneous population that governs the primary tumor might develop metastatic abilities just to escape to a friendlier environment where a higher payoff is initially possible, at least theoretically.

5. Conclusions

Tumor spatial specialization with metastasizing subclones located in the tumor interior, the demonstrated ability of metastases to metastasize, and the sociological tumor cell interactions unveiled by the Hawk–Dove game reinforce the storyline of this perspective in the sense that the search for a better environment by tumor cells is a constant event in malignant tumors.

Author Contributions: A.L., C.M., E.I. and J.I.L. conceived and wrote the article. All authors have read and agreed to the published version of the manuscript.

Funding: This research received no external funding.

Conflicts of Interest: The authors have no conflicts of interest.

References

1. Langley, R.R.; Fidler, I.J. The seed and soil hypothesis revisited. The role of tumor-stroma interactions in metastasis to different organs. *Int. J. Cancer* **2011**, *128*, 2527–2535. [CrossRef]
2. De La Fuente, I.M.; López, J.I. Cell motility and cancer. *Cancers* **2020**, *12*, 2177. [CrossRef]
3. De La Fuente, I.M.; Bringas, C.; Malaina, I.; Fedetz, M.; Carrasco-Pujante, J.; Morales, M.; Knafo, S.; Martínez, L.; Pérez-Samartín, A.; López, J.I.; et al. Evidence of conditioned behavior in amoebae. *Nat. Commun.* **2019**, *10*, 3690. [CrossRef]
4. Kirschner, M.K.; Gerhart, J. Evolvability. *Proc. Nat. Acad. Sci. USA* **1998**, *95*, 8420–8427. [CrossRef]
5. Axelrod, R.; Pienta, K.J. Cancer as a social dysfunction. Why cancer research needs new thinking. *Mol. Cancer Res.* **2018**, *16*, 1346–1347. [CrossRef]
6. Reynolds, B.A.; Oli, M.W.; Oli, M.K. Eco-oncology: Applying ecological principles to understand and manage cancer. *Ecol. Evol.* **2020**, *10*, 8538–8553. [CrossRef]
7. Cancer Genome Atlas Research Network. Comprehensive molecular characterization of clear cell renal cell carcinoma. *Nature* **2013**, *499*, 43–49. [CrossRef]
8. Patel, S.A.; Rodrigues, P.; Wesolowski, L.; Vanharanta, S. Genomic control of metastasis. *Br. J. Cancer* **2021**, *124*, 3–12. [CrossRef] [PubMed]
9. Trpkov, K.; Hes, O.; Williamson, S.; Adeniran, A.J.; Agaimy, A.; Alaghehbandan, R.; Amin, M.B.; Argani, P.; Chen, Y.-B.; Cheng, L.; et al. New developments in existing WHO entities and evolving molecular concepts: The Genitourinary Pathology Society (GUPS) Update on Renal Neoplasia. *Mod. Pathol.* **2021**. [CrossRef]
10. Turajlic, S.; Xu, H.; Litchfield, K.; Rowan, A.; Horswell, S.; Chambers, T.; O'Brien, T.; Lopez, J.I.; Watkins, T.B.K.; Nicol, D.; et al. Deterministic evolutionary trajectories influence primary tumor growth: TRACERx Renal. *Cell* **2018**, *173*, 595–610. [CrossRef] [PubMed]
11. Turajlic, S.; Xu, H.; Litchfield, K.; Rowan, A.; Chambers, T.; López, J.I.; Nicol, D.; O'Brien, T.; Larkin, J.; Horswell, S.; et al. Tracking renal cancer evolution reveals constrained routes to metastases: TRACERx Renal. *Cell* **2018**, *173*, 581–594. [CrossRef] [PubMed]
12. Zhao, Y.; Fu, X.; López, J.I.; Rowan, A.; Au, L.; Fendler, A.; Hazell, S.; Xu, H.; Horswell, S.; Shepherd, S.T.C.; et al. Selection of metastasis competent subclones in the tumour interior. *Nat. Ecol. Evol.* **2021**, *5*, 1033–1045. [CrossRef] [PubMed]
13. Massagué, J.; Ganesh, K. Metastasis-initiating cells and ecosystems. *Cancer Discov.* **2021**, *11*, 971–994. [CrossRef] [PubMed]
14. Angulo, J.C.; Manini, C.; López, J.I.; Pueyo, A.; Colás, B.; Ropero, S. The role of epigenetics in the progression of clear cell renal cell carcinoma and the basis for future epigenetic treatments. *Cancers* **2021**, *13*, 2071. [CrossRef] [PubMed]
15. Gundem, G.; Van Loo, P.; Kremeyer, B.; Alexandrov, L.B.; Tubio, J.M.C.; Papaemmanull, E.; Brewer, D.S.; Kallio, H.M.L.; Högnäs, G.; Annala, M.; et al. The evolutionary history of lethal metastatic prostate cancer. *Nature* **2015**, *520*, 353–357. [CrossRef] [PubMed]
16. Turajlic, S.; Swanton, C. Metastasis as an evolutionary process. *Science* **2016**, *352*, 169–175. [CrossRef]
17. Angulo, J.C.; Shapiro, O. The changing therapeutic landscape of metastatic renal cancer. *Cancers* **2019**, *11*, 1227. [CrossRef] [PubMed]
18. Nunes-Xavier, C.E.; Angulo, J.C.; Pulido, R.; López, J.I. A critical insight into the clinical translation of PD-1/PD-L1 blockade therapy in clear cell renal cell carcinoma. *Curr. Urol. Rep.* **2019**, *20*, 1. [CrossRef]
19. Larringa, G.; Solano-Iturri, J.D.; Errarte, P.; Unda, M.; Loizaga-Iriarte, A.; Pérez-Fernández, A.; Echevarría, E.; Asumendi, A.; Manini, C.; Angulo, J.C.; et al. Soluble PD-L1 is an independent prognostic factor in clear cell renal cell carcinoma. *Cancers* **2021**, *13*, 667. [CrossRef]

20. Brugarolas, J.; Rajaram, S.; Christie, A.; Kapur, P. The evolution of angiogenic and inflamed tumors: The renal cancer paradigm. *Cancer Cell* **2020**, *38*, 771–773. [CrossRef]
21. Flem-Karlsen, K.; Fodstad, O.; Tan, M.; Nunes-Xavier, C.E. B7-H3 in cancer. Beyond immune regulation. *Trends Cancer* **2018**, *4*, 401–404. [CrossRef]
22. Wu, Y.; Du, K.; Guan, W.; Wu, D.; Tang, H.; Wang, N.; Qi, J.; Gu, Z.; Yang, J.; Ding, J. A novel definition of microvessel density in renal cell carcinoma: Angiogenesis plus vasculogenic mimicry. *Oncol. Lett.* **2020**, *20*, 192. [CrossRef]
23. Angulo, J.C.; Lawrie, C.H.; López, J.I. Sequential treatment of metastatic renal cancer in a complex evolving landscape. *Ann. Transl. Med.* **2019**, *7*, S272. [CrossRef]
24. Singla, N.; Xie, Z.; Zhang, Z.; Gao, M.; Yousuf, Q.; Onabolu, O.; McKenzie, T.; Tcheuyap, V.T.; Ma, Y.; Choi, J.; et al. Pancreatic trophism of metastatic renal cell carcinoma. *JCI Insight* **2020**, *5*, e134564. [CrossRef]
25. Errarte, P.; Larrinaga, G.; López, J.I. The role of cancer-associated fibroblasts in renal cell carcinoma. An example of tumor modulation through tumor/non-tumor cell interactions. *J. Adv. Res.* **2019**, *21*, 103–108. [CrossRef] [PubMed]
26. Cirri, P.; Chiarugi, P. Cancer-associated-fibroblasts and tumor cells: A diabolic liaison driving cancer progression. *Cancer Metastasis Rev.* **2012**, *31*, 195–208. [CrossRef] [PubMed]
27. López, J.I.; Errarte, P.; Erramuzpe, A.; Guarch, R.; Cortés, J.M.; Angulo, J.C.; Pulido, R.; Irazusta, J.; Llarena, R.; Larrinaga, G. Fibroblast activation protein predicts prognosis in clear cell renal cell carcinoma. *Hum. Pathol.* **2016**, *54*, 100–105. [CrossRef] [PubMed]
28. Errarte, P.; Guarch, R.; Pulido, R.; Blanco, L.; Nunes-Xavier, C.E.; Beitia, M.; Gil, J.; Angulo, J.C.; López, J.I.; Larrinaga, G. The expresión of fibroblast activation protein in clear cell renal cell carcinomas is associated with synchronous lymph node metastases. *PLoS ONE* **2016**, *11*, e0169105. [CrossRef]
29. Kalluri, R.; Zeisberg, M. Fibroblasts in cancer. *Nat. Rev. Cancer* **2006**, *6*, 392–401. [CrossRef]
30. Von Neumann, J.; Morgenstern, O. *Theory of Games and Economic Behavior*; Princeton University Press: Princeton, NJ, USA, 1944.
31. Maynard Smith, J.; Price, G.R. The logic of animal conflict. *Nature* **1973**, *246*, 15–18. [CrossRef]
32. Archetti, M.; Pienta, K.J. Cooperation among cancer cells: Applying game theory to cancer. *Nat. Rev. Cancer* **2019**, *19*, 110–117. [CrossRef] [PubMed]
33. Inarra, E.; Laruelle, A. Artificial distinction and real discrimination. *J. Theor. Biol.* **2012**, *505*, 110–117. [CrossRef] [PubMed]

Article

Restaging the Biochemical Recurrence of Prostate Cancer with [⁶⁸Ga]Ga-PSMA-11 PET/CT: Diagnostic Performance and Impact on Patient Disease Management

Aloÿse Fourquet [1], Lucien Lahmi [2], Timofei Rusu [1], Yazid Belkacemi [3], Gilles Créhange [4], Alexandre de la Taille [5], Georges Fournier [6], Olivier Cussenot [7] and Mathieu Gauthé [1,8,*]

1. Department of Nuclear Medicine, Hôpital Tenon-AP-HP, Sorbonne Université, 75020 Paris, France; aloyse.fourquet@aphp.fr (A.F.); timofei.rusu@aphp.fr (T.R.)
2. Department of Radiation Oncology, Hôpital Tenon-AP-HP, Sorbonne Université, 75020 Paris, France; Lucien.lahmi@gustaveroussy.fr
3. Department of Radiation Oncology and Henri Mondor Breast Center, Hôpitaux Universitaires Henri Mondor, Université Paris-Est Créteil (UEPC) et IMRB—INSERM U955 Team 21, 94000 Créteil, France; yazid.belkacemi@aphp.fr
4. Department of Radiation Oncology, Institut Curie, 75005 Paris, France; gilles.crehange@curie.fr
5. Department of Urology, Hôpitaux Universitaires Henri Mondor, Université Paris-Est Créteil (UEPC), 94000 Créteil, France; alexandre.de-la-taille@aphp.fr
6. Department of Urology, Hôpital de la Cavale Blanche, Université de Brest, 29200 Brest, France; gfournier@afu.fr
7. Department of Urology, Hôpital Tenon-AP-HP, Sorbonne Université, 75020 Paris, France; olivier.cussenot@aphp.fr
8. AP-HP Health Economics Research Unit, INSERM-UMR1153, 75004 Paris, France
* Correspondence author: mathieugauthe@yahoo.fr; Tel.: +33-156017842; Fax: +33-156016171

Simple Summary: We aimed to evaluate the diagnostic performance, impact on patient disease management, and therapy efficacy prediction of [⁶⁸Ga]Ga-PSMA-11 PET/CT on 294 patients with biochemical recurrence of prostate cancer. We established a composite standard of truth for the imaging based on all clinical data available collected during the follow-up period with a median duration of follow-up of 17 months. Using this methodology, we found that the overall per-patient sensitivity and specificity were both 70%, the patient disease management was changed in 68% of patients, and that [⁶⁸Ga]Ga-PSMA-11 PET/CT impacted this change in 86% of patients. The treatment carried out on the patient was considered effective in 78% of patients; in 89% of patients when guided by [⁶⁸Ga]Ga-PSMA-11 PET/CT versus 61% of patients when not guided by [⁶⁸Ga]Ga-PSMA-11 PET/CT.

Abstract: Background: Detection rates of [⁶⁸Ga]Ga-PSMA-11 PET/CT on the restaging of prostate cancer (PCa) patients presenting with biochemical recurrence (BCR) have been well documented, but its performance and impact on patient management have not been evaluated as extensively. Methods: Retrospective analysis of PCa patients presenting with BCR and referred for [⁶⁸Ga]Ga-PSMA-11 PET/CT. Pathological foci were classified according to six anatomical sites and evaluated with a three-point scale according to the uptake intensity. The impact of [⁶⁸Ga]Ga-PSMA-11 PET/CT was defined as any change in management that was triggered by [⁶⁸Ga]Ga-PSMA-11 PET/CT. The existence of a PCa lesion was established according to a composite standard of truth based on all clinical data available collected during the follow-up period. Results: We included 294 patients. The detection rate was 69%. Per-patient sensitivity and specificity were both 70%. Patient disease management was changed in 68% of patients, and [⁶⁸Ga]Ga-PSMA-11 PET/CT impacted this change in 86% of patients. The treatment carried out on patient was considered effective in 89% of patients when guided by [⁶⁸Ga]Ga-PSMA-11 PET/CT versus 61% of patients when not guided by [⁶⁸Ga]Ga-PSMA-11 PET/CT ($p < 0.001$). Conclusions: [⁶⁸Ga]Ga-PSMA-11 PET/CT demonstrated high performance in locating PCa recurrence sites and impacted therapeutic management in nearly two out of three patients.

Keywords: prostatic neoplasms; positron-emission tomography; decision making

1. Introduction

Prostate cancer (PCa) is the most prevalent cancer in men worldwide, accounting for approximately 21% of all diagnosed cancers [1]. Up to 40% of patients with PCa initially treated with curative intent will experience biochemical recurrence (BCR) [2,3], which is defined following radical prostatectomy by two consecutive rising prostate-specific antigen (PSA) values >0.2 ng/mL, or after primary radiation therapy by any PSA increase ≥2 ng/mL higher than the PSA nadir value, regardless of the serum concentration of the nadir [4,5]. Accurately locating the recurrence site(s) is essential for optimizing patient management, as localized or oligometastatic recurrences could be eligible for salvage targeted treatments with curative intent, such as local therapy [6] or stereotactic radiation therapy [7]. Conventional imaging modalities, such as bone scan and computed tomography (CT), have limited utility in this setting, especially when PSA serum levels are below 10 ng/mL [8]. [^{18}F]fluorocholine positron emission tomography associated with computed tomography (PET/CT) was demonstrated to have better performance than conventional imaging but may also fail to locate recurrence at low PSA levels [9].

The prostate-specific membrane antigen (PSMA) is a transmembrane protein that is over-expressed by up to 1000-fold in almost all PCa cells [10,11]. The recent introduction of PET/CT using a PSMA radioligand for imaging of PCa BCR has shown promising results due to its performance in detecting lesions, even at very low PSA levels, impacting on the therapeutic management of PCa patients [12–14].

Although the detection rates of ^{68}Ga-PSMA PET/CT have been well documented, its sensitivity, specificity, impact on patient management, and therapy efficacy prediction have not been evaluated as extensively.

The purpose of this study was to evaluate the diagnostic performance, impact on patient disease management, and therapy efficacy prediction of PET/CT using a PSMA ligand radiolabelled with gallium-68, the [^{68}Ga]Ga-PSMA-11, on the restaging of PCa patients presenting with BCR.

2. Materials and Methods

2.1. Population

Patients presenting with BCR of PCa who were addressed to our department for [^{68}Ga]Ga-PSMA-11 PET/CT were consecutively included and retrospectively analyzed. These patients had shown no sign of distant metastases at [^{18}F]fluorocholine PET/CT; for this reason, they were referred to [^{68}Ga]Ga-PSMA-11 PET/CT, based on the French regulation for compassionate use of pharmaceutical, which is authorized on an individual basis by the National Medicine Agency.

Inclusion criteria for patients were as follows: 1—histologically confirmed PCa previously treated with curative intent; 2—no known history of PCa distant metastases (invaded locoregional pelvic lymph node at diagnosis was not considered as metastatic according to the 2009 TNM classification for staging PCa [15]); 3—currently presenting a biochemical recurrence defined as two consecutive rising PSA values above 0.2 ng/mL following radical prostatectomy or any PSA increase greater than or equal to 2 ng/mL higher than the PSA nadir value, regardless of the nadir value, for non-surgical first-line definitive treatments [4].

Exclusion criteria were as follows: 1—PCa with known distant metastases; 2—patients with persistent PSA after prostatectomy (PSA ≥ 0.1 ng/mL) [15] or radiation therapy (nadir PSA < 2 ng/mL with testosterone recovered if previous androgen deprivation therapy (ADT) [5]); 3—patients who were never treated with curative intent for PCa; 4—the presence of a second active neoplasm other than PCa.

This research implied no intervention on the patient. According to French regulations, the approval of an institutional review board was not necessary for performing this retrospective analysis of already available data. Patients were informed that their data collected for the [^{68}Ga]Ga-PSMA-11 PET/CT would be analyzed and published anonymously, and did not object.

2.2. [^{68}Ga]Ga-PSMA-11 PET/CT Imaging Procedure

Gallium-68 was obtained from a [^{68}Ge]Ge/ [^{68}Ga]Ga radionuclide generator (GalliaPharm, Eckert & Ziegler Radiopharma GmBH, Berlin, Germany) and used for radiolabelling of PSMA-11 according to the manufacturer's instructions (IASON GmbH). Patients did not require specific preparation before the injection. Patients received 1–2 MBq/kg of the radiotracer injected in saline via an infusion line.

Images were acquired using a Gemini TF16 (Philips Medical Systems, Cleveland, OH, USA) or a Biograph mCTflow (Siemens Healthcare, Erlangen, Germany) PET/CT. Both PET/CT scanners included time-of-flight technology. Dynamic images were acquired on the pelvis immediately after [^{68}Ga]Ga-PSMA-11 injection (10 images of one-minute duration each) and from vertex to mid-thigh 60 to 90 min after injection. On the Gemini TF16 PET/CT, the pelvis was imaged for 3 min, and every other bed position was imaged for 2 min in 3D mode with a 576 mm FOV and a 144 × 144 matrix. Images were reconstructed from 3 iterations and 33 subsets using the OSEM weighted method. Low-dose CT without contrast-enhancement was performed prior to PET acquisition (120 kVp, 80 mA.s, slice thickness 2.5 mm, pitch 0.813, rotation time 0.5 s, FOV 600 mm). On the Biograph mCTflow PET/CT, the scanning speed was set to 0.7 cm/min over the pelvis and 0.9 cm/min for the rest of the acquisition field. Images were taken in 3D mode with a 780 mm FOV and a 200 × 200 matrix. Images were reconstructed from 2 iterations and 21 subsets using the OSEM weighted method. Low-dose CT without contrast-enhancement was performed prior to PET acquisition (CareDose® automatic modulation for keV and mA.s, slice thickness 2 mm, pitch 0.813, rotation time 0.5 s, FOV 500 mm). A harmonization in PET images between the two scanners by using EQ.PET, a NEMA-referenced SUV across technologies was performed [16].

2.3. [^{68}Ga]Ga-PSMA-11 PET/CT Image Analysis

[^{68}Ga]Ga-PSMA-11 PET/CTs were read on-site on the day of image acquisition (routine unmasked reading) by local nuclear physicians with at least 4 years of experience in reading PET/CT and 6 months of experience in reading [^{68}Ga]Ga-PSMA-11 PET/CT. An expert nuclear medicine physician with 10 years of experience in reading PET/CT and 4 years of experience in reading [^{68}Ga]Ga-PSMA-11 PET/CT, who was blinded to all clinical data, performed a retrospective reading of the [^{68}Ga]Ga-PSMA-11 PET/CTs of the patients who met the inclusion criteria (masked retrospective reading). Anonymized images presented in a random order were independently reviewed on a dedicated workstation (Syngo.via, Siemens Healthcare).

The expert reader assessed uptakes across six anatomical sites and attributed them value on a 3-point qualitative scale according their intensity: 0—no suspicious uptake (at best equal to muscle background); 1—equivocal uptake (between background in muscles and vessels); 2—malignant uptake (higher than background in vessels) [17]. CT images were used for anatomic allocation of a suspicious focus and to facilitate diagnosis. We considered six anatomical sites: prostate/prostatic lodge, pelvic lymph nodes (up to the common iliac lymph nodes), paraaortic lymph nodes, lymph nodes above the diaphragm, bone, and viscera. If at least one suspicious uptake (equivocal or malignant) was detected in an anatomical site, the entire areas was quoted as equivocal or malignant. The intensity of the most intense abnormal uptake was determined by the maximum standardized uptake value (SUVmax) for each anatomical site during the retrospective reading. Based on the results of [^{68}Ga]Ga-PSMA-11 PET/CT, we categorized patients as oligometastatic if they present between 1 and 3, 4, or 5 distant malignant uptakes excluding the prostate/prostatic lodge (oligo-3, oligo-4, and oligo-5, respectively); polymetastatic if more than 5 distant malignant uptakes were detected [18].

2.4. Follow-Up and Evaluation of [^{68}Ga]Ga-PSMA-11 PET/CT Impact on Patient Disease Management

After imaging, the clinical follow-up was performed for each patient by his referring physician. Clinicians decided for each patient management plan during multidisciplinary meetings dedicated to urological cancers. These multidisciplinary meeting boards were constituted by a urologist, a radiation oncologist, a medical oncologist, a pathologist, a radiologist, and a nuclear medicine physician. They analyzed all clinical data available before and after [^{68}Ga]Ga-PSMA-11 PET/CT and then decided the management of the patients. We defined the impact of [^{68}Ga]Ga-PSMA-11 PET/CT as any change in management decided by clinicians during the multidisciplinary meeting triggered by [^{68}Ga]Ga-PSMA-11 PET/CT.

The treatment carried out on the patient was considered to be effective if the PSA declined by more than 50% (compared to the baseline value) following treatment modification or if the PSA remained stable (maximum variation of 10% compared to baseline) on at least 2 assays performed at least 3 weeks apart when surveillance was decided [19].

2.5. Standard of Truth

Existence of a PCa lesion was established for each patient according to a composite standard of truth (SOT) based on all clinical data that were available during the follow-up period: histological findings, results of other imaging, follow-up imaging and PSA evolution. Histological findings when available were considered as the strongest criteria. When histological confirmation was not available, SOT criteria were as follows:

1. True-positive if at least 3 criteria were met: the imaging was positive for a location; the patient received targeted treatment for imaging findings; the PSA decreased in response to the targeted treatment; the number or the size of the lesions decreased on follow-up imaging;
2. True-negative if at least 3 criteria were met: the imaging was negative for a location; the patient received targeted treatment on another location; the PSA decreased in response to the targeted treatment; no evolution on follow-up imaging;
3. False-positive if at least 3 criteria were met: the imaging was positive for a location; the location was atypical for a PCa metastasis; the patient received targeted treatment for atypical imaging findings leading to an absence of PSA decrease in response to treatment; the patient received targeted treatment on another location leading to a PSA decrease in response to treatment; persistence and stability of the abnormality on follow-up imaging;
4. False-negative if at least 3 criteria were met: the imaging was negative for a location; the patient received targeted treatment on that location leading (as PCa patients presenting with first BCR after prostatectomy and in whom there is no evidence of distant metastatic disease can be offered salvage radiation therapy according to guidelines [20]); PSA decrease in response to treatment; appearance of a typical abnormality in that location on follow-up imaging.

2.6. Statistical Analysis

Data were analyzed using the IBM SPSS software. We considered a p-value less than 0.05 to be statistically significant. We performed logistic regression to search for a relationship between [^{68}Ga]Ga-PSMA-11 PET/CT positivity (patient-based and per anatomical site) and initial ISUP grade group, initial d'Amico group risk, PSA in surgical patients (closest assay to the [^{68}Ga]Ga-PSMA-11 PET/CT), PSA doubling time in months, and PSA velocity in ng/mL/year in all patients. Receiver operating characteristic (ROC) analysis was used to determine area under the curve and cut-off values of PSA parameters in relation to [^{68}Ga]Ga-PSMA-11 PET/CT positivity. Comparisons of [^{68}Ga]Ga-PSMA-11 PET/CT detection rates (at least one suspicious abnormality suggestive of PCa) and impact on patient management to parameters in relation with [^{68}Ga]Ga-PSMA-11 PET/CT positivity were performed by chi-squared test, a Fisher's exact test, or a Student's t-test according to the type of variable. Detection rates and accuracies between the two PET/CT scanners were compared by chi-squared test. Therapy efficacies when guided or not by [^{68}Ga]Ga-

PSMA-11 PET/CT imaging were compared via Fisher's exact test. The agreement between retrospective masked and routine unmasked [^{68}Ga]Ga-PSMA-11 PET/CT readings, overall and per anatomical site, were assessed using Cohen's kappa coefficient (0–0.20: very weak; 0.21–0.40: weak; 0.41–0.60: moderate; 0.61–0.80: strong; 0.81–1.0: very strong).

3. Results

3.1. BCR Patient Characteristics

Between June 2016 and November 2018, 294 consecutive patients who met the inclusion criteria were retrospectively included (Table 1). No eligible patient was excluded. One-hundred and ninety-three patients were presenting with first PCa BCR among whom 159 had prostatectomy.

Table 1. Patient characteristics.

Parameter	
n	294
Mean age in years	
At prostate cancer diagnosis (range)	61 (42–83)
The day of [^{68}Ga]Ga-PSMA-11 PET/CT (range)	68 (43–88)
Initial group according to d'Amico classification	
Low risk	32 (11%)
Intermediate risk	170 (58%)
High risk	70 (24%)
Unknown	22 (7%)
International Society of Urological Pathologists (ISUP) 2014 grade group	
1	47 (16%)
2	106 (36%)
3	98 (33%)
4	23 (8%)
5	17 (6%)
Unknown	2 (1%)
Initial treatment	
Surgery (prostatectomy ± lymph node dissection)	210 (71.5%)
Surgery + adjuvant radiation therapy	42 (14%)
Definitive radiation therapy ± androgen deprivation therapy	27 (9%)
Brachytherapy	14 (5%)
High Intensity Focused Ultrasound	1 (0.5%)
PSA parameters at [^{68}Ga]Ga-PSMA-11 PET/CT (closest assay to the examination)	
Mean delay between PSA assay and [^{68}Ga]Ga-PSMA-11 PET/CT in weeks	10.5 [9.7–11.3]
Mean serum level in ng/mL in operated patients ($n = 252$)	2.97 [1.96–3.98]
0.20–0.49	57 (23%)
0.50–0.99	45 (18%)
1–1.99	59 (23%)
Greater than 2	91 (36%)
Mean serum level in ng/mL in non-operated patients ($n = 42$)	4.96 [3.60–6.31]
Mean doubling time in months *	12.9 [11.4–14.7]
Under 6	102 (36%)
Between 6 and 12	80 (28%)
Above 12	103 (36%)
Mean velocity in ng/mL/year *	2.95 [2.17–3.74]

* Evaluated on 285 patients; 95% confidence intervals are presented between brackets.

The mean time from PCa diagnosis to the first BCR was 42 months [95%CI: 37–46], longer for ISUP 1–2 patients (49 months [95%CI: 43–56]) than for ISUP 3–5 patients (33 months [95%CI: 28–39]) ($p < 0.001$; Student t test).

Sixteen patients were considered lost to follow-up (no follow-up data available). The median duration of follow up after [^{68}Ga]Ga-PSMA-11 PET/CT for the 278 assessable patients was 17 months [95%CI: 14–19]. A patient-based SOT was feasible in 176 patients (60%) among whom histological confirmation was available in 27 patients.

3.2. [^{68}Ga]Ga-PSMA-11 PET/CT Positivity Rates and Performance

On the 294 patients, 140 (48%) where scanned on the Gemini TF16 and 154 (52%) on the Biograph CTflow.

At least one abnormal focus was found in 237 patients (81%) on routine unmasked reading and in 229 patients (78%) on masked retrospective reading. The overall and per anatomical site detection rates, irrespective of PSA, for readings, as well as the SUVmax of detected foci, are presented in Table 2.

Table 2. [^{68}Ga]Ga-PSMA-11 PET/CT positivity rates in prostate cancer patients investigated due to a biochemical recurrence (irrespective of total prostate-specific antigen serum values). Results of routine unmasked and retrospective masked readings, both by anatomical site and overall, are presented. Median maximum standard uptake values (SUVmax) per anatomical site are presented with their range brackets. Agreement was evaluated with Cohen's kappa coefficient κ.

n = 294	Malignant	Equivocal	Negative	SUVmax [Range]	κ
Overall					
Routine unmasked	202 (69%)	35 (12%)	57 (19%)		
Retrospective masked	202 (69%)	27 (9%)	65 (22%)	-	0.68
Prostate/prostatic lodge					
Routine unmasked	60 (20%)	18 (6%)	216 (74%)		
Retrospective masked	60 (20%)	8 (3%)	226 (77%)	5.3 [1.7–20.9]	0.54
Pelvic lymph nodes					
Routine unmasked	110 (38%)	6 (2%)	178 (61%)		
Retrospective masked	111 (38%)	5 (2%)	178 (61%)	5.9 [1.7–58.3]	0.90
Paraaortic lymph nodes					
Routine unmasked	47 (16%)	3 (1%)	244 (83%)		
Retrospective masked	47 (16%)	2 (1%)	245 (83%)	5.5 [1.8–71.7]	0.84
Lymph nodes above the diaphragm					
Routine unmasked	17 (6%)	12 (4%)	265 (90%)		
Retrospective masked	25 (9%)	7 (2%)	262 (89%)	3.9 [2–19.6]	0.73
Bone					
Routine unmasked	53 (18%)	14 (5%)	227 (77%)		
Retrospective masked	57 (19%)	26 (9%)	211 (72%)	3.4 [1.1–38.6]	0.74
Viscera					
Routine unmasked	18 (6%)	9 (3%)	267 (91%)		
Retrospective masked	20 * (7%)	12 ** (4%)	262 (89%)	6.2 [2.2–18.6]	0.56

*: 7 carcinomatosis, 7 pleura/lung, 2 testis, 1 liver, 1 penile, 1 intramedullary spinal, 1 rectal. **: 4 carcinomatosis, 3 liver, 3 testis, 1 penile, 1 pancreas.

Based on the masked retrospective reading results, we identified a relationship between [^{68}Ga]Ga-PSMA-11 PET/CT detection rates and PSA serum level in surgical patients: 37%, 56%, 71%, and 87% for PSA serum levels of 0.20–0.49 ng/mL (n = 57), 0.50–0.99 ng/mL (n = 45); 1.00–1.99 ng/mL (n = 59), and ≥2.00 ng/mL (n = 91), respectively, if equivocal findings were considered negative for malignancy; and 59%, 67%, 81%, and 88%, respectively, if equivocal findings were considered positive for malignancy.

From the ROC analysis, the best cut-off value of PSA to perform a [^{68}Ga]Ga-PSMA-11 PET/CT in surgical patients (n = 252) was 1 ng/mL (area under curve of 0.59), whether considering equivocal findings as positive or negative for malignancy.

Overall, the mean PSA serum value of patients with negative [^{68}Ga]Ga-PSMA-11 PET/CT was significantly lower than that of the patients with positive examinations, whether considering equivocal findings as positive for malignancy (1.4 vs. 3.8 ng/mL: $p = 0.03$) or negative for malignancy (1.8 vs. 3.9 ng/mL; $p = 0.03$).

A relationship was found between [^{68}Ga]Ga-PSMA-11 PET/CT positivity on bone and initial ISUP grade group. However, the rate of bone foci positive was statistically higher for ISUP 3–5 patients than for ISUP 1–2 patients only when considering equivocal findings as negative for malignancy (14% vs. 25%; $p = 0.03$).

We did not find any relationship between [^{68}Ga]Ga-PSMA-11 PET/CT positivity and the other tested parameters.

Furthermore, we did not identify a difference in [^{68}Ga]Ga-PSMA-11 PET/CT positivity between patients presenting with a first BCR and patients presenting with a second or third episode of BCR ($p = 0.1$).

According to [^{68}Ga]Ga-PSMA-11 PET/CT results (retrospective masked reading results, equivocal findings considered positive for malignancy), 65 patients (22%) had no detectable disease (60 surgeries), 34 patients (12%) presented an isolated focus in the prostate/prostatic lodge (24 surgeries), 132 patients (45%) were categorized oligo-3 (116 surgeries), 142 patients (48%) oligo-4 (125 surgeries), 149 (51%) oligo-5 (130 surgeries), and 46 (16%) had more than five distant malignant foci (37 surgeries).

The dynamic images acquired on the pelvis provided an additional diagnostic information in 6/294 patients (2%), all of whom had surgery, since an abnormal focus in the prostatic lodge was detected on this acquisition but masked by the physiologic urinary uptake in the bladder on the 60 to 90 min after injection PET acquisition.

The overall and per anatomical site diagnostic performances of [^{68}Ga]Ga-PSMA-11 PET/CT based on the 176 patients on whom a SOT was feasible are presented in Table 3.

Table 3. [^{68}Ga]Ga-PSMA-11 PET/CT performances in prostate cancer patients investigated due to a biochemical recurrence (irrespective of total prostate-specific antigen serum values). Results with equivocal findings considered positive for malignancy and with equivocal results negative for malignancy are both presented. Patient-based and region-based analyses with the number of cases on which the standard of truth was feasible.

	Se	Sp	Acc
Overall ($n = 176$)			
Equivocal positive for malignancy	73%	57%	71%
Equivocal negative for malignancy	70%	70%	70%
Prostate/prostatic lodge ($n = 121$)			
Equivocal positive for malignancy	76%	91%	85%
Equivocal negative for malignancy	69%	94%	87%
Pelvic lymph nodes ($n = 116$)			
Equivocal positive for malignancy	90%	98%	94%
Equivocal negative for malignancy	88%	100%	95%
Paraaortic lymph nodes ($n = 103$)			
Equivocal positive for malignancy	100%	99%	99%
Equivocal negative for malignancy	100%	100%	100%
Lymph nodes above the diaphragm ($n = 101$)			
Equivocal positive for malignancy	78%	97%	95%
Equivocal negative for malignancy	56%	98%	94%
Bone ($n = 109$)			
Equivocal positive for malignancy	88%	92%	91%
Equivocal negative for malignancy	88%	95%	94%
Viscera ($n = 101$)			
Equivocal positive for malignancy	78%	97%	95%
Equivocal negative for malignancy	56%	98%	94%

We found no differences in detection rates or in accuracies between the two scanners.

3.3. Impact of [^{68}Ga]Ga-PSMA-11 PET/CT on BCR Patient Management and Therapy Efficacy Prediction

The impact of [^{68}Ga]Ga-PSMA-11 PET/CT was assessable for the 278 patients for whom follow-up data were available. Patient disease management changed in 189/278 (68%) cases, and [^{68}Ga]Ga-PSMA-11 PET/CT impacted this change in 162 cases (86%), 21 being minor changes (Table 4).

Table 4. Impact on patient management. Management scheduled before and in view of [68Ga]Ga-PSMA-11 PET/CT, overall, for surgical patients and for non-operated patients. Changes induced by [68Ga]Ga-PSMA-11 PET/CT results are highlighted in bold and underlined.

		Scheduled (n = 278)			
Overall Management		Undecided n = 82	Treatment with Curative Intent n = 50	ADT n = 23	Surveillance n = 123
Indicated after [68Ga]Ga-PSMA-11 PET/CT	Treatment with curative intent n = 140	48 + 3 #	18 + 19 *	4	45 + 3
	ADT n = 68	15 + 3 ##	11 + 1	1 + 17 **	18 + 2
	Surveillance n = 70	13	1	1	2 + 53 ***

		Scheduled (n = 240)			
Surgical Patients Management		Undecided n = 71	Treatment with Curative Intent n = 48	ADT n = 18	Surveillance n = 103
Indicated after [68Ga]Ga-PSMA-11 PET/CT	Treatment with curative intent n = 127	41 + 3	18 + 19	3	40 + 3
	ADT n = 57	14 + 3	10 + 1	1 + 14	13 + 1
	Surveillance n = 56	10	0	0	46

		Scheduled (n = 38)			
Non-Operated Patients Management		Undecided n = 11	Treatment with Curative Intent n = 2	ADT n = 5	Surveillance n = 20
Indicated after [68Ga]Ga-PSMA-11 PET/CT	Treatment with curative intent n = 13	7	0	1	5
	ADT n = 11	1	1	3	5 + 1
	Surveillance n = 14	3	1	1	2 + 7

ADT: androgen-deprivation therapy; #: in 3 cases imaging was negative and patients were treated by salvage radiation therapy according to guidelines; ##: in 3 cases imaging was negative and patients were treated by ADT because of a rapid PSA doubling time (less than 3 months); *: in 18 cases, imaging triggered a modification of radiotherapy fields by finding lymph node metastases (n = 17) or an isolated bone metastasis (n = 1); **: in 1 case, imaging found multiple bone metastasis and triggered a modification of the planned ADT regimen (switch for a second-generation ADT); ***: in 2 cases, imaging triggered a biopsy of an abnormal uptake, for which pathology demonstrated normal prostatic tissue (false positive of the imaging on the prostate) and a solitary fibrous tumor (false positive of imaging on viscera).

This impact was statistically higher when PSA was greater than 1 ng/mL in surgical patients (97/145 = 67% vs. 43/95 = 45%; $p < 0.001$), but only when PSA was superior to 2 ng/mL in non-surgical patients (21/31 = 68% vs. 1/7 = 14%). Overall, the impact was statistically higher when PSA doubling time was less than one year (113/176 = 65% vs. 46/95 = 48%: $p = 0.01$), and tended to be higher for ISUP 3–5 patients (83/131 = 63% vs. 77/145 = 53%; $p = 0.08$).

The therapy efficacy was assessable for 257 patients (87%) for whom sufficient follow-up data were available. Treatments with curative intent consisted in 107 radiation therapies focused on abnormalities detected by [^{68}Ga]Ga-PSMA-11 PET/CT, 12 salvage lymphadenectomies, two focal irreversible electroporations, one cryosurgery, and one left orchidectomy (isolated CaP metastasis of the testis histologically proven). Eleven patients with negative [^{68}Ga]Ga-PSMA-11 PET/CT were treated by radiation therapy of the prostatic lodge in accordance with the guidelines for salvage radiation therapy after prostatectomy [20]. ADT was started for 67 patients among whom three benefited from novel androgen axis drugs (such abiraterone or enzalutamide). Surveillance was finally decided for 75 patients, while it was only indicated in 70 after multidisciplinary meeting, as five patients, in whom PSA presented a long doubling time, refused the offered treatment.

The treatment carried out on the patient was considered effective according to the defined criteria in 78% (200/257) of patients overall, 89% (138/155) when guided by [^{68}Ga]Ga-PSMA-11 PET/CT versus 61% (62/102) when not guided by [^{68}Ga]Ga-PSMA-11 PET/CT ($p < 0.001$). It was considered effective in 84% (112/133) when a treatment with curative intent was performed, 94% (60/64) when ADT was started, and 47% (28/60) when surveillance was decided. The treatment carried out on the patient was considered effective in 85%, 86%, and 87% when a treatment with curative intent was performed in oligo-3, oligo-4, and oligo-5 patients, respectively.

3.4. Agreement between Routine Unmasked and Retrospective Masked [^{68}Ga]Ga-PSMA-11 PET/CT Readings

The agreements between routine unmasked and retrospective masked readings are presented in Table 2. Overall agreement was strong (k = 0.68). The agreement was moderate for the prostate/prostatic lodge (k = 0.54) and viscera (k = 0.56), strong for lymph nodes above the diaphragm (k = 0.73) and bone (k = 0.74), and very strong for pelvic lymph nodes (k = 0.90) and paraaortic lymph nodes (k = 0.84). Both readings were similar in 92% (272/294) of cases. In the 22 cases for which readings were different, the findings on masked readings were considered more accurate according to the follow-up in 6% (17/294: 10 on prostate/prostatic lodge, 2two on bone, two on lymph nodes above the diaphragm, one on pelvic lymph nodes, one on paraaortic lymph nodes, and one on lung) versus 2% (5/294: three on the prostate/prostatic lodge and two on pelvic lymph nodes) for routine reading findings.

4. Discussion

4.1. [^{68}Ga]Ga-PSMA-11 PET/CT Performances

PSMA-11 labelled with gallium-68 a is the most studied ligand for imaging of PCa, especially patients with BCR, for whom detection rates were largely reported, but this study is one of the largest series presenting [^{68}Ga]Ga-PSMA-11 PET/CT performances (i.e., sensitivity, specificity, and accuracy), based on a composite SOT. In the present study, [^{68}Ga]Ga-PSMA-11 PET/CT demonstrated an overall positivity rate of 78% for restaging PCa patients with BCR, which is consistent with that of 76% reported in a recent meta-analysis [21]. Our per anatomical site positivity rates were also in agreement with an updated version of this meta-analysis [13], except for the extrapelvic lymph nodes positivity rate as we chose to analyze this location in two separate areas (paraortic lymph nodes and lymph nodes above the diaphragm). We decided to do this because patient management may significantly differ for PCa recurrence between these locations. We also confirmed the relationship between [^{68}Ga]Ga-PSMA-11 PET/CT positivity rate and PSA serum levels that has already been reported several times [22–24], observing positivity rates in surgical patients comparable to those reported by Perera et al. [13].

In our study, we established a composite SOT for PCa on 176 patients overall and at least a hundred patients for each anatomical area. As a histological confirmation of the detected abnormalities was only available in 27 patients, the SOT was primarily based on clinical follow-up including the PSA response to targeted treatment for imaging findings during a median follow-up period of 17 months. Using those criteria, we found overall sensitivity and specificity of both 70%, lower than that of 86% reported by Perera et al. [21], which were only based on histopathologic correlation with ^{68}Ga-PSMA PET/CT abnormal findings and, therefore, did not take into account anatomical areas without abnormal PSMA uptake. [^{68}Ga]Ga-PSMA-11 PET/CT accuracies for anatomical areas were above 90%, except for the prostate/prostatic lodge area, for which image reading was impaired by the urinary physiologic uptake in the bladder. To improve reading in this location, we performed dynamic PET acquisition over the pelvis immediately after [^{68}Ga]Ga-PSMA-11 injection as it was published that this imaging sequence increases the detection rate of local recurrence [25]. In our study these early images provided additional information in 2% (6/294) of patients, all of whom received surgery, by detecting pathological foci in prostate lodge that were then masked by the urinary physiologic uptake in the bladder during the 60 min post-IV PET images. Considering these results, adding dynamic PET images to [^{68}Ga]Ga-PSMA-11 PET/CT imaging protocol should be considered for surgical patients.

4.2. [^{68}Ga]Ga-PSMA-11 PET/CT's Impact on PCa Management and Therapy Efficacy Prediction

We found that [^{68}Ga]Ga-PSMA-11 PET/CT impacted patient disease management in 58% of cases, resulting in an increased proportion of treatments with curative intent. These findings are consistent with the 54% reported by a recent meta-analysis [14]. Moreover, we found that [^{68}Ga]Ga-PSMA-11 PET/CT's impact was significantly higher when PSA was greater than 1 ng/mL in surgical patients but only when PSA was greater than 2 ng/mL in non-surgical patients. All patients considered, the impact was significantly higher when PSA doubling time was less than one year and tended to be higher for ISUP 3–5 patients.

In this study, the treatment carried out on patient was considered effective in 91% of cases when guided by [^{68}Ga]Ga-PSMA-11 PET/CT versus 40% when not guided by [^{68}Ga]Ga-PSMA-11 PET/CT ($p < 0.001$). We previously reported similar findings in a small cohort of 30 castration-resistant PCa patients restaged by [^{68}Ga]Ga-PSMA-11 PET/CT [26].

4.3. [^{68}Ga]Ga-PSMA-11 PET/CT Reading Agreement

In our study we analyzed [^{68}Ga]Ga-PSMA-11 PET/CTs according to a three-point scale to introduce diagnostic uncertainty in imaging reading, as pitfalls and equivocal findings are inseparable from any diagnostic procedure. Thus, we found equivocal results in approximately 10% for both routine unmasked and retrospective masked readings, which is relatively low. We also noted that the proportion of equivocal findings decreased when PSA increased. Systematic approaches to the interpretation of PSMA imaging studies, using a five-point scale, were recently proposed to classify imaging findings and better reflects the likelihood of the presence of PCa [27,28]. However, we could not use those approaches for the routine readings that were already performed and chose not to use it for the retrospective readings as we wanted to ensure that evaluation between readings would be comparable. Furthermore, we could not use the 2021 EANM standardized guidelines for PSMA-PET as we conducted this research in 2019–2020.

We found an overall strong agreement between [^{68}Ga]Ga-PSMA-11 PET/CT routine unmasked and retrospective masked readings ($k = 0.68$), which is comparable to a previously report on a more heterogeneous series of 50 patients ($k = 0.62$) [29].

Agreement was moderate ($k = 0.54$) for the prostate/prostatic lodge, slightly lower than previously reported ($k = 0.62$) [29], likely because our series reports a large proportion of surgical patients. The reading in the prostate/prostate bed is impaired by the physiological uptake of the urine in the bladder, especially in operated patients. Dynamic images aim to improve the reading in the prostate/prostatic lodge. The unmasked reader may look less attentively at the dynamic images because he relies on the clinical data, such as mpMRI or PSA serum value or doubling time that may influence its diagnosis (more equivocal findings in the prostate/prostatic lodge). The experienced masked reader relies only on the dynamic images to improve its reading in the prostate bed.

We found that agreement was strong to very strong for all lymph nodes areas (k = 0.73 for lymph nodes above the diaphragm, k = 0.84 for paraaortic lymph nodes, and k = 0.90 for pelvic lymph nodes), similar to previously reported values of k = 0.74, considering all lymph node areas [29]. However, we distinguished invasion of the lymph nodes between pelvic and paraaortic regions and above the diaphragm, as we assumed that therapeutic management for involved lymph nodes differed between these areas. We found a strong agreement in readings for bone (k = 0.74), which is comparable to that previously reported [29]. Finally, we found a moderate agreement in readings for viscera (k = 0.56), likely due to the low incidence of visceral metastases in patients with BCR [30] and to atypical locations (like penile, testis, intramedullary spinal cord) or challenging location for imaging (30% of visceral lesions were peritoneal carcinomatosis) [31,32].

We assumed that the disagreements between readings might be explain by the higher number of equivocal foci found by the routine unmasked reading. Indeed, the knowledge of clinical parameters such as PSA serum level or doubling time may influence the reading.

4.4. Limitations

This study has several limitations. The primary one, shared by most imaging studies addressing the search for metastatic disease, is the lack of sufficient histological proof for most of the suspected metastases, which were primarily characterized based on follow-up data. Indeed, obtaining a histopathological evidence for asymptomatic and possibly benign lesions, or locations that are negative on imaging, is ethically questionable and hardly feasible in practice. We chose to base our SOT on the variation of PSA, excluding patients with a change in their ADT regimen after [^{68}Ga]Ga-PSMA-11 PET/CT and on histological findings, if available. In this work, a SOT was feasible for 60% of patients, with a 17-month median duration of follow-up, which allowed us to calculate overall performances of [^{68}Ga]Ga-PSMA-11 PET/CT, as well as performances per anatomical areas. Furthermore, we were able to determine from the ROC analysis that the best cut-off values of PSA level to perform a [^{68}Ga]Ga-PSMA-11 PET/CT in operated patients was 1 ng/mL. These findings need to be confirmed by other large studies, as most available data report ^{68}Ga-PSMA PET/CT detection rates [30,33,34].

The second major limitation of this work was its retrospective design. However, this study is one of the larger homogenous cohorts of BCR patients which has evaluated imaging performance based on a composite SOT.

In this work, we reported [^{68}Ga]Ga-PSMA-11 PET/CT performance according to PSA serum levels which were not evaluated the day of the PET, in the same laboratory, but corresponded to the value of the closet assays to the examinations. Thus, we assume that we may have overestimated detection rates, especially for low PSA levels.

In this work, [^{68}Ga]Ga-PSMA-11 PET/CT acquisition time varied from 60 to 90 min after [^{68}Ga]Ga-PSMA-11 injection, which may be important as increased lesion detection was reported with delayed imaging times up to four hours [35]. However, acquisition times were within the acceptable range of 50 to 100 min that is recommended by the current guidelines for ^{68}Ga-PSMA PET/CT [35]. Thus, we assume that the limited variation in acquisition times in our study did not significantly affect the results.

Finally, another notable feature is that all patients in our study who underwent [^{68}Ga]Ga-PSMA-11 PET/CT had previously show no sign of distant metastases at [^{18}F]fluorocholine PET/CT and were referred for this reason to [^{68}Ga]Ga-PSMA-11 PET/CT based on the French regulation for compassionate use of pharmaceutical, which is authorized on an individual basis by the National Medicine Agency. Therefore, our study population may not reflect that of other international studies, and results may differ due to this selection method. However, we found comparable positivity rates than that reported on larger series and metanalyses [13,36].

Because of these limitations, the promising performances and impact rate on patient disease management of [^{68}Ga]Ga-PSMA-11 PET/CT need to be confirmed in a larger prospective study.

5. Conclusions

In conclusion, [^{68}Ga]Ga-PSMA-11 PET/CT demonstrated reliable performance in locating recurrence sites of prostate cancer and motivated disease management changes in almost two out of three patients. Those performances and impact rates were better when PSA serum level were above 1 ng/mL.

Comparison between routine unmasked and retrospective masked readings demonstrated that this imaging modality is highly reproducible, especially for the detection of pelvic and paraaortic lymph nodes.

The use of PSMA radioligands with PET/CT should be considered, when available, as a first line imaging modality for biochemical recurrence of prostate cancer.

Author Contributions: Conception and design: M.G.; acquisition of data: M.G., A.F., L.L. and G.C.; analysis and interpretation of data: M.G., A.F. and L.L.; drafting the manuscript: M.G., A.F., Y.B., A.d.l.T., O.C., G.C. and G.F.; critical revision of the manuscript: M.G., Y.B., A.d.l.T., O.C. and G.F.; statistical analysis: M.G.; technical and material support: M.G. and T.R. All authors have read and agreed to the published version of the manuscript.

Funding: This research received no external funding.

Institutional Review Board Statement: According to French regulation, the approval of an institutional review board was not necessary for performing this retrospective analysis of already available data.

Informed Consent Statement: According to French regulation, the written consent was not necessary for performing this retrospective analysis of already available data.

Data Availability Statement: The data that support the findings of this study are available from the corresponding author, M.G., upon reasonable request.

Acknowledgments: The authors are grateful to the following physicians for their confidence in referring their patients: Chauveinc, Radiothérapie—Institut de radiothérapie Hartmann, Levallois-Perret; Simon, Radiothérapie—Hôpital de la Pitié-Salpétrière, Paris; Quero, Radiothérapie—Hôpital Saint-Louis, Paris; Graff Caillaud, Radiothérapie—IUCT Oncopole, Toulouse.

Conflicts of Interest: The authors declare no conflict of interest.

References

1. Siegel, R.L.; Miller, K.D.; Jemal, A. Cancer Statistics, 2020. *CA Cancer J. Clin.* **2020**, *70*, 7–30. [CrossRef] [PubMed]
2. Zietman, A.L.; DeSilvio, M.L.; Slater, J.D.; Rossi, C.J.; Miller, D.W.; Adams, J.A.; Shipley, W.U. Comparison of Conventional-Dose vs High-Dose Conformal Radiation Therapy in Clinically Localized Adenocarcinoma of the Prostate: A Randomized Controlled Trial. *JAMA* **2005**, *294*, 1233–1239. [CrossRef]
3. Remmers, S.; Verbeek, J.F.M.; Nieboer, D.; van der Kwast, T.; Roobol, M.J. Predicting Biochemical Recurrence and Prostate Cancer-Specific Mortality after Radical Prostatectomy: Comparison of Six Prediction Models in a Cohort of Patients with Screening- and Clinically Detected Prostate Cancer. *BJU Int.* **2019**. [CrossRef] [PubMed]
4. Cornford, P.; Bellmunt, J.; Bolla, M.; Briers, E.; Santis, M.D.; Gross, T.; Henry, A.M.; Joniau, S.; Lam, T.B.; Mason, M.D.; et al. EAU-ESTRO-SIOG Guidelines on Prostate Cancer. Part II: Treatment of Relapsing, Metastatic, and Castration-Resistant Prostate Cancer. *Eur. Urol.* **2017**, *71*, 630–642. [CrossRef]
5. Sanda, M.G.; Chen, R.C.; Crispino, T.; Freedland, S.; Greene, K.; Klotz, L.H.; Makarov, D.V.; Reston, J.; Rodrigues, G.; Sandler, H.M.; et al. Clinically Localized Prostate Cancer: AUA/ASTRO/SUO Guideline. Available online: https://www.astro.org/Patient-Care-and-Research/Clinical-Practice-Statements/ASTRO-39;s-evidence-based-guideline-on-clinically (accessed on 29 March 2021).
6. Artibani, W.; Porcaro, A.B.; De Marco, V.; Cerruto, M.A.; Siracusano, S. Management of Biochemical Recurrence after Primary Curative Treatment for Prostate Cancer: A Review. *Urol. Int.* **2018**, *100*, 251–262. [CrossRef] [PubMed]

7. Ost, P.; Jereczek-Fossa, B.A.; As, N.V.; Zilli, T.; Muacevic, A.; Olivier, K.; Henderson, D.; Casamassima, F.; Orecchia, R.; Surgo, A.; et al. Progression-Free Survival Following Stereotactic Body Radiotherapy for Oligometastatic Prostate Cancer Treatment-Naive Recurrence: A Multi-Institutional Analysis. *Eur. Urol.* **2016**, *69*, 9–12. [CrossRef] [PubMed]
8. Kane, C.J.; Amling, C.L.; Johnstone, P.A.S.; Pak, N.; Lance, R.S.; Thrasher, J.B.; Foley, J.P.; Riffenburgh, R.H.; Moul, J.W. Limited Value of Bone Scintigraphy and Computed Tomography in Assessing Biochemical Failure after Radical Prostatectomy. *Urology* **2003**, *61*, 607–611. [CrossRef]
9. Graute, V.; Jansen, N.; Ubleis, C.; Seitz, M.; Hartenbach, M.; Scherr, M.K.; Thieme, S.; Cumming, P.; Klanke, K.; Tiling, R.; et al. Relationship between PSA Kinetics and [18F]Fluorocholine PET/CT Detection Rates of Recurrence in Patients with Prostate Cancer after Total Prostatectomy. *Eur. J. Nucl. Med. Mol. Imaging* **2012**, *39*, 271–282. [CrossRef]
10. Emmett, L.; Willowson, K.; Violet, J.; Shin, J.; Blanksby, A.; Lee, J. Lutetium 177 PSMA Radionuclide Therapy for Men with Prostate Cancer: A Review of the Current Literature and Discussion of Practical Aspects of Therapy. *J. Med. Radiat. Sci.* **2017**, *64*, 52–60. [CrossRef]
11. Kasperzyk, J.L.; Finn, S.P.; Flavin, R.; Fiorentino, M.; Lis, R.; Hendrickson, W.K.; Clinton, S.K.; Sesso, H.D.; Giovannucci, E.L.; Stampfer, M.J.; et al. Prostate-Specific Membrane Antigen Protein Expression in Tumor Tissue and Risk of Lethal Prostate Cancer. *Cancer Epidemiol. Biomark. Prev.* **2013**, *22*, 2354–2363. [CrossRef]
12. Gauthé, M.; Belissant, O.; Girard, A.; Zhang Yin, J.; Ohnona, J.; Cottereau, A.-S.; Nataf, V.; Balogova, S.; Pontvert, D.; Lebret, T.; et al. TEP/TDM et récidive biologique d'adénocarcinome prostatique: Apport du 68Ga-PSMA-11 lorsque la 18F-fluorocholine n'est pas contributive. *Prog. Urol.* **2017**, *27*, 474–481. [CrossRef] [PubMed]
13. Perera, M.; Papa, N.; Roberts, M.; Williams, M.; Udovicich, C.; Vela, I.; Christidis, D.; Bolton, D.; Hofman, M.S.; Lawrentschuk, N.; et al. Gallium-68 Prostate-Specific Membrane Antigen Positron Emission Tomography in Advanced Prostate Cancer-Updated Diagnostic Utility, Sensitivity, Specificity, and Distribution of Prostate-Specific Membrane Antigen-Avid Lesions: A Systematic Review and Meta-Analysis. *Eur. Urol.* **2019**. [CrossRef]
14. Han, S.; Woo, S.; Kim, Y.J.; Suh, C.H. Impact of 68Ga-PSMA PET on the Management of Patients with Prostate Cancer: A Systematic Review and Meta-Analysis. *Eur. Urol.* **2018**, *74*, 179–190. [CrossRef]
15. Mottet, N.; Bellmunt, J.; Bolla, M.; Briers, E.; Cumberbatch, M.G.; De Santis, M.; Fossati, N.; Gross, T.; Henry, A.M.; Joniau, S.; et al. EAU-ESTRO-SIOG Guidelines on Prostate Cancer. Part 1: Screening, Diagnosis, and Local Treatment with Curative Intent. *Eur. Urol.* **2017**, *71*, 618–629. [CrossRef]
16. Lowe, S.; Spottiswoode, B.; Declerck, J.; Sullivan, K.; Sharif, M.S.; Wong, W.-L.; Sanghera, B. Positron Emission Tomography PET/CT Harmonisation Study of Different Clinical PET/CT Scanners Using Commercially Available Software. *BJR Open* **2020**, *2*. [CrossRef] [PubMed]
17. Rauscher, I.; Maurer, T.; Beer, A.J.; Graner, F.-P.; Haller, B.; Weirich, G.; Doherty, A.; Gschwend, J.E.; Schwaiger, M.; Eiber, M. Value of 68Ga-PSMA HBED-CC PET for the Assessment of Lymph Node Metastases in Prostate Cancer Patients with Biochemical Recurrence: Comparison with Histopathology after Salvage Lymphadenectomy. *J. Nucl. Med.* **2016**, *57*, 1713–1719. [CrossRef]
18. Foster, C.C.; Weichselbaum, R.R.; Pitroda, S.P. Oligometastatic Prostate Cancer: Reality or Figment of Imagination? *Cancer* **2019**, *125*, 340–352. [CrossRef]
19. Scher, H.I.; Eisenberger, M.; D'Amico, A.V.; Halabi, S.; Small, E.J.; Morris, M.; Kattan, M.W.; Roach, M.; Kantoff, P.; Pienta, K.J.; et al. Eligibility and Outcomes Reporting Guidelines for Clinical Trials for Patients in the State of a Rising Prostate-Specific Antigen: Recommendations from the Prostate-Specific Antigen Working Group. *J. Clin. Oncol.* **2004**, *22*, 537–556. [CrossRef]
20. Thompson, I.M.; Valicenti, R.; Albertsen, P.C.; Goldenberg, S.L.; Hahn, C.A.; Klein, E.A.; Michalski, J.; Iii, M.R.; Sartor, O.; Wolf, J.S., Jr.; et al. Adjuvant and Salvage Radiotherapy after Prostatectomy: ASTRO/AUA Guideline. *J. Urol.* **2013**, *190*, 441–449. [CrossRef]
21. Perera, M.; Papa, N.; Christidis, D.; Wetherell, D.; Hofman, M.S.; Murphy, D.G.; Bolton, D.; Lawrentschuk, N. Sensitivity, Specificity, and Predictors of Positive 68Ga–Prostate-Specific Membrane Antigen Positron Emission Tomography in Advanced Prostate Cancer: A Systematic Review and Meta-Analysis. *Eur. Urol.* **2016**, *70*, 926–937. [CrossRef] [PubMed]
22. Verburg, F.A.; Pfister, D.; Heidenreich, A.; Vogg, A.; Drude, N.I.; Vöö, S.; Mottaghy, F.M.; Behrendt, F.F. Extent of Disease in Recurrent Prostate Cancer Determined by [^{68}Ga]PSMA-HBED-CC PET/CT in Relation to PSA Levels, PSA Doubling Time and Gleason Score. *Eur. J. Nucl. Med. Mol. Imaging* **2016**, *43*, 397–403. [CrossRef]
23. Ceci, F.; Uprimny, C.; Nilica, B.; Geraldo, L.; Kendler, D.; Kroiss, A.; Bektic, J.; Horninger, W.; Lukas, P.; Decristoforo, C.; et al. 68Ga-PSMA PET/CT for Restaging Recurrent Prostate Cancer: Which Factors Are Associated with PET/CT Detection Rate? *Eur. J. Nucl. Med. Mol. Imaging* **2015**, *42*, 1284–1294. [CrossRef]
24. Afshar-Oromieh, A.; Holland-Letz, T.; Giesel, F.L.; Kratochwil, C.; Mier, W.; Haufe, S.; Debus, N.; Eder, M.; Eisenhut, M.; Schäfer, M.; et al. Diagnostic Performance of 68Ga-PSMA-11 (HBED-CC) PET/CT in Patients with Recurrent Prostate Cancer: Evaluation in 1007 Patients. *Eur. J. Nucl. Med. Mol. Imaging* **2017**, *44*, 1258–1268. [CrossRef]
25. Uprimny, C.; Kroiss, A.S.; Decristoforo, C.; Fritz, J.; Warwitz, B.; Scarpa, L.; Roig, L.G.; Kendler, D.; von Guggenberg, E.; Bektic, J.; et al. Early Dynamic Imaging in 68Ga- PSMA-11 PET/CT Allows Discrimination of Urinary Bladder Activity and Prostate Cancer Lesions. *Eur. J. Nucl. Med. Mol. Imaging* **2016**, 1–11. [CrossRef]
26. Fourquet, A.; Aveline, C.; Cussenot, O.; Créhange, G.; Montravers, F.; Talbot, J.-N.; Gauthé, M. 68Ga-PSMA-11 PET/CT in Restaging Castration-Resistant Nonmetastatic Prostate Cancer: Detection Rate, Impact on Patients' Disease Management and Adequacy of Impact. *Sci. Rep.* **2020**, *10*, 2104. [CrossRef]

27. Rowe, S.P.; Pienta, K.J.; Pomper, M.G.; Gorin, M.A. PSMA-RADS Version 1.0: A Step Towards Standardizing the Interpretation and Reporting of PSMA-Targeted PET Imaging Studies. *Eur. Urol.* **2018**, *73*, 485–487. [CrossRef]
28. Ceci, F.; Oprea-Lager, D.E.; Emmett, L.; Adam, J.A.; Bomanji, J.; Czernin, J.; Eiber, M.; Haberkorn, U.; Hofman, M.S.; Hope, T.A.; et al. E-PSMA: The EANM Standardized Reporting Guidelines v1.0 for PSMA-PET. *Eur. J. Nucl. Med. Mol. Imaging* **2021**. [CrossRef] [PubMed]
29. Fendler, W.P.; Calais, J.; Allen-Auerbach, M.; Bluemel, C.; Eberhardt, N.; Emmett, L.; Gupta, P.; Hartenbach, M.; Hope, T.A.; Okamoto, S.; et al. 68Ga-PSMA-11 PET/CT Interobserver Agreement for Prostate Cancer Assessments: An International Multicenter Prospective Study. *J. Nucl. Med.* **2017**, *58*, 1617–1623. [CrossRef] [PubMed]
30. Eiber, M.; Maurer, T.; Souvatzoglou, M.; Beer, A.J.; Ruffani, A.; Haller, B.; Graner, F.-P.; Kübler, H.; Haberhorn, U.; Eisenhut, M.; et al. Evaluation of Hybrid 68Ga-PSMA Ligand PET/CT in 248 Patients with Biochemical Recurrence after Radical Prostatectomy. *J. Nucl. Med.* **2015**, *56*, 668–674. [CrossRef] [PubMed]
31. González-Moreno, S.; González-Bayón, L.; Ortega-Pérez, G.; González-Hernando, C. Imaging of Peritoneal Carcinomatosis. *Cancer J.* **2009**, *15*, 184–189. [CrossRef] [PubMed]
32. Chang, M.-C.; Chen, J.-H.; Liang, J.-A.; Huang, W.-S.; Cheng, K.-Y.; Kao, C.-H. PET or PET/CT for Detection of Peritoneal Carcinomatosis: A Meta-Analysis. *Clin. Nucl. Med.* **2013**, *38*, 623–629. [CrossRef] [PubMed]
33. Bluemel, C.; Linke, F.; Herrmann, K.; Simunovic, I.; Eiber, M.; Kestler, C.; Buck, A.K.; Schirbel, A.; Bley, T.A.; Wester, H.-J.; et al. Impact of 68 Ga-PSMA PET/CT on Salvage Radiotherapy Planning in Patients with Prostate Cancer and Persisting PSA Values or Biochemical Relapse after Prostatectomy. *EJNMMI Res.* **2016**, *6*, 78. [CrossRef] [PubMed]
34. Grubmüller, B.; Baltzer, P.; D'Andrea, D.; Korn, S.; Haug, A.R.; Hacker, M.; Grubmüller, K.H.; Goldner, G.M.; Wadsak, W.; Pfaff, S.; et al. 68Ga-PSMA 11 Ligand PET Imaging in Patients with Biochemical Recurrence after Radical Prostatectomy—Diagnostic Performance and Impact on Therapeutic Decision-Making. *Eur. J. Nucl. Med. Mol. Imaging* **2018**, *45*, 235–242. [CrossRef] [PubMed]
35. Fendler, W.P.; Eiber, M.; Beheshti, M.; Bomanji, J.; Ceci, F.; Cho, S.; Giesel, F.; Haberkorn, U.; Hope, T.A.; Kopka, K.; et al. 68Ga-PSMA PET/CT: Joint EANM and SNMMI Procedure Guideline for Prostate Cancer Imaging: Version 1.0. *Eur. J. Nucl. Med. Mol. Imaging* **2017**, *44*, 1014–1024. [CrossRef]
36. Afshar-Oromieh, A.; da Cunha, M.L.; Wagner, J.; Haberkorn, U.; Debus, N.; Weber, W.; Eiber, M.; Holland-Letz, T.; Rauscher, I. Performance of [^{68}Ga]Ga-PSMA-11 PET/CT in Patients with Recurrent Prostate Cancer after Prostatectomy—A Multi-Centre Evaluation of 2533 Patients. *Eur. J. Nucl. Med. Mol. Imaging* **2021**. [CrossRef]

Article

Treatment Pattern and Outcomes with Systemic Therapy in Men with Metastatic Prostate Cancer in the Real-World Patients in the United States

Umang Swami [1,†], Jennifer Anne Sinnott [1,2,3,4,†], Benjamin Haaland [1,5], Nicolas Sayegh [1], Taylor Ryan McFarland [1], Nishita Tripathi [1], Benjamin L. Maughan [1], Nityam Rathi [1], Deepika Sirohi [6], Roberto Nussenzveig [1], Manish Kohli [1], Sumanta K. Pal [7] and Neeraj Agarwal [1,*]

[1] Division of Oncology, Department of Internal Medicine, Huntsman Cancer Institute, University of Utah, Salt Lake City, UT 84112, USA; Umang.Swami@hci.utah.edu (U.S.); jsinnott@stat.osu.edu (J.A.S.); ben.haaland@hci.utah.edu (B.H.); Nicolas.Sayegh@hci.utah.edu (N.S.); Taylor.McFarland@hci.utah.edu (T.R.M.); Nishita.Tripathi@hci.utah.edu (N.T.); Benjamin.Maughan@hci.utah.edu (B.L.M.); Nityam.Rathi@hci.utah.edu (N.R.); Roberto.Nussenzveig@hci.utah.edu (R.N.); manish.kohli@hci.utah.edu (M.K.)
[2] Department of Internal Medicine, University of Utah, Salt Lake City, UT 84132, USA
[3] Department of Pediatrics, University of Utah, Salt Lake City, UT 84108, USA
[4] Department of Statistics, The Ohio State University, Columbus, OH 43210, USA
[5] Department of Population Health Sciences, University of Utah, Salt Lake City, UT 84108, USA
[6] Department of Pathology, University of Utah and ARUP Laboratories, Salt Lake City, UT 84112, USA; Deepika.Sirohi@hsc.utah.edu
[7] Department of Medical Oncology & Experimental Therapeutics, City of Hope Comprehensive Cancer Center, Duarte, CA 91010, USA; spal@coh.org
* Correspondence: Neeraj.agarwal@hci.utah.edu; Tel.: +1-801-213-5658
† These authors contributed equally to this paper as first authors.

Simple Summary: Novel hormonal therapies (such as abiraterone and enzalutamide) and docetaxel are approved treatments for metastatic prostate cancer. Upfront use of these agents has been shown to improve overall survival. However, we do not know the real-world treatment patterns of these agents or the comparative effectiveness of these agents after treatment with a prior novel hormonal therapy in patients with metastatic prostate cancer. In this large study, we found that most patients with metastatic prostate cancer received only androgen deprivation therapy as upfront therapy without novel hormonal therapies or docetaxel. In patients treated with one novel hormonal therapy, alternate novel hormonal therapy was the most common next therapy and was associated with improved overall survival over docetaxel with the caveat of this being a non-randomized comparison. The study's limitations also include its retrospective design.

Abstract: Background: Both novel hormonal therapies and docetaxel are approved for treatment of metastatic prostate cancer (mPC; in castration sensitive or refractory settings). Present knowledge gaps include lack of real-world data on treatment patterns in patients with newly diagnosed mPC, and comparative effectiveness of novel hormonal therapies (NHT) versus docetaxel after treatment with a prior NHT. Methods: Herein we extracted patient-level data from a large real-world database of patients with mPC in United States. Utilization of NHT or docetaxel for mPC and comparative effectiveness of an alternate NHT versus docetaxel after one prior NHT was evaluated. Comparative effectiveness was examined via Cox proportional hazards model with propensity score matching weights. Each patient's propensity for treatment was modeled via random forest based on 22 factors potentially driving treatment selection. Results: The majority of patients (54%) received only androgen deprivation therapy for mPC. In patients treated with an NHT, alternate NHT was the most common next therapy and was associated with improved median overall survival over docetaxel (abiraterone followed by docetaxel vs. enzalutamide (8.7 vs. 15.6 months; adjusted hazards ratio; aHR 1.32; $p = 0.009$; and enzalutamide followed by docetaxel vs. abiraterone (9.7 vs. 13.2 months aHR 1.40; $p = 0.009$). Limitations of the study include retrospective design.

Keywords: abiraterone; enzalutamide; docetaxel; novel hormonal therapies; comparative effectiveness; real-world treatment pattern; metastatic prostate cancer

1. Introduction

Based on significant improvement in overall survival in randomized controlled trials, docetaxel and novel hormonal therapies (NHTs) such as abiraterone or enzalutamide, were first approved for metastatic castration-resistant prostate cancer (mCRPC), and subsequently for men with metastatic castration-sensitive prostate cancer (mCSPC) [1–5]. However, contemporary data of treatment patterns are lacking to demonstrate whether these successes are reaching the real-world patients with metastatic prostate cancer (mPC) in the United States.

Another knowledge gap pertains to treatment selection in the real world after the progression of mPC on one NHT. In this setting, both NHTs and docetaxel are available and considered standard of care [6]. However, the efficacy of docetaxel and NHT after prior NHT use in mPC has not been compared in a randomized controlled trial, and/or in any large real-world patient-level dataset. In clinical trials, the time on alternate NHT after disease progression on one NHT has been 3.6 to 5.7 months [7,8]. Clinical trials have not evaluated the efficacy of docetaxel after one prior NHT but the median duration of docetaxel as the first subsequent therapy in the registration trial of abiraterone in chemotherapy-naïve mCRPC setting (COU-AA-302 trial) was 3.02 months (interquartile range 0.95–5.72) [9].

The current lack of real-world treatment patterns and treatment-related outcomes hamper efforts on improving patients' access to these therapeutic agents as well as designing of clinical trials in these men. Furthermore, a lack of comparative effectiveness data prevents patients from making best-informed treatment decisions. Drug development is hindered, due to lack of information on estimates of progression-free survival and overall survival (OS) on subsequent therapies after treatment with one NHT, as well as due to challenges in terms of selection of the best control arm in the randomized trials in this setting. In this study, our objective was to fill in these knowledge gaps by evaluating the treatment patterns in patients with new mPC and comparing the efficacy of an alternate NHT (abiraterone or enzalutamide) versus docetaxel after prior therapy with only one NHT (abiraterone or enzalutamide) in real-world patients with mPC.

2. Materials and Methods

2.1. Study Population

De-identified patient-level data from patients with mPC were extracted from the Flatiron Health Electronic Health Record database. The Flatiron database consists of nationally representative real-world data from community practices and academic medical centers from 2011 through the present and contains structured and unstructured data curated via technology-enabled abstraction and supplemented with third-party death information. Details of the Flatiron database and its comparison with other real-world databases have been discussed elsewhere [10]. This study was approved by the Institutional Review Board at the University of Utah (IRB_00067518, last approved 9 July 2021).

Eligibility criteria: patients with mPC diagnosed from January 2011 and treated through 30 September 2019; the patient had some evidence of contact within 180 days of diagnosis of metastatic disease to ensure that the patient was actively engaged in care at the data providing institution, and availability of treatment data after diagnosis of metastatic disease. For comparative effectiveness analysis, patients also needed to receive one NHT for mPC followed by an alternate NHT or docetaxel. Patients with systemic therapy with any anti-cancer drug (except ADT) prior to first NHT were excluded. For the purpose of comparative effectiveness analysis use of first NHT was considered first line (1L) therapy. Receipt of a second NHT or docetaxel after one prior NHT was considered second-line (2L) treatment. The follow-up period was until 9 September 2019. All patients were required to have begun 2L therapy at least 6 months before the end of the follow-up period. Patients were excluded if they received any other anti-cancer agent in 1L or 2L, or if they had prior exposure to NHT or docetaxel in the non-metastatic setting.

2.2. Outcome Definitions for Comparative Effective Analysis

OS was defined as the time from the start of 2L therapy to death from any cause. Among men who did not die during follow-up, censoring time was defined as the time of most recent contact in the data, which could have been a therapy end date or a visit, drug episode, or medication order. Time to initiation of 3L therapy or death (TTTTD) was defined as the time from the start of 2L therapy to the start of third-line (3L) therapy, or death. Among men who did not initiate a 3L therapy and died, death within 180 days after the end of 2L therapy was considered an event. Patients who died after the 180-day window were censored at the time of the last contact since we would anticipate that many of these patients were pursuing a 3L treatment at a different institution. Among men not pursuing 3L who did not die, censoring time was defined as the time of most recent contact in the data.

2.3. Statistical Analysis

Receipt of systemic therapy at the time of diagnosis of metastatic prostate cancer, and subsequent therapies and treatment patterns from 2011 to 2019 were summarized descriptively. For a comparative effectiveness study, the analyses were performed separately among 1L abiraterone and 1L enzalutamide patients. Patients treated with 2L NHT and docetaxel were compared at baseline and pre-2L characteristics using Wilcoxon rank-sum tests for quantitative variables and chi-squared tests for categorical variables. The survival outcomes of interest (TTTTD and OS) were compared visually without any adjustment using Kaplan–Meier survival curve estimates, and median survival times were estimated overall and in groups defined by time on 1L therapy. To account for patient characteristics that may affect both treatment selection and outcomes, Cox proportional hazards models with propensity score weighting were used. The probability of receiving docetaxel vs. alternate NHT was estimated using a random forest approach [11], with candidate variables including: Gleason score at initial diagnosis; prostate-specific antigen (PSA) at the diagnosis of metastatic disease and at start of 1L (an indirect measure of disease volume); insurance status, which may influence the selection of 2L therapy (most recent reported payer prior to the 2L start date, including commercial health plan, Medicaid, Medicare, other government, other payer, and patient assistance program); age and year at the time of starting 2L; race; time on ADT-only therapy after metastasis; time on 1L therapy and whether the patient was considered hormone sensitive at time of 1L initiation; the number of diagnoses in the medical records; indicators for diagnosis codes for visceral metastasis, any other specific metastasis, diabetes, heart failure, or neuropathy; Eastern Cooperative Oncology Group (ECOG) performance status in the 3 months prior to 2L start; and PSA, lactate dehydrogenase (LDH), alkaline phosphatase, and hemoglobin in the 3 months prior to starting 2L therapy. In all cases, a separate category was coded for missing values. The propensity scores were used to calculate matching weights, targeting the same estimand as 1:1 matching of treatment groups on combinations of potential confounders [11]. Covariate balance was assessed via weighted tests and by examining standardized mean differences. The weights were then used in Cox proportional hazards models to evaluate the effect of docetaxel compared with other NHT, balanced on potential confounders.

In addition to the main analyses, subgroup analyses were performed based on age, Gleason score, time on 1L therapy, performance status, alkaline phosphatase, PSA, LDH, and hemoglobin at 2L initiation. Propensity scores were recalculated within each subgroup. Additionally, selected post-2L characteristics, including post-2L ECOG and numbers of post-2L therapies, were compared across the groups. All analyses were performed in R version 4.0.2, using packages ggplot2, randomForest, survey, survival, and tableone.

3. Results

A flow diagram illustrating the selection of patients for both the full cohort and 2L analyses is presented in Figure 1.

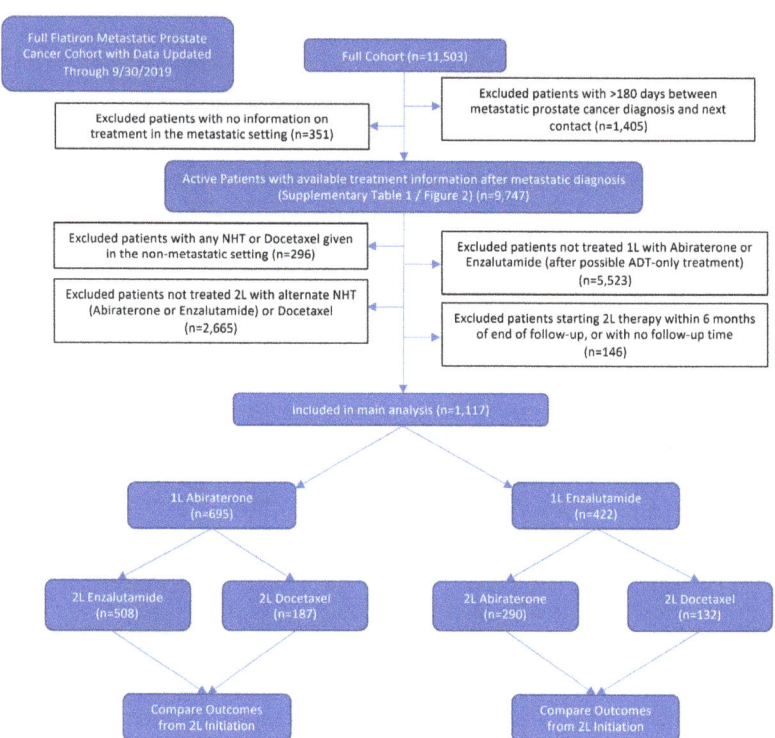

Figure 1. Flow diagram depicting stepwise patient selection.

3.1. Treatment Patterns in Patients with Metastatic Prostate Cancer

Table S1 summarizes patient characteristics for the full cohort of mPC patients (N = 9747 after exclusions from initial N = 11,503) and Figure 2 summarizes treatment patterns after diagnosis of mPC. After the diagnosis of new mPC disease, 54.2% of patients were treated with ADT only. Abiraterone (15.5%) was the most frequently used intensifying agent, followed by docetaxel (13.8%) and enzalutamide (8.3%). The yearly trend of use of therapeutic agents for new diagnosis of mPC is presented in Figure 3 which demonstrates a gradual but encouraging increase in the use of NHTs at the time of onset of metastatic prostate cancer.

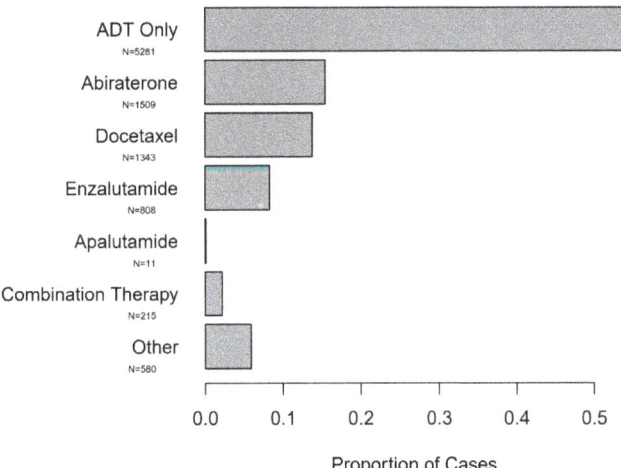

Figure 2. First Treatment After Metastatic Diagnosis ($N = 9747$).

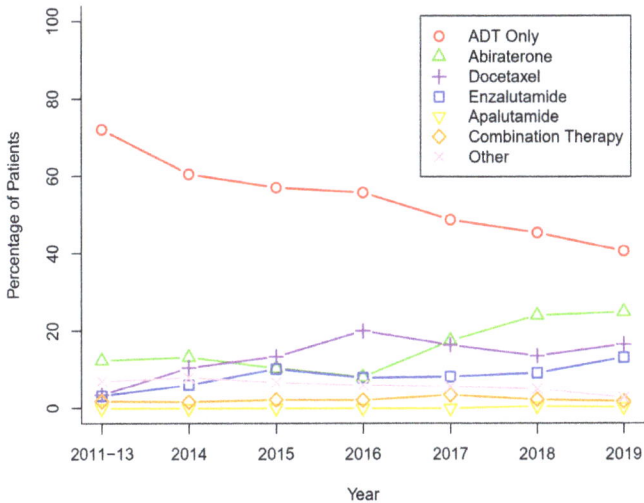

Figure 3. First Treatment After Metastatic Diagnosis, By Year ($N = 9747$).

The predominant subsequent treatments in the entire cohort of these men with mPC consisted of NHTs, although other approved life-prolonging therapies were also utilized. These treatment patterns are summarized in Figures 4 and 5.

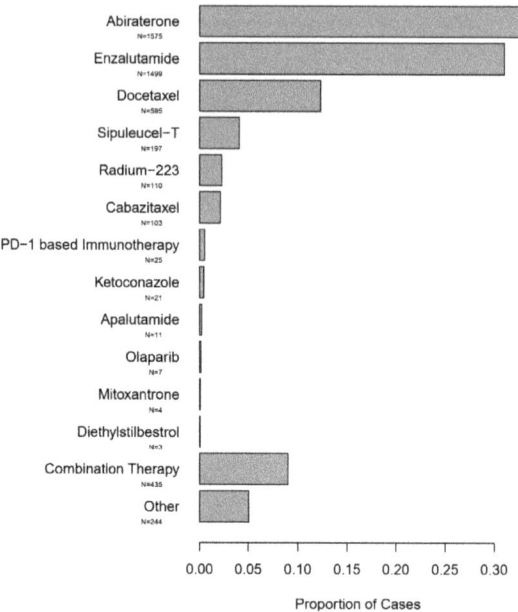

Figure 4. Second Treatment After Metastatic Diagnosis (N = 4829).

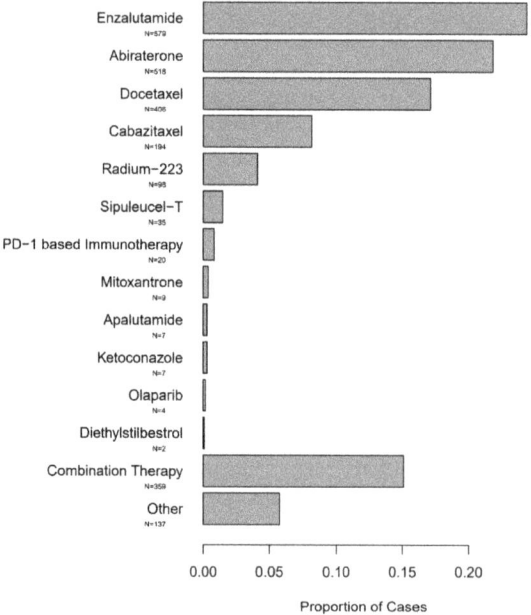

Figure 5. Third Treatment After Metastatic Diagnosis (N = 2375).

3.2. Comparison of Effectiveness of NHT Versus Docetaxel after a Prior NHT

Thereafter we aimed to evaluate the comparative effectiveness of alternate NHTs vs. docetaxel after treatment with an NHT in this real-world patient population. Out of the 9747 patients with mPC in the dataset, 1117 patients met all eligibility criteria for this analysis. The most common reason for exclusion were lack of information on any treatment other than ADT ($N = 2733$), 1L treatment other than abiraterone or enzalutamide (including $N = 1588$ 1L docetaxel and $N = 329$ combination therapy), lack of information on 2L treatment ($N = 2219$) or 2L treatment other than abiraterone, enzalutamide, or docetaxel. Of these 1117 included patients, in the 1L therapy setting, 695 men received abiraterone, and 422 men received enzalutamide. In the 1L abiraterone group, 2L treatment consisted of enzalutamide in 508 and docetaxel in 187 patients. In the 1L enzalutamide group, 2L treatment consisted of abiraterone in 290 and docetaxel in 132 patients. Median follow-up for the study cohort was 9.8 months (range 0.1–64.4) and median follow-up among patients alive at the cutoff date for analysis was 12.5 months (range 0.2–64.4 months).

Table S2 presents an extensive comparison of patient characteristics in alternate NHT and docetaxel groups in 2L. Propensity scores were estimated and used to calculate matching weights, and propensity score overlap (evaluated graphically Figure S1) and covariate balance (evaluated by standardized mean differences) were deemed satisfactory; more details are in the supplementary material. This suggests that analyses adjusted by using weighting based on the propensity score should largely eliminate potential confounding from the measured variables.

3.3. Primary TTTTD and OS Analysis

Figure 6 displays Kaplan–Meier curves for the two survival outcomes of interest, TTTTD, and OS, for 2L NHT vs. docetaxel, separately for the 1L abiraterone (Figure 6a) and 1L enzalutamide (Figure 6b) patient groups. In both groups, 2L NHT showed evidence of superior survival experiences as compared with 2L docetaxel. Table 1 presents (unadjusted) median survival in all groups, as well as groups defined by time on 1L therapy ($<$ or ≥ 6 months; $<$ or ≥ 12 months). Median TTTTD was between 4.4 and 8.5 months across the 2L sub-groups and was longer in nearly all alternate NHT subgroups as compared with docetaxel. Median OS from the start of 2L therapy was consistently longer with alternate NHT as compared to docetaxel in both groups. In the 1L abiraterone group, the median OS with enzalutamide was 15.6 months as compared to 8.7 months with docetaxel. Similarly, in the 1L enzalutamide group, the median OS with abiraterone was 13.2 months as compared to 9.7 months with docetaxel.

Table 2 presents HRs from the Cox proportional hazards model adjusted using matching weights from the propensity score model for the overall population, and the results are consistent with the unadjusted results. The TTTTD HR for 2L docetaxel vs. alternate NHT was 1.26 (95% CI 1.04, 1.53) in the 1L abiraterone group and 1.32 (95% CI 1.07, 1.64) in the 1L enzalutamide group. The analogous HRs for OS were 1.36 (95% CI 1.09, 1.70) in the 1L abiraterone group and 1.40 (95% CI 1.09, 1.80) in the 1L enzalutamide group.

(a) 1L Abiraterone

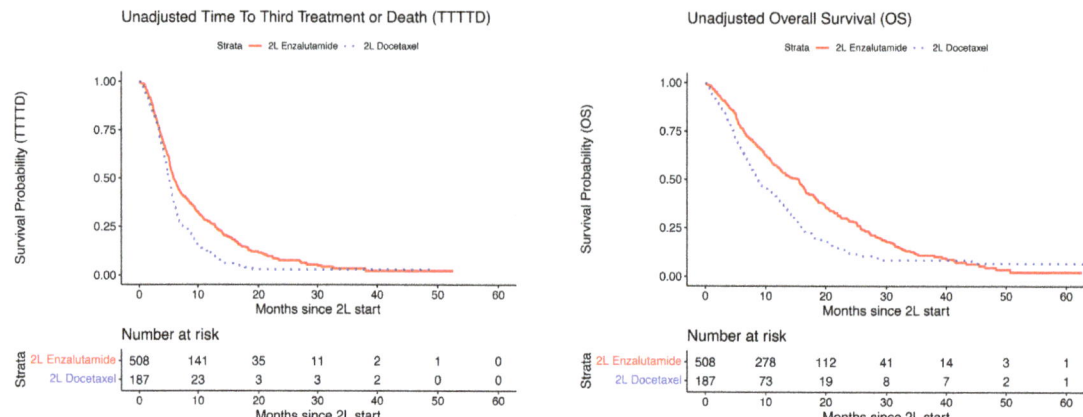

(b) 1L Enzalutamide

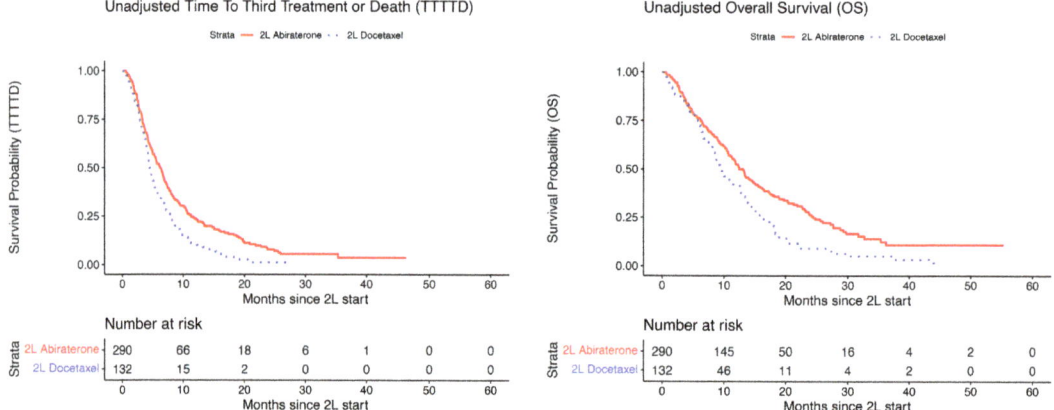

Figure 6. Unadjusted time to third treatment or death (TTTTD) and overall survival (OS) from 2L, with (**a**) 1L abiraterone (Figure 1a) and (**b**) 1L enzalutamide (Figure 1b).

Table 1. Median TTTTD and OS times, starting from the initiation of second-line (2L) therapy, overall and in subgroups defined by time on first-line (1L) therapy. Median survival estimates and 95% confidence intervals (CIs) are estimated using the Kaplan–Meier method, and are not adjusted for any covariates. Confidence limits that could not be estimated due to limited sample size are denoted by "-". TTTTD is time to third-line treatment or death, and OS is overall survival.

1L Therapy	Population	Outcome	2L Therapy	N	Events	Median Survival, Months (95% CI)
Abiraterone	Overall	TTTTD	Enzalutamide	508	422	5.9 (5.6, 6.7)
			Docetaxel	187	161	5.1 (4.7, 5.7)
		OS	Enzalutamide	508	352	15.6 (12.7, 16.7)
			Docetaxel	187	144	8.7 (7.7, 11.6)
	1L NHT < 6 months	TTTTD	Enzalutamide	184	154	5.9 (5.4, 7.1)
			Docetaxel	115	104	5.0 (4.7, 5.7)
		OS	Enzalutamide	184	128	15.1 (11.0, 18.4)
			Docetaxel	115	91	9.1 (7.7, 12.1)
	1L NHT ≥ 6 months	TTTTD	Enzalutamide	324	268	6.0 (5.5, 7.0)
			Docetaxel	72	57	5.5 (4.3, 6.6)
		OS	Enzalutamide	324	224	15.6 (12.8, 16.7)
			Docetaxel	72	53	8.4 (6.3, 13.2)
	1L NHT < 12 months	TTTTD	Enzalutamide	362	306	5.6 (5.3, 6.4)
			Docetaxel	164	148	5.0 (4.6, 5.5)
		OS	Enzalutamide	362	263	12.5 (10.9, 15.6)
			Docetaxel	164	132	8.1 (7.3, 10.3)
	1L NHT ≥ 12 months	TTTTD	Enzalutamide	146	116	8.2 (5.9, 10.1)
			Docetaxel	23	13	8.5 (4.7, -)
		OS	Enzalutamide	146	89	18.9 (16.5, 24.6)
			Docetaxel	23	12	16.0 (12.4, -)
Enzalutamide	Overall	TTTTD	Abiraterone	290	226	6.3 (5.5, 7.0)
			Docetaxel	132	117	4.7 (4.4, 5.4)
		OS	Abiraterone	290	182	13.2 (11.4, 15.0)
			Docetaxel	132	103	9.7 (8.6, 12.6)
	1L NHT < 6 months	TTTTD	Abiraterone	92	72	4.8 (4.2, 6.3)
			Docetaxel	69	63	4.4 (3.7, 5.3)
		OS	Abiraterone	92	66	9.8 (6.7, 13.5)
			Docetaxel	69	58	7.0 (6.3, 9.3)
	1L NHT ≥ 6 months	TTTTD	Abiraterone	198	154	6.8 (6.2, 8.0)
			Docetaxel	63	54	5.3 (4.4, 8.2)
		OS	Abiraterone	198	116	14.0 (11.9, 17.6)
			Docetaxel	63	45	12.8 (10.9, 16.6)
	1L NHT < 12 months	TTTTD	Abiraterone	184	140	5.7 (5.0, 7.0)
			Docetaxel	107	96	4.5 (4.1, 5.3)
		OS	Abiraterone	184	125	10.4 (8.4, 12.0)
			Docetaxel	107	87	8.8 (7.7, 11.5)
	1L NHT ≥ 12 months	TTTTD	Abiraterone	106	86	6.9 (5.9, 8.5)
			Docetaxel	25	21	7.2 (4.9, 14.8)
		OS	Abiraterone	106	57	19.1 (14.7, 24.2)
			Docetaxel	25	16	15.1 (10.0, -)

Table 2. Hazard ratios (HRs) and 95% confidence intervals (CIs) comparing survival outcomes on second-line (2L) docetaxel vs. 2L alternate NHT. The HRs are from Cox proportional hazards models weighted using matching weights from propensity scores to adjust for potential confounding.

1L Therapy	Outcome	2L Therapy	N	HR (95% CI)	p
Abiraterone	TTTTD	Enzalutamide	508	1.00 (ref)	
		Docetaxel	187	1.26 (1.04, 1.53)	0.018
	OS	Enzalutamide	508	1.00 (ref)	
		Docetaxel	187	1.32 (1.07, 1.64)	0.009
Enzalutamide	TTTTD	Abiraterone	290	1.00 (ref)	
		Docetaxel	132	1.36 (1.09, 1.70)	0.008
	OS	Abiraterone	290	1.00 (ref)	
		Docetaxel	132	1.40 (1.09, 1.80)	0.009

3.4. Subgroup Analyses

In addition to the main analyses, subgroup analyses based on characteristics of interest were performed. Results from these are presented visually in Figure 7a,b and numerically in Table S3. With only a few exceptions, the HR point estimates for docetaxel vs. NHT were above 1 across subgroups and are not significantly less than 1 for any subgroup. Some subgroups with OS HRs significantly above 1 (at the 0.05 level) included age ≤ 70 (1L abiraterone); age 70–75 (1L enzalutamide); and ECOG 0–1 (1L abiraterone and 1L enzalutamide).

3.5. Post-2L Characteristics

Finally, we investigated selected post-2L characteristics, to evaluate whether any differences in these could help elucidate the main associations. Comparisons of these characteristics are presented in Table S4 and should be interpreted with some caution because the end of follow-up due to death or censoring could potentially impact the observed summaries. First, the median number of docetaxel cycles among patients in 2L docetaxel groups was 6 in both groups considered, which is lower than the recommended dose of docetaxel in men with mCRPC. Second, we evaluated ECOG after 4 months on 2L therapy, and before the end of 2L therapy; this was available for 35.8–40.0% patients across the arms. There was no evidence suggesting differences in this post-2L ECOG between 2L NHT and 2L docetaxel groups.

The rates of 3L initiation were similar among the two arms in the 1L abiraterone group, but differed in the 1L enzalutamide group with 42.4% patients in the 2L abiraterone subgroup starting a 3L therapy as compared to 57.6% in the 2L docetaxel subgroup, despite a similar follow-up time between the groups. The total number of post-2L lines of treatment, the number of post-2L lines of treatment that include at least one of the approved life-extending drugs [6], and the number of unique post-2L life-extending drugs used [6] were similar among the 1L abiraterone patients, though nominally lower in the 2L docetaxel patients; these metrics were also similar among the 1L enzalutamide patients, though they tended to be nominally higher among 2L docetaxel patients.

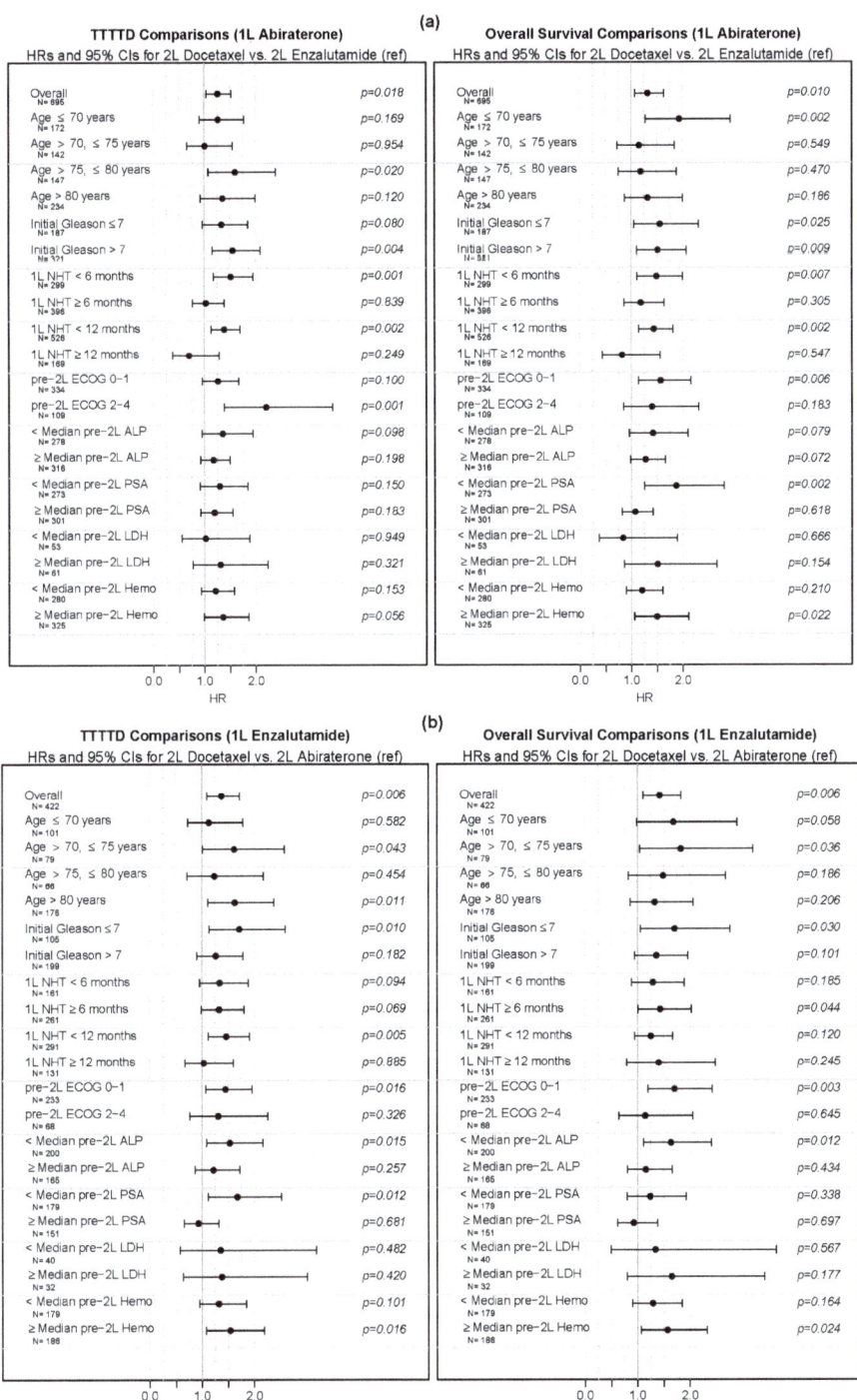

Figure 7. Time to third treatment (TTTTD) and overall survival (OS) comparisons between alternate novel hormonal therapy and docetaxel in patients treated with (**a**) first-line abiraterone (Figure 2a) and (**b**) first line-enzalutamide (Figure 2b).

4. Discussion

Once metastatic, prostate cancer is traditionally described either as mCSPC during which it can be treated by depleting the serum testosterone to castrate levels (<50 ng/dl), and mCRPC when prostate cancer continues to progress even in the presence of castrate levels of testosterone [12]. However, this difference is now losing its significance as the therapies which were utilized for mCRPC have been also approved for mCSPC. Upfront utilization of these therapies (NHT and docetaxel) has shown to improve OS and are recommended by current guidelines [6,12]. However, as discussed above, the real-world adaptation of these life-prolonging therapies in clinical practice, or the survival outcomes with docetaxel versus a NHT after treatment with one NHT are not well characterized. There has not been a randomized controlled trial in this setting either.

There are four main contributions of this paper. First is the data on treatment patterns and utilization of systemic therapies in the real-world men with mPC in the United States. Until now, there are only a few published studies that have investigated the treatment patterns and use of upfront intensification in a real-world population. Upfront intensification with NHTs or docetaxel is currently recommended as initial treatment for patients with mCSPC [6]. Upfront intensification not only prolongs life but does so without compromising quality-of-life as observed in randomized clinical trials [13]. However, retrospective studies (mostly as abstracts) from multiple different databases including Optum, Medicare and ConcertAI Oncology Dataset have shown a consistent underutilization of intensification ranging from <10% to up to 30% of mCSPC patients and even those with visceral disease and in those with insurance [14–17]. In our dataset, we confirm that upfront intensification was low but a gradual and encouraging trend towards increased intensification was observed over the last 5 years (Figure 3).

Second, we observe that in a large real-world dataset, more than two-thirds of patients with mPC treated with an NHT subsequently received an alternate NHT, and <30% of patients received docetaxel as a subsequent therapy. This suggests that an alternate NHT is widely used and is the preferred therapy over docetaxel in this real-world population in the United States.

Third, we provide estimates of TTTTD and OS after disease progression on one NHT. These data are not currently available from prospective datasets. In our view, these estimates may be useful in the counseling of patients and for clinical trial design.

Fourth, with the caveat of a non-randomized comparison and the retrospective nature of the study, we observed that the alternate NHT, when compared to docetaxel, was associated with superior OS after treatment with a prior NHT in mPC setting. It should be noted that efficacy data comparing docetaxel to alternate NHT after treatment with one NHT either from a clinical trial or a retrospective experience are currently non-existent.

Abiraterone, enzalutamide, and docetaxel have been shown to improve survival in men with metastatic castration-resistant prostate cancer [18–23]. Abiraterone is an irreversible inhibitor of 17 α-hydroxylase/17, 20-lyase (cytochrome P450c17 [CYP17A1]), blocks intratumoral production of testosterone, and one of its metabolites, Δ^4-abiraterone is a direct androgen receptor (AR) inhibitor AR [24–27]. Enzalutamide impairs androgen binding to the AR, AR nuclear translocation, DNA binding, and coactivator recruitment [28]. Docetaxel exerts its anticancer effect at least in part by impairing AR signaling by inhibiting its nuclear translocation by stabilizing microtubules [29]. Preclinical studies have shown cross-resistance between these agents despite apparently differing mechanisms of action [8,30–32].

In a prospective single-arm trial (n = 215) in men with progressive mCRPC on abiraterone, the median time to PSA progression and radiographic progression with enzalutamide was 5.7 months and 8.1 months, respectively [7]. In a prospective, randomized, cross-over trial (n = 212) of abiraterone versus enzalutamide followed by cross-over to enzalutamide or abiraterone in progressive mCRPC, the median time to PSA progression with enzalutamide after abiraterone was 3.5 months while with abiraterone after enzalutamide was 1.7 months. This trial did not report radiographic progression-free survival with 2L

NHT [7,8]. While the efficacy of an alternate NHT has been tested after a prior NHT in mCRPC in the two above-mentioned prospective trials, such prospective evaluation of the efficacy of docetaxel after prior treatment on an NHT has not yet been reported. However, in a post hoc analysis of the COU-AA-302 trial, the median duration of docetaxel treatment was ~3 months in men who had prior disease progression on abiraterone which is similar to that seen in our study [9]. Multiple other studies in this setting have shown similar results and are summarized in Table S5. At present, no evidence exists based on a randomized trial to support the use of an alternate NHT over docetaxel or vice versa after prior disease progression on one NHT for mPC. Furthermore, the estimates for TTTTD and OS are not available with 2L NHT or docetaxel after a prior 1L NHT as these agents have never been tested in any prospective clinical trial. Recent data from the CARD trial, which was conducted in the 3L systemic therapy setting, demonstrated significant improvement in clinical outcomes with cabazitaxel as compared to an alternate NHT in men with mCRPC who had previously received docetaxel and experienced disease progression within 12 months of being on the treatment of an NHT [33]. However, these data may not apply to 2L systemic therapy setting, i.e., to the patients who only received treatment with only one NHT in 1L, and a 2L therapy selection needs to be made between docetaxel and an alternate NHT. This lack of evidence on the comparative efficacy of alternate NHT versus docetaxel and survival estimates with these agents in the 2L setting poses serious challenges to treating providers on treatment selection, prognostication, and patient counseling. It is unlikely that we will have these data available from a randomized trial of 2L NHT or docetaxel in the near future, though the results of this study suggest that such a trial is warranted.

To better explain the lower OS with docetaxel, we evaluated multiple hypotheses. We investigated whether patients in the real-world were receiving fewer cycles of docetaxel, whether receipt of docetaxel was leading to early deterioration of performance status, or if there were differences in 3L therapies. In this real-world population, 2L docetaxel patients were indeed receiving fewer than recommended cycles of docetaxel (medians 5.5–6 cycles in the 2 groups) as opposed to ~10 cycles generally recommended in the clinical trial setting based on the results from the seminal TAX327 study which led to the approval of docetaxel in the mCRPC setting [23]. The proportions of patients initiating 3L treatment in the docetaxel and alternate NHT arms were similar in the 1L abiraterone group but greater for docetaxel vs. alternate NHT in the 1L enzalutamide group (Table S4). Similarly, the metrics comparing the number of life-extending agents were similar between docetaxel and alternate NHT in the 1L abiraterone group, but greater for docetaxel than alternate NHT in the 1L enzalutamide group. These findings may be a reflection of an earlier switch to subsequent therapy in those receiving 2L docetaxel possibly due to poor tolerability or earlier disease progression.

The strengths of our study include real-world patient-level data from a large cohort of patients with the propensity-weighted matching of multiple potential confounders. We report both TTTTD and OS with 2L docetaxel or alternate NHT, along with comprehensive subgroup analyses.

The central weakness of this study is that our data are observational, and associations in observational data may be impacted by confounding. A priori, we would expect that physicians were selecting docetaxel for men whose cancer appeared to be more severe, in which case we would expect unadjusted estimates of TTTTD and OS to be shorter in that group, which we do indeed observe. If docetaxel was superior to alternate NHT after treatment with one NHT, we would expect that docetaxel would show similar or better performance than alternate NHT in the analyses where we adjust for measures of disease severity and treatment applicability; however, we did not observe this. Our approach to adjustment is designed to essentially match men based on key covariates that capture disease severity and the rapidity of progression, such as PSA, time on ADT, time on the first NHT, and key comorbidities. We did indeed see some indication of an imbalance in covariates between the treatment arms on some of these variables, but when using matching weights based on the propensity scores, the treatment arms became balanced, suggesting that confounding by the variables we have measured does not have a major impact on our results. However, our results may be impacted by confounding by variables we do not have measured, such as patient treatment preferences, healthcare access information, or genetic phenotypes that are not available to us. They may also be impacted by missingness, particularly if that missingness is differential—e.g., if patients in one treatment arm are typically missing PSA values only if they are very high, and that is not true of patients in the other treatment arm. Although these concerns are significant and urge caution in relying too strongly on results from this observational study, if docetaxel is in fact superior to alternate NHT in all men, it would require fairly strong effects of the unmeasured variables on both treatment selection and outcomes. Our results showing improvement in survival outcomes in the alternate NHT arms vs. docetaxel were very consistent across adjusted analyses and within subgroups. We therefore think it is worthwhile to share the results from this observational study, and strongly advocate for further studies investigating these treatment regimens to understand whether alternate NHT may be a better alternative than docetaxel for some patients. If a randomized clinical trial is unlikely, further observational studies with more careful annotation of potential confounders could better elucidate which treatment plan is superior and for which patients.

5. Conclusions

In this large observational real-world study, most men with new mPC did not receive NHT or docetaxel despite large, randomized trials showing significantly improved survival outcomes with these agents. The next step needs to be understanding the reasons for underutilization including lack of patient and physician awareness, barriers to access to these life-prolonging therapies including insurance/cost/access, fear of toxicities (drug or financial), or other reasons including co-morbidities, age, social, demographic or racial disparities. Once the causes are identified a combined and cohesive effort can be undertaken by various stakeholders to resolve them. In patients with mPC with prior treatment with only one NHT, the use of alternate NHT was more common and associated with superior survival compared to docetaxel in this retrospective dataset, which warrants a randomized controlled trial in this setting. These data also show survival estimates on NHT and docetaxel after progression on one NHT which may assist with designing clinical trials in this setting, as well as counseling and prognostication in the clinic.

Supplementary Materials: The following are available online at https://www.mdpi.com/article/10.3390/cancers13194951/s1, Table S1: Patient characteristics among all patients with evidence of contact within 180 days of metastatic prostate cancer diagnosis and with information on treatment in the metastatic setting, Table S2: Patient Characteristics Among 2L Alternate NHT vs. 2L Docetaxel, separately by 1L Abiraterone and 1L Enzalutamide, Table S3: Subgroup analyses comparing 2L Docetaxel vs. 2L NHT, separately by 1L Abiraterone and 1L Enzalutamide, Table S4: Comparison of patient characteristics after initiation of second line (2L) treatment, Table S5: Selected studies evaluating effectiveness of alternate novel hormonal therapies (NHT; abiraterone (A) or enzalutamide (E)) or docetaxel (D) after progression on an NHT (E or A), Figure S1: Smoothed histograms of propensity scores. Propensity scores model the probability of treatment with 2L Docetaxel vs. 2L Alternate NHT, and are presented for both 2L groups for the 1L Abiraterone patients (Supplementary Figure S1a) and 1L Enzalutamide patients (Supplementary Figure S1b).

Author Contributions: Conception and design: U.S., J.A.S., B.H., N.A. Acquisition of data: U.S., J.A.S., B.H., N.A. Analysis and interpretation of data: U.S., J.A.S., B.H., B.L.M., N.R., T.R.M., D.S., R.N., M.K., S.K.P., N.A. Drafting of the manuscript: U.S., J.A.S., B.H., B.L.M., N.S., N.T., N.R., T.R.M., D.S., R.N., M.K., S.K.P., N.A. Critical revision of the manuscript for important intellectual content: U.S., J.A.S., B.H., B.L.M., N.S., N.T., N.R., T.R.M., D.S., R.N., M.K., S.K.P., N.A. Statistical analysis: J.A.S., B.H. Obtaining funding: NA. Administrative, technical, or material support: U.S., J.A.S., B.H., B.L.M., N.R., T.R.M., D.S., R.N., M.K., S.K.P., N.A. Supervision: U.S., J.A.S., B.H., R.N., N.A. Other (specify): N.A. All authors have read and agreed to the published version of the manuscript.

Funding: This research received no external funding.

Institutional Review Board Statement: The study was conducted according to the guidelines of the Declaration of Helsinki, and approved by the Institutional Review Board of the University of Utah (IRB_00067518, last approved 9/7/2021).

Informed Consent Statement: Patient consent was waived because it is retrospective, non-interventional, and used anonymized data provided by Flatiron.

Data Availability Statement: The data that support the findings of this study have been originated by Flatiron Health, Inc. These de-identified data may be made available upon request, and are subject to a license agreement with Flatiron Health; interested researchers should contact DataAccess@flatiron.com to determine licensing terms.

Acknowledgments: Research reported in this publication utilized the Cancer Biostatistics Shared Resource at Huntsman Cancer Institute at the University of Utah and was supported by the National Cancer Institute of the National Institutes of Health under Award Number P30CA042014. The content is solely the responsibility of the authors and does not necessarily represent the official views of the NIH.

Conflicts of Interest: U.S. reports consultancy fees from Seattle Genetics and research funding paid to his institution from Seattle Genetics/Astellas and Janssen. B.L.M. is a paid consultant/advisor to Pfizer, AVEO oncology, Janssen, Astellas, Bristol-Myers Squibb, Clovis, Tempus, Merck, Exelixis, Bayer Oncology and Peloton Therapeutics; he has received research funding to his institution from Exelixis, Bavarian-Nordic, Clovis, Genentech and Bristol-Myers Squibb. N.A. reports consultancy to Astellas, Astra Zeneca, Aveo, Bayer, Bristol Myers Squibb, Calithera, Clovis, Eisai, Eli Lilly, EMD Serono, Exelixis, Foundation Medicine, Genentech, Gilead, Janssen, Merck, MEI Pharma, Nektar, Novartis, Pfizer, Pharmacyclics, and Seattle Genetics and research funding to his institution from Astra Zeneca, Bavarian Nordic, Bayer, Bristol Myers Squibb, Calithera, Celldex, Clovis, Eisai, Eli Lilly, EMD Serono, Exelixis, Genentech, Glaxo Smith Kline, Immunomedics, Janssen, Medivation, Merck, Nektar, New Link Genetics, Novartis, Pfizer, Prometheus, Rexahn, Roche, Sanofi, Seattle Genetics, Takeda, and Tracon. J.A.S., B.H., N.S., T.R.M., N.T., N.R., D.S., R.N., M.K. and S.K.P. report no COI.

References

1. James, N.D.; de Bono, J.S.; Spears, M.R.; Clarke, N.W.; Mason, M.D.; Dearnaley, D.P.; Ritchie, A.W.S.; Amos, C.L.; Gilson, C.; Jones, R.J.; et al. Abiraterone for prostate cancer not previously treated with hormone therapy. *N. Engl. J. Med.* **2017**, *377*, 338–351. [CrossRef]
2. Davis, I.D.; Martin, A.J.; Stockler, M.R.; Begbie, S.; Chi, K.N.; Chowdhury, S.; Coskinas, X.; Frydenberg, M.; Hague, W.E.; Horvath, L.; et al. Enzalutamide with standard first-line therapy in metastatic prostate cancer. *N. Engl. J. Med.* **2019**, *381*, 121–131. [CrossRef] [PubMed]
3. Chi, K.N.; Agarwal, N.; Bjartell, A.; Chung, B.H.; Pereira de Santana Gomes, A.J.; Given, R.; Soto, A.J.; Merseburger, A.S.; Özgüroglu, M.; Uemura, H.; et al. Apalutamide for metastatic, castration-sensitive prostate cancer. *N. Engl. J. Med.* **2019**, *381*, 13–24. [CrossRef]
4. Kyriakopoulos, C.E.; Chen, Y.-H.; Carducci, M.A.; Liu, G.; Jarrard, D.F.; Hahn, N.M.; Shevrin, D.H.; Dreicer, R.; Hussain, M.; Eisenberger, M.; et al. Chemohormonal therapy in metastatic hormone-sensitive prostate cancer: Long-term survival analysis of the randomized phase III E3805 CHAARTED trial. *J. Clin. Oncol.* **2018**, *36*, 1080–1087. [CrossRef] [PubMed]
5. James, N.D.; Sydes, M.R.; Clarke, N.W.; Mason, M.D.; Dearnaley, D.P.; Spears, M.R.; Ritchie, A.W.S.; Parker, C.C.; Russell, J.M.; Attard, G.; et al. Addition of docetaxel, zoledronic acid, or both to first-line long-term hormone therapy in prostate cancer (STAMPEDE): Survival results from an adaptive, multiarm, multistage, platform randomised controlled trial. *Lancet* **2016**, *387*, 1163–1177. [CrossRef]
6. National Comprehensive Cancer Network. Prostate Cancer (Version 2.2021). Available online: http://www.nccn.org/professionals/physician_gls/pdf/prostate.pdf (accessed on 23 February 2021).
7. De Bono, J.S.; Chowdhury, S.; Feyerabend, S.; Elliott, T.; Grande, E.; Melhem-Bertrandt, A.; Baron, B.; Hirmand, M.; Werbrouck, P.; Fizazi, K. Antitumour activity and safety of enzalutamide in patients with metastatic castration-resistant prostate cancer previously treated with abiraterone acetate plus prednisone for ≥24 weeks in Europe. *Eur. Urol.* **2018**, *74*, 37–45. [CrossRef]
8. Khalaf, D.J.; Annala, M.; Taavitsainen, S.; Finch, D.L.; Oja, C.; Vergidis, J.; Zulfiqar, M.; Sunderland, K.; Azad, A.A.; Kollmannsberger, C.K.; et al. Optimal sequencing of enzalutamide and abiraterone acetate plus prednisone in metastatic castration-resistant prostate cancer: A multicentre, randomised, open-label, phase 2, crossover trial. *Lancet Oncol.* **2019**, *20*, 1730–1739. [CrossRef]
9. De Bono, J.S.; Smith, M.R.; Saad, F.; Rathkopf, D.E.; Mulders, P.F.; Small, E.J.; Shore, N.D.; Fizazi, K.; de Porre, P.; Kheoh, T.; et al. Subsequent chemotherapy and treatment patterns after abiraterone acetate in patients with metastatic castration-resistant prostate cancer: Post hoc analysis of COU-AA-302. *Eur. Urol.* **2017**, *71*, 656–664. [CrossRef]
10. Ma, X.; Long, L.; Moon, S.; Adamson, B.J.S.; Baxi, S.S. Comparison of population characteristics in real-world clinical oncology databases in the US: Flatiron health, SEER, and NPCR. *medRxiv* **2020**, medRxiv:2020.03.16.20037143. [CrossRef]

11. Breiman, L. Random forests. *Mach. Learn.* **2001**, *45*, 5–32. [CrossRef]
12. Swami, U.; McFarland, T.R.; Nussenzveig, R.; Agarwal, N. Advanced prostate cancer: Treatment advances and future directions. *Trends Cancer* **2020**, *6*, 702–715. [CrossRef]
13. Agarwal, N.; McQuarrie, K.; Bjartell, A.; Chowdhury, S.; Pereira de Santana Gomes, A.J.; Chung, B.H.; Özgüroglu, M.; Soto, Á.J.; Merseburger, A.S.; Uemura, H.; et al. Health-related quality of life after apalutamide treatment in patients with metastatic castration-sensitive prostate cancer (TITAN): A randomised, placebo-controlled, phase 3 study. *Lancet Oncol.* **2019**, *20*, 1518–1530. [CrossRef]
14. Ryan, C.J.; Ke, X.; Lafeuille, M.-H.; Romdhani, H.; Kinkead, F.; Lefebvre, P.; Petrilla, A.; Pulungan, Z.; Kim, S.; D'Andrea, D.M.; et al. Management of patients with metastatic castration-sensitive prostate cancer in the real-world setting in the United States. *J. Urol.* **2021**. [CrossRef]
15. Swami, U.; Hong, A.; El-Chaar, N.N.; Nimke, D.; Ramaswamy, K.; Bell, E.J.; Sandin, R.; Agarwal, N. Real-world first-line (1L) treatment patterns in patients (pts) with metastatic castration-sensitive prostate cancer (mCSPC) in a US health insurance database. *J. Clin. Oncol.* **2021**, *39*, 5072. [CrossRef]
16. Freedland, S.J.; Agarwal, N.; Ramaswamy, K.; Sandin, R.; Russell, D.; Hong, A.; Yang, H.; Gao, W.; Hagan, K.; George, D.J. Real-world utilization of advanced therapies and racial disparity among patients with metastatic castration-sensitive prostate cancer (mCSPC): A Medicare database analysis. *J. Clin. Oncol.* **2021**, *39*, 5073. [CrossRef]
17. George, D.J.; Agarwal, N.; Rider, J.R.; Li, B.; Shirali, R.; Sandin, R.; Hong, A.; Russell, D.; Ramaswamy, K.; Freedland, S.J. Real-world treatment patterns among patients diagnosed with metastatic castration-sensitive prostate cancer (mCSPC) in community oncology settings. *J. Clin. Oncol.* **2021**, *39*, 5074. [CrossRef]
18. De Bono, J.S.; Logothetis, C.J.; Molina, A.; Fizazi, K.; North, S.; Chu, L.; Chi, K.N.; Jones, R.J.; Goodman, O.B., Jr.; Saad, F.; et al. Abiraterone and increased survival in metastatic prostate cancer. *N. Engl. J. Med.* **2011**, *364*, 1995–2005. [CrossRef]
19. Ryan, C.J.; Smith, M.R.; de Bono, J.S.; Molina, A.; Logothetis, C.J.; de Souza, P.; Fizazi, K.; Mainwaring, P.; Piulats, J.M.; Vogelzang, N.J.; et al. Abiraterone in metastatic prostate cancer without previous chemotherapy. *N. Engl. J. Med.* **2013**, *368*, 138–148. [CrossRef] [PubMed]
20. Scher, H.I.; Fizazi, K.; Saad, F.; Taplin, M.-E.; Sternberg, C.N.; Miller, K.; de Wit, R.; Mulders, P.; Chi, K.N.; Shore, N.D.; et al. Increased survival with enzalutamide in prostate cancer after chemotherapy. *N. Engl. J. Med.* **2012**, *367*, 1187–1197. [CrossRef] [PubMed]
21. Beer, T.M.; Armstrong, A.J.; Rathkopf, D.E.; Loriot, Y.; Sternberg, C.N.; Higano, C.S.; Iversen, P.; Bhattacharya, S.; Carles, J.; Chowdhury, S.; et al. Enzalutamide in metastatic prostate cancer before chemotherapy. *N. Engl. J. Med.* **2014**, *371*, 424–433. [CrossRef] [PubMed]
22. Petrylak, D.P.; Tangen, C.M.; Hussain, M.H.; Lara, P.N.; Jones, J.A.; Taplin, M.E.; Burch, P.A.; Berry, D.; Moinpour, C.; Kohli, M.; et al. Docetaxel and estramustine compared with mitoxantrone and prednisone for advanced refractory prostate cancer. *N. Engl. J. Med.* **2004**, *351*, 1513–1520. [CrossRef] [PubMed]
23. Tannock, I.F.; de Wit, R.; Berry, W.R.; Horti, J.; Pluzanska, A.; Chi, K.N.; Oudard, S.; Theodore, C.; James, N.D.; Turesson, I.; et al. Docetaxel plus prednisone or mitoxantrone plus prednisone for advanced prostate cancer. *N. Engl. J. Med.* **2004**, *351*, 1502–1512. [CrossRef] [PubMed]
24. Potter, G.A.; Barrie, S.E.; Jarman, M.; Rowlands, M.G. Novel steroidal inhibitors of human cytochrome P45017. alpha.-Hydroxylase-C17, 20-lyase): Potential agents for the treatment of prostatic cancer. *J. Med. Chem.* **1995**, *38*, 2463–2471. [CrossRef]
25. Barrie, S.; Potter, G.; Goddard, P.; Haynes, B.; Dowsett, M.; Jarman, M. Pharmacology of novel steroidal inhibitors of cytochrome P45017α (17α-hydroxylase/C17–20 lyase). *J. Steroid Biochem. Mol. Biol.* **1994**, *50*, 267–273. [CrossRef]
26. Odonnell, A.G.; Judson, I.; Dowsett, M.; Raynaud, F.; Dearnaley, D.P.; Mason, M.G.; Harland, S.J.; Robbins, A.S.; Halbert, G.; Nutley, B.; et al. Hormonal impact of the 17 α-hydroxylase/C 17, 20-lyase inhibitor abiraterone acetate (CB7630) in patients with prostate cancer. *Br. J. Cancer* **2004**, *90*, 2317–2325. [CrossRef]
27. Li, Z.; Alyamani, M.; Li, J.; Rogacki, K.; Abazeed, M.; Upadhyay, S.K.; Balk, S.P.; Taplin, M.-E.; Auchus, R.J.; Sharifi, N. Redirecting abiraterone metabolism to fine-tune prostate cancer anti-androgen therapy. *Nature* **2016**, *533*, 547–551. [CrossRef]
28. Tran, C.; Ouk, S.; Clegg, N.J.; Chen, Y.; Watson, P.A.; Arora, V.; Wongvipat, J.; Smith-Jones, P.M.; Yoo, D.; Kwon, A.; et al. Development of a second-generation antiandrogen for treatment of advanced prostate cancer. *Science* **2009**, *324*, 787–790. [CrossRef]
29. Zhu, M.-L.; Horbinski, C.M.; Garzotto, M.; Qian, D.Z.; Beer, T.M.; Kyprianou, N. Tubulin-targeting chemotherapy impairs androgen receptor activity in prostate cancer. *Cancer Res.* **2010**, *70*, 7992–8002. [CrossRef]
30. Van Soest, R.J.; van Royen, M.E.; de Morrée, E.S.; Moll, J.M.; Teubel, W.; Wiemer, E.A.C.; Mathijssen, R.H.; de Wit, R.; van Weerden, W.M. Cross-resistance between taxanes and new hormonal agents abiraterone and enzalutamide may affect drug sequence choices in metastatic castration-resistant prostate cancer. *Eur. J. Cancer* **2013**, *49*, 3821–3830. [CrossRef] [PubMed]
31. Van Soest, R.J.; de Morrée, E.S.; Kweldam, C.F.; de Ridder, C.M.; Wiemer, E.A.C.; Mathijssen, R.H.; de Wit, R.; van Weerden, W.M. Targeting the androgen receptor confers in vivo cross-resistance between enzalutamide and docetaxel, but not cabazitaxel, in castration-resistant prostate cancer. *Eur. Urol.* **2015**, *67*, 981–985. [CrossRef] [PubMed]

32. Lombard, A.P.; Liu, L.; Cucchiara, V.; Liu, C.; Armstrong, C.M.; Zhao, R.; Yang, J.C.; Lou, W.; Evans, C.P.; Gao, A.C. Intra vs inter cross-resistance determines treatment sequence between taxane and AR-targeting therapies in advanced prostate cancer. *Mol. Cancer Ther.* **2018**, *17*, 2197–2205. [CrossRef] [PubMed]
33. De Wit, R.; de Bono, J.; Sternberg, C.N.; Fizazi, K.; Tombal, B.; Wülfing, C.; Kramer, G.; Eymard, J.-C.; Bamias, A.; Carles, J.; et al. Cabazitaxel versus abiraterone or enzalutamide in metastatic prostate cancer. *N. Engl. J. Med.* **2019**, *381*, 2506–2518. [CrossRef] [PubMed]

Article

Response and Toxicity to the Second Course of 3 Cycles of ^{177}Lu-PSMA Therapy Every 4 Weeks in Patients with Metastatic Castration-Resistant Prostate Cancer

Sazan Rasul [1], Tim Wollenweber [1], Lucia Zisser [1], Elisabeth Kretschmer-Chott [1], Bernhard Grubmüller [2], Gero Kramer [2], Shahrokh F. Shariat [2,3,4,5,6], Harald Eidherr [1], Markus Mitterhauser [1,7], Chrysoula Vraka [1], Werner Langsteger [1], Marcus Hacker [1] and Alexander R. Haug [1,8,*]

1. Division of Nuclear Medicine, Department of Biomedical Imaging and Image-guided Therapy, Medical University of Vienna, 1090 Vienna, Austria; sazan.rasul@meduniwien.ac.at (S.R.); tim.wollenweber@meduniwien.ac.at (T.W.); lucia.zisser@meduniwien.ac.at (L.Z.); elisabeth.kretschmer-chott@meduniwien.ac.at (E.K.-C.); harald.eidherr@meduniwien.ac.at (H.E.); markus.mitterhauser@meduniwien.ac.at (M.M.); chrysoula.vraka@meduniwien.ac.at (C.V.); werner.langsteger@meduniwien.ac.at (W.L.); marcus.hacker@meduniwien.ac.at (M.H.)
2. Department of Urology, Medical University of Vienna, 1090 Vienna, Austria; bernhard.grubmueller@meduniwien.ac.at (B.G.); gero.kramer@meduniwien.ac.at (G.K.); shahrokh.shariat@meduniwien.ac.at (S.F.S.)
3. Department of Urology, Weill Cornell Medical College, New York, NY 10065, USA
4. Department of Urology, Second Faculty of Medicine, Charles University, 15006 Prague, Czech Republic
5. Institute for Urology and Reproductive Health, I.M. Sechenov First Moscow State Medical University, 119991 Moscow, Russia
6. Department of Urology, University of Texas Southwestern Medical Center, Dallas, TX 75390, USA
7. Ludwig Boltzmann Institute Applied Diagnostics, 1090 Vienna, Austria
8. Christian Doppler Laboratory for Applied Metabolomics (CDL AM), Medical University of Vienna, 1090 Vienna, Austria
* Correspondence: alexander.haug@meduniwien.ac.at; Tel.: +43-1-40400-39360; Fax: +43-1-40400-5520

Simple Summary: The [^{177}Lu]Lu-PSMA radioligand therapy (PSMA-RLT) has emerged as a successful treatment option in patients with metastatic castration-resistant prostate cancer (mCRPC). Nevertheless, the therapeutic protocol of this treatment is still heterogeneous in many centers, in terms of the number of cycles and the interval between the cycles. Recently, we published the clinical impact of a homogeneous PSMA-RLT protocol that has been applied in our clinic since we started offering this treatment to patients with mCRPC. The outcomes were supportive and promising for analyzing the efficacy and toxicity of using the same treatment regimen in patients who benefited from the first treatment course. Based on the results, we concluded that a second course of three cycles of standardized PSMA-RLT with only a 4-week interval between the cycles is safe and offers favorable tolerability, response rates, overall survival, and progression-free survival, rendering it a promising alternative for the retreatment of mCRPC patients who have formerly responded well to PSMA-RLT.

Abstract: Background: We investigated the response rate and degree of toxicity of a second course of three cycles of [^{177}Lu]Lu-PSMA radioligand therapy (PSMA-RLT) every 4 weeks in mCRPC patients. Methods: Forty-three men (71.5 ± 6.6 years, median PSA 40.8 (0.87–1358 µg/L)) were studied. The response was based on the PSA level 4 weeks after the third cycle. The laboratory parameters before and one month after the last cycle were compared. Kaplan–Meier methods were used to estimate the progression-free survival (PFS) and overall survival (OS), and the Cox regression model was performed to find predictors of survival. Results: Twenty-six patients (60.5%) exhibited a PSA reduction (median PSA declined from 40.8 to 20.2, range 0.6–1926 µg/L, p = 0.002); 18 (42%) and 8 (19%) patients showed a PSA decline of ≥50% and ≥80%, respectively. The median OS and PFS were 136 and 31 weeks, respectively. The patients with only lymph node metastases survived longer (p = 0.02), whereas the patients with bone metastases had a shorter survival (p = 0.03). In the multivariate analysis, only the levels of PSA prior to the therapy remained significant for OS

Citation: Rasul, S.; Wollenweber, T.; Zisser, L.; Kretschmer-Chott, E.; Grubmüller, B.; Kramer, G.; Shariat, S.F.; Eidherr, H.; Mitterhauser, M.; Vraka, C.; et al. Response and Toxicity to the Second Course of 3 Cycles of ^{177}Lu-PSMA Therapy Every 4 Weeks in Patients with Metastatic Castration-Resistant Prostate Cancer. *Cancers* **2021**, *13*, 2489. https://doi.org/10.3390/cancers13102489

Academic Editor: José I. López

Received: 28 April 2021
Accepted: 15 May 2021
Published: 20 May 2021

Publisher's Note: MDPI stays neutral with regard to jurisdictional claims in published maps and institutional affiliations.

Copyright: © 2021 by the authors. Licensee MDPI, Basel, Switzerland. This article is an open access article distributed under the terms and conditions of the Creative Commons Attribution (CC BY) license (https://creativecommons.org/licenses/by/4.0/).

($p < 0.05$, hazard ratio 2.43, 95% CI 1.01–5.87). The levels of hemoglobin (11.5 ± 1.7 g/dL vs. xm, $p = 0.006$) and platelets (208 ± 63 g/L vs. 185 ± 63 g/L, $p = 0.002$) significantly decreased one month after cycle three, though only two grade 3 anemia and one grade 3 thrombocytopenia were recorded. Conclusion: A further intensive PSMA-RLT course is well tolerated in mCRPC patients and associated with promising response rates and OS.

Keywords: PSMA-RLT; ^{177}Lu-PSMA; PSA; mCRPC; prostate cancer

1. Introduction

Prostate cancer is one of the most common cancers and one of the leading oncologic causes of death in men in western countries. In these patients, particularly in those with aggressive, metastatic or castration-resistant prostate cancer (mCRPC), the levels of prostate-specific membrane antigen, also called glutamate carboxypeptidase type II and abbreviated as PSMA, are elevated up to 1000 times the normal value and are inversely correlated with the levels of androgens [1]. These receptors are a highly potent target in the diagnosis and treatment of patients with prostate tumors. Therefore, radionuclides targeting these peptides, such as [^{68}Ga]Ga-PSMA-11 ligand positron emission tomography (^{68}Ga-PSMA PET), which is widely applied as a non-invasive molecular method for imaging prostate cancer, and [^{177}Lu]Lu-PSMA-617 radio-ligand therapy (PSMA-RLT), which has emerged as a valuable treatment in patients with mCRPC, are currently available. Although this novel therapy has not yet been approved for clinical use, it has been successfully administered to patients with mCRPC, based on the results of numerous studies [2–5]. Despite encouraging favorable outcomes of these trials on the efficacy and safety of PSMA-RLT [2,6], this therapy is presently used in mCRPC patients only as a last therapy option when other available standard medical procedures have failed to show clinical improvement. The therapeutic protocol in many centers, however, is quite heterogenous with treatments differing between two to six cycles of 3.7–9.3 GBq PSMA-RLT every 6 to 8 weeks [7]. Most recently, the TheraP study, a multicenter, unblinded, randomized phase 2 trial involving 11 centers in Australia, demonstrated a more frequent PSA response in mCRPC men treated with PSMA-RLT than in patients receiving cabazitaxel at the same stage of the disease. In addition, the results of this study reported fewer serious adverse events in men treated with PSMA-RLT than with cabazitaxel [8].

We lately published the clinical impact of a homogeneous PSMA-RLT protocol consisting of three cycles of 7400 MBq PSMA-RLT with 4-week intervals, which has been used in our clinic since we started offering this treatment to patients with mCRPC [2]. The results of this standardized treatment protocol were very favorable concerning the rate of response, overall survival (OS) as well as progression-free survival (PFS) and therapy-related toxicity, also in comparison to the findings of previous studies [4,6,9,10]. These observations and outcomes were highly supportive to analyze the efficacy and toxicity of applying the same treatment regimen in patients who gained benefit from the first treatment course. Hence, in this study, we aimed to elucidate the response rate and toxicity in mCRPC patients who underwent a second course consisting of three cycles of highly standardized PSMA-RLT every 4 weeks.

2. Patients and Methods

2.1. Patients

This retrospective study included data from all mCRPC patients being referred to the Department of Nuclear Medicine, Medical University of Vienna, Vienna General Hospital, between September 2015 and May 2020 who were eligible and scheduled for the second course of PSMA-RLT. The median distance between the 1st and the 2nd therapy course was 16 weeks (range 4–96 weeks) and the median duration of follow-up of these patients was 30 months (ranged 4–50). In an interdisciplinary tumor board, the initiation of PSMA-RLT was endorsed for all the mCRPC patients studied. The decisive requirements for applying a second course of treatment were a sufficient response rate, high tumor burden, and good tolerability of the first PSMA-RLT course in the absence of clinical and laboratory signs of severe therapy toxicities among the treated patients. Furthermore, the presence of PSMA-positive lesions in a ^{68}Ga-PSMA PET scan conducted for each patient before the start of the 2nd treatment course was mandatory for receiving the therapy. The protocol of the performed ^{68}Ga-PSMA PET/computed tomography (CT) or PET/magnetic resonance imaging (MRI) scan for these patients was previously described in the study by Grubmüller et al. [11].

2.2. Medical Care of the Patients and the Applied PSMA-RLT Protocol

As described previously [2,12], each therapy course routinely applied in our clinic was composed of 3 cycles of PSMA-RLT acquired from ABX GmbH (Radeberg, Germany) with 4-week intervals between every cycle. For each of them, every patient was hospitalized, received medical care and was monitored for at least 72 h. Their Karnofsky performance score, and Eastern Cooperative Oncology Group (ECOG) index were then determined by an experienced medical doctor. Laboratory parameters such as complete blood count, biochemistry, and PSA levels were assessed for each patient at each this visit and 1 month after the last 3rd cycle. Based on the laboratory results, the common terminology criteria for adverse events (CTCAE), version 4.0, were considered to evaluate treatment toxicities. The therapy was intravenously administered following paragraph § 8 of the Austrian Medicinal Products Act (AMG). Thirty minutes prior to the application of the PSMA-RL, every patient obtained 1 liter of normal saline infusion at 300 mL/h. Subsequently, in order to enable imaging assessments of treatment responses, all patients received a second [^{68}Ga]Ga-PSMA-11 whole-body PET scan 4–6 weeks after the last cycle of the therapy.

2.3. Statistical Analysis

All statistical methods mentioned in this study were performed using IBM SPSS Statistics for Windows, version 24.0 (IBM Corp., Armonk, NY, USA). The Kolmogorov–Smirnov test was conducted to estimate the distribution of all data used in this study. Normally distributed data were shown as mean ± standard deviation, whereas non-normally distributed data were displayed as median and range, and log-10 transformed for analysis. All categorical variables were presented in percentages and number of recorded cases, and the comparison of laboratory parameters before and 1 month after acquiring the last 3rd cycle of the PSMA-RLT was conducted using the paired t-test. In all patients, PFS and OS were estimated using Kaplan–Meier estimates and a Cox proportional hazard model. Additionally, log-rank analyses (Mantel–Cox test) were performed to examine the impact of factors such as type and location of metastasis and history of previously receiving other therapies like hormonal as well as chemo- and Ra-223 (Xofigo®) therapy before the start of PSMA RLT on the survival and PFS in this studied cohort. The PFS was defined as the time from the first cycle of the second course of therapy until the detection of PSA progression. OS was ascertained from the date of the first cycle of the first course of therapy as well as from the date of the first cycle of the second course of therapy until the date of death or until the date of the last hospital follow-up. For all results, a p-value < 0.05 was considered statistically significant.

3. Results

Collectively, 43 mCRPC patients (aged 71.4 ± 6.6 years) were valid to acquire the second course of PSMA-RLT, which was composed of three cycles of standardized [^{177}Lu]Lu-PSMA-617 (7351 ± 647 MBq) every 4 weeks. The clinical characteristics of these patients prior receiving the second PSMA-RLT therapy are presented in Table 1. Among this cohort, 26 patients (60.5%) responded to the first PSMA-RLT course with a PSA reduction of more than 50%. The Karnofsky score was lower than 80% in only 16 (37.2%) patients, and equal and higher than 80% in 27 (62.8%) patients. The ECOG index was 0 in 8 (18.6%), 1 in

26 (60.5%) and 2 in 9 (20.9%) patients. Twenty-seven (62.8%) patients had a history of enzalutamide or abiraterone therapy, while 30 (69.7%) patients were previously treated with chemotherapy (docetaxel and/or cabazitaxel) and only 12 (27.9%) patients were treated with Ra-223 (Xofigo®). Between the first and second therapy courses, the patients did not obtain newly initiated treatments with chemo- and Ra-223 therapies. However, in the men already treated with abiraterone or enzalutamide, these therapies were continued in individual patients between the two PSMA-RLT courses without starting new antiandrogenic therapies.

Table 1. Clinical characteristics of the studied mCRPC patients prior to obtaining the second PSMA-RLT therapy.

Parameters	Values
Patients (n)	43
Age (mean ± SD) years	71.4 ± 6.6
Weight (mean ± SD) kilogram	83.1 ± 11.4
[^{177}Lu]Lu-PSMA-617 MBq	7351 ± 647
≥50% PSA decline after 1st PSMA-RLT (n) %	(26) 60.5
Karnofsky score (n) %	
<80%	(16) 37.2
≥80%	(27) 62.8
ECOG index (n) %	
0	(8) 18.6
1	(26) 60.5
2	(9) 20.9
Previous treatments (n) %	
Enzalutamide/abiraterone	(27) 62.8
Docetaxel/cabazitaxel	(30) 69.7
Ra-223 (Xofigo®)	(12) 27.9
Metastatic lesions (n) %	
M1a	(8) 18.6
M1b	(27) 62.8
M1c	(8) 18.6

(n): number; MBq: megabecquerel; ECOG: Eastern Cooperative Oncology Group. M1a: lymph node only; M1b: bone ± lymph node without visceral metastasis; M1c: visceral metastasis.

The distributions of metastatic lesions on the basis of the [^{68}Ga]Ga-PSMA-11 scan were as follows: lymph node only (M1a) in 8 (18.6%), bone ± lymph node without visceral metastasis (M1b) in 27 (62.8%), and any visceral metastasis (M1c) in 8 (18.6%) patients, all shown in Table 1.

3.1. Response Rate and Clinical Effects of Second PSMA-RLT Course

In Table 2, the laboratory parameters of the entire studied mCRPC patients before and 1 month after the third last cycle of the second course of PSMA-RLT applied every 4 weeks have been compared. The PSA levels of the treated patients decreased significantly after three cycles of PSMA-RLT compared with baseline, median PSA 40.8 (range 0.87–1358 µg/L) vs. 20.2 (range 0.6–1926 µg/L), $p = 0.002$. Overall, 26 out of 43 (60.5%) patients demonstrated any decrease in PSA levels, 18 out of 43 (42%) had a PSA decline of ≥50%, and 8 of 43 (19%) patients showed a PSA decrease of ≥80%. The percentage of the

PSA decline after both treatment courses of highly standardized PSMA-RLT, each three cycles with a 4-week interval, in all patients studied are depicted in Figure 1.

Table 2. Comparison of laboratory parameters of the studied mCRPC patients before and after the second course of three cycles of PSMA-RLT every 4 weeks.

Parameters	Before Therapy	After Therapy	q-Value
* PSA µg/L	40.8 (0.87–1358)	20.2 (0.60–1962)	0.002
Hemoglobin g/dL (mean ± SD)	11.5 ± 1.7	11 ± 1.6	0.006
Thrombocyte g/L (mean ± SD)	208 ± 63	185 ± 63	0.002
* Leucocyte g/L	5.4 (1.17–14.3)	4.8 (2.1–14.1)	n.s.
* Creatinine mg/dL	0.96 (0.54–2.24)	0.94 (0.61–2.6)	n.s.
* Alkaline phosphatase U/L	78 (42–995)	84 (47–1345)	n.s.
* LDH U/L	205 (96–278)	194 (86–551)	n.s.

PSA: prostate-specific antigen; LDH: lactate dehydrogenase; n.s.: not significant; (*) data not normally distributed, presented in median and range and $\log_{10}$ transferred for analysis.

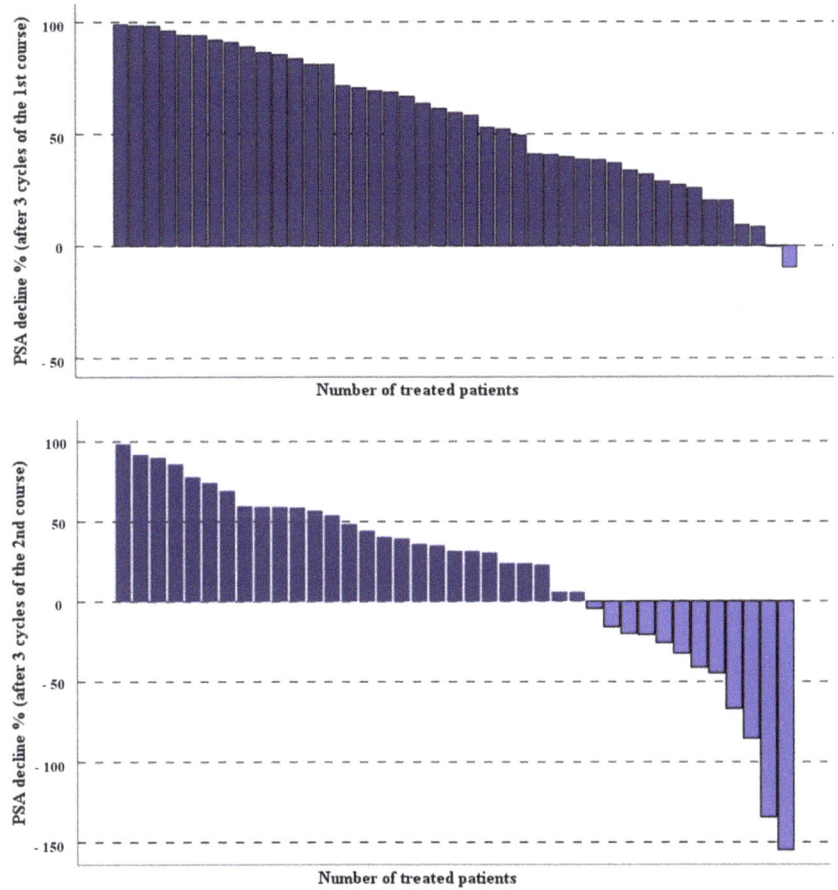

Figure 1. Percentage of PSA change in the studied patients after two courses of the highly standardized PSMA-RLT, each three cycles with 4-week interval.

Moreover, levels of hemoglobin (Hb) (11.5 ± 1.7 g/dL vs. 11 ± 1.6 g/dL, $p = 0.006$) and platelets (208 ± 63 g/L vs. 185 ± 63 g/L, $p = 0.002$) one month after the third cycle were significantly lower compared to the baseline (Table 2). However, only two cases of grade 3 anemia and one case of grade 3 thrombocytopenia were observed among all the treated patients (Table 3). In addition, no statistically significant changes in the levels of leukocyte, creatinine, alkaline phosphatase, and lactate dehydrogenase were observed when we compared their basal values with those one month after treatment with three cycles of PSMA-RLT (Table 2). No patients with severe gastrointestinal adverse events, as well as no patients with acute parotitis and myelodysplastic syndrome, were reported during the second PSMA-RLT course.

Table 3. Evaluation of treatment toxicities based on CTCAE version 4.0.

Parameters	Before Therapy Toxicity (n)			After Therapy Toxicity (n)		
	Grade 1	Grade 2	Grade 3	Grade 1	Grade 2	Grade 3
Hemoglobin g/dL	18	7	0	15	9	2
Thrombocyte g/L	8	0	0	8	0	1
Leucocyte g/L	2	0	1	3	1	0
Creatinine mg/dL	3	0	0	1	0	0

(n): number of reported cases.

3.2. Overall Survival of Patients Treated with Second PSMA-RLT Course

Kaplan–Meier plots of the entire treated patients revealed a median OS of 188 weeks from the beginning of the first cycle of the first course, and a median OS of 136 weeks from the start of the first cycle of the second course of the treatment, shown in Figure 2. Among the collective populations studied, the median PFS from the time of the beginning of the second therapy course was 31 weeks (95% CI 26–36), while from the first cycle of the first course to the PSA progression was 27 weeks (95% CI 22–32). In Table 4, we have presented the OS as well as the PFS calculated from the time of the first cycles of both PSMA-RLT courses depending on the type of metastases present in the treated patients. As shown in that table, after receiving the first and second therapy courses, the shortest OS was observed among the patients with M1c, whereas the shortest PFS was seen in the patients with M1b.

The results of log-rank analyses to ascertain the overall survival of the patients by the type of metastasis indicated a significantly shorter survival of the patients who had metastatic bone lesions (M1b) compared with those with other types of metastases (M1a or M1c), they had a median survival of 123 weeks vs. not reached, $p = 0.03$, 95% CI 2.42–243, shown in Figure 3. Additionally, the existence of only lymph node metastases (M1a) was significantly associated with a longer survival compared with the availability of prostate metastatic lesions of other types (147 weeks vs. median survival not reached, $p = 0.02$). Figure 4 illustrates the ^{68}Ga-PSMA-11 PET scan images of a patient with M1a demonstrated a highly favorable response to two courses of PSMA-RLT.

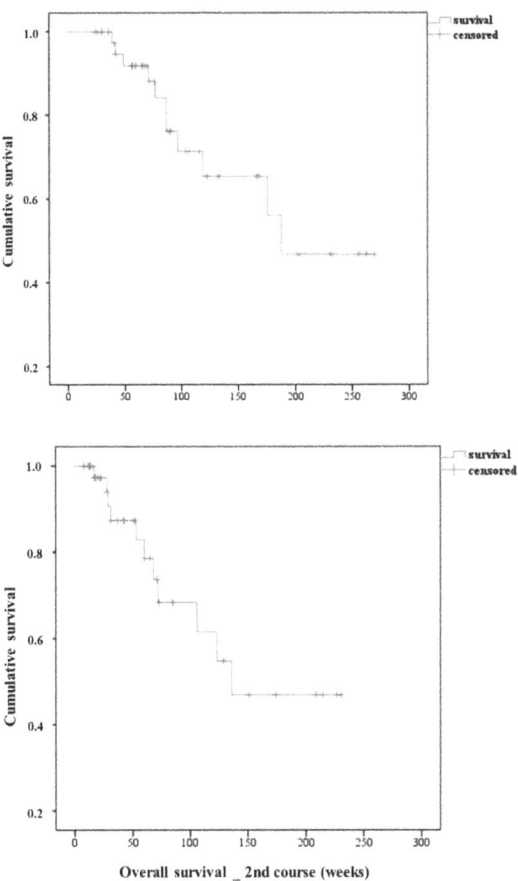

Figure 2. Kaplan–Meier plots of the treated patients received collectively two courses of PSMA-RLT, each course composed of three cycles of 7400 MBq PSMA-RLT every 4 weeks. The median overall survival from the time of the first cycle of the first PSMA-RTL course was 188 weeks, whereas the median survival from the time of the first cycle of the second PSMA-RLT course was 136 weeks.

Table 4. Median overall survival and median progression-free survival in weeks calculated from the time of the first and second PSMA-RLT course in relation to the type of metastases present.

Type of Metastasis	OS Calculated from 1st Course (Weeks)	OS Calculated from 2nd Course (Weeks)	PFS after 1st Course (Weeks)	PFS after 2nd Course (Weeks)
Total population	188	136	27	31
M1a	>169 *	>147 *	41	32
M1b	176	123	25	24
M1c	119	106	44	40

M1a: patients with lymph node metastasis; M1b: patients with bone ± lymph node without visceral metastasis; M1c: patients with visceral metastasis; OS: overall survival; PFS: progression-free survival; (*): no patient died in this group during follow-up.

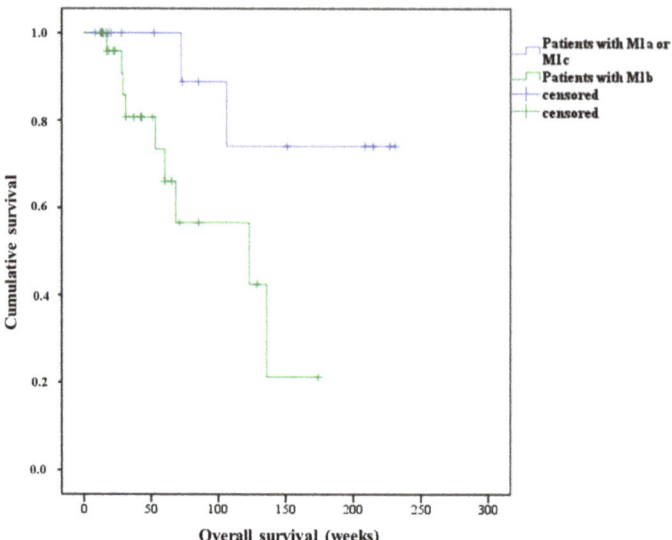

Figure 3. Kaplan–Meier plots using log-rank (Mantel–Cox) test illustrate overall survival of mCRPC patients treated with a second course of three cycles of PSMA-RLT. Male patients with bone metastasis (M1b) had significantly shorter survival compared to patients with lymph node (M1a) or visceral metastases (M1c), median survival 123 weeks vs. not reached, $p = 0.03$, 95% CI 2.42–243.

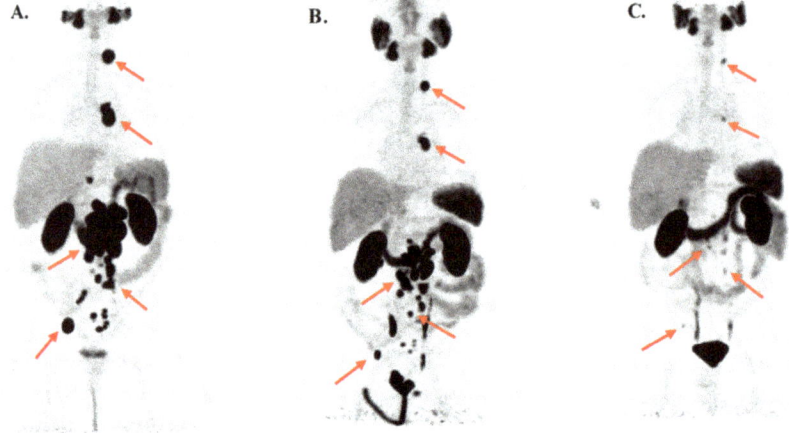

Figure 4. The [68Ga]-Ga-PSMA PET examination of an mCRPC patient with only lymph node metastasis who received two courses (each three cycles with 4-week interval) of PSMA-RLT. The [^{68}Ga]GaPSMA-11 PET images of a 76-year-old mCRPC patient. (**A**): prior the first cycle of first course of PSMA-RLT with clearly PSMA-overexpressed lymph node (LN) metastases in upper and lower diaphragm and a PSA value of 597 ng/mL. (**B**): one month after the third cycle of the first course of PSMA-RLT the metastatic LN were measured smaller with computed tomography (CT) and the PSA level declined to 81.2 ng/mL (reduction of 85%). (**C**): One month after the third cycle of the second course of PSMA-RLT, these LN were either tiny or completely disappeared in the CT and the PSA level dropped further to 0.93 ng/mL (reduction of >95%). The overall survival of this patient was 263 weeks from the beginning of the first cycle of first course and 227 weeks from the beginning of the first cycle of second course of the PSMA-RLT.

While the results of the univariate analysis with the Cox regression model showed levels of Hb as well as serum alkaline phosphatase and PSA prior to the first cycle of the second PSMA-RLT course as predictors for OS in these retreated patients (all $p < 0.05$), in multivariate analysis, only the PSA levels remained significant for survival ($p < 0.05$, hazard ratio 2.43, 95% CI 1.01–5.87). Nevertheless, we did not observe a significant impact of other therapies, such as hormonal as well as chemo- and Ra-223 (Xofigo®) therapies, before the start of PSMA-RLT on the survival and PFS in our investigated cohort.

4. Discussion

Owing to the poor prognosis of the patients with mCRPC, and their low survival rate of less than 2 years from the time of their diagnosis, new therapeutic approaches and strategies are constantly striven to improve the survival and quality of life of these subjects [13].

In this study, we presented data of a rather selected mCRPC population, who previously benefited from PSMA-RLT and was retreated with another course consisting of three cycles of a standardized radionuclide therapy every 4 weeks, with a median interval of 16 weeks (range 4–96 weeks) between the first and the second therapy courses. Most of these patients were pretreated with enzalutamide and/or abiraterone as well as docetaxel and/or cabazitaxel therapy and the majority exhibited a good response rate and tolerability without having clinical and laboratory signs of severe therapy toxicities to the first PSMA-RLT course.

Consistent with the outcomes of other previous studies where patients were retreated with PSMA-RLT [14,15], approximately 42% of the treated patients who responded to therapy showed a PSA reduction of greater than 50%. Indeed, the levels of PSMA expression in prostatic tumors, their related metastases and levels of serum PSA might not correlate with each other, as the expression of PSA, unlike PSMA, is mainly promoted by androgens and regulated by the androgen receptor [16]. In the prospective TheraP study by Hofman et al., PSA was used as the primary endpoint to evaluate the therapy response to both PSMA-RLT and cabazitaxel in mCRPC patients [8]. Consistently, response assessment based on PSA levels was one of the main endpoints of a single-center phase II prospective trial by Violet et al. [14], which included 50 mCRPC patients retreated with PSMA-RLT. In a study by Grubmüller et al., which involved patients who underwent the first PSMA-RLT course, treatment response was assessed in these mCRPC patients by comparing PSMA uptake in tumors and metastases before and one month after the third cycle of therapy using ^{68}Ga-PSMA PET scan examinations. The results revealed a significant association of changes in the total tumor volume on the PSMA PET scan, but not in the RECIST (response criteria in solid tumors) evaluation with the PSA response [11]. Although it was not the focus of this current study, the PSMA PET parameters were a strong predictor of survival in men treated with PSMA-RLT in a study by Ferdinandus et al. [17].

In addition, the report of only two cases of grade 3 anemia and one case of severe thrombocytopenia suggests that the rate of treatment-related toxicity remained good and consistent with the results of previous studies [15,18,19]. These results, thus, indicated that a further therapy course with an additional three cycles of PSMA-RLT every 4 weeks is well tolerated and accompanied by satisfactory response rates to the treatment. Likewise, a second course of treatment for 30 mCRPC patients was also performed in a study by Yordanova et al. [20], and the outcomes have similarly demonstrated safety and efficacy of rechallenge PSMA-RLT. However, unlike our study, the first and second rechallenge therapy in that study were heterogenous regarding the injected activity (ranged 3.8–6.7 GBq) as well as the number of cycles in each course (ranged 1–6 cycles), and the interval between these cycles was ambiguous. In this respect, the diversity of the patients treated, and the differences in their tumor burden and comorbidities in our study and the study by Yordanova et al. should be taken into consideration.

Furthermore, a median OS of approximately 4 years from the onset of the first cycle of the first course and a median OS of approximately 3 years from the date of initiation of the first cycle of the second therapy course were observed in our studied cohort. The median PFS with 31 weeks after the second therapy course was slightly superior to the median PFS from the last cycle of the first course until PSA progress (= 27 weeks). These findings are not unexpected given the growing recognition of the therapeutic efficacy and, thus, the prolonged survival of patients with advanced mCRPC who have acquired PSMA-RLT, as a PSA decrease of only $\geq$20% is predictive of prolonged survival [2,6,20]. The shortest OS was found in patients with M1c and the shortest time from the last treatment cycle to disease progression was identified in patients with M1b. Moreover, the patients who had only metastatic bone lesions lived significantly shorter than the patients with other types of metastases (median survival not reached vs. 123 weeks, p = 0.03). Although the type and distribution of metastases did not influence patient survival in the results of previous studies [10,12,21], the results of this current analysis indicated an association between the presence of lymph node metastases and longer survival, whereas the presence of bone metastases was significantly linked to a shorter survival in patients retreated with three cycles of PSMA-RLT every 4 weeks. In agreement with these findings, Ahmadzadehfar et al. have shown a negative impact of bone metastasis on the survival of patients treated with PSMA-RLT in 11 different clinics in a multicenter study, including data from more than 400 mCRPC patients [22]. Furthermore, the results of the same study showed 30 patients with only lymph node metastases that had the longest median OS among all other patients studied. In fact, at this advanced tumor stage with bone involvement, they often have diffused bone marrow metastasis, which limits the effectiveness of the therapy. Hence, patients with only lymph node metastasis might have a better outcome and a higher response rate to PSMA-RLT than patients with bone $\pm$ lymph node metastases. This has also been demonstrated in other studies [23,24], particularly in a study by von Eyben et al. in 45 patients with predominant lymph node metastatic prostate cancer [25].

Firstly, the retrospective design is the major limitation of the study. Secondly, the small sample of the included patients with different tumor burdens and diverse pretreatments will restrict the results of this investigation. However, this treated cohort represents the patient population referred to PSMA-RLT in clinical routine quite well. Additionally, although we have previously published part of our dosimetric data in a subgroup of patients who obtained the first PSMA-RLT course [26], the lack of such information in this current study could limit its outcomes. The reason lies in the crucial role of radiation dosimetry in estimating the therapy response and level of absorbed radiation dose, not only for each individual metastasis but also for the organs physiologically exhibiting an uptake of PSMA-RLT [27] and thereby evaluating their degree of therapy toxicity [28]. Moreover, no comparison has been conducted between PSMA-RLT and other hormonal or chemotherapies that might be optioned for these mCRPC patients at this stage of the tumor. Thus, the results of larger prospective studies such as VISION [29], comparing survival outcomes of patients receiving PSMA-RLT with those acquiring the best standard medical care, as well as the interesting results of the multicenter TheraP trial [8], should help to support the forthcoming implementation of this radionuclide therapy into the clinical treatment routine of patients with prostate cancer.

5. Conclusions

A second course of three cycles of standardized PSMA-RLT with only a 4-week interval between the cycles is safe and yields favorable tolerability, response rates, OS and PFS, thereby making it a promising option for the retreatment of mCRPC patients who have previously responded well to PSMA-RLT.

Author Contributions: S.R. wrote the manuscript, collected, and researched the data, contributed to discussion. T.W., L.Z., E.K.-C. collected the data. B.G. and G.K. contributed to patient's recruitment and study design. S.F.S., M.M., W.L. and M.H. reviewed the manuscript and contributed to discussion. H.E. and C.V. contributed to therapy preparation and reviewed the manuscript. A.R.H. designed the study, contributed to discussion, reviewed/edited the manuscript. All authors have read and agreed to the published version of the manuscript.

Funding: This research received no external funding.

Institutional Review Board Statement: The study was authorized by the Ethics Committee of the Medical University of Vienna (EK: 1143/2019). All procedures performed in this study were in accordance with the ethical standards of the institutional and/or national research committee and with the principles of the 1964 Declaration of Helsinki and its later amendments or comparable ethical standards. This paper does not include any animal study conducted by authors.

Informed Consent Statement: Before applying any therapy, a written informed consent was obtained from each patient.

Data Availability Statement: Not applicable.

Acknowledgments: The authors would like to thank Verena Pichler for radiopharmaceutical production management and the radiopharmaceutical production team Mag. pharm. Stefan Schmitl, Rainer Bartosch and Sabine Wunsch, who contributed to preparation, production, quality control of the radiopharmaceuticals and calibrations and service of modules and other devices. Moreover, we thank Michael Weber for the statistical support.

Conflicts of Interest: The authors declare no conflict of interest.

References

1. Silver, D.A.; Pellicer, I.; Fair, W.R.; Heston, W.D.; Cordon-Cardo, C. Prostate-specific membrane antigen expression in normal and malignant human tissues. *Clin. Cancer Res.* **1997**, *3*, 81–85.
2. Rasul, S.; Hacker, M.; Kretschmer-Chott, E.; Leisser, A.; Grubmuller, B.; Kramer, G.; Shariat, S.; Wadsak, W.; Mitterhauser, M.; Hartenbach, M.; et al. Clinical outcome of standardized (177)Lu-PSMA-617 therapy in metastatic prostate cancer patients receiving 7400 MBq every 4 weeks. *Eur. J. Nucl. Med. Mol. Imaging* **2020**, *47*, 713–720. [CrossRef] [PubMed]
3. Ahmadzadehfar, H.; Rahbar, K.; Kurpig, S.; Bogemann, M.; Claesener, M.; Eppard, E.; Gartner, F.; Rogenhofer, S.; Schafers, M.; Essler, M. Early side effects and first results of radioligand therapy with (177)Lu-DKFZ-617 PSMA of castrate-resistant metastatic prostate cancer: A two-centre study. *EJNMMI Res.* **2015**, *5*, 114. [CrossRef]
4. Ahmadzadehfar, H.; Eppard, E.; Kurpig, S.; Fimmers, R.; Yordanova, A.; Schlenkhoff, C.D.; Gartner, F.; Rogenhofer, S.; Essler, M. Therapeutic response and side effects of repeated radioligand therapy with 177Lu-PSMA-DKFZ-617 of castrate-resistant metastatic prostate cancer. *Oncotarget* **2016**, *7*, 12477–12488. [CrossRef]
5. McBean, R.; O'Kane, B.; Parsons, R.; Wong, D. Lu177-PSMA therapy for men with advanced prostate cancer: Initial 18 months experience at a single Australian tertiary institution. *J. Med. Imaging Radiat. Oncol.* **2019**, *63*, 538–545. [CrossRef]
6. Rahbar, K.; Boegemann, M.; Yordanova, A.; Eveslage, M.; Schafers, M.; Essler, M.; Ahmadzadehfar, H. PSMA targeted radioligandtherapy in metastatic castration resistant prostate cancer after chemotherapy, abiraterone and/or enzalutamide. A retrospective analysis of overall survival. *Eur. J. Nucl. Med. Mol. Imaging* **2018**, *45*, 12–19. [CrossRef] [PubMed]
7. Kratochwil, C.; Fendler, W.P.; Eiber, M.; Baum, R.; Bozkurt, M.F.; Czernin, J.; Delgado Bolton, R.C.; Ezziddin, S.; Forrer, F.; Hicks, R.J.; et al. EANM procedure guidelines for radionuclide therapy with (177)Lu-labelled PSMA-ligands ((177)Lu-PSMA-RLT). *Eur. J. Nucl. Med. Mol. Imaging* **2019**, *46*, 2536–2544. [CrossRef] [PubMed]
8. Hofman, M.S.; Emmett, L.; Sandhu, S.; Iravani, A.; Joshua, A.M.; Goh, J.C.; Pattison, D.A.; Tan, T.H.; Kirkwood, I.D.; Ng, S.; et al. [(177)Lu]Lu-PSMA-617 versus cabazitaxel in patients with metastatic castration-resistant prostate cancer (TheraP): A randomised, open-label, phase 2 trial. *Lancet* **2021**, *397*, 797–804. [CrossRef]
9. Fendler, W.P.; Reinhardt, S.; Ilhan, H.; Delker, A.; Boning, G.; Gildehaus, F.J.; Stief, C.; Bartenstein, P.; Gratzke, C.; Lehner, S.; et al. Preliminary experience with dosimetry, response and patient reported outcome after 177Lu-PSMA-617 therapy for metastatic castration-resistant prostate cancer. *Oncotarget* **2017**, *8*, 3581–3590. [CrossRef] [PubMed]
10. Ferdinandus, J.; Eppard, E.; Gaertner, F.C.; Kurpig, S.; Fimmers, R.; Yordanova, A.; Hauser, S.; Feldmann, G.; Essler, M.; Ahmadzadehfar, H. Predictors of Response to Radioligand Therapy of Metastatic Castrate-Resistant Prostate Cancer with 177Lu-PSMA-617. *J. Nucl. Med.* **2017**, *58*, 312–319. [CrossRef] [PubMed]
11. Grubmuller, B.; Senn, D.; Kramer, G.; Baltzer, P.; D'Andrea, D.; Grubmuller, K.H.; Mitterhauser, M.; Eidherr, H.; Haug, A.R.; Wadsak, W.; et al. Response assessment using (68)Ga-PSMA ligand PET in patients undergoing (177)Lu-PSMA radioligand therapy for metastatic castration-resistant prostate cancer. *Eur. J. Nucl. Med. Mol. Imaging* **2019**, *46*, 1063–1072. [CrossRef]
12. Rasul, S.; Hartenbach, M.; Wollenweber, T.; Kretschmer-Chott, E.; Grubmuller, B.; Kramer, G.; Shariat, S.; Wadsak, W.; Mitterhauser, M.; Pichler, V.; et al. Prediction of response and survival after standardized treatment with 7400 MBq (177)Lu-PSMA-617 every 4 weeks in patients with metastatic castration-resistant prostate cancer. *Eur. J. Nucl. Med. Mol. Imaging* **2020**. [CrossRef] [PubMed]
13. Huang, X.; Chau, C.H.; Figg, W.D. Challenges to improved therapeutics for metastatic castrate resistant prostate cancer: From recent successes and failures. *J. Hematol. Oncol.* **2012**, *5*, 35. [CrossRef]
14. Violet, J.; Sandhu, S.; Iravani, A.; Ferdinandus, J.; Thang, S.P.; Kong, G.; Kumar, A.R.; Akhurst, T.; Pattison, D.A.; Beaulieu, A.; et al. Long-Term Follow-up and Outcomes of Retreatment in an Expanded 50-Patient Single-Center Phase II Prospective Trial of (177)Lu-PSMA-617 Theranostics in Metastatic Castration-Resistant Prostate Cancer. *J. Nucl. Med.* **2020**, *61*, 857–865. [CrossRef]

15. Gafita, A.; Rauscher, I.; Retz, M.; Knorr, K.; Heck, M.; Wester, H.J.; D'Alessandria, C.; Weber, W.A.; Eiber, M.; Tauber, R. Early Experience of Rechallenge (177)Lu-PSMA Radioligand Therapy After an Initial Good Response in Patients with Advanced Prostate Cancer. *J. Nucl. Med.* **2019**, *60*, 644–648. [CrossRef]
16. Evans, M.J.; Smith-Jones, P.M.; Wongvipat, J.; Navarro, V.; Kim, S.; Bander, N.H.; Larson, S.M.; Sawyers, C.L. Noninvasive measurement of androgen receptor signaling with a positron-emitting radiopharmaceutical that targets prostate-specific membrane antigen. *Proc. Natl. Acad. Sci. USA* **2011**, *108*, 9578–9582. [CrossRef] [PubMed]
17. Ferdinandus, J.; Violet, J.; Sandhu, S.; Hicks, R.J.; Ravi Kumar, A.S.; Iravani, A.; Kong, G.; Akhurst, T.; Thang, S.P.; Murphy, D.G.; et al. Prognostic biomarkers in men with metastatic castration-resistant prostate cancer receiving [177Lu]-PSMA-617. *Eur. J. Nucl. Med. Mol. Imaging* **2020**, *47*, 2322–2327. [CrossRef] [PubMed]
18. Zechmann, C.M.; Afshar-Oromieh, A.; Armor, T.; Stubbs, J.B.; Mier, W.; Hadaschik, B.; Joyal, J.; Kopka, K.; Debus, J.; Babich, J.W.; et al. Radiation dosimetry and first therapy results with a (124)I/ (131)I-labeled small molecule (MIP-1095) targeting PSMA for prostate cancer therapy. *Eur. J. Nucl. Med. Mol. Imaging* **2014**, *41*, 1280–1292. [CrossRef]
19. Baum, R.P.; Kulkarni, H.R.; Schuchardt, C.; Singh, A.; Wirtz, M.; Wiessalla, S.; Schottelius, M.; Mueller, D.; Klette, I.; Wester, H.J. 177Lu-Labeled Prostate-Specific Membrane Antigen Radioligand Therapy of Metastatic Castration-Resistant Prostate Cancer: Safety and Efficacy. *J. Nucl. Med.* **2016**, *57*, 1006–1013. [CrossRef]
20. Yordanova, A.; Linden, P.; Hauser, S.; Meisenheimer, M.; Kurpig, S.; Feldmann, G.; Gaertner, F.C.; Essler, M.; Ahmadzadehfar, H. Outcome and safety of rechallenge [(177)Lu-PSMA-617 in patients with metastatic prostate cancer. *Eur. J. Nucl. Med. Mol. Imaging* **2019**, *46*, 1073–1080. [CrossRef]
21. Ahmadzadehfar, H.; Schlolaut, S.; Fimmers, R.; Yordanova, A.; Hirzebruch, S.; Schlenkhoff, C.; Gaertner, F.C.; Awang, Z.H.; Hauser, S.; Essler, M. Predictors of overall survival in metastatic castration-resistant prostate cancer patients receiving [(177)Lu]-PSMA-617 radioligand therapy. *Oncotarget* **2017**, *8*, 103108–103116. [CrossRef]
22. Ahmadzadehfar, H.; Rahbar, K.; Baum, R.P.; Seifert, R.; Kessel, K.; Bogemann, M.; Kulkarni, H.R.; Zhang, J.; Gerke, C.; Fimmers, R.; et al. Prior therapies as prognostic factors of overall survival in metastatic castration-resistant prostate cancer patients treated with [(177)Lu]-PSMA-617. A WARMTH multicenter study (the 617 trial). *Eur. J. Nucl. Med. Mol. Imaging* **2021**, *48*, 113–122. [CrossRef] [PubMed]
23. Kulkarni, H.R.; Singh, A.; Schuchardt, C.; Niepsch, K.; Sayeg, M.; Leshch, Y.; Wester, H.J.; Baum, R.P. PSMA-Based Radioligand Therapy for Metastatic Castration-Resistant Prostate Cancer: The Bad Berka Experience Since 2013. *J. Nucl. Med.* **2016**, *57* (Suppl. 3), 97S–104S. [CrossRef]
24. Von Eyben, F.E.; Kiljunen, T.; Joensuu, T.; Kairemo, K.; Uprimny, C.; Virgolini, I. (177)Lu-PSMA-617 radioligand therapy for a patient with lymph node metastatic prostate cancer. *Oncotarget* **2017**, *8*, 66112–66116. [CrossRef]
25. Von Eyben, F.E.; Singh, A.; Zhang, J.; Nipsch, K.; Meyrick, D.; Lenzo, N.; Kairemo, K.; Joensuu, T.; Virgolini, I.; Soydal, C.; et al. (177)Lu-PSMA radioligand therapy of predominant lymph node metastatic prostate cancer. *Oncotarget* **2019**, *10*, 2451–2461. [CrossRef] [PubMed]
26. Barna, S.; Haug, A.R.; Hartenbach, M.; Rasul, S.; Grubmuller, B.; Kramer, G.; Blaickner, M. Dose Calculations and Dose-Effect Relationships in 177Lu-PSMA I&T Radionuclide Therapy for Metastatic Castration-Resistant Prostate Cancer. *Clin. Nucl. Med.* **2020**, *45*, 661–667. [PubMed]
27. Violet, J.; Jackson, P.; Ferdinandus, J.; Sandhu, S.; Akhurst, T.; Iravani, A.; Kong, G.; Kumar, A.R.; Thang, S.P.; Eu, P.; et al. Dosimetry of (177)Lu-PSMA-617 in Metastatic Castration-Resistant Prostate Cancer: Correlations Between Pretherapeutic Imaging and Whole-Body Tumor Dosimetry with Treatment Outcomes. *J. Nucl. Med.* **2019**, *60*, 517–523. [CrossRef] [PubMed]
28. Paganelli, G.; Sarnelli, A.; Severi, S.; Sansovini, M.; Belli, M.L.; Monti, M.; Foca, F.; Celli, M.; Nicolini, S.; Tardelli, E.; et al. Dosimetry and safety of (177)Lu PSMA-617 along with polyglutamate parotid gland protector: Preliminary results in metastatic castration-resistant prostate cancer patients. *Eur. J. Nucl. Med. Mol. Imaging* **2020**, *47*, 3008–3017. [CrossRef] [PubMed]
29. Rahbar, K.; Bodei, L.; Morris, M.J. Is the Vision of Radioligand Therapy for Prostate Cancer Becoming a Reality? An Overview of the Phase III VISION Trial and Its Importance for the Future of Theranostics. *J. Nucl. Med.* **2019**, *60*, 1504–1506. [CrossRef] [PubMed]

Communication

Multiparametric Magnetic Resonance Imaging-Ultrasound Fusion Transperineal Prostate Biopsy: Diagnostic Accuracy from a Single Center Retrospective Study

Andrea Fulco [1,2], Francesco Chiaradia [1], Luigi Ascalone [1], Vincenzo Andracchio [1,2], Antonio Greco [1], Manlio Cappa [1], Marcello Scarcia [3], Giuseppe Mario Ludovico [3], Vincenzo Pagliarulo [4], Camillo Palmieri [2] and Stefano Alba [1,*]

[1] Department of Urology, Romolo Hospital, 88821 Rocca di Neto, Italy; andreafulco.md@gmail.com (A.F.); francescochiaradia@romolohospital.com (F.C.); luigiascalone@romolohospital.com (L.A.); andracchio.v@libero.it (V.A.); antoniogreco@romolohospital.com (A.G.); manlio.cappa@gmail.com (M.C.)
[2] Department of Clinical and Experimental Medicine, University Magna Grecia of Catanzaro, 88100 Catanzaro, Italy; cpalmieri@unicz.it
[3] Department of Urology, Ente Ecclesiastico Ospedale Generale Regionale "F. Miulli", 70021 Acquaviva delle Fonti, Italy; m.scarcia@miulli.it (M.S.); g.ludovico@miulli.it (G.M.L.)
[4] Department of Urology, Ospedale Vito Fazzi, 73100 Lecce, Italy; enzopagliarulo@yahoo.com
* Correspondence: stefanoalba@romolohospital.com; Tel.: +39-0962-80322

Simple Summary: The introduction of imaging techniques has improved the diagnostic pathway for prostate cancer. In this study we compared the diagnostic accuracy of multiparametric MRI with fusion ultrasound-guided prostate biopsy and standard biopsy, both performed through the transperineal route. Our results support the combined targeted and standard biopsy pathway to reduce the risk of missing clinically significant prostate cancer.

Abstract: The management of prostate biopsy in men with clinical suspicion of prostate cancer has changed in the last few years, especially with the introduction of imaging techniques, to overcome the low efficacy of risk stratification based on PSA levels. Here, we aimed to compare the diagnostic accuracy of multiparametric MRI with fusion ultrasound-guided prostate biopsy and standard biopsy, both performed through the transperineal route. To this end, we retrospectively analyzed 272 patients who underwent combined transperineal targeted and standard biopsy during the same session. The primary outcome was to compare the cancer detection rate between targeted and standard biopsy. The secondary outcome was to evaluate the added value of combined targeted and standard biopsy approach as compared to only targeted or standard biopsy. Results showed that a rate of 16.7% clinically significant tumors (International Society of Urological Pathology (ISUP) grade ≥ 2) would have been lost if only the standard biopsy had been used. The combined targeted and standard biopsy showed an added value of 10.3% and 9.9% in reducing the risk of prostate cancer missing after targeted or standard biopsy alone, respectively. The combined targeted and standard biopsy pathway is recommended to reduce the risk of missing clinically significant prostate cancer.

Keywords: image-guided; magnetic resonance imaging; ultrasonography; prostatic neoplasms; biopsy

1. Introduction

Prostate cancer (PCa) is a leading cause of cancer death in men. The diagnostic pathway of PCa in men presenting with symptoms referable to a possible prostate disease includes the combined use of digital rectal examination (DRE), serum biomarkers, imaging techniques, and biopsy. The introduction of robust prostate-specific antigen (PSA) assay has long fostered the possibility of screening for early disease prediction in asymptomatic men. In addition to PSA, other serum and urinary biomarkers (reviewed in [1]) have been identified and approved by the Food and Drug Administration (FDA) and Clinical

Laboratory Improvement Amendments (CLIA) for improving PCa diagnosis and prognosis, and helping in biopsy decision.

Over the past decade, the introduction of multiparametric magnetic resonance imaging (mpMRI) and mpMRI-Ultrasound Fusion-Targeted (TBx) has raised great expectations about the diagnostic pathway of PCa. The large-scale Patient-Reported Outcomes Measurement Information System (PROMIS) multicenter study indicated that mpMRI, when performed as a triage test prior to biopsy, has a two-fold advantage: on the one hand, it significantly reduces overdiagnosis and overtreatment of clinically indolent tumors; on the other hand, it significantly increases the diagnosis of clinically significant tumors compared to ultrasound (US)-guided random biopsy [2]. Many studies comparing the diagnostic accuracy between TBx and standard biopsy (SBx) in detecting clinically significant PCa (ISUP grade ≥ 2) demonstrated the superiority of TBx in the repeat-biopsy setting [3]. Even in biopsy-naïve patients, TBx out-performed SBx, but the difference appears to be less pronounced and insignificant [2,4]. Furthermore, TBx appeared to detect fewer patients with clinically insignificant PCa (ISUP grade 1, maximum core length < 6 mm) than SBx. In consequence, TBx was superior to SBx in reducing overdiagnosis of low-risk disease [3–6].

Many studies have evaluated the combined diagnostic pathway, in which SBx and TBx biopsy was performed in the same patients with a positive mpMRI. The data from the Cochrane meta-analysis of these studies indicated that the absolute added value of TBx for detecting ISUP grade > 2 cancers is higher than that of SBx [3].

Prostate lesions found on mpMRI are graded from 1 to 5, according to the Prostate Imaging-Reporting and Data System (PIRADS) version 2, where higher imaging suspicion scores are associated with a higher risk of clinically significant PCs (csPCa) [7]. A score of "3" indicates equivocal results, "4" results are likely to be prostate cancer, and "5" results are highly likely to be prostate cancer.

The TBx can be achieved through a transrectal or transperineal route. Both approaches are equivalent for patient tolerability and PCa detection rates in an SBx setting, with slight differences in infectious and retention complications [8–10]. However, the comparison in cancer detection rate (CDR) in a TBx configuration remains unclear.

The purpose of this study was to compare the CDR of TBx and SBx, both performed through the transperineal route. An additional outcome was to evaluate the added value of the combined TBx and SBx approach as compared to only TBx or SBx biopsy.

2. Materials and Methods

We analyzed data from a database of 272 patients who underwent primary or repeated prostate biopsy at a single institution in the period from December 2017 to February 2020. All patients gave their informed consent to use the data obtained from medical records; biopsies were performed according to standard procedures, and no additional biopsy was performed for this study; in addition, patient identification information was anonymized before analysis. All procedures performed in this study were in accordance with the Helsinki declaration and its later amendments for comparable ethical standards.

Data were from men aged at least 18 years, referred with a clinical suspicion of prostate cancer based on elevated levels of prostate-specific antigen (PSA, ≥ 4 ng/mL) and/or suspicious DRE results or family history of prostate cancer that were fit to undergo all protocol procedures, including a transrectal ultrasound. Patients were excluded if they used 5-alpha-reductase inhibitors during the previous six months, had a history of prostate cancer, or had evidence of urinary tract infection or acute prostatitis. All patients had a recent prostate-mpMRI (<45 days) with at least one lesion with PIRADS v2 score ≥ 3 (Study Flow Chart, Figure S1).

The mpMRI was performed with a 1.5-T MRI scanner using a 32-channel phased-array coil combined with an endorectal coil and included three orthogonal triplanar T2-weighted (T2w), diffusion-weighted imaging (DWI) with calculated *b*-value images, axial apparent diffusion coefficient (ADC) map, and dynamic contrast-enhanced (DCE) image sequences. All patients were scanned with the same MRI protocol, MRI scanner, and software version.

Details of the MRI protocol are provided in Supplemental Materials. The MRIs were reviewed by two experienced uro-radiologists and scored according to the PIRADS v2.

Each participant underwent TBx and SBx during the same session, with TBx taking place prior to SBx. The same person performed both TBx and SBx. The number of biopsies on targeted areas of the mpMRI-US and random biopsies were performed according to guidelines [11]. For SBx, a total of 10 biopsy cores were obtained from the peripheral zone of the prostate at the base, mid gland, and apex. For TBx, three to five biopsy cores per target were obtained for participant; the transition zone was not subjected to routine biopsy, except in the case of positive mpMRI. All biopsies were performed using the transperineal approach under monitored anesthesia, or local anesthesia, by two operators with experience of more than 1000 procedures. Biopsies performed during the learning curve were excluded from the analysis. Fusion biopsies were carried out with the Koelis Trinity system (Koelis, Meylan, France). Koelis Trinity system creates a precise and highly detailed 3D map of the prostate, showing the biopsy cores locations and suspicious areas delineated on mpMRI sequences. Trinity integrates 3D ultrasound, multimodal elastic fusion, and Organ-Based Tracking, which allows the device to follow the prostate's position and not that of the probe, automatically compensating for patient movement and prostate deformation.

All biopsy cores were analyzed by a centralized pathological anatomy laboratory and by a single operator with experience in uropathology and reported according to ISUP 2014 criteria [12,13]. Clinically significant (csPCa) was defined as having an ISUP score ≥ 2, and clinically insignificant (ciPCa) prostate cancer was defined as having an ISUP score < 1.

Descriptive statistics was evaluated using GraphPad Prism version 9.0.0 (GraphPad Software, San Diego, CA, USA). Difference of detection rates between TBx snd SBx was evaluated by McNemar's test, Chi-square, and Fisher' exact test. Statistical significance was established for $p < 0.05$. The added values were calculated by considering the cancer prevalence in the entire cohort.

3. Results

Our cohort consisted of 272 patients who had a mpMRI result suggestive of prostate cancer (PIRADS v2 score, ≥ 3), and underwent TBx and SBx during the investigated period. Eighty-two patients (30.1%) had at least one prior negative biopsy, whereas 190 (69.9%) were biopsy-naïve. Our cohort included 115 cases (42.3%) with areas of the prostate classified as PIRADS 3, 129 cases (47.4%) as PIRADS 4, and 28 cases (10.3%) as PIRADS 5. Our cohort consisted of 272 patients who had a mpMRI result suggestive of prostate cancer (PIRADS v2 score, ≥ 3), and underwent TBx and SBx during the investigated period. In total, 82 patients (30.1%) had at least one prior negative biopsy, whereas 190 (69.9%) were biopsy-naïve. Patients' characteristics and PIRADS classification are summarized in Table 1.

Table 1. Characteristics of the participants. [1] IQR, Inter-Quartile Range.

Characteristics	All (n = 272)	Biopsy Naïve (n = 190)	Prior Negative Biopsy (n = 82)
Age (years), median (IQR [1])	68 (62–74)	68 (62–74)	68 (63–72)
PSA (ng/mL), median (IQR)	7.2 (4.8–10.1)	7.1 (4.8–9.6)	7.3 (4.9–11.7)
PIRADS n, (%)			
3	115 (42.3%)	73 (26.8%)	42 (15.4%)
4	129 (47.4%)	96 (35.3%)	33 (12.1%)
5	28 (10.3%)	21 (7.7%)	7 (2.6%)

Overall, prostate cancer was detected in 117 of 272 men (43.0%), including 74 csPCa (27.2%) and 43 ciPCa (15.8%). The highest rate of csPCa was observed in PIRADS 5 index lesions, while the highest rate of negative biopsies was observed in PIRADS 3 (Figure 1).

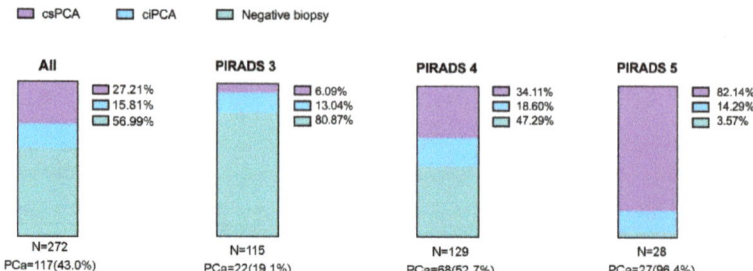

Figure 1. Percentages of patients with clinically significant (csPCa), insignificant (ciPCa), and no cancer in the whole cohort (leftmost bar chart) or according to PIRADS v2 score (remaining bar charts).

PCa was detected more frequently in biopsy-naïve patients than in patients undergoing prior biopsy, while negative biopsies were more frequent in prior biopsy patients ($p = 0.0162$; Table 2). The amount of csPCa was also more frequent in biopsy-naïve patients than in patients undergoing previous biopsy (31.6% versus 17.1% of all PCa, $p = 0.0136$), the latter showing a comparable frequency of csPca and ciPCA (Table 2). There was no significant difference between TBx and SBx biopsy in the detection rate of any PCa (32.7% vs. 33.1%), csPCa (22.4% vs. 22.4.0%), ciPCa (10.3% vs. 10.7%), or negative biopsy (67.3% vs. 66.9%, $p = 1$), as assessed by McNemar's test (Table 3).

Table 2. Overall cancer detection rate in naïve and prior biopsy cases. For each diagnosis, the number of cases and, in brackets, the percentage and 95% confidence interval are reported.

Cases	Biopsy Negative	csPCa (ISUP ≥ 2)	ciPCa (ISUP < 1)
Biopsy naïve 190/272 (69.9%)	99 (52.1%; 59.1–45.0)	60 (31.6%; 38.5–25.4)	31 (16.3%; 22.2–11.7)
Prior biopsy 82/272 (30.1%)	56 (68.3%; 77.4–57.6)	14 (17.1%; 26.6–10.4)	12 (14.6%; 23.9–8.6)

Table 3. Detection rate, diagnostic sensitivity, and specificity of TBx and SBx.

Cases	All n (%)	TBx n (%)	SBx n (%)	TBx/SBx n (%)
[1] Biopsy negative	155 (57.0%)	183 (67.3%)	182 (66.9%)	155 (57.0%)
[1] CDR for ciPCa	43 (15.8%)	28 (10.3%)	29 (10.7%)	43 (15.8%)
[1] CDR for csPCa	74 (27.2%)	61 (22.4%)	61 (22.4%)	74 (27.2%)
[1] False negative PCa		28 (10.3%)	27 (9.9%)	0 (0%)
[2] Sensitivity for csPCa		82.4%	82.4%	100%
Specificity for csPCa		85.%	85.8%	78.3%

[1] Percentage refers to the entire cohort ($n = 272$). [2] False negative includes biopsy negative plus ciPCa; since all diagnosed csPCa are derived from either TBx or SBx, the sensitivity of the combined TBx/SBx is 100%.

Notably, we observed a comparable number of patients showing either TBx-negative and SBx-positive results, or SBx-negative and TBx-positive results (27 and 28, respectively, Table 3), indicating that the combined TBx and SBX biopsy has an added value of 10.3% and 9.9% in reducing the risk of PCa missing after TBx or SBx alone, respectively (examples of mpMRI image for each clinical scenario are provided as Supplemental Figures S2–S4). Furthermore, only 62/117 PCas (53.0%) were positive for both biopsy approaches, meaning that 12/43 ciPCa and 26/74 csPCa would not have been diagnosed if patients had undergone TBx alone, or SBx; these data demonstrate an added value of 10.3% and 9.6% of the combined biopsy in the detection of ciPCa and csPCA, respectively.

The CDRs of TBx and SBx according to PIRADS subgroups were not significantly different (Figure 2A). For PIRADS 3 lesions, only one csPCa was detected by both TBx and SBx, while three out of seven csPCa were detected by TBx or SBx alone (Figure 2B).

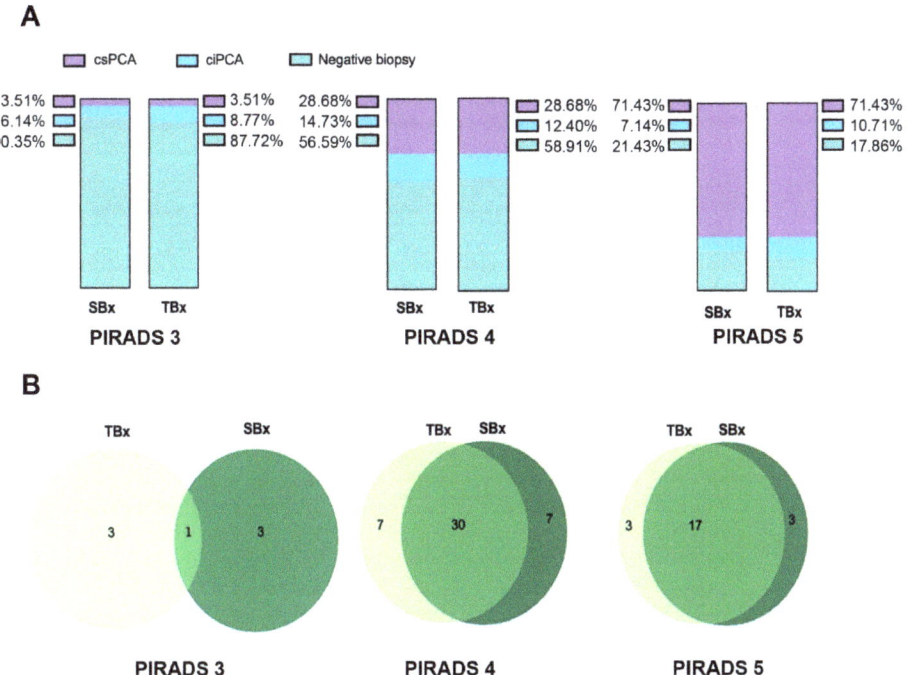

Figure 2. Diagnostic accuracy of TBx and SBx. (**A**) Percentages of patients with csPCa, ciPCa, or with negative biopsy revealed by TBx or SBx according to the PIRADS v2 score. (**B**) Venn diagrams showing the number of csPCa detected by TBx alone, SBx alone, or by the combined TBx/SBx approach.

Similarly, in PIRADS 4 and 5 lesions, a certain number of csPCa were detected either only after TBx, or only after SBx, although most of them were detected by both biopsy approaches. Overall, 13/74 (17.6%) of csPCa would not have been diagnosed if patients had undergone TBx alone, and 13/74 (17.6%) of csPCa would not have been diagnosed if patients had undergone SBx alone. These results further emphasize the non-redundant value for the two approaches in detecting csPCa. No major complications (Clavien Grade 3–4) were observed after 272 biopsies, but only minor complications that did not require hospitalization, such as hematospermia (38%), hematuria (8%), and fever > 38 °C (0.01%).

4. Discussion

The management of prostate biopsy in men with clinical suspicion of PCa has changed in the last 10 years, especially with the introduction of imaging techniques, in order to overcome the low efficacy of risk stratification based on PSA levels [5]. To implement proper treatment, extensive efforts have been made to identify csPCa. Several studies have reported the superiority of targeted and systematic biopsy based on mpMRI compared to US-guided transrectal biopsy [2,4,5]. The ideal detection method for diagnosing csPCa should be minimally invasive, have fewer complications and provide a higher detection rate for diagnosis. In recent years, interest in the transperineal biopsy approach has been growing [14–17], as the benefit appears to be not only in terms of reduced complications but also in terms of cancer characterization [9].

In our retrospective analysis, the detection rate of any PCa on mpMRI-positive cases was similar to the one observed by others [8,14–17]. Overall, PCa was detected more significantly in biopsy-naïve patients than in patients undergoing prior biopsy, while negative biopsies were more frequent in prior biopsy.

The CDR between targeted and standard biopsy were not significantly different and in line with experiences already reported in the literature [8–10]. We did not observe an advantage of CDR by using only one biopsy technique; instead, the combination of the TBx and SBx approaches has shown a significant 9.6% added value in csPCa detection. Our results confirm previous studies showing that 13–16% of csPCa were missed by TBx in comparison with a combination of SBx and TBx [18,19]. The CDR of targeted biopsy has shown superior performance for detection of prostate cancer, but the role of systematic biopsy should not be overlooked, and the combination of targeted and systematic biopsy is essential [9]. Similarly, the combined TBx and SBx approach showed an added value of 10.3% and 9.9% in reducing the risk of any PCa missing after TBx or SBx alone, respectively. Obviously, the advantage of a reduced risk must be weighed with the corresponding increase in biopsy cores and the consequent increase in possible complications related to the biopsy. From this point of view, the transperineal approach has the advantage of a lower risk of sepsis than the transrectal route [8,9], thus representing a good alternative to conventional US-guided transrectal biopsy, which is the current gold standard [11].

Our study highlighted a significant number of negative or indolent tumor biopsies, which could be reduced by a diagnostic pathway that includes more effective risk indicators. Although this area has benefited from mpMRI, the problem of overdiagnosis and the excessive number of unnecessary biopsies is still debated. In addition to PSA and its derivatives, a large number of new blood and urinary biomarkers to assist in biopsy decisions have appeared in clinical trials, largely thanks to advances in genomic technologies, among which the Prostate cancer antigen 3 (PCA3) [20,21], the SelectMDx multiplex biomarker [22–24], the ExoDx prostate Intelliscore [25,26], the Prostarix Risk Score [27,28], and the Prostate-specific G-protein coupled receptor (PSGR) [29,30]. However, their contribution in predicting biopsy outcomes needs to be more rigorously weighed.

Our study is affected by several limitations, and primarily by its retrospective and non-randomized nature. The small size of our single study center in a heterogeneous study population is hampered by potential biases, among which the use of a single fusion-system, and a limited number of surgeons. Another factor that weakens the accuracy analysis is the variable number of cores (three to five) per lesion. Moreover, we do not have a quality review of mpMRI classification. These limitations could benefit from a multicentric, prospective, and randomized study using a combination of biomarkers for a more rigorous evaluation of a diagnostic pathway of csPCa that makes use of transperineal biopsy.

5. Conclusions

The transperineal TBx proved a detection rate of overall PCa and csPca consistent with the literature. The transperineal approach is feasible and can be performed under local anesthesia as conventional US-guided transrectal biopsy. All patients had no complication and no impact on erectile or urinary function. The combined TBx and SBx pathway in patients with a positive MRI is recommended to reduce the risk of csPCa missing. Future prospective larger-scale studies are needed to confirm our findings.

Supplementary Materials: The following are available online at https://www.mdpi.com/article/10.3390/cancers13194833/s1, Supplemental Methods: MRI Protocol and sequence parameters; Figure S1: Study flow chart; Figures S2–S4: Examples of mpMRI images of the study.

Author Contributions: Study concept, design and supervision: S.A.; Data collection: L.A. and V.A.; Analysis and data interpretation: A.G. and M.S.; Drafting of the manuscript: A.F. and F.C.; Critical revision of the manuscript for important intellectual content: M.C., G.M.L., and V.P.; Statistical analysis: C.P. All authors have read and agreed to the published version of the manuscript.

Funding: This research received no external funding.

Institutional Review Board Statement: The study was conducted according to the guidelines of the Declaration of Helsinki; all patients gave their informed consent to use the data obtained from medical records; biopsies were performed according to standard procedures, and no additional biopsy was performed for this study; in addition, patient identification information was anonymized before analysis.

Data Availability Statement: The data presented in this study are available on request from the corresponding author.

Acknowledgments: We acknowledge Mariantonietta Rota, Maristella Genevose, and Vincenzo Aiello for data collection, and Anna Varano and Costanzo Battista for technical support.

Conflicts of Interest: The authors declare no conflict of interest.

References

1. Kohaar, I.; Petrovics, G.; Srivastava, S. A Rich Array of Prostate Cancer Molecular Biomarkers: Opportunities and Challenges. *Int. J. Mol. Sci.* **2019**, *20*, 1813. [CrossRef]
2. Ahmed, H.U.; El-Shater Bosaily, A.; Brown, L.C.; Gabe, R.; Kaplan, R.; Parmar, M.K.; Collaco-Moraes, Y.; Ward, K.; Hindley, R.G.; Freeman, A.; et al. Diagnostic accuracy of multi-parametric MRI and TRUS biopsy in prostate cancer (PROMIS): A paired validating confirmatory study. *Lancet* **2017**, *389*, 815–822. [CrossRef]
3. Drost, F.H.; Osses, D.F.; Nieboer, D.; Steyerberg, E.W.; Bangma, C.H.; Roobol, M.J.; Schoots, I.G. Prostate MRI, with or without MRI-targeted biopsy, and systematic biopsy for detecting prostate cancer. *Cochrane Database Syst. Rev.* **2019**, *4*, CD012663. [CrossRef] [PubMed]
4. Kasivisvanathan, V.; Rannikko, A.S.; Borghi, M.; Panebianco, V.; Mynderse, L.A.; Vaarala, M.H.; Briganti, A.; Budaus, L.; Hellawell, G.; Hindley, R.G.; et al. MRI-Targeted or Standard Biopsy for Prostate-Cancer Diagnosis. *N. Engl. J. Med.* **2018**, *378*, 1767–1777. [CrossRef] [PubMed]
5. Rouviere, O.; Puech, P.; Renard-Penna, R.; Claudon, M.; Roy, C.; Mege-Lechevallier, F.; Decaussin-Petrucci, M.; Dubreuil-Chambardel, M.; Magaud, L.; Remontet, L.; et al. Use of prostate systematic and targeted biopsy on the basis of multiparametric MRI in biopsy-naive patients (MRI-FIRST): A prospective, multicentre, paired diagnostic study. *Lancet. Oncol.* **2019**, *20*, 100–109. [CrossRef]
6. van der Leest, M.; Cornel, E.; Israel, B.; Hendriks, R.; Padhani, A.R.; Hoogenboom, M.; Zamecnik, P.; Bakker, D.; Setiasti, A.Y.; Veltman, J.; et al. Head-to-head Comparison of Transrectal Ultrasound-guided Prostate Biopsy Versus Multiparametric Prostate Resonance Imaging with Subsequent Magnetic Resonance-guided Biopsy in Biopsy-naive Men with Elevated Prostate-specific Antigen: A Large Prospective Multicenter Clinical Study. *Eur. Urol.* **2019**, *75*, 570–578. [CrossRef] [PubMed]
7. Weinreb, J.C.; Barentsz, J.O.; Choyke, P.L.; Cornud, F.; Haider, M.A.; Macura, K.J.; Margolis, D.; Schnall, M.D.; Shtern, F.; Tempany, C.M.; et al. PI-RADS Prostate Imaging—Reporting and Data System: 2015, Version 2. *Eur. Urol.* **2016**, *69*, 16–40. [CrossRef] [PubMed]
8. Marra, G.; Marquis, A.; Tappero, S.; D'Agate, D.; Oderda, M.; Calleris, G.; Falcone, M.; Faletti, R.; Molinaro, L.; Zitella, A.; et al. Transperineal Free-hand mpMRI Fusion-targeted Biopsies Under Local Anesthesia: Technique and Feasibility From a Single-center Prospective Study. *Urology* **2020**, *140*, 122–131. [CrossRef] [PubMed]
9. Marra, G.; Ploussard, G.; Futterer, J.; Valerio, M.; Party, E.-Y.P.C.W. Controversies in MR targeted biopsy: Alone or combined, cognitive versus software-based fusion, transrectal versus transperineal approach? *World J. Urol.* **2019**, *37*, 277–287. [CrossRef]
10. Marra, G.; Zhuang, J.; Beltrami, M.; Calleris, G.; Zhao, X.; Marquis, A.; Kan, Y.; Oderda, M.; Huang, H.; Faletti, R.; et al. Transperineal freehand multiparametric MRI fusion targeted biopsies under local anaesthesia for prostate cancer diagnosis: A multicentre prospective study of 1014 cases. *BJU Int.* **2021**, *127*, 122–130. [CrossRef]
11. Mottet, N.; van den Bergh, R.C.N.; Briers, E.; Van den Broeck, T.; Cumberbatch, M.G.; De Santis, M.; Fanti, S.; Fossati, N.; Gandaglia, G.; Gillessen, S.; et al. EAU-EANM-ESTRO-ESUR-SIOG Guidelines on Prostate Cancer-2020 Update. Part 1: Screening, Diagnosis, and Local Treatment with Curative Intent. *Eur. Urol.* **2021**, *79*, 243–262. [CrossRef]
12. Epstein, J.I.; Egevad, L.; Amin, M.B.; Delahunt, B.; Srigley, J.R.; Humphrey, P.A.; Grading, C. The 2014 International Society of Urological Pathology (ISUP) Consensus Conference on Gleason Grading of Prostatic Carcinoma: Definition of Grading Patterns and Proposal for a New Grading System. *Am. J. Surg. Pathol.* **2016**, *40*, 244–252. [CrossRef]
13. Epstein, J.I.; Zelefsky, M.J.; Sjoberg, D.D.; Nelson, J.B.; Egevad, L.; Magi-Galluzzi, C.; Vickers, A.J.; Parwani, A.V.; Reuter, V.E.; Fine, S.W.; et al. A Contemporary Prostate Cancer Grading System: A Validated Alternative to the Gleason Score. *Eur. Urol.* **2016**, *69*, 428–435. [CrossRef] [PubMed]
14. Mischinger, J.; Kaufmann, S.; Russo, G.I.; Harland, N.; Rausch, S.; Amend, B.; Scharpf, M.; Loewe, L.; Todenhoefer, T.; Notohamiprodjo, M.; et al. Targeted vs systematic robot-assisted transperineal magnetic resonance imaging-transrectal ultrasonography fusion prostate biopsy. *BJU Int.* **2018**, *121*, 791–798. [CrossRef]
15. Borkowetz, A.; Hadaschik, B.; Platzek, I.; Toma, M.; Torsev, G.; Renner, T.; Herout, R.; Baunacke, M.; Laniado, M.; Baretton, G.; et al. Prospective comparison of transperineal magnetic resonance imaging/ultrasonography fusion biopsy and transrectal systematic biopsy in biopsy-naive patients. *BJU Int.* **2018**, *121*, 53–60. [CrossRef] [PubMed]
16. Hakozaki, Y.; Matsushima, H.; Murata, T.; Masuda, T.; Hirai, Y.; Oda, M.; Kawauchi, N.; Yokoyama, M.; Kume, H. Detection rate of clinically significant prostate cancer in magnetic resonance imaging and ultrasonography-fusion transperineal targeted biopsy for lesions with a prostate imaging reporting and data system version 2 score of 3–5. *Int. J. Urol. Off. J. Jpn. Urol. Assoc.* **2019**, *26*, 217–222. [CrossRef]

17. Hansen, N.L.; Barrett, T.; Kesch, C.; Pepdjonovic, L.; Bonekamp, D.; O'Sullivan, R.; Distler, F.; Warren, A.; Samel, C.; Hadaschik, B.; et al. Multicentre evaluation of magnetic resonance imaging supported transperineal prostate biopsy in biopsy-naive men with suspicion of prostate cancer. *BJU Int.* **2018**, *122*, 40–49. [CrossRef]
18. Baco, E.; Rud, E.; Eri, L.M.; Moen, G.; Vlatkovic, L.; Svindland, A.; Eggesbo, H.B.; Ukimura, O. A Randomized Controlled Trial To Assess and Compare the Outcomes of Two-core Prostate Biopsy Guided by Fused Magnetic Resonance and Transrectal Ultrasound Images and Traditional 12-core Systematic Biopsy. *Eur. Urol.* **2016**, *69*, 149–156. [CrossRef]
19. Maxeiner, A.; Kittner, B.; Blobel, C.; Wiemer, L.; Hofbauer, S.L.; Fischer, T.; Asbach, P.; Haas, M.; Penzkofer, T.; Fuller, F.; et al. Primary magnetic resonance imaging/ultrasonography fusion-guided biopsy of the prostate. *BJU Int.* **2018**, *122*, 211–218. [CrossRef] [PubMed]
20. Leyten, G.H.; Hessels, D.; Jannink, S.A.; Smit, F.P.; de Jong, H.; Cornel, E.B.; de Reijke, T.M.; Vergunst, H.; Kil, P.; Knipscheer, B.C.; et al. Prospective multicentre evaluation of PCA3 and TMPRSS2-ERG gene fusions as diagnostic and prognostic urinary biomarkers for prostate cancer. *Eur. Urol.* **2014**, *65*, 534–542. [CrossRef]
21. Marks, L.S.; Fradet, Y.; Deras, I.L.; Blase, A.; Mathis, J.; Aubin, S.M.; Cancio, A.T.; Desaulniers, M.; Ellis, W.J.; Rittenhouse, H.; et al. PCA3 molecular urine assay for prostate cancer in men undergoing repeat biopsy. *Urology* **2007**, *69*, 532–535. [CrossRef] [PubMed]
22. Van Neste, L.; Hendriks, R.J.; Dijkstra, S.; Trooskens, G.; Cornel, E.B.; Jannink, S.A.; de Jong, H.; Hessels, D.; Smit, F.P.; Melchers, W.J.; et al. Detection of High-grade Prostate Cancer Using a Urinary Molecular Biomarker-Based Risk Score. *Eur. Urol.* **2016**, *70*, 740–748. [CrossRef] [PubMed]
23. Busetto, G.M.; Del Giudice, F.; Maggi, M.; De Marco, F.; Porreca, A.; Sperduti, I.; Magliocca, F.M.; Salciccia, S.; Chung, B.I.; De Berardinis, E.; et al. Prospective assessment of two-gene urinary test with multiparametric magnetic resonance imaging of the prostate for men undergoing primary prostate biopsy. *World J. Urol.* **2021**, *39*, 1869–1877. [CrossRef] [PubMed]
24. Maggi, M.; Salciccia, S.; Del Giudice, F.; Busetto, G.M.; Falagario, U.G.; Carrieri, G.; Ferro, M.; Porreca, A.; Di Pierro, G.B.; Fasulo, V.; et al. A Systematic Review and Meta-Analysis of Randomized Controlled Trials With Novel Hormonal Therapies for Non-Metastatic Castration-Resistant Prostate Cancer: An Update From Mature Overall Survival Data. *Front. Oncol.* **2021**, *11*, 700258. [CrossRef]
25. McKiernan, J.; Donovan, M.J.; O'Neill, V.; Bentink, S.; Noerholm, M.; Belzer, S.; Skog, J.; Kattan, M.W.; Partin, A.; Andriole, G.; et al. A Novel Urine Exosome Gene Expression Assay to Predict High-grade Prostate Cancer at Initial Biopsy. *Jama Oncol.* **2016**, *2*, 882–889. [CrossRef] [PubMed]
26. McKiernan, J.; Donovan, M.J.; Margolis, E.; Partin, A.; Carter, B.; Brown, G.; Torkler, P.; Noerholm, M.; Skog, J.; Shore, N.; et al. A Prospective Adaptive Utility Trial to Validate Performance of a Novel Urine Exosome Gene Expression Assay to Predict High-grade Prostate Cancer in Patients with Prostate-specific Antigen 2–10 ng/mL at Initial Biopsy. *Eur. Urol.* **2018**, *74*, 731–738. [CrossRef]
27. Sartori, D.A.; Chan, D.W. Biomarkers in prostate cancer: What's new? *Curr. Opin. Oncol.* **2014**, *26*, 259–264. [CrossRef]
28. Eifler, J.B.; Feng, Z.; Lin, B.M.; Partin, M.T.; Humphreys, E.B.; Han, M.; Epstein, J.I.; Walsh, P.C.; Trock, B.J.; Partin, A.W. An updated prostate cancer staging nomogram (Partin tables) based on cases from 2006 to 2011. *BJU Int.* **2013**, *111*, 22–29. [CrossRef]
29. Rigau, M.; Morote, J.; Mir, M.C.; Ballesteros, C.; Ortega, I.; Sanchez, A.; Colas, E.; Garcia, M.; Ruiz, A.; Abal, M.; et al. PSGR and PCA3 as biomarkers for the detection of prostate cancer in urine. *Prostate* **2010**, *70*, 1760–1767. [CrossRef]
30. Rigau, M.; Ortega, I.; Mir, M.C.; Ballesteros, C.; Garcia, M.; Llaurado, M.; Colas, E.; Pedrola, N.; Montes, M.; Sequeiros, T.; et al. A three-gene panel on urine increases PSA specificity in the detection of prostate cancer. *Prostate* **2011**, *71*, 1736–1745. [CrossRef]

Article

Evaluation of Glutaminase Expression in Prostate Adenocarcinoma and Correlation with Clinicopathologic Parameters

Zin W. Myint [1,2,*], Ramon C. Sun [2,3], Patrick J. Hensley [4], Andrew C. James [2,5], Peng Wang [1,2], Stephen E. Strup [2,5], Robert J. McDonald [6], Donglin Yan [2,7], William H. St. Clair [2,8] and Derek B. Allison [2,5,6]

1. Department of Internal Medicine, Division of Medical Oncology, University of Kentucky, Lexington, KY 40536, USA; p.wang@uky.edu
2. Markey Cancer Center, University of Kentucky, Lexington, KY 40536, USA; ramon.sun@uky.edu (R.C.S.); andrew.james@uky.edu (A.C.J.); Stephen.Strup@uky.edu (S.E.S.); donglin.yan@uky.edu (D.Y.); stclair@email.uky.edu (W.H.S.C.); Derek.Allison@uky.edu (D.B.A.)
3. Department of Neuroscience, University of Kentucky College of Medicine, Lexington, KY 40536, USA
4. Department of Urology, University of Texas MD Anderson Cancer Center, Houston, TX 77030, USA; hpatrick1@mdanderson.org
5. Department of Urology, University of Kentucky, Lexington, KY 40536, USA
6. Department of Pathology and Laboratory Medicine, University of Kentucky, Lexington KY 40536, USA; rmcdonald@uky.edu
7. Department of Internal Medicine-Health Services Research, University of Kentucky, Lexington, KY 40536, USA
8. Department of Radiation Medicine, University of Kentucky, Lexington, KY 40536, USA
* Correspondence: zin.myint@uky.edu; Tel.: +1-859-323-3964

Simple Summary: High expression levels of glutaminase (GLS1) are reported for several cancers, and correlate with parameters of disease status. GLS1, the rate-limiting enzyme in the glutamine pathway, is involved in DNA/RNA and amino acid synthesis and contributes to other pathways (e.g., TCA cycle). Inhibition of GLS1 has shown anti-tumor activity in both solid tumors and hematological malignancies. The CB-839 agent, a novel GLS1 inhibitor, has been under investigation clinically. GLS1 expression by immunohistochemical (IHC) staining in prostate has not been definitively demonstrated. We present a retrospective study evaluating GLS1 expression utilizing The Cancer Genome Atlas (TCGA) RNA-Seq data and by IHC in formalin-fixed paraffin embedded radical prostatectomy samples. The study showed a significant difference in GLS1 levels between cancer and non-cancer, but fell short as a prognostic marker. As the study cohort was skewed to less aggressive localized prostate cancer, we support further studies that incorporate high-risk and very high-risk localized and metastatic prostate cancers.

Abstract: High Glutaminase (GLS1) expression may have prognostic implications in colorectal and breast cancers; however, high quality data for expression in prostate cancer (PCa) are lacking. The purpose of this study is to investigate the status of GLS1 expression in PCa and correlated expression levels with clinicopathologic parameters. This study was conducted in two phases: an exploratory cohort analyzing RNA-Seq data for GLS1 from The Cancer Genome Atlas (TCGA) data portal (246 PCa samples) and a GLS1 immunohistochemical protein expression cohort utilizing a tissue microarray (TMA) (154 PCa samples; 41 benign samples) for correlation with clinicopathologic parameters. In the TCGA cohort, GLS1 mRNA expression did not show a statistically significant difference in disease-free survival (DFS) but did show a small significant difference in overall survival (OS). In the TMA cohort, there was no correlation between GLS1 expression and stage, Gleason score, DFS and OS. GLS1 expression did not significantly correlate with the clinical outcomes measured; however, GLS1 expression was higher in PCa cells compared to benign epithelium. Future studies are warranted to evaluate expression levels in greater numbers of high-grade and advanced PCa samples to investigate whether there is a rational basis for GLS1 targeted therapy in a subset of patients with prostate cancer.

Citation: Myint, Z.W.; Sun, R.C.; Hensley, P.J.; James, A.C.; Wang, P.; Strup, S.E.; McDonald, R.J.; Yan, D.; St. Clair, W.H.; Allison, D.B. Evaluation of Glutaminase Expression in Prostate Adenocarcinoma and Correlation with Clinicopathologic Parameters. *Cancers* **2021**, *13*, 2157. https://doi.org/10.3390/cancers13092157

Academic Editor: José I. López

Received: 23 March 2021
Accepted: 28 April 2021
Published: 29 April 2021

Publisher's Note: MDPI stays neutral with regard to jurisdictional claims in published maps and institutional affiliations.

Copyright: © 2021 by the authors. Licensee MDPI, Basel, Switzerland. This article is an open access article distributed under the terms and conditions of the Creative Commons Attribution (CC BY) license (https://creativecommons.org/licenses/by/4.0/).

Keywords: glutaminase; immunohistochemistry; in situ methods; prostate; prognosis

1. Introduction

The essential role of dysregulated glucose metabolism, also known as "aerobic glycolysis", in cancer was first discovered by Otto Warburg and colleagues in the 1920s [1–3]. Glucose and glutamine are essential nutrients for cell growth and survival [4]. Glutamine supplies nitrogen for purine and pyrimidine synthesis and nonessential amino acids for protein synthesis through glutaminolysis, which converts glutamine to glutamate by the rate-limiting enzyme, glutaminase (GLS). Subsequently, metabolic reactions produce alpha-ketoglutarate that contributes to the tricarboxylic acid (TCA) cycle and energy production in cancer cells through oxidative phosphorylation [5]. As such, glutaminolysis plays a significant role in the metabolic reprogramming of cancer cell growth and proliferation. It is now believed that glutaminolysis is associated with either activation of oncogenes (such as *MYC*) [6,7] and/or loss of tumor suppressor genes (such as *TP53*) [8,9].

Cancer associated fibroblasts (CAFs) are associated with prostate cancer (PCa) growth, progression to metastasis, and the development of castration resistance [10–13]. Glutamine metabolism may play a role in CAFs by providing enhanced secretion of glutamine from the tumor stromal cells as demonstrated in an ovarian cancer animal model [14]. Several studies have shown that androgen receptor signaling increases the utilization of glycolysis and alters glutamine metabolism in prostate cancer (PCa) [15,16]. Thus, it was hypothesized that blocking glutamine metabolism would deprive cancer cells of needed nutrients and lead to cell death. Furthermore, a new first-in-class orally bioavailable glutaminase inhibitor (CB-839), which inhibits glutathione production and blocks tumor glutamine consumption, is in early phase clinical trials as monotherapy or in combination for select tumors; examples include colorectal cancer (ClinicalTrials.gov Identifier: NCT02861300), non-small cell lung cancer (ClinicalTrials.gov Identifier: NCT04250545), hematological malignancy (ClinicalTrials.gov Identifier: NCT03047993), brain cancer (ClinicalTrials.gov Identifier: NCT03528642), and so on. However, there are no current ongoing clinical trials for PCa at the time of this writing.

Several studies reported that increased GLS1 expression was associated with disease aggressiveness and could be used as a prognostic marker in select solid tumors [17–20]. However, there is a lack of high-quality studies analyzing GLS1 expression in PCa. As a result, in the present study, we report an exploratory cohort correlating mRNA GLS1 expression with disease-free and overall survival, followed by a tissue microarray (TMA) cohort comparing GLS1 immunohistochemical (IHC) protein expression in PCa tissue and benign prostatic tissue, as well as a correlation with a more in-depth set of clinicopathologic parameters.

2. Results

In the TCGA cohort, we analyzed mRNA expression levels for GLS1 from 246 prostate cancer specimens. GLS1 mRNA expression barely showed a statistically significant difference in overall survival (Figure 1A,B) and did not show a statistically significant difference in disease-free survival; however, there was a trend toward a worse disease-free survival with high GLS1 expression (Figure 1C,D).

To further analyze GLS1 expression and to validate these findings, we studied a cohort of 154 patients with localized prostate adenocarcinoma and performed GLS1 IHC, an in situ method, to specifically measure GLS1 protein expression in tumor cells and to correlate the expression with a larger number of clinicopathologic parameters. The clinicopathological features of prostate cancer cases in this cohort are presented in Table 1. Briefly, the ages ranged from 43 to 73 years with a mean range of 58 years. Most patients (78%) were Caucasian and 20% were Black. Half of the patients (52.9%) were stage T2; 43.1% were stage T3, and the majority (94.7%) were node negative. Regarding Gleason grade group, 42.2% had grade group 2, 18.8% grade group 3, 5.2% grade group 4, and 13.4% grade group 5. In this cohort, 53% were negative for GLS1 expression, 21.5% had low GLS1 expression, and 25.5% had high GLS1 expression. Ten patients had androgen deprivation therapy (ADT) exposure prior to definitive prostatectomy. Among them, 5 out of 10 were GLS1 0, 4 were GLS1 low, and 1 was GLS1 high. Essentially, 50% were GLS negative and 50% were GLS1 positive. There was no statistical difference in GLS1 expression in patients with prior ADT exposure prior to definitive prostatectomy.

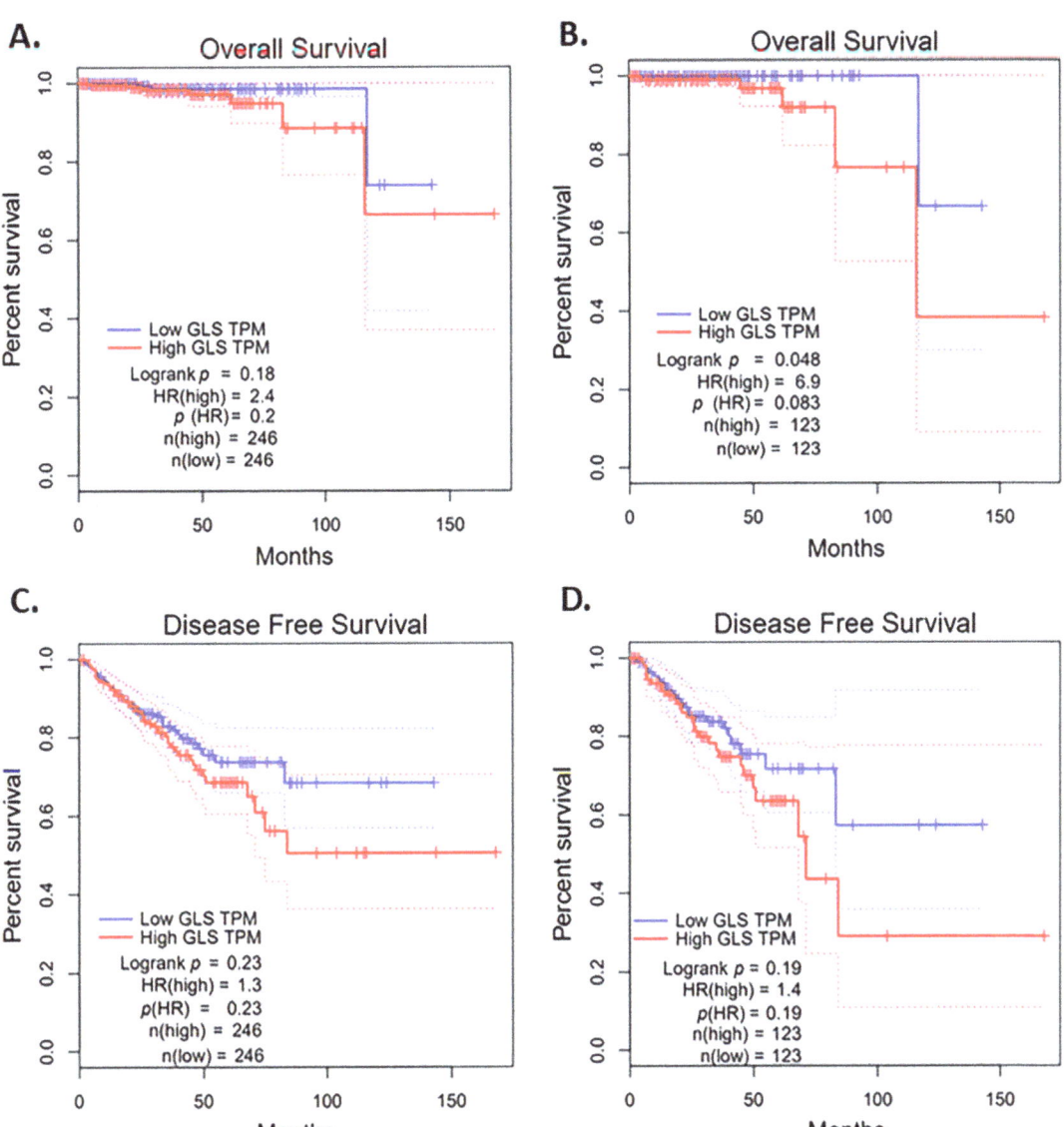

Figure 1. Impact of glutaminase (GLS1) mRNA on survivals using The Cancer Genome Atlas (TCGA) data portal cohort. (**A**) GLS1 mRNA expression and overall survival with a medium cutoff, (**B**) GLS1 mRNA expression and overall survival with a quartile cutoff, (**C**) mRNA expression and disease-free survival with a medium cutoff and (**D**) GLS1 mRNA expression and disease-free survival with a quartile cutoff. Dotted lines represent 95% confidence interval (CI).

Table 1. Clinicopathological characteristics of prostrate adenocarcinoma cases in tissue microarray (TMA) cohort.

Clinicopathological Characteristics	Number of Prostate Cancer Cases, n (%)
	Total n = 154
Age	
<60	84 (55%)
≥60	65 (45%)
Race	
White	121 (78.6%)
Black	31 (20.1%)
Other	2 (1.3%)
Smoking	
Yes	56 (36.4%)
No	50 (32.5%)
Unknown	48 (31.2%)
Pathological Tumor (pT) stage	
2	81 (52.9%)
3	66 (43.1%)
4	6 (3.9%)
Missing	1
Pathological Node (pN) status	
Yes	7 (5.3%)
No	124 (94.7%)
Missing	23
Gleason Grade Group	
Grade Group 1	21 (13.6%)
Grade Group 2	65 (42.2%)
Grade Group 3	29 (18.8%)
Grade Group 4	8 (5.2%)
Grade Group 5	21 (13.6%)
Missing	10 (6.4%)
GLS1 Score	
High expression	38 (25.5%)
Low expression	32 (21.5%)
Negative	79 (53%)
Missing *	5

* Missing means there was no remaining tumor cells available for scoring in the prostate cancer (PCa) sample.

To evaluate the clinical significance of GLS1 IHC protein expression in PCa, we compared the expression levels to benign glandular prostatic cells. A total of 41 cases of benign prostate cases were included; 17 out of 41 (41%) cases had low GLS1 protein expression and the remaining 24 cases (59%) had no observable GLS1 protein expression. IHC staining for GLS1 expression for both groups is shown in Figure 2. Additionally, we used the t test to compare the difference in GLS1 protein expression between malignant and benign prostate cases. We saw a statistically significant difference in GLS1 protein expression between PCa cells and benign glandular epithelium ($p < 0.003$) by the t test (Figure 3).

Furthermore, we compared the correlation between GLS1 expression and clinicopathological parameters as an indication of prognostic value in this cohort. There was no difference between GLS1 low vs. high protein expression for age, race, Gleason score, stage, node status, and smoking status by univariate analysis (Table 2).

In the TMA cohort, 55 out of 154 (36%) patients had biochemical progression, 65% underwent salvage radiation, and 35% had long-term hormonal therapy. Five patients

(3%) had systemic chemotherapy in addition to hormonal therapy. The estimated 5-year biochemical recurrence-free survival rate for high, low, and no GLS1 expression was 67.2%, 64.0%, and 74.1%, respectively ($p = 0.8$). The median overall survival (OS) was 12 years (0–23 years). The median OS time for high, low, and no GLS1 expression were 10.8, 11.8, 14.5 years, respectively ($p = 0.76$). There was no biochemical PFS difference between no, low, and high GLS1 expression ($p = 0.48$) (Figure 4A). No statistically significant between GLS1 expression and OS ($p = 0.76$) was found (Figure 4B).

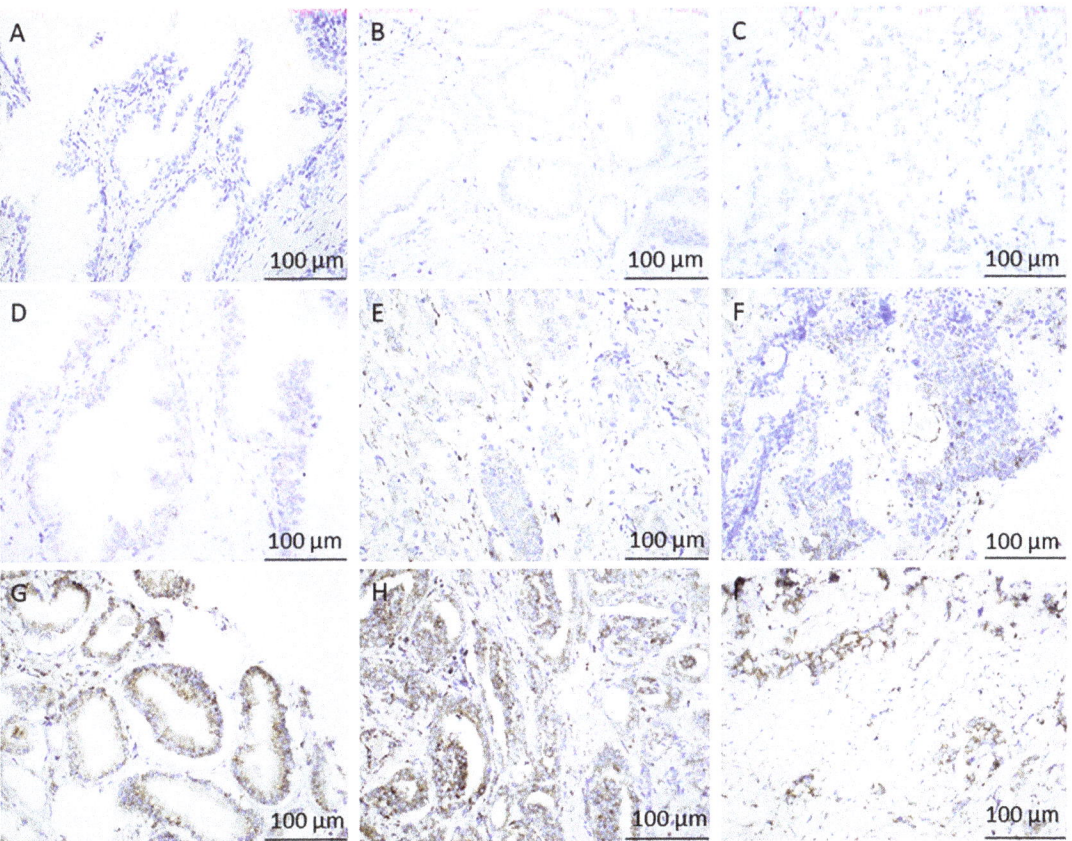

Figure 2. Immunohistochemical Analysis of GLS1 in Prostate Tissues. (**A**) Benign central zone of the prostate with no glutaminase (GLS1) staining, 40× magnification. (**B**) Grade Group 1 Prostate Adenocarcinoma showing no staining with GLS1, 40× magnification. (**C**) Grade Group 3 Prostate Adenocarcinoma showing no GLS1 expression, 40× magnification. (**D**) Benign Central zone of the prostate with faint but diffuse granular cytoplasmic staining (low expression) with GLS1, 40× magnification. (**E**) Grade Group 2 Prostate Adenocarcinoma showing faint but diffuse granular cytoplasmic staining (low expression) with GLS1, 40× magnification. (**F**) Grade Group 5 prostate adenocarcinoma showing moderate but focal granular cytoplasmic staining (low expression) with GLS1, 40× magnification. (**G**) Grade Group 1 Prostate Adenocarcinoma showing strong and diffuse granular cytoplasmic staining (high expression) with GLS1, 40×. (**H**) Grade Group 3 Prostate Adenocarcinoma showing variably strong but diffuse cytoplasmic granular staining (high expression) with GLS1, 40×. (**I**) Grade Group 5 Prostate Adenocarcinoma showing single infiltrating tumor cells and poorly formed glands with strong, diffuse granular cytoplasmic staining (high expression) with GLS1, 40×.

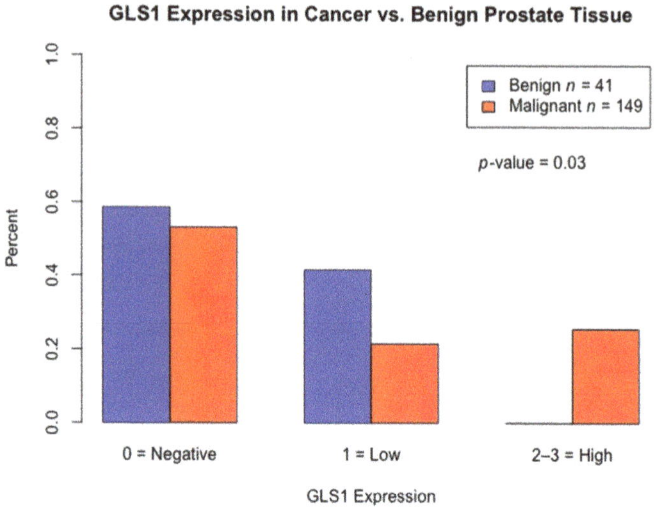

Figure 3. Distribution of glutaminase (GLS1) expression: Comparison of GLS1 protein expression between malignant prostate cancer vs. benign prostate tissue (control) by *t*-test.

Table 2. Glutaminase (GLS1) immunohistochemistry protein expression and clinicopathological parameters of prostate adenocarcinoma cases in TMA cohort.

Variables	GLS1 Score				*p*-Value
	High Expression (*n* = 38)	Low Expression (*n* = 32)	No (*n* = 79)	Missing (*n* = 5)	
	Age				
<60	20 (24%)	17 (20%)	47 (56%)		0.29
≥60	18 (28%)	15 (23%)	32 (49%)		
	Race				
White	33 (86.8%)	27 (84.4%)	57 (72.2%)	4	0.3
Black	5 (13.2%)	5 (15.6%)	20 (25.3%)	1	
Other	0	0	2 (2.5%)		
	pT stage				
pT2	19 (50%)	18 (56.3%)	40 (51.3%)	4	0.6
pT3	19 (50%)	12 (37.5%)	34 (43.6%)	1	
pT4	0	2 (6.3%)	4 (5.1%)	0	
Missing	0	0	1		
	pN				
Yes	1 (2.9%)	0	5 (7.9%)	1	0.2
No	34 (97.1%)	29 (100%)	58 (92.1%)	3	
	Gleason Grade Group				
1	9 (23.6%)	6 (18%)	5 (6.3%)	1	
2	16 (42%)	11 (34%)	37 (46.8%)	1	
3	4 (10.5%)	6 (18.7%)	18 (22.8)	1	0.5
4	3 (7.8%)	1 (3%)	4 (5%)		
5	5 (13%)	5 (15.6%)	11(13.9%)		
Unknown	1(2.6%)	3 (9.3%)	4 (5%)	2	
	Smoking				
Yes	12 (31.6%)	13 (40.6%)	28 (35.4%)	3	0.9
No	13 (34.2%)	10 (31.3%)	27 (34.2%)	0	
Unknown	13 (34.2%)	9 (28.1%)	24 (30.4%)	2	

pT = pathological stage, pN = pathological node.

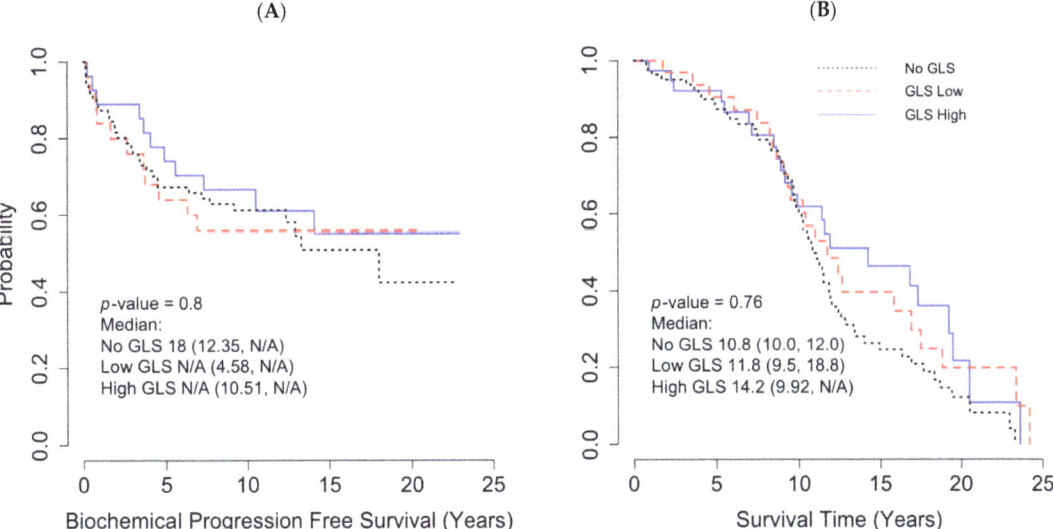

Figure 4. Impact of glutaminase (GLS1) mRNA on survivals with using a tissue microarray (TMA) Cohort. (**A**) Pearson correlation between GLS1 protein expression and biochemical progression-free survival and (**B**) Pearson Correlation between GLS1 protein expression and overall survival. N/A = estimates not available due to small sample size/censoring.

3. Discussion

High GLS1 expression has been reported as a potential prognostic marker for poor response in solid tumors such as colorectal and triple negative breast cancers [17,18]. In prostate cancer, previously published studies have reported that high GLS1 expression was associated with a higher Gleason score and higher tumor stage [21,22]. In order to explore whether GLS1 is a negative prognostic marker, we analyzed GLS1 RNA-Seq data from TCGA database. In this cohort, GLS1 mRNA expression did not show a statistically significant difference in disease-free survival but did show a small statistically significant difference in overall survival. We next sought to further validate these findings utilizing GLS1 IHC to restrict the analysis of GLS1 expression more specifically to tumor cells, and to study the relationship with a wider range of clinicopathologic parameters from our patient population. In summary, we found a lack of correlation between GLS1 protein expression by IHC and various clinicopathological parameters. However, we did find a statistically significant difference between GLS1 protein expression in PCa cells versus benign glands, with high GLS1 protein expression limited to prostate cancer. There are important differences between our study and previous investigations: (1) techniques/methods of measuring GLS1 expression; (2) tissue sample types (preclinical cell lines vs. patient radical prostatectomy samples); (3) patients' baseline characteristics.

With respect to the first difference, Zhang et al. demonstrated that high expression of GLS mRNA was significantly associated with high Gleason score (≥8) and higher tumor stage (≥T3) [22]. They utilized qRT-PCR and Western blot analyses to measure GLS mRNA while we analyzed RNA-Seq data. Furthermore, no in situ methods were performed, such as IHC, which is important for prostate cancer considering tumors are often rich in stroma and intimately associated with benign glands. In contrast, Dorai et al. studied GLS IHC on FFPE samples of PCa [23]. However, a careful examination of their GLS IHC figures shows that staining was primarily localized to the stroma and not the actual tumor cells, which questions the reliability of the data [23]. Prior to the present study, a high-quality in situ investigation that localizes GLS1 expression to tumor cells, the biologically relevant compartment, has been lacking. In our study, we used a combination of RNA-Seq data, as well as a high quality in situ IHC protein method, to study GLS1 expression (Figure 3).

With respect to the second difference, the Zhang et al. study also reported that the expression of GLS1 mRNA levels by qRT-PCR and Western blot methods were higher in different prostate cancer cell lines (DU145, PC-3 and LNCaP) compared with a normal prostate epithelial cell line (RWPE-1) [22]. This finding is consistent with our IHC results, which showed a statistically significant difference in GLS1 protein expression in prostate cancer cells versus benign glandular epithelium.

Finally, with respect to the third difference, many of our samples in the TMA cohort were from patients with low/intermediate risk, localized prostate cancer: T2 (52.9%), Gleason grade group 2 (42.2%), Gleason grade group 5 (13.4%) and N0 (94.7%). Furthermore, the TCGA cohort is similarly enriched for localized, low-to-intermediate grade PCa. In contrast to our study, the Zhang et al. study included more high-risk patients; Gleason score ≥ 8 (63%), $\geq$T3 (56%) and PSA ≥ 10 ng/mL (59%) [22]. Given the smaller sample size for aggressive disease, our cohorts may not be sufficiently powered to detect a difference in this population. Furthermore, the cell lines DU145, PC-3, and LNCap are all derived from Stage IV metastatic prostate cancer. As a result, it is not surprising that GLS1 expression may be different in these aggressive tumors than in our cohorts. To this point, Zacharias et al. evaluated the metabolic differences between the aggressive prostate cancer cell line (PC3) and the even more aggressive metastatic cell line (P3M) by using hyperpolarized in vivo pyruvate studies, nuclear magnetic resonance spectroscopy, and carbo-13 feeding studies, showing that the P3M cell line utilized a higher amount of glutamine than PC3 [24]. These findings are interesting, and more studies are warranted in high-risk and aggressive PCa.

In addition, GLS exists in two major isoforms, kidney-type glutaminase (KGA) and glutaminase C (GAC) [25], which are collectively referred to as GLS1; the antibody used for these analyses detects both isoforms [26]. Since GAC may be more active than KGA, further studies that utilize in situ methods of detection with these two splice variants may be warranted. However, interestingly, Zacharias et al. performed Western blot analyses of KGA and GAC and showed similar expression levels in all prostate cancer cell lysates for both the PC3 and PC3M cell lines [24]. As a result, since our antibody detects both KGA and GAC, it is unclear that additional IHC would provide any further information. To this point, GLS1 expression (detecting both KGA and GAC) was found to be undetectable in 53% of our prostate cancer samples. This finding suggests that KGA specific and GAC specific IHC would be negative in the majority of our samples. Despite lack of any additional IHC markers, our data suggest that the glutaminolysis pathway may not play a major role in many clinically encountered, localized prostate cancers. However, additional studies are warranted, particularly in more high-risk and advanced prostate cancer cohorts.

Another limitation in our TMA cohort is related to the age of the FFPE samples, which ranged from 4–18 years old. Antigenicity (antibody to antigen specificity) can change over time; however, it is crucial to include samples from patients for whom long-term follow-up is available. This factor may result in less intense staining; however, in comparison to the published literature, our rate of GLS1 protein expression in prostate cancer is not dramatically dissimilar from mRNA data. One study reported that GLS1 expression was found in 64% (68 out of 107 patients) in the malignant prostate specimens by mRNA analysis [21], while in our study, GLS1 IHC protein expression was found in 47% (70 out of 154 patients) malignant prostate specimens.

Despite the lack of support for GLS1 IHC protein expression as a prognostic marker in early stage prostate cancer in our study, the findings from preclinical models and the trend in our TCGA analysis warrant further investigation into whether GLS1 expression could be a predictive marker for response to the CB-839, a small molecule allosteric inhibitor of GLS. Furthermore, the fact that high protein expression by IHC was restricted to cancer specimens may indicate that this pathway could be a potential target for a subset of patients. Importantly, pre-clinical studies have shown anti-tumor activity in PCa cell lines treated with CB-839 as monotherapy [24], and it showed synergistic activity in combination with the PARP inhibitors in multiple solid tumors including PCa cells [27], in combination with metformin in osteosarcoma [28] in vitro, in combination with cabozantinib in renal cell carcinoma [29], in combination of CDK4/6 inhibitor in colorectal, and breast cancers preclinical models [27] as well as CB-839 could be utilized as a radiosensitizer in lung cancer shown in vivo and vitro [30]. These encouraging in vitro results support the confirmation of CB-839 anti-tumor activity in xenograft models, but these findings need to be further translated. Essentially, additional studies enriched for high-risk and aggressive PCa are warranted.

4. Materials and Methods

This study was conducted in two phases: an exploratory cohort analyzing RNA-Seq data for GLS1 from The Cancer Genome Atlas (TCGA) data portal and a GLS1 immunohistochemical (IHC) protein expression cohort utilizing a TMA for correlation with additional clinicopathologic parameters.

4.1. TCGA Cohort

246 PCa samples with RNA-Seq data available for GLS1 were identified from TCGA (https://cancergenome.nih.gov/, accessed on 4 March 2021) and included in an exploratory phase of this study.

4.2. TMA Cohort

A TMA was constructed from formalin-fixed paraffin-embedded tissue from 154 patients who underwent radical prostatectomy for localized prostate adenocarcinoma between 2002 and 2016 at University of Kentucky Markey Cancer Center. Survival intervals were calculated from the time of prostate surgery until death or last clinic follow-up. Clinicopathological information including age, race, tumor stage, histology, grade, lymph node status, Gleason score, smoking status, subsequent treatment history (radiation therapy and or hormonal therapy), and survival data were collected using the Kentucky Cancer Registry. The study was approved by the local Institutional Review Boards (IRBs) at University of Kentucky (53854).

The TMAs were used to evaluate GLS1 protein expression by IHC. Slides were sectioned at 4 microns and baked at 58 °C for a minimum of one hour. Staining was conducted on the Ventana Discovery Ultra using Standard CC2 (Roche, Tucson, AZ, USA) antigen retrieval, followed by incubation with an anti-GLS1 antibody (ab156876, abcam, Cambridge, MA, USA) at 1:400 for 1 h at 37 °C and then incubation with OmniMap anti-Rabbit-HRP (Roche, Tucson, AZ, USA) and visualization with DAB (Roche), according to the manufacturer's recommendations. Slides were lightly counterstained in Mayer's hematoxylin before permanent mounting. Two board-certified anatomic pathologists participated in this study (Derek B. Allison and Robert J. McDonald) and worked together to optimize the IHC staining conditions and agree upon staining thresholds. Cases where there was a question about scoring were reviewed by both pathologists. The expression scores were calculated as intensity score (0 = no staining; 1 = weak granular cytoplasmic staining; 2 = moderate granular cytoplasmic staining; 3 = strong granular cytoplasmic staining) X proportion score (0 = no positive cells; 1= <10% positive cells; 2= 10–40% positive cells; 3 = 40–70% positive cells; 4 = >70% positive cells, as modified from Xiang et al [31]. For the PCa samples, interpretation was restricted to the PCa tumor cells while interpretation in the benign tissue samples was restricted to the benign glandular epithelium. Based on the distribution of the scores, the cases were then classified into one of three tiers: no expression, low expression, or high expression. Forty-one benign prostate samples were included for comparison and scored using the same method. Associations between GLS1 levels and clinicopathological parameters and survival were analyzed by Pearson's chi-squared and Log-rank tests.

4.3. Statistical Analysis

4.3.1. TCGA Cohort

Kaplan–Meier estimate or survival analysis was performed using the publicly available software GEPIA (http://gepia.cancer-pku.cn, accessed on 4 March 2021) that utilizes RNA sequencing expression data from tumor and normal samples from the TCGA database using a standard processing pipeline. For both disease-free survival and overall survival, upper and lower quartile were used for high and low expression cut-offs and the nonparametric Mantel-Cox test was used for the calculation of Log-rank score.

4.3.2. TMA Cohort

A total of 154 patients were included in the TMA cohort IHC analysis. Basic characteristics, including age, race, smoking history, pathological tumor stage, pathological node status, Gleason grade, and GLS expression were summarized by descriptive statistics. Univariate analysis was conducted to examine differences in age, race, Gleason score, stage, node status, and smoking status by GLS expression (no expression, low expression, or high expression). Associations between categorical endpoints were examined by Chi-square test, and continuous endpoints were compared by Student's t-test. Disease-free (biochemical recurrence-free) survival and overall survival were estimated and plotted using the Kaplan–Meier method. Log-rank tests were performed to detect survival differences by GLS expression. Analyses were conducted using SAS 9.4 (SAS Institute Inc., Cary, NC, USA) and R Studio 1.4 (RStudio, PBC, Boston, MA, USA). All hypotheses testing and confidence intervals were conducted at 95% significance level.

5. Conclusions

In our study, PCa cells were more likely to have increased GLS1 protein expression compared to benign glandular epithelium. Although GLS1 protein expression did not appear to be a statistically significant prognostic marker, high GLS1 mRNA expression showed a trend toward worse disease-free and overall survival. Our cohorts, however, were enriched for cases with localized disease and low-to-intermediate grade PCa. As a result, future studies are warranted to evaluate GLS1 expression in high grade and advanced PCa cases to determine whether there is a rational basis for GLS1 targeted therapy with CB-839 in a subset of patients with PCa.

Author Contributions: Conceptualization, Z.W.M., R.C.S., P.J.H., A.C.J., P.W., S.E.S., R.J.M., D.Y., W.H.S.C., D.B.A.; Methodology, Z.W.M., R.C.S., D.Y., D.B.A.; Validation, Z.W.M., R.J.M., D.B.A.; Analysis, R.C.S., D.Y.; Writing, Original Draft Preparation, Z.W.M., R.C.S., P.J.H., A.C.J., P.W., S.E.S., R.J.M., D.Y., W.H.S.C., D.B.A.; Writing, Review and Editing & Final approval, Z.W.M., R.C.S., P.J.H., A.C.J., P.W., S.E.S., R.J.M., D.Y., W.H.S.C., D.B.A.; Supervision, Z.W.M. and D.B.A.; Administrative, technical or material support, Z.W.M. All authors have read and agreed to the published version of the manuscript.

Funding: This study received no external funding.

Institutional Review Board Statement: The study was conducted according to the guidelines of the Declaration of Helsinki, and approved by the local Institutional Review Boards (IRBs) at University of Kentucky. The approval number is 53854 and date of approval began on 22 June 2020.

Informed Consent Statement: Patient consent was waived due to retrospective nature.

Data Availability Statement: The data presented in this study are available on request from the corresponding author.

Acknowledgments: The University of Kentucky Markey Cancer Center's Research Communications Office assisted with manuscript preparation, the Biostatistics and Bioinformatics Shared Resource Facility provided statistical support, and the Biospecimen Procurement & Translational Pathology Shared Resource Facility provided IHC preparation and staining support; these Shared Resource Facilities are supported by NCI Cancer Center Support Grant (P30 CA177558).

Conflicts of Interest: The authors declare no conflict of interest.

References

1. Warburg, O. The Metabolism of Carcinoma Cells. *J. Cancer Res.* **1925**, *9*, 148–163. [CrossRef]
2. Warburg, O.; Wind, F.; Negelein, E. The Metabolism of Tumors in the Body. *J. Gen. Physiol.* **1927**, *8*, 519–530. [CrossRef]

3. Heiden, M.G.V.; Cantley, L.C.; Thompson, C.B. Understanding the Warburg effect: The metabolic requirements of cell proliferation. *Science* **2009**, *324*, 1029–1033. [CrossRef] [PubMed]
4. Dang, C.V. Links between metabolism and cancer. *Genes Dev.* **2012**, *26*, 877–890. [CrossRef]
5. Altman, B.J.; Stine, Z.E.; Dang, B.J.A.Z.E.S.C.V. From Krebs to clinic: Glutamine metabolism to cancer therapy. *Nat. Rev. Cancer* **2016**, *16*, 619–634. [CrossRef] [PubMed]
6. Gao, P.; Tchernyshyov, I.; Chang, T.-C.; Lee, Y.-S.; Kita, K.; Ochi, T.; Zeller, K.I.; De Marzo, A.M.; Van Eyk, J.E.; Mendell, J.T.; et al. c-Myc suppression of miR-23a/b enhances mitochondrial glutaminase expression and glutamine metabolism. *Nat. Cell Biol.* **2009**, *458*, 762–765. [CrossRef]
7. Wise, D.R.; Ward, P.S.; Shay, J.E.S.; Cross, J.R.; Gruber, J.J.; Sachdeva, U.M.; Platt, J.M.; DeMatteo, R.G.; Simon, M.C.; Thompson, C.B. Hypoxia promotes isocitrate dehydrogenase-dependent carboxylation of α-ketoglutarate to citrate to support cell growth and viability. *Proc. Natl. Acad. Sci. USA* **2011**, *108*, 19611–19616. [CrossRef]
8. Anderson, N.M.; Mucka, P.; Kern, J.G.; Feng, H. The emerging role and targetability of the TCA cycle in cancer metabolism. *Protein Cell* **2018**, *9*, 216–237. [CrossRef]
9. Liang, Y.; Liu, J.; Feng, Z. The regulation of cellular metabolism by tumor suppressor p53. *Cell Biosci.* **2013**, *3*, 9. [CrossRef] [PubMed]
10. Olumi, A.F.; Grossfeld, G.D.; Hayward, S.W.; Carroll, P.R.; Tlsty, T.D.; Cunha, G.R. Carcinoma-associated fibroblasts direct tumor progression of initiated human prostatic epithelium. *Cancer Res.* **1999**, *59*, 5002–5011. [CrossRef]
11. Thalmann, G.N.; Rhee, H.; Sikes, R.A.; Pathak, S.; Multani, A.; Zhau, H.E.; Marshall, F.F.; Chung, L.W. Human Prostate Fibroblasts Induce Growth and Confer Castration Resistance and Metastatic Potential in LNCaP Cells. *Eur. Urol.* **2010**, *58*, 162–172. [CrossRef] [PubMed]
12. Bonollo, F.; Thalmann, G.N.; Julio, M.K.-D.; Karkampouna, S. The Role of Cancer-Associated Fibroblasts in Prostate Cancer Tumorigenesis. *Cancers* **2020**, *12*, 1887. [CrossRef] [PubMed]
13. Chiarugi, P.; Paoli, P.; Cirri, P. Tumor Microenvironment and Metabolism in Prostate Cancer. *Semin. Oncol.* **2014**, *41*, 267–280. [CrossRef] [PubMed]
14. Yang, L.; Achreja, A.; Yeung, T.-L.; Mangala, L.S.; Jiang, D.; Han, C.; Baddour, J.; Marini, J.C.; Ni, J.; Nakahara, R.; et al. Targeting Stromal Glutamine Synthetase in Tumors Disrupts Tumor Microenvironment-Regulated Cancer Cell Growth. *Cell Metab.* **2016**, *24*, 685–700. [CrossRef]
15. Massie, C.E.; Lynch, A.; Ramos-Montoya, A.; Boren, J.; Stark, R.; Fazli, L.; Warren, A.; Scott, H.; Madhu, B.; Sharma, N.; et al. The androgen receptor fuels prostate cancer by regulating central metabolism and biosynthesis. *EMBO J.* **2011**, *30*, 2719–2733. [CrossRef]
16. Tennakoon, J.B.; Shi, Y.; Han, J.J.; Tsouko, E.; White, M.A.; Burns, A.R.; Zhang, A.; Xia, X.; Ilkayeva, O.R.; Xin, L.; et al. Androgens regulate prostate cancer cell growth via an AMPK-PGC-1α-mediated metabolic switch. *Oncogene* **2014**, *33*, 5251–5261. [CrossRef]
17. Huang, F.; Zhang, Q.; Ma, H.; Lv, Q.; Zhang, T. Expression of glutaminase is upregulated in colorectal cancer and of clinical significance. *Int. J. Clin. Exp. Pathol.* **2014**, *7*, 1093–1100.
18. Vidula, N.; Yau, C.; Rugo, H.S. Glutaminase (GLS) expression in primary breast cancer (BC): Correlations with clinical and tumor characteristics. *J. Clin. Oncol.* **2019**, *37*, 558. [CrossRef]
19. Cao, J.; Zhang, C.; Jiang, G.; Jin, S.; Gao, Z.; Wang, Q.; Yu, D.; Ke, A.; Fan, Y.; Li, D.; et al. Expression of GLS1 in intrahepatic cholangiocarcinoma and its clinical significance. *Mol. Med. Rep.* **2019**, *20*, 1915–1924. [CrossRef]
20. Kim, H.M.; Lee, Y.K.; Koo, J.S. Expression of glutamine metabolism-related proteins in thyroid cancer. *Oncotarget* **2016**, *7*, 53628–53641. [CrossRef]
21. Pan, T.; Gao, L.; Wu, G.; Shen, G.; Xie, S.; Wen, H.; Yang, J.; Zhou, Y.; Tu, Z.; Qian, W. Elevated expression of glutaminase confers glucose utilization via glutaminolysis in prostate cancer. *Biochem. Biophys. Res. Commun.* **2015**, *456*, 452–458. [CrossRef] [PubMed]
22. Zhang, J.; Mao, S.Y.; Guo, Y.D.; Wu, Y.; Yao, X.D.; Huang, Y. Inhibition of GLS suppresses proliferation and promotes apoptosis in prostate cancer. *Biosci. Rep.* **2019**, *39*. [CrossRef]
23. Dorai, T.; Dorai, B.; Pinto, J.T.; Grasso, M.; Cooper, A.J.L. High Levels of Glutaminase II Pathway Enzymes in Normal and Cancerous Prostate Suggest a Role in 'Glutamine Addiction'. *Biomolecules* **2019**, *10*, 2. [CrossRef]
24. Zacharias, N.M.; McCullough, C.; Shanmugavelandy, S.; Lee, J.; Lee, Y.; Dutta, P.; McHenry, J.; Nguyen, L.; Norton, W.; Jones, L.W.; et al. Metabolic Differences in Glutamine Utilization Lead to Metabolic Vulnerabilities in Prostate Cancer. *Sci. Rep.* **2017**, *7*, 1–11. [CrossRef]
25. Cassago, A.; Ferreira, A.P.S.; Ferreira, I.M.; Fornezari, C.; Gomes, E.R.M.; Greene, K.S.; Pereira, H.D.; Garratt, R.C.; Dias, S.M.G.; Ambrosio, A.L.B. Mitochondrial localization and structure-based phosphate activation mechanism of Glutaminase C with implications for cancer metabolism. *Proc. Natl. Acad. Sci. USA* **2012**, *109*, 1092–1097. [CrossRef] [PubMed]
26. de la Rosa, V.; Campos-Sandoval, J.A.; Martín-Rufián, M.; Cardona, C.; Matés, J.M.; Segura, J.A.; Alonso, F.J.; Márquez, J. A novel glutaminase isoform in mammalian tissues. *Neurochem. Int.* **2009**, *55*, 76–84. [CrossRef]
27. Emberley, E.; Bennett, M.; Chen, J.; Gross, M.; Huang, T.; Makkouk, A.; Parlati, F. The glutaminase inhibitor CB-839 synergizes with CDK4/6 and PARP inhibitors in pre-clinical tumor models. *Cancer Res.* **2018**, *78*, 3509.
28. Ren, L.; Ruiz-Rodado, V.; Dowdy, T.; Huang, S.; Issaq, S.H.; Beck, J.; Wang, H.; Hoang, C.T.; Lita, A.; Larion, M.; et al. Glutaminase-1 (GLS1) inhibition limits metastatic progression in osteosarcoma. *Cancer Metab.* **2020**, *8*, 4–13. [CrossRef] [PubMed]

29. Meric-Bernstam, F.; Lee, R.J.; Carthon, B.C.; Iliopoulos, O.; Mier, J.W.; Patel, M.R.; Tannir, N.M.; Owonikoko, T.K.; Haas, N.B.; Voss, M.H.; et al. CB-839, a glutaminase inhibitor, in combination with cabozantinib in patients with clear cell and papillary metastatic renal cell cancer (mRCC): Results of a phase I study. *J. Clin. Oncol.* **2019**, *37*, 549. [CrossRef]
30. Boysen, G.; Jamshidi-Parsian, A.; Davis, M.A.; Siegel, E.R.; Simecka, C.M.; Kore, R.A.; Dings, R.P.M.; Griffin, R.J. Glutaminase inhibitor CB-839 increases radiation sensitivity of lung tumor cells and human lung tumor xenografts in mice. *Int. J. Radiat. Biol.* **2019**, *95*, 436–442. [CrossRef]
31. Xiang, L.; Mou, J.; Shao, B.; Wei, Y.; Liang, H.; Takano, N.; Semenza, G.L.; Xie, G. Glutaminase 1 expression in colorectal cancer cells is induced by hypoxia and required for tumor growth, invasion, and metastatic colonization. *Cell Death Dis.* **2019**, *10*, 40. [CrossRef] [PubMed]

Article

Integration of Urinary EN2 Protein & Cell-Free RNA Data in the Development of a Multivariable Risk Model for the Detection of Prostate Cancer Prior to Biopsy

Shea P. Connell [1], Robert Mills [2], Hardev Pandha [3], Richard Morgan [4], Colin S. Cooper [1], Jeremy Clark [1], Daniel S. Brewer [1,5,*] and The Movember GAP1 Urine Biomarker Consortium [1,†]

[1] Norwich Medical School, University of East Anglia, Norwich Research Park, Norwich NR4 7TJ, UK; sheaconnell@gmail.com (S.P.C.); colin.cooper@uea.ac.uk (C.S.C.); Jeremy.Clark@uea.ac.uk (J.C.)
[2] Norfolk and Norwich University Hospitals NHS Foundation Trust, Norwich, Norfolk NR4 7UY, UK; robert.mills@nnuh.nhs.uk
[3] Faculty of Health and Medical Sciences, The University of Surrey, Guildford GU2 7XH, UK; h.pandha@surrey.ac.uk
[4] School of Pharmacy and Medical Sciences, University of Bradford, Bradford BD7 1DP, UK; r.morgan3@bradford.ac.uk
[5] The Earlham Institute, Norwich Research Park, Norwich, Norfolk NR4 7UZ, UK
* Correspondence: d.brewer@uea.ac.uk; Tel.: +44-(0)-1603-593761
[†] The Movember GAP1 Urine Biomarker Consortium: Bharati Bapat, Rob Bristow, Andreas Doll, Jeremy Clark, Colin Cooper, Hing Leung, Ian Mills, David Neal, Mireia Olivan, Hardev Pandha, Antoinette Perry, Chris Parker, Martin Sanda, Jack Schalken, Hayley Whitaker.

Simple Summary: Prostate cancer is a disease responsible for a large proportion of all male cancer deaths but there is a high chance that a patient will die with the disease rather than from. Therefore, there is a desperate need for improvements in diagnosing and predicting outcomes for prostate cancer patients to minimise overdiagnosis and overtreatment whilst appropriately treating men with aggressive disease, especially if this can be done without taking an invasive biopsy. In this work we develop a test that predicts whether a patient has prostate cancer and how aggressive the disease is from a urine sample. This model combines the measurement of a protein-marker called EN2 and the levels of 10 genes measured in urine and proves that integration of information from multiple, non-invasive biomarker sources has the potential to greatly improve how patients with a clinical suspicion of prostate cancer are risk-assessed prior to an invasive biopsy.

Abstract: The objective is to develop a multivariable risk model for the non-invasive detection of prostate cancer prior to biopsy by integrating information from clinically available parameters, Engrailed-2 (EN2) whole-urine protein levels and data from urinary cell-free RNA. Post-digital-rectal examination urine samples collected as part of the Movember Global Action Plan 1 study which has been analysed for both cell-free-RNA and EN2 protein levels were chosen to be integrated with clinical parameters (n = 207). A previously described robust feature selection framework incorporating bootstrap resampling and permutation was applied to the data to generate an optimal feature set for use in Random Forest models for prediction. The fully integrated model was named ExoGrail, and the out-of-bag predictions were used to evaluate the diagnostic potential of the risk model. ExoGrail risk (range 0–1) was able to determine the outcome of an initial trans-rectal ultrasound guided (TRUS) biopsy more accurately than clinical standards of care, predicting the presence of any cancer with an area under the receiver operator curve (AUC) = 0.89 (95% confidence interval(CI): 0.85–0.94), and discriminating more aggressive Gleason $\geq 3 + 4$ disease returning an AUC = 0.84 (95% CI: 0.78–0.89). The likelihood of more aggressive disease being detected significantly increased as ExoGrail risk score increased (Odds Ratio (OR) = 2.21 per 0.1 ExoGrail increase, 95% CI: 1.91–2.59). Decision curve analysis of the net benefit of ExoGrail showed the potential to reduce the numbers of unnecessary biopsies by 35% when compared to current standards of care. Integration of information from multiple, non-invasive biomarker sources has the potential to greatly improve how patients with a clinical suspicion of prostate cancer are risk-assessed prior to an invasive biopsy.

Keywords: prostate cancer; biomarker; urine; machine learning; TRIPOD; liquid biopsy

1. Introduction

Prostate cancer is responsible for 13% of all male cancer deaths in the UK, yet this is contrasted by 10-year survival rates approaching 84% [1]. This dichotomy has led to uncertainty for clinicians in how best to diagnose and predict the outcome for prostate cancer patients to minimise overdiagnosis and overtreatment whilst appropriately treating men with aggressive disease [2]. More accurate discrimination of disease state in biopsy naïve men would mark a significant development compared to current standards and impact large numbers of patients suspected of harbouring prostate cancer. The development of such a pre-biopsy screening test would provide a convenient checkpoint along the clinical pathway for patients to exit without the need for further invasive and stressful follow-up.

Under current guidelines patients are selected for further clinical investigations for prostate cancer if they have an elevated prostate specific antigen (PSA) (≥ 4 ng/mL) and/or an adverse finding on digital rectal examination (DRE) or lower urinary tract symptoms, whilst other factors such as age and ethnicity are also considered alongside patient preference [3–5]. More recently multiparametric MRI (mpMRI) has been used as a triage tool to reduce negative biopsy rates since its validation in the PROMIS clinical trial [6]. However, as it has gained more widespread adoption, mpMRI has shown a higher rate of inter-operator and inter-machine variability than reported in controlled clinical trials; up to 28% of clinically significant disease is missed in practice [5,7–9]. Coupled with the relative expense, time and expertise required to undertake an mpMRI meeting the current clinical guidelines, there is a need to improve on current clinical practices.

Biomarkers utilising tissue samples taken at the time of diagnosis for the detection of aggressive or significant prostate cancer requiring clinical attention are relatively plentiful [10–13]. Many of these markers are good tests, whether that be for discerning the most aggressive disease [11,14], or for predicting disease-free survival following radical prostatectomy [15]. However, requiring tissue means a biopsy must already have been performed, making these tests incompatible with reducing the rates of unnecessary biopsy that come at considerable economic, psychological and societal cost to patients and healthcare systems alike [2,16,17].

As a secretory organ directly interacting with the male urinary tract, the prostate is well-placed as a candidate for non-invasive liquid biopsy from urine samples [18]. Single- or few-biomarker panels such as Engrailed-2 (EN2) protein expression [19], the SelectMDx [20] and ExoDx Prostate (IntelliScore) [21] tests have published promising results for the non-invasive detection of significant disease (Gleason score (Gs) ≥ 7). However, they are in various stages of clinical validation and none are currently implemented in the UK healthcare system [5]. Most urinary biomarkers developed to date for the prediction of biopsy outcome are unimodal; considering a singular fraction of urine (such as the cell-pellet or cell-free fractions) or biological aspect of cancer to appraise disease status. Whilst these tests have shown promising clinical use and accuracy, for the majority it has not yet been explored whether extra predictive value could be derived by integrating multiple streams of information from other sources.

Since initial development, the SelectMDx model has been updated to include clinically available parameters of serum PSA, patient age and DRE alongside urinary *HOXC6* and *DLX1* mRNA, adding significant predictive ability for patients with a PSA < 10 ng/mL [22]. We have also recently shown the benefit of such a holistic approach, presenting the development of the multivariable ExoMeth risk prediction model integrating clinical parameters, hypermethylation within the urinary cell pellet and urinary cell-free RNA expression data that displayed improved clinical utility over any single mode [23].

EN2 is a homeodomain-containing transcription factor that has an essential function in early development, which in mammals includes the delineation of the midbrain/hindbrain border [24]. For a transcription factor it has a number of unusual properties, including the ability to be secreted from cells and taken up by others [25]. Indeed, a recent study indicated that prostate cancer cells can secrete EN2 protein through vesicles which are then taken up by other non-EN2 expressing cells, where it can directly influence the transcription of target genes [25].

This secretory behaviour of EN2 makes it a potential biomarker for prostate cancer, and indeed EN2 protein can be detected in the urine of men with prostate tumours [19]. The original and subsequent studies have generally supported a diagnostic role for urinary EN2, including a relationship between urinary EN2 concentration and tumour volume [19,26]. More recently, a lateral flow-based test for EN2 has been described that could potentially allow point-of-care testing [27].

In this study, we report the utility of a predictive model produced by the integration of clinically available parameters, urinary EN2 protein levels and targeted cell-free RNA transcriptomics. The data were collected within the Movember Global Action Plan 1 (GAP1) study that explored a range of biomarkers in urine for PCa diagnosis and prognosis. The clinical utility of this model is determined by the ability to predict the presence of Gs ≥ 7 and Gs $\geq 4 + 3$ disease on biopsy, both critical distinctions in clinical settings, where patients with Gs ≥ 7 are recommended radical therapy [5], whilst patients with Gs $4 + 3$ have significantly worse outcomes than Gs $3 + 4$ patients [28]. Aware that most cancer biomarkers and predictive models fail to reach clinical adoption, we have adhered to the guidelines for the transparent reporting of a multivariable prediction model for individual prognosis or diagnosis (TRIPOD) whilst developing the models and results presented here [29].

2. Materials and Methods

2.1. Patient Population and Characteristics

The full Movember GAP1 urine cohort comprises 1257 first-catch post-DRE urine samples collected between 2009 and 2015 from urology clinics at multiple sites, as described in Connell et al. (2019). As a diverse range of techniques was applied to samples from this cohort and restricted amounts of urine, the number of experiments that could be performed on any one sample was limited. Samples within the Movember cohort that were quantified for both EN2 levels by ELISA and cell-free-RNA (cf-RNA) expression by NanoString (Seattle, WA, USA) were eligible for selection for model development in the current study ($n = 218$).

Exclusion criteria for model development included a recent prostate biopsy or transurethral resection of the prostate (<6 weeks) and metastatic disease (confirmed by a positive bone-scan or PSA > 100 ng/mL), resulting in a cohort of 207 samples, deemed the ExoGrail cohort (Table 1). All samples analysed in the ExoGrail cohort were collected from the Norfolk and Norwich University Hospital (NNUH, Norwich, UK). Sample collections and processing were ethically approved by the East of England REC.

2.2. Sample Processing and Analysis

Urine samples were processed according to the Movember GAP1 standard operating procedure (Supplementary Methods). In brief, within 30 min of collection, urine was centrifuged (1200× g 10 min, 6 °C) to remove cellular material. Supernatant extracellular vesicles were harvested by microfiltration and cell-free mRNA extracted (RNeasy micro kit, #74004, Qiagen, Hilden, Germany) on the same day that they were provided by the patient. RNA was amplified as cDNA with an Ovation PicoSL WTA system V2 (Nugen, Redwood City, CA, USA, #3312-48). Urinary EN2 protein concentration was quantified by ELISA from whole urine using a monoclonal anti-mouse EN2 antibody, as described by Morgan et al. (2011) [19]. Cell-free mRNA was quantified from urinary extracellular vesicles using NanoString technology, with 167 gene-probes (Table S1), as described in Connell et al. (2019), with the modification that NanoString data were normalised according to NanoString guidelines using NanoString internal positive controls, and $\log_2$ transformed.

Clinical variables serum PSA, age at sample collection, DRE finding, and urine volume collected were considered.

Table 1. Characteristics of the ExoGrail development cohort, stratified according to a record of cancer or not, either on an initial biopsy or for a no cancer finding if a biopsy was not performed.

	No Cancer Finding:	Biopsy Positive Cancer Finding
Collection Centre:		
NNUH, n (%)	77 (100)	130 (100)
Age:		
minimum	45.00	53.00
median (IQR)	65.00 (59.00, 71.00)	68.50 (65.00, 76.00)
mean (sd)	65.22 ± 8.10	69.71 ± 7.67
maximum	82.00	91.00
PSA:		
minimum	0.30	4.10
median (IQR)	6.10 (3.70, 8.80)	10.35 (6.82, 16.48)
mean (sd)	7.89 ± 8.72	17.08 ± 18.33
maximum	63.80	95.90
Prostate Size (DRE Estimate):		
Small, n (%)	13 (17)	13 (10)
Medium, n (%)	34 (44)	64 (49)
Large, n (%)	21 (27)	38 (29)
Unknown, n (%)	9 (12)	15 (12)
Gleason Score:		
0, n (%)	77 (100)	0 (0)
6, n (%)	0 (0)	30 (23)
3 + 4, n (%)	0 (0)	48 (37)
4 + 3, n (%)	0 (0)	24 (18)
≥8, n (%)	0 (0)	28 (22)
Biopsy Outcome:		
No Biopsy, n (%)	25 (32)	0 (0)
Biopsy Negative, n (%)	52 (68)	0 (0)
Biopsy Positive, n (%)	0 (0)	130 (100)

2.3. Statistical Analysis

All analyses, model construction and data preparation were undertaken in R version 3.5.3 [30], and unless otherwise stated, utilised base R and default parameters. All data and code required to reproduce these analyses can be found at the UEA Cancer Genetic GitHub repository [31].

2.4. Feature Selection

In total, 172 variables were available for prediction (cf-RNA ($n = 167$), clinical variables ($n = 4$) and urinary EN2 ($n = 1$); for full list see Table S1), making feature selection a key task for minimising model overfitting and increasing the robustness of trained models. To avoid dataset-specific features being positively selected [32], we implemented a robust feature selection workflow utilising the Boruta algorithm [33] and bootstrap resampling. Boruta is a random forest-based algorithm that iteratively compares feature importance against random predictors, deemed "shadow features." Features that perform significantly worse compared to the maximally performing shadow feature at each permutation, ($p \leq 0.01$, calculated by Z-score difference in mean accuracy decrease) are consecutively dropped until only confirmed, stable features remain.

Boruta was applied on 1000 datasets generated by resampling with replacement. Features were only positively selected for model construction when confirmed as stable features in ≥90% of resampled Boruta runs.

2.5. Comparator Models

To evaluate potential clinical utility, additional models were trained as comparators using subsets of the available variables across the patient population: a clinical standard of care (SoC) model was trained by incorporating age, PSA, T-staging and clinician DRE impression; a model using only the values from the EN2 ELISA (EN2, $n = 1$); and a model only using NanoString gene-probe information (NanoString, $n = 167$). The fully integrated ExoGrail model was trained by incorporating information from all of the above variables ($n = 177$). Each set of variables for comparator models were independently selected via the bootstrapped Boruta feature selection process described above to select the most optimal subset of variables possible for each predictive model.

2.6. Model Construction

All models were trained via the random forest algorithm [34], using the *randomForest* package [35] with default parameters except for resampling without replacement and 401 trees being grown per model. Risk scores from trained models are presented as the out-of-bag predictions; the aggregated outputs from decision trees within the forest where the sample in question has not been included within the resampled dataset [34]. Bootstrap resamples were identical for feature selection and model training for all models and used the same seed for random number generator.

Models were trained on a modified continuous label, based on biopsy outcome and constructed as follows: samples were scored on a continuous scale (range: 0–1) according to the dominant Gleason pattern: where 0 represented no evidence of cancer, Gleason scores 6 & 3 + 4 were assigned to 0.5 and Gleason scores $\geq 4 + 3$ are set to 1. Following this categorisation, the score is treated as a continuous variable by the Random Forest algorithm described above. This process was designed to recognise that two patients with the same TRUS-biopsy Gleason score will not share the exact same proportions of tumour pattern, or overall disease burden. This scale was solely used for model training and was not represented in any endpoint measurements, or for determining the predictive ability and clinical utility.

2.7. Statistical Evaluation of Models

Area Under the Receiver-Operator Characteristic curve (AUC) metrics were produced using the *pROC* package [36], with confidence intervals calculated via 1000 stratified bootstrap resamples. Density plots of model risk scores, and all other plots were created using the *ggplot2* package [37]. Partial dependency plots were calculated using the *pdp* package [38]. Cumming estimation plots and calculations were produced using the *dabestr* package [39] and 1000 bootstrap resamples were used to visualise robust effect size estimates of model predictions.

Decision curve analysis (DCA) [40] examined the potential net benefit of using PUR-signatures in the clinic. Standardised net benefit (sNB) was calculated with the *rmda* package [41] and presented throughout our decision curve analyses as it is a more directly interpretable metric compared to net benefit [42]. In order to ensure DCA was representative of a more general population, the prevalence of Gleason scores within the ExoGrail cohort were adjusted via bootstrap resampling to match those observed in a population of 219,439 men that were in the control arm of the Cluster Randomised Trial of PSA Testing for Prostate Cancer (CAP) Trial [43], as described in Connell et al. (2019). Briefly, of the biopsied men within this CAP cohort, 23.6% were Gs 6, 8.7% Gs 7 and 7.1% Gs ≥ 8, with 60.6% of biopsies showing no evidence of cancer. These ratios were used to perform stratified bootstrap sampling with a replacement of the Movember cohort to produce a "new" dataset of 197 samples with risk scores from each comparator model. sNB was then calculated for this resampled dataset, and the process repeated for a total of 1000 resamples with replacement. The mean sNB for each risk score and the "treat-all" options over all of the iterations were used to produce the presented figures to account for variance in resampling. Net reduction in biopsies, based on the adoption of models versus the default treatment option of undertaking biopsy in all men with PSA ≥ 4 ng/mL was calculated as:

$$Biopsy_{NetReduction} = (NB_{Model} - NB_{All}) \times \frac{1 - Threshold}{Threshold} \quad (1)$$

where the decision threshold (*Threshold*) is determined by accepted patient/clinician risk [40]. For example, a clinician may accept up to a 25% perceived risk of cancer before recommending biopsy to a patient, equating to a decision threshold of 0.25.

3. Results

3.1. The ExoGrail Development Cohort

Urinary EN2 protein and cf-RNA data were available for 207 patients within the Movember GAP1 cohort, with all samples originating from the NNUH to form the ExoGrail development cohort (Table 1). The proportion of Gleason ≥ 7 disease in the ExoGrail cohort was 48%, whilst 25 patients were deemed to have no evidence of cancer (NEC, PSA < 4 ng/mL), and did not receive a biopsy.

3.2. Feature Selection and Model Development

Using the robust feature selection framework, four models were produced in total: a standard of care (SoC) model incorporating only clinically available parameters (age and PSA), a model using urinary EN2 protein levels as the sole predictor variable (Engrailed), a model using only cf-RNA information (ExoRNA, 11 gene-probes) and the integrated model, named ExoGrail that incorporated variables from all three sources (12 variables) (Table 2). The ExoGrail model is a multivariable risk prediction model incorporating clinical parameters, urinary EN2 protein levels and cf-RNA expression information. When the resampling strategy was applied for feature reduction using Boruta, 12 variables were selected for the ExoGrail model. Each of the retained variables were positively selected in every resample and notably included information from clinical and cf-RNA variables, as well as urinary EN2 (Figure 1; full resample-derived Boruta variable importance for the SoC, Engrailed and ExoRNA comparator models can be seen in Figures S1–S3, respectively).

Table 2. Features positively selected for each model by bootstrap resampling and the Boruta algorithm. Features are selected for each model by being confirmed as important for predicting biopsy outcome, categorised as a modified ordinal variable (see Methods) by Boruta in ≥90% of bootstrap resamples.

	SoC	Engrailed	ExoRNA	ExoGrail
Clinical Parameters	Serum PSA Age	- -	- -	Serum PSA -
ELISA Targets		EN2 (ELISA)	-	EN2 (ELISA)
NanoString cf-RNA targets			ERG exons 4-5 ERG exons 6-7 GJB1 HOXC6 HPN NKAIN1 PCA3 PPFIA2 RPLP2 - TMEM45B TMPRSS2/ERG fusion	ERG exons 4-5 ERG exons 6-7 GJB1 HOXC6 HPN - PCA3 PPFIA2 - SLC12A1 TMEM45B TMPRSS2/ERG fusion

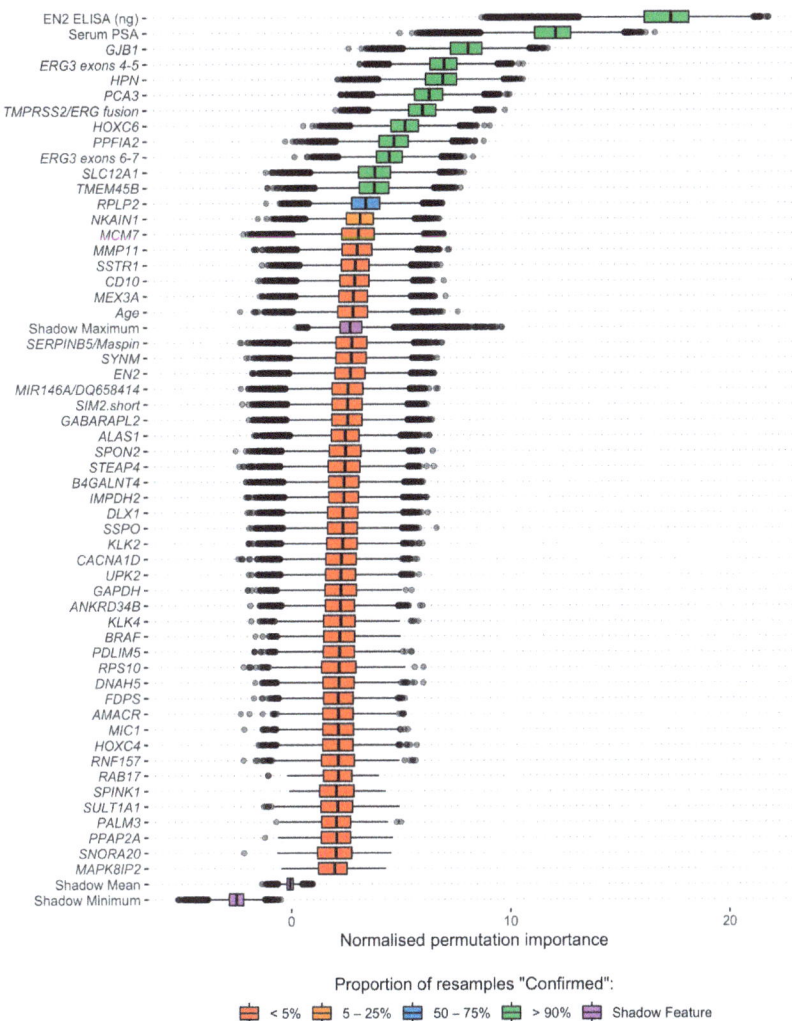

Figure 1. Analysis of variables available for the training of the ExoGrail model through the application of the Boruta algorithm via bootstrap resampling. 1000 resamples with replacement of the available data were made, with the normalised permutation importance of each variable recorded at each iteration, along with the decision of Boruta within that resample. Fill colour shows the proportion of resamples that a feature was positively retained by Boruta. Those features selected in ≥90% of resamples were selected for fitting predictive models. Variables rejected in all of the 1000 resamples are not shown here but are fully detailed in Table S1.

In the SoC comparator model, only PSA and age were selected as important predictors. Urinary EN2 levels were confirmed as important in the independent Engrailed model as the sole variable, and also within the ExoGrail model (Table 2). For the cf-RNA model, ExoRNA, 11 transcripts were selected, notably including both variants of the ERG gene-probe and *TMPRSS2/ERG* fusion gene-probe. ExoGrail incorporated an additional cf-RNA transctript, *SLC1A1*, which was not previously selected in the ExoRNA comparator model. When this was examined by partial dependency plots, an additive interaction effect was observed between quantified levels of urinary EN2 and counts of SLC12A1 on the predicted ExoGrail risk signature output (Figure S4).

3.3. ExoGrail Predictive Ability

As ExoGrail Risk Score (range 0–1) increased, the likelihood of high-grade disease detection on TRUS-biopsy was significantly greater (Proportional odds ratio = 2.21 per 0.1 ExoGrail increase, 95% CI: 1.91–2.59; ordinal logistic regression, Figure 2). The median ExoGrail risk score for metastatic patients was 0.76 (n = 11). These patients were excluded from model training and can be considered as a positive control for model calibration.

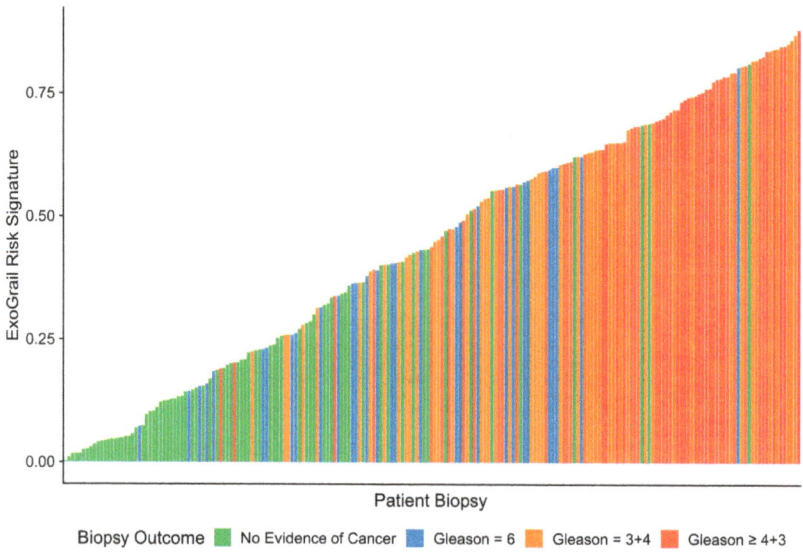

Figure 2. Representation of the ExoGrail risk score for each patient within a waterfall plot, where each coloured bar represents an individual's biopsy outcome (fill colour) and predicted ExoGrail risk score (bar height). Green—No evidence of cancer, Blue—Gs 6, Orange—Gs 3 + 4, Red—Gs ≥ 4 + 3.

ExoGrail was superior to all other models for the detection of Gleason ≥ 3 + 4 (AUC = 0.90 (95% CI: 0.86–0.94), p < 0.001, bootstrap test with 1000 resamples) and for any cancer (AUC = 0.89 (95% CI: 0.85–0.94), p < 0.001, bootstrap test with 1000 resamples) (Table 3). When Gleason ≥ 4 + 3 was considered, ExoGrail returned an AUC = 0.84 (95% CI: 0.78–0.89), outperforming the SoC and cf-RNA models (p < 0.001, bootstrap test with 1000 resamples), whilst the Engrailed model displayed similar performance by AUC metrics (Table 3). A model consisting of the combination of EN2 and PSA showed a similar ability in the detection of Gleason ≥ 4 + 3 compared to ExoGrail (AUCs of 0.85 compared to 0.84), whilst ExoGrail showed a small improvement in the detection of Gleason ≥ 3 + 4 disease and any cancer (Table S2).

Table 3. Area under the receiver operator curve (AUC) of all trained models for detecting outcomes of an initial biopsy for varying clinically significant thresholds. Numbers within brackets detail 95% confidence intervals of the AUC, calculated from 1000 stratified bootstrap resamples. Input variables for each model are detailed in Table 1.

Initial Biopsy Outcome:	SoC	Engrailed	ExoRNA	ExoGrail
Gleason ≥ 4 + 3:	0.77 (0.69–0.84)	0.81 (0.74–0.88)	0.67 (0.59–0.75)	0.84 (0.78–0.89)
Gleason ≥ 3 + 4:	0.72 (0.65–0.79)	0.83 (0.77–0.88)	0.77 (0.70–0.83)	0.90 (0.86–0.94)
Any Cancer	0.75 (0.68–0.82)	0.81 (0.74–0.86)	0.81 (0.74–0.87)	0.89 (0.85–0.94)

As revealed by the distributions of risk scores and AUC, ExoGrail achieved clearer discrimination of disease status Gleason $\geq 3 + 4$ disease from other biopsy outcomes when compared to any of the other models (ExoGrail all comparisons $p < 0.01$ bootstrap test, 1000 resamples, Figure 3).

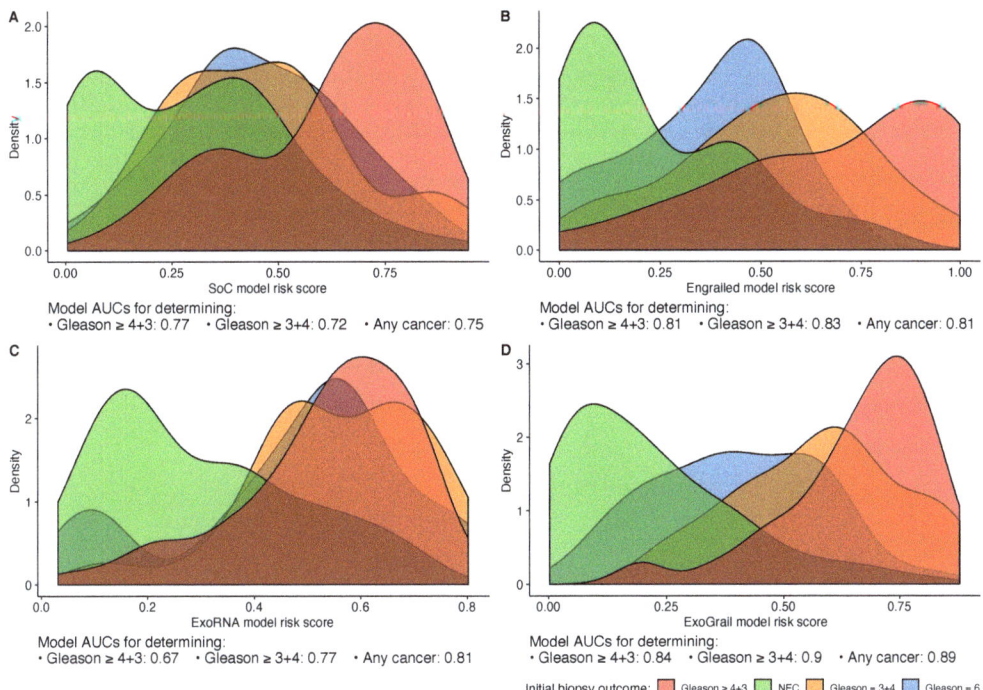

Figure 3. Risk score distributions of the four trained models, calculated as the out-of-bag predictions and represented as density plots. Area under the receiver operator curves (AUCs) for each model's predictive ability for clinically relevant outcomes are detailed underneath each panel. Each random forest model was fit using different input variables; (A) SoC clinical risk model, including Age and serum prostate specific antigen (PSA), (B) Engrailed model, (C) ExoRNA model and (D) ExoGrail model, combining predictors from all three modes of analysis. The full list of variables in each model is available in Table 1. Fill colour shows the risk score distribution of patients with respect to biopsy outcome: No evidence of cancer (Green), Gleason 6 (Blue), Gleason 3 + 4 (Orange), Gleason $\geq 4 + 3$ (Red).

Investigation of risk score distributions found that whilst the SoC model returned respectable AUCs and detection of the higher grade disease (Gleason $\geq 3 + 4$), it displayed a relative inability to clearly stratify intermediate disease states. This uncertainty would cause large numbers of patients to be inappropriately selected for further investigation (Figure 3A). For example, to classify 90% of patients with Gleason 7 disease correctly, an SoC risk score of 0.251 would misclassify 64.5% of men with less significant, or no disease. The Engrailed model detailed clearer discrimination, though featured a bimodal distribution of patients without prostate cancer (Figure 3B, green density plot), misidentifying 51.4% of patients with low-grade disease as similar to those with more clinically significant disease (Figure 3B). A similar bimodal distribution was seen for the EN2 plus PSA model (Figure S5). Whilst the AUCs returned for the ExoRNA model were lower, the distribution of risk scores shows that ExoRNA could more accurately discriminate cancer from non-cancer than either the SoC or EN2 models, a key clinical step in the triage of patients prior to biopsy (Figure 3C).

Examination of ExoGrail scores displayed similar distributions for NEC patients as the ExoRNA model whilst also being able to more accurately separate different cancer outcomes from biopsy, resulting in fewer misclassifications of patients without cancer if binary detection of 95% of Gleason ≥ 3 + 4 were considered (28% of NEC patients misclassified). The greater discriminatory ability of the ExoGrail model when biopsy outcomes are considered as a binary Gleason ≥ 3 + 4 threshold can also be seen in Figure S6.

Comparisons of mean ExoGrail scores between groups were performed with resampling and Cumming estimation plots (1000 bias-corrected and accelerated bootstrap resamples, Figure 4). The mean ExoGrail differences between patients with no evidence of cancer on biopsy were: Gleason 6 = 0.3 (95% CI: 0.22–0.37), Gleason 3 + 4 = 0.48 (95% CI: 0.41–0.53) and Gleason ≥ 4 + 3 = 0.56 (95% CI: 0.51–0.61). Of note, patients with no evidence of cancer had a lower ExoGrail risk score (mean difference = 0.17 (95% CI: 0.11–0.24)) than those with a raised PSA but no findings of cancer on biopsy (Figure 4).

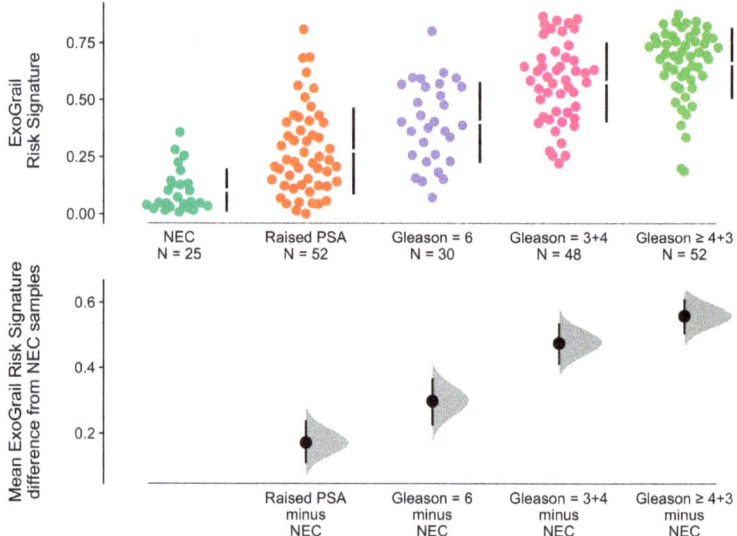

Figure 4. Mean ExoGrail risk score differences between biopsy outcomes, as represented by Cumming estimation plots. Individual patient risk scores (y-axis) are presented as points in the top panel, separated according to Gleason score (x-axis) with gapped vertical lines detailing the mean and standard deviation of each clinical group's ExoGrail risk score. Mean ExoGrail risk score differences relative to the no evidence of cancer (NEC) group are shown in the bottom panel. Mean difference and 95% confidence intervals are shown as a point estimate and vertical bar, respectively, with density plots generated from 1000 bias-corrected and accelerated bootstrap resamples.

Decision curve analyses examined the net benefit of ExoGrail adoption in a population of patients with a clinical suspicion of prostate cancer and a PSA level suitable to trigger biopsy (≥4 ng/mL). The biopsy of men based upon their ExoGrail risk score provided a net benefit over current standards of care across all decision thresholds examined and was the most consistent amongst all comparator models across a range of clinically relevant endpoints for biopsy (Figure 5).

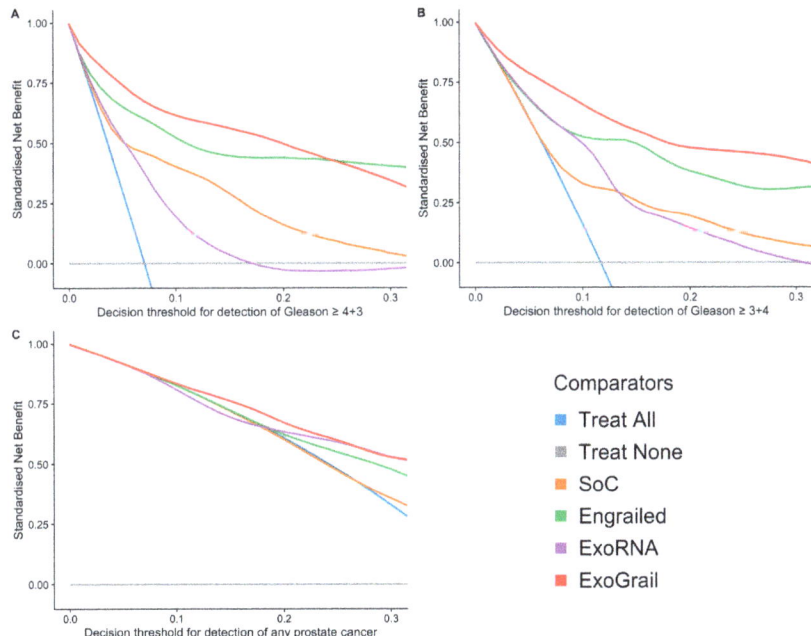

Figure 5. Exploration of the standardised net benefit (sNB) by decision curve analysis (DCA) for adopting risk models to aid the decision to undertake an initial biopsy for patients presenting with a serum PSA $\geq$ 4 ng/mL, where current clinical practice is to biopsy all patients. The accepted patient/clinician risk threshold for accepting biopsy is detailed on the x-axis. Different biopsy outcomes are shown in each of the three panels; (**A**) detection of Gleason $\geq$ 4 + 3, (**B**) detection of Gleason $\geq$ 3 + 4, (**C**) any cancer; Blue—biopsy all patients with a PSA >4 ng/mL, Orange—biopsy patients according to the SoC model, Green—biopsy patients based on the Engrailed model, Purple—biopsy patients based on the exoRNA model, Red—biopsy patients based on a the ExoGrail model. To assess the benefit of adopting these risk models in a clinically relevant population, we used data available from the control arm of the The Cluster Randomized Trial of PSA Testing for Prostate Cancer (CAP) study [42] for proportionally resampling the ExoGrail cohort. DCA curves were calculated from 1000 bootstrap resamples of the available data to match the distribution of disease reported in the CAP trial population. Mean sNB from these resampled DCA results are plotted here. See Methods for full details.

Using the SoC model as the baseline with which to compare the potential for biopsy reduction of each model, we found that ExoGrail could reduce unnecessary biopsy rates by upwards of 40%, depending on accepted patient-clinician risk. For example, if a decision threshold of 0.1 were accepted, representing a perceived risk of 1 in 10 for Gleason $\geq$ 3 + 4 on biopsy, ExoGrail could result in up to a 35% reduction in unnecessary biopsies of men presenting with a suspicion of prostate cancer, whilst also correctly identifying patients with more aggressive disease. If Gleason $\geq$ 4 + 3 were considered the threshold of clinical significance, a more conservative decision threshold of 0.05 could save 32% of men from receiving an unnecessary biopsy (Figure 6).

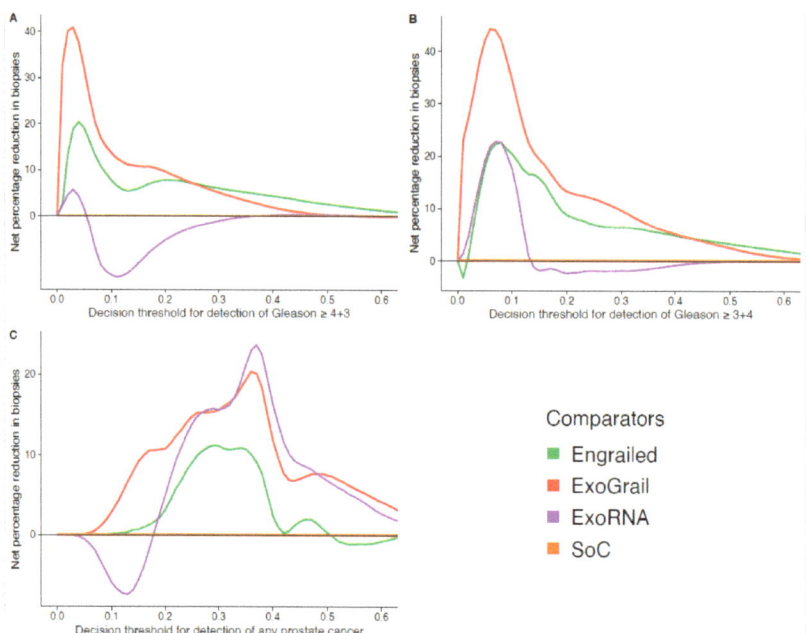

Figure 6. Estimation of biopsy reduction, as calculated by comparing the DCA-calculated net benefit of each risk model to the net benefit of the standard of care (SoC) model. The accepted patient/clinician risk threshold for accepting biopsy is detailed on the x-axis. Different biopsy outcomes are shown in each of the three panels: (**A**) detection of Gleason $\geq 4 + 3$, (**B**) detection of Gleason $\geq 3 + 4$ and (**C**) any cancer. Coloured lines show differing comparator models; Orange—biopsy patients according to the SoC model, Green—biopsy patients based on the Engrailed model, Purple—biopsy patients based on the ExoRNA model, Red—biopsy patients based on the ExoGrail model. To assess the benefit of adopting these risk models in a clinically relevant population we used data available from the control arm of the CAP study [42] for proportionally resampling the ExoGrail cohort. DCA curves were calculated from 1000 bootstrap resamples of the available data to match the distribution of disease reported in the CAP trial population. Net benefit averaged over all resamples was used to calculate the potential reductions in biopsy rates here. See Methods for full details.

4. Discussion

Discriminating disease status in patients before a diagnostic biopsy with higher accuracy than current standards could bring about a sizeable change in treatment pathways and reduce the number of men sent forward for ultimately unnecessary biopsy. Given that up to 75% of patients are negative for prostate cancer when presenting with serum PSA levels ≥ 4 ng/mL [5,43,44], a concentration of research efforts has been made to address this problem. To date, several biomarker panels have been successfully developed to non-invasively detect prostate cancer using urine samples, Gleason $\geq 3 + 4$ disease with superior accuracy to current clinically implemented methods, including the PUR model developed by ourselves [20,21,45,46]. However, as only a single aspect of urine, assay method or biological process are assessed by these examples, the heterogeneity of prostate cancer may not be entirely accounted for [47], requiring an approach to be taken that provides a more holistic insight into disease status.

Recent analyses, including those presented here, have demonstrated the added value of integrating multiple prognostic biomarkers within the process of fitting risk models for determining patient risk upon an initial biopsy [23,48]. Urine clearly contains a wealth of useful information concerning the disease status of the prostate through the quantification of cf-RNA transcripts, circulating and cell-free DNA, hypermethylation of DNA, and protein biomarker levels [19,46,49–52].

Our results show that an improved multivariable risk prediction model can be developed from the careful consideration of information from multiple different urine fractions in men suspected to have prostate cancer. Urinary levels of EN2 protein were quantified by ELISA, whilst the transcript levels of 167 cell-free mRNAs were quantified using NanoString technology. The final model integrating information from those assays with serum PSA levels was deemed ExoGrail. Markers selected for the model include well-known genes associated with prostate cancer and proven in other diagnostic tests, such as PCA3 [45], HOXC6 [20], and the TMPRSS2/ERG gene fusion [53]. An interaction between urinary EN2 protein levels and quantified transcripts of SLC12A1 was observed, further demonstrating the benefit of considering information from multiple biological sources (Figure S4).

ExoGrail was able to accurately predict the presence of significant (Gs $\geq$ 7) prostate cancer on biopsy with an AUC of 0.89, comparing favourably to other published tests (AUCs for Gs $\geq$ 7: PUR = 0.77 [46], ExoMeth = 0.89 [23], ExoDX Prostate IntelliScore = 0.77 [21], SelectMDX = 0.78 [20], epiCaPture Gs $\geq$ 4 + 3 AUC = 0.73 [49]). Furthermore, ExoGrail resulted in accurate predictions even when serum PSA levels alone proved inaccurate; patients with a raised PSA but negative biopsy result possessed ExoGrail scores significantly different from both clinically benign patients and those with low-grade Gleason 6 disease, whilst still able to discriminate between more clinically significant Gleason $\geq$ 7 cancers (Figure 4). The adoption of ExoGrail into current clinical pathways for reducing unnecessary biopsies was considered, showing the potential for up to 32% of patients to safely forgo an invasive biopsy without incurring excessive risk (Figure 6).

ExoGrail was developed with the explicit goal of being robust to potential overfitting and bias, using strong internal validation methods in bootstrap resampling and out-of-bag predictions. Nonetheless, ExoGrail was developed in a relatively small dataset and so requires external validation in an independent cohort before it can be considered for use as a clinical risk model. To this end, we are currently collecting samples from multiple sites in the UK, EU and Canada using an updated 'At-Home' Collection system [54]. The At-Home collection system enables biomarker analysis to be performed on urine samples provided by patients at home, which they send in the post to a centralised laboratory. This collection and analysis system will sidestep the need for a visit to the clinic and lead to a postal screening system for prostate cancer diagnosis and prognosis. In this study, we will also assess the potential utility of supplementing MP-MRI with ExoGrail, as MP-MRI can misrepresent disease status, even with rigorous controls in place [6]. The NanoString expression analysis system used in the ExoGrail signature is a rapid and cost-effective analysis system that is also used in the FDA-approved Prosigna Pam50 test for breast cancer aggressiveness [55], making ExoGrail well-positioned for implementation for patient benefit.

5. Conclusions

ExoGrail was able to accurately predict the presence of significant (Gs $\geq$ 7) prostate cancer on biopsy and showed the potential for an important number of patients to safely forgo an invasive biopsy. If validated in future studies, ExoGrail has the potential to positively impact the clinical experience of patients being investigated for prostate cancer that ultimately have no disease or indolent prostate cancer.

6. Patents

A patent application has been filed by the authors for the present work and work related to this.

Supplementary Materials: The following are available online at https://www.mdpi.com/article/10.3390/cancers13092102/s1, Figure S1: Boruta analysis of variables available for the training of the SoC model, Figure S2: Boruta analysis of variables available for the training of the Engrailed model, Figure S3: Boruta analysis of variables available for the training of the ExoRNA model, Figure S4:

Partial dependency plots detailing the marginal effects and interactions of SLC12A1 and urinary EN2 on predicted ExoGrail Risk Score, Figure S5: Risk score distributions of the trained models including the EN2 and PSA model, Figure S6: Density plots detailing risk score distributions generated from four trained models. Table S1: List of all features available for selection as input variables for each model prior to bootstrapped Boruta feature selection, Table S2: AUC of all trained models, including a combination of EN2 and PSA, for detecting outcomes of an initial biopsy for varying clinically significant thresholds. Supplementary Methods.

Author Contributions: S.P.C. and D.S.B drafted the manuscript and conceived, designed, and performed the statistical analyses. H.P. and R.M. (Richard Morgan) were involved in sample collection and ELISA analyses at their respective institutes. J.C. and R.M. were involved in sample collection and NanoString analyses, as well as the development of clinical methodologies. D.S.B., J.C., R.M. (Robert Mills), H.P., and C.S.C. had joint and equal contributions to senior authorship and were contributors in writing the manuscript. C.S.C., J.C., H.P. and D.S.B. provided the original idea for this study. All authors have read and agreed to the published version of the manuscript. All authors critiqued the manuscript for intellectual content.

Funding: This research was funded by Movember Foundation GAP1 Urine Biomarker project, The Masonic Charitable Foundation, The Bob Champion Cancer Trust, the King family, The Andy Ripley Memorial Fund and the Stephen Hargrave Trust.

Institutional Review Board Statement: The study was conducted according to the guidelines of the Declaration of Helsinki, and approval was granted for the collection and processing of samples by the Ethics Committees at the East of England REC.

Informed Consent Statement: Informed consent was obtained from all subjects involved in the study.

Data Availability Statement: All data and code required to reproduce these analyses can be found at https://github.com/UEA-Cancer-Genetics-Lab/ExoGrail_paper.

Acknowledgments: The research presented in this paper was carried out on the High Performance Computing Cluster supported by the Research and Specialist Computing Support service at the University of East Anglia.

Conflicts of Interest: A patent application has been filed by the authors for the present work and work related to this. There are no other conflicts of interest to disclose.

References

1. Cancer Research UK Prostate Cancer Incidence Statistics. 2019. Available online: http://www.cancerresearchuk.org/health-professional/cancer-statistics/statistics-by-cancer-type/prostate-cancer/incidence (accessed on 29 June 2019).
2. Loeb, S.; Bjurlin, M.A.; Nicholson, J.; Tammela, T.L.; Penson, D.F.; Carter, H.B.; Carroll, P.; Etzioni, R. Overdiagnosis and Overtreatment of Prostate Cancer. *Eur. Urol.* **2014**, *65*, 1046–1055. [CrossRef]
3. Sanda, M.G.; Cadeddu, J.A.; Kirkby, E.; Chen, R.C.; Crispino, T.; Fontanarosa, J.; Freedland, S.J.; Greene, K.; Klotz, L.H.; Makarov, D.V.; et al. Clinically Localized Prostate Cancer: AUA/ASTRO/SUO Guideline. Part I: Risk Stratification, Shared Decision Making, and Care Options. *J. Urol.* **2018**, *199*, 683–690. [CrossRef]
4. Cornford, P.; Bellmunt, J.; Bolla, M.; Briers, E.; De Santis, M.; Gross, T.; Henry, A.M.; Joniau, S.; Lam, T.B.; Mason, M.D.; et al. EAU-ESTRO-SIOG Guidelines on Prostate Cancer. Part II: Treatment of Relapsing, Metastatic, and Castration-Resistant Prostate Cancer. *Eur. Urol.* **2017**, *71*, 630–642. [CrossRef]
5. National Institute for Health and Care Excellence. *Prostate Cancer: Diagnosis and Management (Update)*; NICE: London, UK, 2015.
6. Ahmed, H.U.; Bosaily, A.E.-S.; Brown, L.C.; Gabe, R.; Kaplan, R.; Parmar, M.K.; Collaco-Moraes, Y.; Ward, K.; Hindley, R.G.; Freeman, A.; et al. Diagnostic accuracy of multi-parametric MRI and TRUS biopsy in prostate cancer (PROMIS): A paired validating confirmatory study. *Lancet* **2017**, *389*, 815–822. [CrossRef]
7. Pepe, P.; Pennisi, M. Gleason score stratification according to age at diagnosis in 1028 men. *Współczesna Onkol.* **2015**, *19*, 471–473. [CrossRef]
8. Sonn, G.A.; Fan, R.E.; Ghanouni, P.; Wang, N.N.; Brooks, J.D.; Loening, A.M.; Daniel, B.L.; To'O, K.J.; Thong, A.E.; Leppert, J.T. Prostate Magnetic Resonance Imaging Interpretation Varies Substantially Across Radiologists. *Eur. Urol. Focus* **2019**, *5*, 592–599. [CrossRef] [PubMed]
9. Walz, J. The "PROMIS" of Magnetic Resonance Imaging Cost Effectiveness in Prostate Cancer Diagnosis? *Eur. Urol.* **2018**, *73*, 31–32. [CrossRef] [PubMed]
10. Moschini, M.; Spahn, M.; Mattei, A.; Cheville, J.; Karnes, R.J. Incorporation of tissue-based genomic biomarkers into localized prostate cancer clinics. *BMC Med.* **2016**, *14*, 67. [CrossRef] [PubMed]

11. Luca, B.-A.; Brewer, D.S.; Edwards, D.R.; Edwards, S.; Whitaker, H.C.; Merson, S.; Dennis, N.; Cooper, R.A.; Hazell, S.; Warren, A.Y.; et al. DESNT: A Poor Prognosis Category of Human Prostate Cancer. *Eur. Urol. Focus* **2018**, *4*, 842–850. [CrossRef]
12. Knezevic, D.; Goddard, A.D.; Natraj, N.; Cherbavaz, D.B.; Clark-Langone, K.M.; Snable, J.; Watson, D.; Falzarano, S.M.; Magi-Galluzzi, C.; Klein, E.A.; et al. Analytical validation of the Oncotype DX prostate cancer assay—A clinical RT-PCR assay optimized for prostate needle biopsies. *BMC Genom.* **2013**, *14*, 690. [CrossRef] [PubMed]
13. Cuzick, J.; Berney, D.M.; Fisher, G.J.; Mesher, D.; Moller, H.; Reid, J.; Perry, M.B.A.; Park, J.; Younus, A.; on behalf of the Transatlantic Prostate Group; et al. Prognostic value of a cell cycle progression signature for prostate cancer death in a conservatively managed needle biopsy cohort. *Br. J. Cancer* **2012**, *106*, 1095–1099. [CrossRef]
14. Luca, B.-A.; Moulton, V.; Ellis, C.; Edwards, D.R.; Campbell, C.; Cooper, R.A.; Clark, J.; Brewer, D.S.; Cooper, C.S. A novel stratification framework for predicting outcome in patients with prostate cancer. *Br. J. Cancer* **2020**, *122*, 1467–1476. [CrossRef] [PubMed]
15. Cooperberg, M.R.; Davicioni, E.; Crisan, A.; Jenkins, R.B.; Ghadessi, M.; Karnes, R.J. Combined Value of Validated Clinical and Genomic Risk Stratification Tools for Predicting Prostate Cancer Mortality in a High-risk Prostatectomy Cohort. *Eur. Urol.* **2015**, *67*, 326–333. [CrossRef] [PubMed]
16. Eklund, M.; Nordström, T.; Aly, M.; Adolfsson, J.; Wiklund, P.; Brandberg, Y.; Thompson, J.; Wiklund, F.; Lindberg, J.; Presti, J.C.; et al. The Stockholm-3 (STHLM3) Model can Improve Prostate Cancer Diagnostics in Men Aged 50–69 yr Compared with Current Prostate Cancer Testing. *Eur. Urol. Focus* **2018**, *4*, 707–710. [CrossRef] [PubMed]
17. Tosoian, J.J.; Carter, H.B.; Lepor, A.; Loeb, S. Active surveillance for prostate cancer: Current evidence and contemporary state of practice. *Nat. Rev. Urol.* **2016**, *13*, 205–215. [CrossRef] [PubMed]
18. Frick, J.; Aulitzky, W. Physiology of the prostate. *Infection* **1991**, *19* (Suppl. 3), S115–S118. [CrossRef] [PubMed]
19. Morgan, R.; Boxall, A.; Bhatt, A.; Bailey, M.; Hindley, R.; Langley, S.; Whitaker, H.C.; Neal, D.E.; Ismail, M.; Whitaker, H.; et al. Engrailed-2 (EN2): A Tumor Specific Urinary Biomarker for the Early Diagnosis of Prostate Cancer. *Clin. Cancer Res.* **2011**, *17*, 1090–1098. [CrossRef]
20. Van Neste, L.; Hendriks, R.J.; Dijkstra, S.; Trooskens, G.; Cornel, E.B.; Jannink, S.A.; de Jong, H.; Hessels, D.; Smit, F.P.; Melchers, W.J.; et al. Detection of High-grade Prostate Cancer Using a Urinary Molecular Biomarker–Based Risk Score. *Eur. Urol.* **2016**, *70*, 740–748. [CrossRef] [PubMed]
21. McKiernan, J.; Donovan, M.J.; O'Neill, V.; Bentink, S.; Noerholm, M.; Belzer, S.; Skog, J.; Kattan, M.W.; Partin, A.; Andriole, G.; et al. A Novel Urine Exosome Gene Expression Assay to Predict High-grade Prostate Cancer at Initial Biopsy. *JAMA Oncol.* **2016**, *2*, 882–889. [CrossRef] [PubMed]
22. Haese, A.; Trooskens, G.; Steyaert, S.; Hessels, D.; Brawer, M.; Vlaeminck-Guillem, V.; Ruffion, A.; Tilki, D.; Schalken, J.; Groskopf, J.; et al. Multicenter Optimization and Validation of a 2-Gene mRNA Urine Test for Detection of Clinically Significant Prostate Cancer before Initial Prostate Biopsy. *J. Urol.* **2019**, *202*, 256–263. [CrossRef]
23. Connell, S.P.; O'Reilly, E.; Tuzova, A.; Webb, M.; Hurst, R.; Mills, R.; Zhao, F.; Bapat, B.; Cooper, C.S.; Perry, A.S.; et al. Development of a multivariable risk model integrating urinary cell DNA methylation and cell-free RNA data for the detection of significant prostate cancer. *Prostate* **2020**, *80*, 547–558. [CrossRef] [PubMed]
24. Morgan, R. Engrailed: Complexity and economy of a multi-functional transcription factor. *FEBS Lett.* **2006**, *580*, 2531–2533. [CrossRef]
25. Punia, N.; Primon, M.; Simpson, G.R.; Pandha, H.S.; Morgan, R. Membrane insertion and secretion of the Engrailed-2 (EN2) transcription factor by prostate cancer cells may induce antiviral activity in the stroma. *Sci. Rep.* **2019**, *9*, 5138. [CrossRef] [PubMed]
26. Pandha, H.; Sorensen, K.D.; Orntoft, T.F.; Langley, S.; Hoyer, S.; Borre, M.; Morgan, R. Urinary engrailed-2 (EN2) levels predict tumour volume in men undergoing radical prostatectomy for prostate cancer. *BJU Int.* **2012**, *110*, E287–E292. [CrossRef] [PubMed]
27. Elamin, A.A.; Klunkelfuß, S.; Kämpfer, S.; Oehlmann, W.; Stehr, M.; Smith, C.; Simpson, G.R.; Morgan, R.; Pandha, H.; Singh, M. A Specific Blood Signature Reveals Higher Levels of S100A12: A Potential Bladder Cancer Diagnostic Biomarker Along with Urinary Engrailed-2 Protein Detection. *Front. Oncol.* **2020**, *9*, 1484. [CrossRef]
28. Stark, J.R.; Perner, S.; Stampfer, M.J.; Sinnott, J.A.; Finn, S.; Eisenstein, A.S.; Ma, J.; Fiorentino, M.; Kurth, T.; Loda, M.; et al. Gleason Score and Lethal Prostate Cancer: Does 3 + 4 = 4 + 3? *J. Clin. Oncol.* **2009**, *27*, 3459–3464. [CrossRef]
29. Collins, G.S.; Reitsma, J.B.; Altman, D.G.; Moons, K.G. Transparent Reporting of a Multivariable Prediction Model for Individual Prognosis or Diagnosis (TRIPOD): The TRIPOD Statement. *Eur. Urol.* **2015**, *67*, 1142–1151. [CrossRef]
30. R Core Team. *R: A Language and Environment for Statistical Computing*; R Core Team: Vienna, Austria, 2019.
31. UEA Cancer Genetic GitHub Repository. Available online: https://github.com/UEA-Cancer-Genetics-Lab/ExoGrail (accessed on 21 April 2021).
32. Guyon, I.; Elisseeff, A. An Introduction to Variable and Feature Selection. *J. Mach. Learn. Res.* **2003**, *3*, 1157–1182.
33. Kursa, M.B.; Rudnicki, W.R. Feature Selection with theBorutaPackage. *J. Stat. Softw.* **2010**, *36*, 1–13. [CrossRef]
34. Breiman, L. Random Forests. *Mach. Learn.* **2001**, *45*, 5–32. [CrossRef]
35. Liaw, A.; Wiener, M. Classification and Regression by randomForest. *R News* **2002**, *2*, 18–22.
36. Robin, X.A.; Turck, N.; Hainard, A.; Tiberti, N.; Lisacek, F.; Sanchez, J.-C.; Muller, M.J. pROC: An open-source package for R and S+ to analyze and compare ROC curves. *BMC Bioinform.* **2011**, *12*, 77. [CrossRef] [PubMed]
37. Wickham, H. *ggplot2: Elegant Graphics for Data Analysis*; Springer: New York, NY, USA, 2016; ISBN 978-3-319-24277-4.

38. Greenwell, B.M. Pdp: An r Package for Constructing Partial Dependence Plots. *R J.* **2017**, *9*, 421–436. [CrossRef]
39. Ho, J.; Tumkaya, T.; Aryal, S.; Choi, H.; Claridge-Chang, A. Moving beyond P values: Data analysis with estimation graphics. *Nat. Methods* **2019**, *16*, 565–566. [CrossRef] [PubMed]
40. Vickers, A.J.; Elkin, E.B. Decision Curve Analysis: A Novel Method for Evaluating Prediction Models. *Med. Decis. Mak.* **2006**, *26*, 565–574. [CrossRef] [PubMed]
41. Brown, M. *rmda: Risk Model Decision Analysis*; Fred Hutchinson Cancer Research Center: Seattle, WA, USA, 2018.
42. Kerr, K.F.; Brown, M.D.; Zhu, K.; Janes, H. Assessing the Clinical Impact of Risk Prediction Models with Decision Curves: Guidance for Correct Interpretation and Appropriate Use. *J. Clin. Oncol.* **2016**, *34*, 2534–2540. [CrossRef]
43. Martin, R.M.; Donovan, J.L.; Turner, E.L.; Metcalfe, C.; Young, G.J.; Walsh, E.I.; Lane, J.A.; Noble, S.; Oliver, S.E.; Evans, S.; et al. Effect of a Low-Intensity PSA-Based Screening Intervention on Prostate Cancer Mortality: The CAP randomized clinical trial. *JAMA* **2018**, *319*, 883–895. [CrossRef]
44. Lane, J.A.; Donovan, J.L.; Davis, M.; Walsh, E.; Dedman, D.; Down, L.; Turner, E.L.; Mason, M.D.; Metcalfe, C.; Peters, T.J.; et al. Active monitoring, radical prostatectomy, or radiotherapy for localised prostate cancer: Study design and diagnostic and baseline results of the ProtecT randomised phase 3 trial. *Lancet Oncol.* **2014**, *15*, 1109–1118. [CrossRef]
45. Hessels, D.; Gunnewiek, J.M.K.; van Oort, I.; Karthaus, H.F.; van Leenders, G.J.; van Balken, B.; Kiemeney, L.A.; Witjes, J.; Schalken, J.A. DD3PCA3-based Molecular Urine Analysis for the Diagnosis of Prostate Cancer. *Eur. Urol.* **2003**, *44*, 8–16. [CrossRef]
46. Connell, S.P.; Yazbek-Hanna, M.; McCarthy, F.; Hurst, R.; Webb, M.; Curley, H.; Walker, H.; Mills, R.; Ball, R.Y.; Sanda, M.G.; et al. A four-group urine risk classifier for predicting outcomes in patients with prostate cancer. *BJU Int.* **2019**, *124*, 609–620. [CrossRef]
47. Ciccarese, C.; Massari, F.; Iacovelli, R.; Fiorentino, M.; Montironi, R.; Di Nunno, V.; Giunchi, F.; Brunelli, M.; Tortora, G. Prostate cancer heterogeneity: Discovering novel molecular targets for therapy. *Cancer Treat. Rev.* **2017**, *54*, 68–73. [CrossRef]
48. Strand, S.H.; Bavafaye-Haghighi, E.; Kristensen, H.; Rasmussen, A.K.; Hoyer, S.; Borre, M.; Mouritzen, P.; Besenbacher, S.; Orntoft, T.F.; Sorensen, K.D. A novel combined miRNA and methylation marker panel (miMe) for prediction of prostate cancer outcome after radical prostatectomy. *Int. J. Cancer* **2019**, *145*, 3445–3452. [CrossRef]
49. O'Reilly, E.; Tuzova, A.V.; Walsh, A.L.; Russell, N.M.; O'Brien, O.; Kelly, S.; Ni Dhomhnallain, O.; DeBarra, L.; Dale, C.M.; Brugman, R.; et al. epiCaPture: A Urine DNA Methylation Test for Early Detection of Aggressive Prostate Cancer. *JCO Precis. Oncol.* **2019**, *2019*, 1–18. [CrossRef]
50. Zhao, F.; Olkhov-Mitsel, E.; Kamdar, S.; Jeyapala, R.; Garcia, J.; Hurst, R.; Hanna, M.Y.; Mills, R.; Tuzova, A.V.; O'Reilly, E.; et al. A urine-based DNA methylation assay, ProCUrE, to identify clinically significant prostate cancer. *Clin. Epigenet.* **2018**, *10*, 147. [CrossRef]
51. Xia, Y.; Huang, C.-C.; Dittmar, R.; Du, M.; Wang, Y.; Liu, H.; Shenoy, N.; Wang, L.; Kohli, M. Copy number variations in urine cell free DNA as biomarkers in advanced prostate cancer. *Oncotarget* **2016**, *7*, 35818–35831. [CrossRef]
52. Killick, E.; Morgan, R.; Launchbury, F.; Bancroft, E.; Page, E.; Castro, E.; Kote-Jarai, Z.; Aprikian, A.; Blanco, I.; Clowes, V.; et al. Role of Engrailed-2 (EN2) as a prostate cancer detection biomarker in genetically high risk men. *Sci. Rep.* **2013**, *3*, 2059. [CrossRef]
53. Tomlins, S.A.; Day, J.R.; Lonigro, R.J.; Hovelson, D.H.; Siddiqui, J.; Kunju, L.P.; Dunn, R.L.; Meyer, S.; Hodge, P.; Groskopf, J.; et al. Urine TMPRSS2:ERG Plus PCA3 for Individualized Prostate Cancer Risk Assessment. *Eur. Urol.* **2016**, *70*, 45–53. [CrossRef] [PubMed]
54. Webb, M.; Manley, K.; Olivan, M.; Guldvik, I.; Palczynska, M.; Hurst, R.; Connell, S.P.; Mills, I.G.; Brewer, D.S.; Mills, R.; et al. Methodology for the at-home collection of urine samples for prostate cancer detection. *Biotechniques* **2020**, *68*, 65–71. [CrossRef] [PubMed]
55. Wallden, B.; Storhoff, J.; Nielsen, T.; Dowidar, N.; Schaper, C.; Ferree, S.; Liu, S.; Leung, S.; Geiss, G.; Snider, J.; et al. Development and Verification of the PAM50-Based Prosigna Breast Cancer Gene Signature Assay. *BMC Med. Genom.* **2015**, *8*, 54. [CrossRef] [PubMed]

Article

Intraoperative Digital Analysis of Ablation Margins (DAAM) by Fluorescent Confocal Microscopy to Improve Partial Prostate Gland Cryoablation Outcomes

Oscar Selvaggio [1], Ugo Giovanni Falagario [1,*], Salvatore Mariano Bruno [1], Marco Recchia [1], Maria Chiara Sighinolfi [2,3], Francesca Sanguedolce [4], Paola Milillo [5], Luca Macarini [5], Ardeshir R. Rastinehad [6], Rafael Sanchez-Salas [7], Eric Barret [7], Franco Lugnani [8], Bernardo Rocco [3], Luigi Cormio [1,9] and Giuseppe Carrieri [1]

[1] Department of Urology and Organ Transplantation, University of Foggia, 71122 Foggia, Italy; oscarsel@libero.it (O.S.); mariano.bruno91@gmail.com (S.M.B.); marco.recchia291292@gmail.com (M.R.); luigi.cormio@unifg.it (L.C.); giuseppe.carrieri@unifg.it (G.C.)
[2] Department of Urology, Azienda Ospedaliero-Universitaria di Modena, 41121 Modena, Italy; sighinolfic@yahoo.com
[3] Department of Urology, ASST Santi Paolo e Carlo Dipartimento di Scienze della Salute, Università degli Studi di Milano, 20122 Milano, Italy; bernardo.rocco@gmail.com
[4] Department of Pathology, University of Foggia, 71122 Foggia, Italy; fradolce@hotmail.com
[5] Department of Radiology, University of Foggia, 71122 Foggia, Italy; paola.milillo@yahoo.com (P.M.); luca.macarini@unifg.it (L.M.)
[6] Department of Urology, Lenox Hill Urology, Northwell Health System, New York, NY 10075, USA; arastine@northwell.edu
[7] Urology Department, Institut Mutualiste Montsouris, 75014 Paris, France; rafael.sanchez-salas@imm.fr (R.S.-S.); Eric.Barret@imm.fr (E.B.)
[8] Hippocrates D.O.O. Center, 6215 Divaca, Slovenia; info@lugnani.com
[9] Department of Urology, Bonomo Teaching Hospital, 76123 Andria, Italy
* Correspondence: ugofalagario@gmail.com; Tel.: +39-08-8173-2111

Simple Summary: This study tested the feasibility and reliability of a novel digital microscopy technique in assessing ablation margins during partial prostate gland cryoablation. Though preliminary, findings suggest that this novel technique may increase the efficacy of focal treatments, by reducing the risk of untreated prostate cancer areas not visible to an MRI, as well as safety, by more precisely sparing uninvolved areas and surrounding structures.

Abstract: Partial gland cryoablation (PGC) aims at destroying prostate cancer (PCa) foci while sparing the unaffected prostate tissue and the functionally relevant structures around the prostate. Magnetic Resonance Imaging (MRI) has boosted PGC, but available evidence suggests that ablation margins may be positive due to MRI-invisible lesions. This study aimed at determining the potential role of intraoperative digital analysis of ablation margins (DAAM) by fluoresce confocal microscopy (FCM) of biopsy cores taken during prostate PGC. Ten patients with low to intermediate risk PCa scheduled for PGC were enrolled. After cryo-needles placement, 76 biopsy cores were taken from the ablation margins and stained by the urologist for FCM analysis. Digital images were sent for "real-time" pathology review. DAAM, always completed within the frame of PGC treatment (median time 25 min), pointed out PCa in 1/10 cores taken from 1 patient, thus prompting placement of another cryo-needle to treat this area. Standard HE evaluation confirmed 75 cores to be cancer-free while displayed a GG 4 PCa in 7% of the core positive at FCM. Our data point out that IDAAM is feasible and reliable, thus representing a potentially useful tool to reduce the risk of missing areas of PCa during PGC.

Keywords: prostate cancer; fluorescence confocal microscopy; prostate biopsy; ablation margins; focal therapy

1. Introduction

EAU guidelines mention partial gland cryoablation (PGC) as a minimally invasive investigational option for the management of organ-confined prostate cancer (PCa) [1]. One of the technical challenges during PGC is to deploy multiple cryo-needles to form an ablation zone sufficiently covering the target volume while sparing surrounding critical structures. Historically, physicians relied on digital rectal examination (DRE) and template prostate biopsy results to plan the cryo-needles placement. The multifocal nature of PCa and the suboptimal prostate sampling obtained using template biopsies, however, limited the widespread use of PGC and supported treatment of the entire gland or at least half of it (hemigland ablation) to achieve wide safety margins. Even so, up to 15% of patients having undergone hemigland cryoablation experienced treatment failure at 5 years of follow-up [2]. The availability of multiparametric Magnetic Resonance Imaging (MRI) has boosted PGC. Indeed, the combination of MRI and target biopsy have been proved to be superior to template prostate biopsy alone in identifying lobes with significant PCa for the application of hemi-ablative focal therapies [3].

Recent studies reported encouraging mid-term results in cohorts of patients undergoing MRI-guided focal therapies [4,5]. Still, MRI reader experience [6], MRI-invisible lesions, and targeting errors in placing the cryo-needles [7] can lead to positive ablation margins impacting PGC outcomes.

Recently, a novel technology has shown promise for the real-time microscopic evaluation of prostate tissue [8]. Fluorescent confocal microscopy (FCM) allows the immediate acquisition of digital images in a hematoxylin-eosin (HE)-like fashion without conventional processing and its attendant time and resource requirements. In the preliminary studies, FCM provided an excellent discrimination performance compared to HE for prostate biopsy cores and peri-prostatic tissue evaluation [9,10].

This pilot study aimed at determining feasibility and reliability of intraoperative FCM in assessing ablation margins during PGC.

2. Materials and Methods

2.1. Study Population

Following ethics committee approval (University of Foggia, approval number 143/CE/2020), ten patients with clinically localized low to intermediate risk PCa (PSA $\leq$ 20.0 ng/mL, Gleason Group (GG) $\leq$ 3) scheduled for PGC between September and November 2020 were enrolled in this prospective study evaluating intraoperative digital analysis of ablation margins (DAAM) by FCM. Patients with high-risk PCa (PSA > 20.0 ng/mL, GG > 3) and those who received any prior treatments for malign and benign prostatic disease were excluded.

All patients meeting inclusion and exclusion criteria signed an informed consent form.

2.2. Multiparametric Magnetic Resonance Imaging and Prostate Biopsy Technique

A prebiopsy mpMRI was performed and interpreted by a single dedicated radiologist (7 years' experience in prostate MRI) according to PIRADS V2.1 recommendation [11]. Specifically, the mpMRI protocol consisted of: (a) Turbo-Spin-Echo (TSE) T2-weighed imaging in axial, coronal, and sagittal planes (repetition time (TR) 5300, echo time (TE) 150 ms, slice thickness 3 mm, field of view (FOV) 180 $\times$ 180, number of signals averaged (NSA 8); (b) TSE T1-weighed imaging in the axial plane (TR/TE 400–650/12 ms, thickness 3 mm, FOV 180$\times$180, NSA 3); (c) Diffusion-weighted imaging sequence (DWI) in the axial plane (TR/TE 3481/92 ms, slice thickness 3 mm, FOV 180 $\times$ 220, NSA 4, b-values 0–500–1000–1500/2000 sec/mm^2); (d) Dynamic contrast enhanced prostate MRI was performed using a T1-weighted high-resolution isotropic volume examination (THRIVE) on the axial plane (TR/TE 4.5/2.2 ms, slice thickness 3 mm, FOV 184 $\times$ 220, NSA 1) following injection of 0.1 mL/kg of gadobutrol followed by 20 mL of saline solution using an automatic injector at a rate of 2 mL/s. All patients underwent prostate biopsy at our institution with 2 to 4 target cores to any MRI suspicious lesion in addition to our 18 cores systematic template [12]. An Electromagnetically Tracked MRI/US Fusion system (Navigo, UC-Care, Tampa, FL, USA) was used to performed target sampling.

2.3. Prostate Gland Cryoablation with Intraoperative Digital Analysis of Ablation Margins (DAAM)

Based on mpMRI findings and biopsy results, two to four 2.4 mm cryo-needles were inserted transperineally using a template grid and with the help of an Electromagnetically Tracked MRI/US Fusion system (Navigo, UC-Care, Tampa, FL, USA). Prostate Cryoablation was performed using an argon/helium gas-based system (Endocare, HeathTonics Inc., Austin, TX, USA). After the creation of the ice balls, 7 to 10 biopsy cores, depending on prostate volume, were taken transperineally from ablation margins and untreated areas of the prostate (Figure 1). In order to avoid potential tissue damage, cores were taken outside ice balls. A single surgeon performed PGC and intraoperative prostate biopsies (OS).

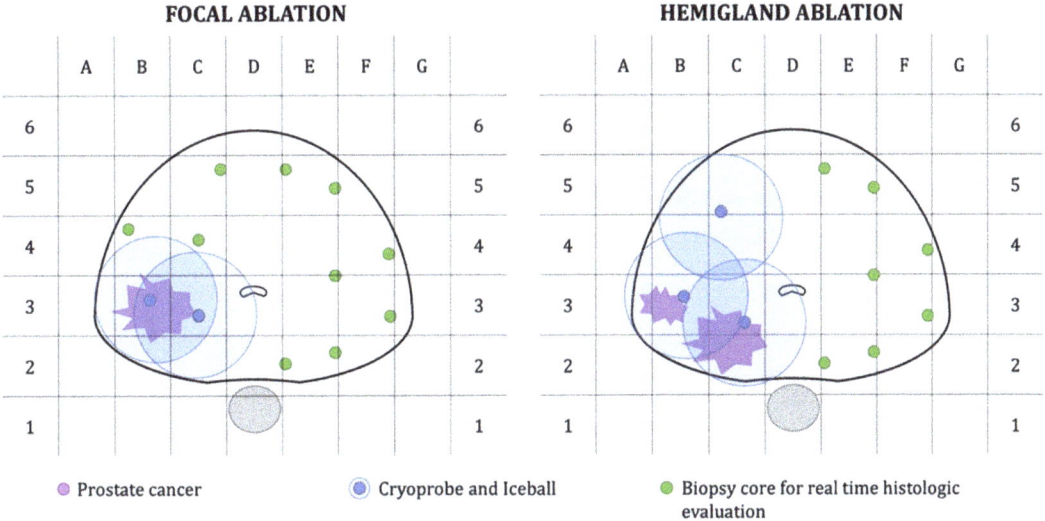

Figure 1. Graph representation of treatment plan and biopsy core template for patients undergoing focal or hemigland cryoablation.

The cores were prepared for FMC (VivaScope® 2500 M-G4, VivaScope GmbH, Munich, Germany) in the operating room by a urologist (UF) following the manufacturer guidelines. A glass slide is loaded with three biopsy cores and then scanned within a maximum time of 2 min. Sample preparation and image acquisition do not require special training. Samples are first cleaned using 70% Ethanol solution and then colored using Acridine Orange (1/50 solution in distilled water for 20–30 seconds). There are no specific requirements for the cores (e.g., length, diameter, thickness, integrity); however, to reduce acquisition time, care needs to be taken to place each biopsy core parallel or perpendicular to the slide axis. The FCM device combines two different lasers that enable tissue examination according to reflectance (785 nm) and fluorescence (488 nm) modalities. Images are rendered by the microscope software as pseudo-Hematoxylin-eosin (HE) images, relying on the combination of two images acquired at each wavelength. Specifically, nuclei appear purple, and collagen and cytoplasm appear pink.

The obtained digital FCM images were sent for "real-time" pathology review to a dedicated uropathologist (FS). Prostate Cryoablation was completed with no adjustments in patients with negative ablation margins. Conversely, if one or more cores were positive at FCM evaluation, the surgical plan was modified in order to ensure treatment of all cancer areas.

2.4. Final Pathology Examination

Biopsy cores were then fixed, stained using standard hematoxylin & eosin (HE), and sent to the pathology department for final diagnosis. A second dedicated uropathologist, blinded to clinical information and digital biopsy results, reported all PB specimens according to the 2019 ISUP recommendations [13] and diagnostic criteria for high-grade prostatic intraepithelial neoplasia and atypical small acinar proliferation of prostate [14]. ISUP Gleason-grade groups (ISUP GG) were reported per each core.

2.5. Study Endpoints and Statistical Analysis

The primary endpoint of the present study was time for FCM diagnosis. The secondary endpoint was the efficacy of FCM as measured by the agreement between digital biopsies and standard HE evaluation in the assessment of PCa. Surgical and intraoperative biopsy complications were reported using the Clavien-Dindo Classification [15]. Finally, PSA values at three months after therapy were used to compute the percentage of PSA reduction (derived from the ratio between PSA at three months and preoperative PSA) [16].

Descriptive statistics were performed using Stata 14 (StataCorp LP, College Station, TX, USA).

3. Results

Descriptive characteristics of the study population are listed in Table 1. Four patients with biopsy-proven MRI-visible GG1-2 PCa were treated with focal Cryoablation. The remaining patients with unilateral non-MRI-visible PCa underwent prostate hemiablation according to the SPARED guidelines [17]. In total, 76 cores were taken and analyzed using FCM (Figure 2). The median time for FCM diagnosis was 25 (IQR: 25, 27) min, with DAAM being always completed before the conclusion of the two treatment cycles. DAAM was negative in 9 patients, whereas in one patient with a preoperative diagnosis of low volume (two cores) GG 2 PCa, it pointed out PCa in 1 of 10 cores from the ablation margin area (left medial posterior core). A third cryo-needle was then placed to treat this positive core. The post-operative stay was uneventful, and all patients were discharged on post-operative day 1 with a urethral catheter removed on post-operative day 7. The median PSA drop at three months was 5.7 (IQR: 4.7, 6.6) ng/mL. The median percentage of PSA reduction was 79.0 (78.0, 85.0).

Table 1. Descriptive characteristics of the study population.

	Study Population (N = 10)
Age, years	70 (68, 73)
PSA, ng/mL	6.9 (6.2, 8.9)
DRE, n (%)	
Negative	4 (40.0%)
Positive	6 (60.0%)
PIRADS, n (%)	
2	2 (20.0%)
3	2 (20.0%)
4	4 (40.0%)
5	2 (20.0%)
Prostate Volume, cc	41.5 (40.0, 50.0)
Positive cores	3.5 (3.0, 5.0)
Bx GG, n (%)	
1	4 (40.0%)
2	4 (40.0%)
3	2 (20.0%)

Table 1. Cont.

	Study Population (N = 10)
EAU risk, n (%)	
Low risk	3 (30.0%)
Intermediate Risk	7 (70.0%)
Treatment, n (%)	
Focal ablation	4 (40.0%)
Hemigland Ablation	6 (60.0%)
N of Probes	3.5 (2.0, 4.0)
Intraoperative cores, n (%)	
7	8 (80.0%)
10	2 (20.0%)
Time to FCM diagnosis	25 (25, 27)
3-month PSA, ng/mL	1.4 (1.0, 1.6)
PSA drop, ng/mL	5.7 (4.7, 6.6)
Percentage of PSA reduction	79.0 (78.0, 85.0)

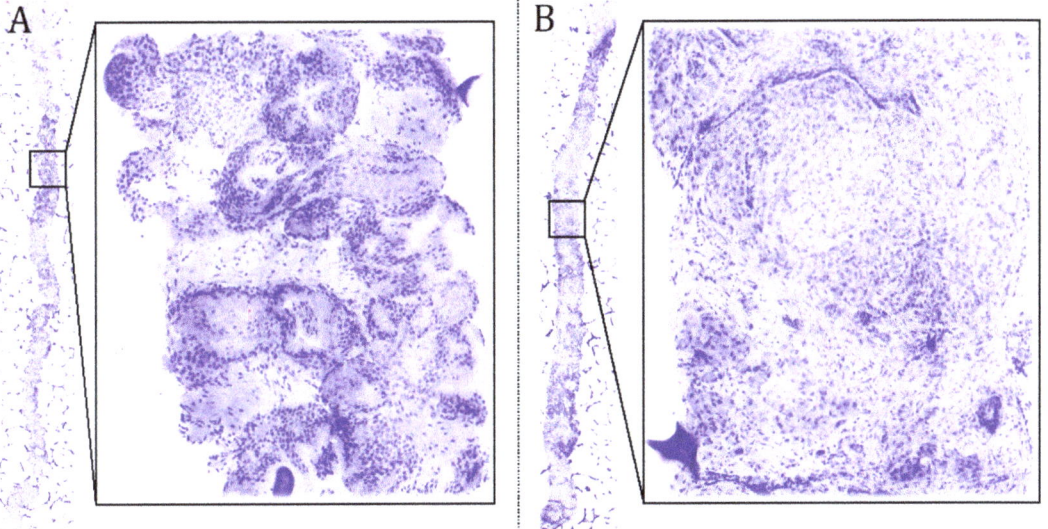

Figure 2. Digital Biopsy images. High-resolution pictures allow zooming into focal areas to better evaluate the architecture of the prostatic gland. (**A**) Negative core and zoom into normal gland architecture. (**B**) Positive core, Gleason Grade Group 4, pattern 4.

Standard HE evaluation confirmed 75 cores to be cancer-free while displayed a GG 4 PCa in 7% of the core positive at FCM.

4. Discussion

To our knowledge, this is the first study testing the potential role of intraoperative digital analysis of ablation margins (DAAM) by fluorescence confocal microscopy during prostate PGC. This novel technique holds promises to revolutionize PCa diagnosis and treatment by potentially replacing traditional frozen sections.

Other optical technologies allowing real-time microscopic evaluation of cancer tissue have been recently described and can be divided into two main groups. Conventional confocal microscopy is the most cost-effective option. UV light-emitting diode and other lasers with different wavelengths have been tested for real-time microscopic examination of fresh tissue taken from different cancers (mainly skin, prostate, and breast cancer) [18–21]. The technical process for image acquisition is the same; all light sources provided optimal quality images; however, comparative studies are lacking.

On the other hand, multiphoton microscopy involves an ultrafast (typically femtosecond pulse duration) laser source to achieve the extremely high photon density at the focal plane needed to excite two-photon absorption-based fluorescence [22,23]. This technique achieves a considerably higher imaging depth (using near-infrared wavelengths) with comparable image resolution, but it is limited by longer acquisition times and much higher costs of the microscope.

While the best technique for image acquisition is still a matter of debate, the relevance of the intraoperative analysis of prostatic tissue was clearly pointed out by the NeuroSAFE dissection technique, which was developed to maximize the preservation of periprostatic tissue during nerve-sparing radical prostatectomies [24]. While the patient is still under general anesthesia, the posterolateral aspect of the prostate is prepared using a cryostat, stained with HE, and analyzed under the microscope by a dedicated pathologist on-site. If persisting malignant glands are noted, a secondary resection of the ipsilateral bundle is performed. Though promising, it required a dedicated setup to obtain a diagnosis in a reasonable time [25].

To overcome these limitations, Rocco et al. proposed using FCM for real-time assessment of surgical margins during radical prostatectomy, reporting excellent results [10]. Indeed, the pathologist provided results from remote in less than 25 min, and there was perfect agreement between FCM and subsequent HE findings.

FCM may be even more crucial in the context of focal therapies. Indeed, frozen sections are not a viable option, and FCM is the fastest technique allowing the preservation of biopsy cores for further HE staining, immunohistochemistry/genomic studies.

The two studies tested the feasibility and diagnostic accuracy of FCM onto prostate biopsy cores rather than surgical margins. Rocco et al. obtained digital images of biopsy cores in 54 patients; perfect agreement between FCM and HE diagnosis was obtained for 95.1% of the 427 tested cores [9]. Similarly, Marenco et al. tested by FCM 182 MRI-targeted biopsy cores obtained in 57 biopsy-naive patients. The median time for FCM processing and analysis was 5 minutes; the positive and negative predictive values were 85% and 95%, respectively [26].

The present study confirmed the perfect agreement between FCM and standard HE findings, with FCM not affecting at all the quality of HE images in all 76 taken cores.

Our findings open novel perspectives on focal therapies since real-time detection of PCa in the ablation margins is likely to dramatically reduce the risk of undertreatment and, therefore, of disease recurrence. All patients included in the present study had a reduction of PSA > 75% at three months of follow-up. Even if the percentage of PSA reduction represents only a surrogate of the oncological outcome, Stabile et al. recently showed it is inversely associated with disease recurrence and the need for additional treatment, thus potentially useful as a proxy of treatment efficacy [16].

The small sample size and short-term follow-up are obvious limitations of this pilot study which, however, provides solid grounds for further evaluation of DAAM in the setting of focal PCa treatments.

5. Conclusions

Intraoperative DAAM by FCM proved to be feasible and reliable. This strategy potentially allows to reduce disease recurrence by extending treatment to previously unsuspected areas of PCa as well as to reduce functional complications by limiting treatment to the affected areas while sparing surrounding structures.

Author Contributions: Conceptualization, O.S., U.G.F., B.R., L.C. and G.C.; Data curation, U.G.F., S.M.B., M.R., F.S., P.M. and L.M.; Formal analysis, U.G.F.; Methodology, O.S., S.M.B., M.C.S., F.S., E.B. and F.L.; Supervision, L.M., B.R., L.C. and G.C.; Validation, M.R.; Writing–original draft, O.S. and U.G.F.; Writing–review & editing, M.C.S., F.S., A.R.R., R.S.-S., E.B. and L.C. All authors have read and agreed to the published version of the manuscript.

Funding: This paper has been published as open access with the financial support of the Dept. of Medical and Surgical Sciences of the University of Foggia. Our funding source was not involved in any of the steps of the study.

Institutional Review Board Statement: The study was conducted according to the guidelines of the Declaration of Helsinki and approved by the Institutional Review Board of the University of Foggia, 143/CE/2020, 30 November 2020.

Informed Consent Statement: Informed consent was obtained from all subjects involved in the study.

Data Availability Statement: Data supporting reported results can be obtained upon reasonable request to the corresponding author.

Conflicts of Interest: The authors declare no conflict of interest.

References

1. Mottet, N.; van den Bergh, R.C.N.; Briers, E.; Cornford, P.; De Santis, M.; Fanti, S.; Gillessen, S.; Grummet, J.; Henry, A.M.; Lam, T.B.; et al. EAU-ESTRO-ESUR-SIOG Guidelines on Prostate Cancer 2020. In *European Association of Urology Guidelines*; European Association of Urology Guidelines Office: Arnhem, The Netherlands, 2020.
2. Oishi, M.; Gill, I.S.; Tafuri, A.; Shakir, A.; Cacciamani, G.E.; Iwata, T.; Iwata, A.; Ashrafi, A.; Park, D.; Cai, J.; et al. Hemigland Cryoablation of Localized Low, Intermediate and High Risk Prostate Cancer: Oncologic and Functional Outcomes at 5 Years. *J. Urol.* **2019**, *202*, 1188–1198. [CrossRef] [PubMed]
3. Tran, M.; Thompson, J.; Bohm, M.; Pulbrook, M.; Moses, D.; Shnier, R.; Brenner, P.; Delprado, W.; Haynes, A.M.; Savdie, R.; et al. Combination of multiparametric MRI and transperineal template-guided mapping biopsy of the prostate to identify candidates for hemi-ablative focal therapy. *BJU Int.* **2016**, *117*, 48–54. [CrossRef] [PubMed]
4. Shah, T.T.; Peters, M.; Eldred-Evans, D.; Miah, S.; Yap, T.; Faure-Walker, N.A.; Hosking-Jervis, F.; Thomas, B.; Dudderidge, T.; Hindley, R.G.; et al. Early-Medium-Term Outcomes of Primary Focal Cryotherapy to Treat Nonmetastatic Clinically Significant Prostate Cancer from a Prospective Multicentre Registry. *Eur. Urol.* **2019**, *76*, 98–105. [CrossRef]
5. Chuang, R.; Kinnaird, A.; Kwan, L.; Sisk, A.; Barsa, D.; Felker, E.; Delfin, M.; Marks, L. Hemigland Cryoablation of Clinically Significant Prostate Cancer: Intermediate-Term Followup via Magnetic Resonance Imaging Guided Biopsy. *J. Urol.* **2020**, *204*, 941–949. [CrossRef]
6. Wajswol, E.; Winoker, J.S.; Anastos, H.; Falagario, U.; Okhawere, K.; Martini, A.; Treacy, P.J.; Voutsinas, N.; Knauer, C.J.; Sfakianos, J.P.; et al. A cohort of transperineal electromagnetically tracked magnetic resonance imaging/ultrasonography fusion-guided biopsy: Assessing the impact of inter-reader variability on cancer detection. *BJU Int.* **2019**, *125*, 531–540. [CrossRef]
7. Moreira, P.; Tuncali, K.; Tempany, C.M.; Tokuda, J. The Impact of Placement Errors on the Tumor Coverage in MRI-Guided Focal Cryoablation of Prostate Cancer. *Acad. Radiol.* **2020**, *28*, 841–848. [CrossRef]
8. Bertoni, L.; Puliatti, S.; Reggiani Bonetti, L.; Maiorana, A.; Eissa, A.; Azzoni, P.; Bevilacqua, L.; Spandri, V.; Kaleci, S.; Zoeir, A.; et al. Ex vivo fluorescence confocal microscopy: Prostatic and periprostatic tissues atlas and evaluation of the learning curve. *Virchows Arch.* **2020**, *476*, 511–520. [CrossRef]
9. Rocco, B.; Sighinolfi, M.C.; Sandri, M.; Spandri, V.; Cimadamore, A.; Volavsek, M.; Mazzucchelli, R.; Lopez-Beltran, A.; Eissa, A.; Bertoni, L.; et al. Digital Biopsy with Fluorescence Confocal Microscope for Effective Real-time Diagnosis of Prostate Cancer: A Prospective, Comparative Study. *Eur. Urol. Oncol.* **2020**, S2588–S9311. [CrossRef]
10. Rocco, B.; Sighinolfi, M.C.; Cimadamore, A.; Reggiani Bonetti, L.; Bertoni, L.; Puliatti, S.; Eissa, A.; Spandri, V.; Azzoni, P.; Dinneen, E.; et al. Digital frozen section of the prostate surface during radical prostatectomy: A novel approach to evaluate surgical margins. *BJU Int.* **2020**, *126*, 336–338. [CrossRef]
11. Turkbey, B.; Rosenkrantz, A.B.; Haider, M.A.; Padhani, A.R.; Villeirs, G.; Macura, K.J.; Tempany, C.M.; Choyke, P.L.; Cornud, F.; Margolis, D.J.; et al. Prostate Imaging Reporting and Data System Version 2.1: 2019 Update of Prostate Imaging Reporting and Data System Version 2. *Eur. Urol* **2019**, *76*, 340–351. [CrossRef] [PubMed]
12. Cormio, L.; Lucarelli, G.; Selvaggio, O.; Di Fino, G.; Mancini, V.; Massenio, P.; Troiano, F.; Sanguedolce, F.; Bufo, P.; Carrieri, G. Absence of Bladder Outlet Obstruction Is an Independent Risk Factor for Prostate Cancer in Men Undergoing Prostate Biopsy. *Medicine* **2016**, *95*, e2551. [CrossRef] [PubMed]
13. van Leenders, G.; van der Kwast, T.H.; Grignon, D.J.; Evans, A.J.; Kristiansen, G.; Kweldam, C.F.; Litjens, G.; McKenney, J.K.; Melamed, J.; Mottet, N.; et al. The 2019 International Society of Urological Pathology (ISUP) Consensus Conference on Grading of Prostatic Carcinoma. *Am. J. Surg. Pathol.* **2020**, *44*, e87–e99. [CrossRef]

14. Sanguedolce, F.; Cormio, A.; Musci, G.; Troiano, F.; Carrieri, G.; Bufo, P.; Cormio, L. Typing the atypical: Diagnostic issues and predictive markers in suspicious prostate lesions. *Crit. Rev. Clin. Lab. Sci.* **2017**, *54*, 309–325. [CrossRef]
15. Mitropoulos, D.; Artibani, W.; Biyani, C.S.; Bjerggaard Jensen, J.; Roupret, M.; Truss, M. Validation of the Clavien-Dindo Grading System in Urology by the European Association of Urology Guidelines Ad Hoc Panel. *Eur. Urol. Focus* **2018**, *4*, 608–613. [CrossRef] [PubMed]
16. Stabile, A.; Orczyk, C.; Giganti, F.; Moschini, M.; Allen, C.; Punwani, S.; Cathala, N.; Ahmed, H.U.; Cathelineau, X.; Montorsi, F.; et al. The Role of Percentage of Prostate-specific Antigen Reduction After Focal Therapy Using High-intensity Focused Ultrasound for Primary Localised Prostate Cancer. Results from a Large Multi-institutional Series. *Eur. Urol.* **2020**, *78*, 155–160. [CrossRef] [PubMed]
17. Gross, M.D.; Sedrakyan, A.; Bianco, F.J.; Carroll, P.R.; Daskivich, T.J.; Eggener, S.E.; Ehdaie, B.; Fisher, B.; Gorin, M.A.; Hunt, B.; et al. SPARED Collaboration: Patient Selection for Partial Gland Ablation in Men with Localized Prostate Cancer. *J. Urol.* **2019**, *202*, 952–958. [CrossRef]
18. Fereidouni, F.; Harmany, Z.T.; Tian, M.; Todd, A.; Kintner, J.A.; McPherson, J.D.; Borowsky, A.D.; Bishop, J.; Lechpammer, M.; Demos, S.G.; et al. Microscopy with ultraviolet surface excitation for rapid slide-free histology. *Nat. Biomed. Eng.* **2017**, *1*, 957–966. [CrossRef]
19. Glaser, A.K.; Reder, N.P.; Chen, Y.; McCarty, E.F.; Yin, C.; Wei, L.; Wang, Y.; True, L.D.; Liu, J.T.C. Light-sheet microscopy for slide-free non-destructive pathology of large clinical specimens. *Nat. Biomed. Eng.* **2017**, *1*, 1–10. [CrossRef] [PubMed]
20. Wong, T.T.W.; Zhang, R.; Hai, P.; Zhang, C.; Pleitez, M.A.; Aft, R.L.; Novack, D.V.; Wang, L.V. Fast label-free multilayered histology-like imaging of human breast cancer by photoacoustic microscopy. *Sci. Adv.* **2017**, *3*, e1602168. [CrossRef]
21. Assayag, O.; Antoine, M.; Sigal-Zafrani, B.; Riben, M.; Harms, F.; Burcheri, A.; Grieve, K.; Dalimier, E.; Le Conte de Poly, B.; Boccara, C. Large field, high resolution full-field optical coherence tomography: A pre-clinical study of human breast tissue and cancer assessment. *Technol. Cancer Res. Treat.* **2014**, *13*, 455–468. [CrossRef]
22. Tao, Y.K.; Shen, D.; Sheikine, Y.; Ahsen, O.O.; Wang, H.H.; Schmolze, D.B.; Johnson, N.B.; Brooker, J.S.; Cable, A.E.; Connolly, J.L.; et al. Assessment of breast pathologies using nonlinear microscopy. *Proc. Natl. Acad. Sci. USA* **2014**, *111*, 15304–15309. [CrossRef] [PubMed]
23. You, S.; Sun, Y.; Chaney, E.J.; Zhao, Y.; Chen, J.; Boppart, S.A.; Tu, H. Slide-free virtual histochemistry (Part II): Detection of field cancerization. *Biomed. Opt. Express* **2018**, *9*, 5253–5268. [CrossRef] [PubMed]
24. Schlomm, T.; Tennstedt, P.; Huxhold, C.; Steuber, T.; Salomon, G.; Michl, U.; Heinzer, H.; Hansen, J.; Budaus, L.; Steurer, S.; et al. Neurovascular structure-adjacent frozen-section examination (NeuroSAFE) increases nerve-sparing frequency and reduces positive surgical margins in open and robot-assisted laparoscopic radical prostatectomy: Experience after 11,069 consecutive patients. *Eur. Urol.* **2012**, *62*, 333–340. [CrossRef]
25. Mirmilstein, G.; Rai, B.P.; Gbolahan, O.; Srirangam, V.; Narula, A.; Agarwal, S.; Lane, T.M.; Vasdev, N.; Adshead, J. The neurovascular structure-adjacent frozen-section examination (NeuroSAFE) approach to nerve sparing in robot-assisted laparoscopic radical prostatectomy in a British setting—A prospective observational comparative study. *BJU Int.* **2018**, *121*, 854–862. [CrossRef] [PubMed]
26. Marenco, J.; Calatrava, A.; Casanova, J.; Claps, F.; Mascaros, J.; Wong, A.; Barrios, M.; Martin, I.; Rubio, J. Evaluation of Fluorescent Confocal Microscopy for Intraoperative Analysis of Prostate Biopsy Cores. *Eur. Urol. Focus* **2020**. [CrossRef]

Article

Prognostic Index for Predicting Prostate Cancer Survival in a Randomized Screening Trial: Development and Validation

Subas Neupane [1,*], Jaakko Nevalainen [1], Jani Raitanen [1,2], Kirsi Talala [3], Paula Kujala [4], Kimmo Taari [5], Teuvo L. J. Tammela [6], Ewout W. Steyerberg [7,8] and Anssi Auvinen [1]

[1] Unit of Health Sciences, Faculty of Social Sciences, Tampere University, FI-33014 Tampere, Finland; jaakko.nevalainen@tuni.fi (J.N.); jani.raitanen@tuni.fi (J.R.); anssi.auvinen@tuni.fi (A.A.)
[2] UKK Institute for Health Promotion Research, FI-33014 Tampere, Finland
[3] Finnish Cancer Registry, FI-00130 Helsinki, Finland; kirsi.talala@cancer.fi
[4] Department of Pathology, FIMLAB laboratory services, FI-33014 Tampere, Finland; paula.kujala@fimlab.fi
[5] Department of Urology, Helsinki University Hospital, University of Helsinki, FI-00014 Helsinki, Finland; kimmo.taari@helsinki.fi
[6] Department of Urology, Tampere University Hospital, University of Tampere, FI-33521 Tampere, Finland; teuvo.tammela@tuni.fi
[7] Department of Public Health, Erasmus MC-University Medical Center Rotterdam, 3015 GD Rotterdam, The Netherlands; e.steyerberg@erasmusmc.nl
[8] Department of Biomedical Data Sciences, Leiden University Medical Center, 2333 ZC Leiden, The Netherlands
* Correspondence: subas.neupane@tuni.fi; Tel.: +358-40-1909709; Fax: +358-3-35516057

Simple Summary: A prognostic index for predicting survival of localized prostate cancer (PCa) up to 15 and 20 years was developed. The prognostic index performed well for predicting PCa survival among screened and non-screened men. The performance of the prediction model was superior to the European Association of Urology (EAU) risk groups as well as a modified cancer of prostate risk assessment (CAPRA) risk score. We further constructed a simplified risk score in an unscreened population, using the three most relevant predictors. The simplified risk score was applied to predict PCa survival at 10 years from diagnosis to provide more accurate risk estimation as the basis for decision making.

Abstract: We developed and validated a prognostic index to predict survival from prostate cancer (PCa) based on the Finnish randomized screening trial (FinRSPC). Men diagnosed with localized PCa (N = 7042) were included. European Association of Urology risk groups were defined. The follow-up was divided into three periods (0–3, 3–9 and 9–20 years) for development and two corresponding validation periods (3–6 and 9–15 years). A multivariable complementary log–log regression model was used to calculate the full prognostic index. Predicted cause-specific survival at 10 years from diagnosis was calculated for the control arm using a simplified risk score at diagnosis. The full prognostic index discriminates well men with PCa with different survival. The area under the curve (AUC) was 0.83 for both the 3–6 year and 9–15 year validation periods. In the simplified risk score, patients with a low risk score at diagnosis had the most favorable survival, while the outcome was poorest for the patients with high risk scores. The prognostic index was able to distinguish well between men with higher and lower survival, and the simplified risk score can be used as a basis for decision making.

Keywords: prognostic index; prediction model; prostate cancer; mortality; screening trial

1. Introduction

Prostate cancer (PCa) presents a wide spectrum of behavior, from indolent to highly aggressive [1]. Treatment decisions are required at several phases during the course of the disease [2]. Optimal disease management should avoid both excessively aggressive treatment in patients who are not at high risk of disease progression and ineffective management of aggressive disease leading to treatment failure and development of metastatic disease. However, the dilemma expressed by Dr. Willet Whitmore persists for PCa: "Is cure possible in those for whom it is necessary—and is cure necessary in those for whom it is possible".

Several prediction methods for the prognosis of localized PCa have been presented as tabulations [3–5], nomograms [6–8], risk groups [9,10] and decision trees [11,12]. However, these methods have mainly divided patients into 3–4 broad risk groups and used biochemical recurrence (BCR) as the end-point rather than PCa death [3,4,9,13,14]; furthermore, few are based on a modern setting with largely prostate-specific antigen (PSA)-detected cases. Prognostic prediction models based on a limited number of relevant clinical characteristics can offer evidence-based input to inform medical practice [15].

We developed and validated a full prognostic index for predicting survival of localized PCa up to 15 and 20 years. We also developed a simplified risk score tool for use at diagnosis and applied it to predict survival at 10 years.

2. Results

Prognostic factors associated with PCa death included age at diagnosis, trial arm, PSA at diagnosis, European Association of Urology (EAU) risk group, treatment modality, mode of detection and biochemical recurrence (Table 1). All prognostic factors except comorbidity index showed a statistically significant difference between men who died from PCa and others.

Older age at diagnosis was marginally associated with lower PCa mortality at 9–20 years (Table S1). PSA at diagnosis was associated with higher PCa mortality in the first and the last development periods. PCa mortality was higher in the intermediate- to high-risk groups compared to the low-risk group in all three follow-up periods. Men treated with radical prostatectomy had the most favorable survival, with the exceptions of radiotherapy and observation in the early follow-up. Biochemical recurrence also predicted increased probability of PCa death, with the largest effect after the first three years.

The distributions of the prognostic index differed markedly across the development periods (Figure 1a–c). The graphs illustrate lower prognostic index (PI) values (indicating worse survival) for men who died from PCa than those who did not die from PCa (cumulative frequency for the former group shown as the dotted blue line above the latter group, shown as the solid red line). PCa mortality increased with increasing values of prognostic index in the initial follow-up, but after 9 years, a clear excess mortality was limited to the two highest quintiles.

The prognostic indices were associated with PCa mortality in all EAU risk groups, though the difference was not obvious in the initial three-year period with low mortality (Figure S1a–c). The prognostic index provided incremental information, especially in the intermediate- and high-risk groups, and its contribution was accentuated with follow-up. Furthermore, the prognostic index also predicted survival within the low-risk group in the longer follow-up.

Table 1. Demographic and clinical characteristics of 7042 localized prostate cancer (PCa) patients in the cohort stratified by survival status at 20-year follow-up from the date of diagnosis.

Characteristic	Total (N = 7042)	No PCa Death [†] (n = 6737)	PCa Death (n = 305)	p-Value [‡]
Age at entry (years)				<0.001
55	1844	1806 (97.9%)	38 (2.1%)	
59	1898	1840 (96.9%)	58 (3.1%)	
63	1782	1681 (94.3%)	101 (5.7%)	
67	1518	1410 (92.9%)	108 (7.1%)	
Age at diagnosis (years)				<0.001
Median (IQR)	69 (65–73)	69 (65–73)	68 (64–71)	
Study arm				
Control	3823	3667 (95.9%)	156 (4.1%)	
Screening	3219	3070 (95.4%)	149 (4.6%)	
PSA at diagnosis (ng/mL)				<0.001
Median (IQR)	7.8 (5.2–11.7)	7.7 (5.2–11.5)	10.3 (6.4–18.6)	
Biopsy Gleason sum				<0.001
2–6	4031	3913 (97.1%)	118 (2.9%)	
7	2249	2148 (95.5%)	101 (4.5%)	
8–10	705	630 (89.4%)	75 (10.6%)	
Missing	57	46 (80.7%)	11 (19.3%)	
EAU risk group				<0.001
Low	2769	2712 (97.9%)	57 (2.1%)	
Intermediate	2988	2864 (95.9%)	124 (4.2%)	
High	1285	1161 (90.4%)	124 (9.7%)	
Missing	236	225 (95.3%)	11 (4.7%)	
Comorbidity index				0.309
0	6320	6041 (95.6%)	279 (4.4%)	
1+	722	696 (96.4%)	26 (3.6%)	
Primary treatment				<0.001
Radical Prostatectomy	1812	1751 (96.6%)	61 (3.4%)	
Radiation	2718	2583 (95.0%)	135 (5.0%)	
Endocrine	643	570 (88.7%)	73 (11.4%)	
Observation	1788	1756 (98.2%)	32 (1.8%)	
No treatment	78	74 (94.9.6%)	4 (5.1%)	
Missing	3	3 (100.0%)	0	
Method of presentation				0.011
Screen-detected	1462	1381 (94.5%)	181 (5.5%)	
Not screen-detected	5578	5354 (96.0%)	224 (4.0%)	
Missing	2	2 (100.0%)	0	
Biochemical recurrence				<0.001
No	4749	4672 (98.4%)	77 (1.6%)	
Yes	2212	1988 (89.9%)	224 (10.1%)	
Missing	81	77 (95.1%)	4 (4.9%)	

IQR: Interquartile range. [†] Includes men alive and deaths due to causes other than PCa. [‡] p-values for categorical variables were derived from a chi-square test, whereas for continuous variable, using ANOVA test.

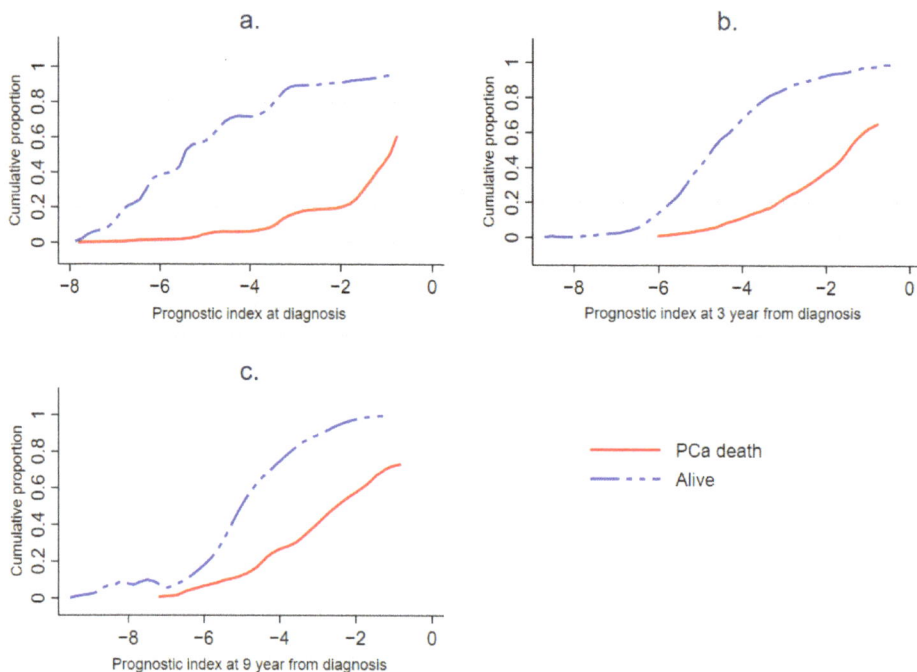

Figure 1. Cumulative distribution of prognostic index (PI) values for men who died due to prostate cancer (PCa) and those who did not during (**a**) the development period of 0–3 year (**b**) the development period of 3–9 years and (**c**) the development period of 9–20 years. The distribution of men is described as a cumulative frequency across values of the PI from low (indicating worse survival) to high (indicating favorable survival).

The observed mortality matched the expected one very well at all levels of the PI during each development and validation period (Table 2). In all follow-up periods, including the validation periods, the highest quintiles of the PI showed the highest observed and expected PCa mortalities. The differences between the lower quintiles were, however, relatively small.

Less than one third of the patients remained in the initial quintile from the 0–3-year development period to the 3–6-year validation period; particularly, progression from Q1 to Q2 and Q4 to Q5 was common (Table S2). The most frequent transition was by one step up, likely due to biochemical recurrence. However, downward transitions also occurred, reflecting changes in the regression coefficients of the variables used in the model.

The predictive ability of the prognostic indices (Figure 2) did not substantially differ between the development and validation periods: area under the curve (AUC) 0.84 (95% confidence interval, CI 0.77–0.90) for the initial development period (0–3 years) and 0.83 (0.79–0.88) for the corresponding validation period (3–6 years). Similarly, for the second development period, the AUC was 0.84 (0.81–0.88), and it was 0.83 (0.79–0.88) for the subsequent validation period. For the 9–20-year development period, the AUC was 0.83 (0.79–0.86).

Table 2. Expected and observed probability and number of PCa deaths in the development and validation periods for quintiles of prognostic index.

Prognostic Index Quintiles	Development Period						Validation Period					
	Men	Number of Deaths [†]	Observed		Expected		Men	Observed		Expected		
			Probability [‡]	PCa Deaths	Probability [‡]	PCa Deaths		Probability [‡]	PCa Deaths	Probability [‡]	PCa Deaths	
			0–3 year					3–6 years				
Q1	1436	67	0.001	1	0.001	1	1323	0.001	1	0.001	1	
Q2	1446	87	0.003	4	0.002	3	1352	0.005	7	0.003	4	
Q3	1438	44	0.003	4	0.003	4	1354	0.003	4	0.004	5	
Q4	1436	66	0.002	3	0.006	9	1299	0.010	13	0.008	10	
Q5	1465	175	0.031	46	0.028	41	1396	0.044	61	0.044	62	
Total	7221	439	0.008	58	0.008	58	6724	0.013	86	0.012	82	
			3–9 years					9–15 years				
Q1	1319	118	0.002	3	0.002	3	1133	0.002	2	0.001	1	
Q2	1353	165	0.007	9	0.004	6	1155	0.005	6	0.003	4	
Q3	1356	150	0.005	7	0.007	10	1147	0.004	4	0.004	5	
Q4	1319	176	0.018	24	0.014	19	1152	0.009	10	0.011	13	
Q5	1377	95	0.074	102	0.075	103	1248	0.054	69	0.047	61	
Total	6724	704	0.022	145	0.021	141	5875	0.016	91	0.014	84	
			9–20 years									
Q1	1122	26	0.002	2	0.002	2						
Q2	1148	55	0.004	5	0.004	5						
Q3	1133	93	0.007	8	0.006	7						
Q4	1180	132	0.019	22	0.014	17						
Q5	1292	126	0.057	74	0.054	70						
Total	5875	432	0.019	111	0.017	101						

[†] Number of deaths due to causes other than PCa; [‡] probability of PCa death for an individual man from the prognostic model.

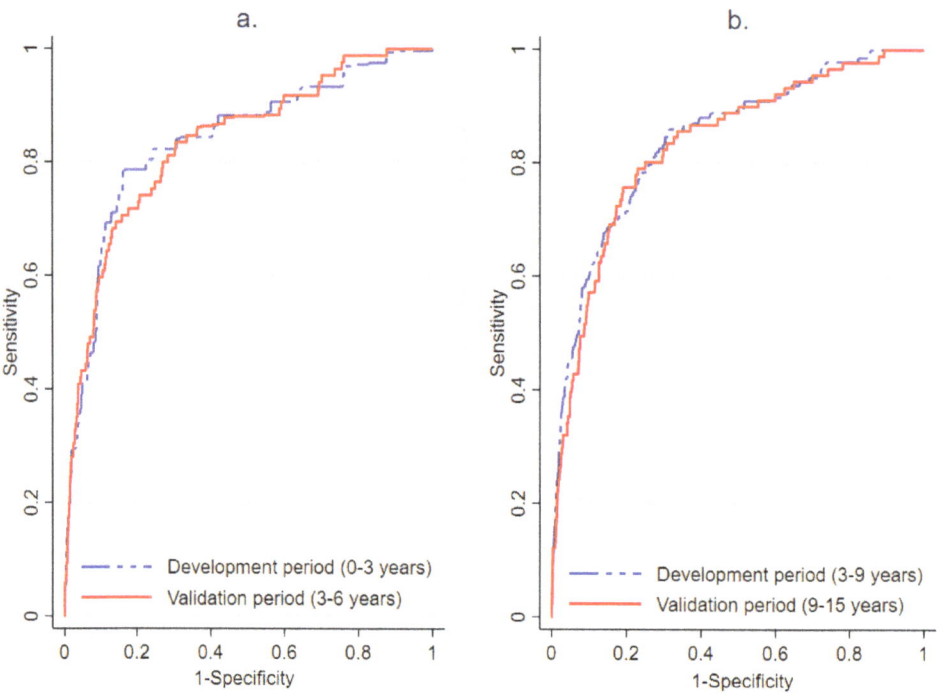

Figure 2. (a) Receiver operating characteristic (ROC) curves for the development period of 0–3 year and validation period of 3–6 years and the (b) development period of 3–9 years and validation period of 9–15 years.

A simplified risk score at diagnosis was calculated among patients in the control arm based on the regression coefficients of three categorical parameters (age at diagnosis, PSA at diagnosis and EAU risk group) to allow easy clinical application. The simplified risk score uses a granular scale of 0 to 100, with higher score indicating increasing risk. The predictive ability of the simplified risk score at diagnosis was 0.68 (0.63–0.73) (Figure S2).

The full prognostic model displayed superior discrimination ($p < 0.001$) compared to the EAU risk group alone in all three periods (AUC for EAU risk group: 0.61, 0.53 and 0.39 during the follow-up periods of 0–3, 3–9 and 9–20 years, respectively). The simplified risk score showed superior discrimination only in the 9–20-year period ($p < 0.001$).

The simplified risk score at diagnosis (Table 3) was used to calculate the predicted PCa survival at 10 years (Figure 3). Overall, men with a high-risk score at diagnosis had poorer survival.

Table 3. Scoring rules for constructing the simplified risk score at diagnosis for PCa survival based on the complementary log–log regression model among PCa-diagnosed cases in the control arm.

Characteristics	Categories	β [†]	Risk Score [§]
Age at diagnosis (years)			
	≤60	1.12	40
	61–70	1.00	35
	71–75	0.72	25
	≥76	Ref	0
PSA at diagnosis (ng/mL)			
	≤19.9	Ref	0
	≥20.0	0.22	10
EAU risk group			
	Low	Ref	0
	Intermediate	0.75	30
	High	1.46	50

[†] Regression coefficients from complementary log–log model. [§] Risk score at baseline calculated by dividing each beta coefficient by the sum of the highest beta coefficient of each variable and then multiplied by 100 (scores are rounded).

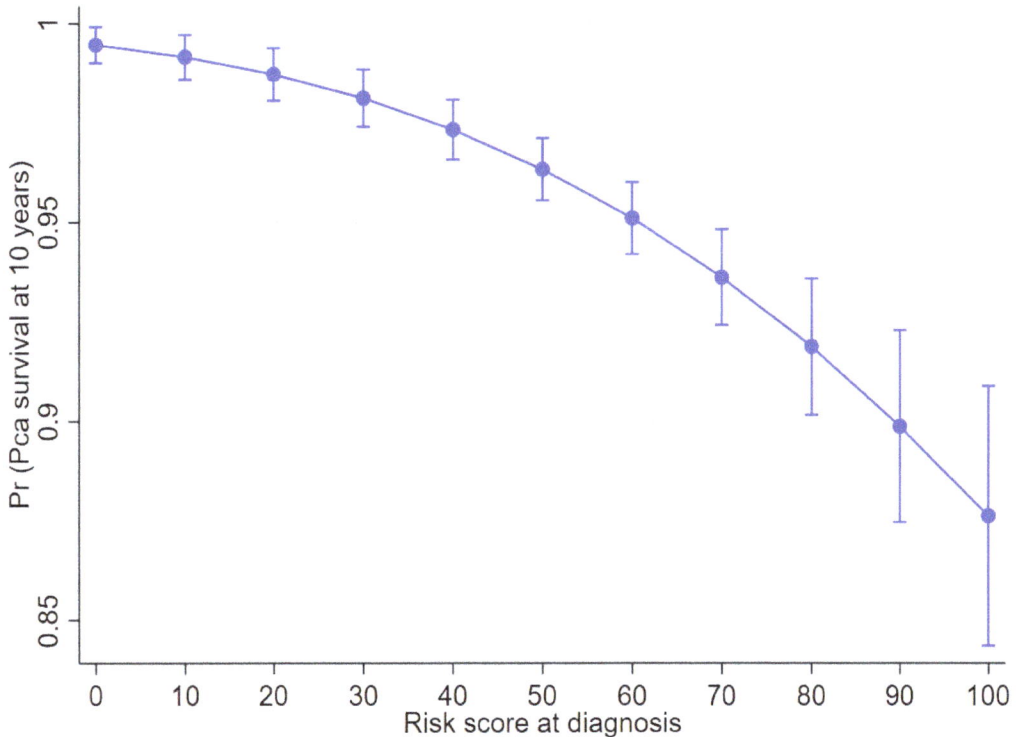

Figure 3. Predicted probability of PCa survival at 10 years from diagnosis among controls in the Finnish randomized screening trial.

We calculated the risk score at diagnosis and 10-year survival probability among patients in both study arms, as well as performing a complete case analysis in the control arm only as a sensitivity analysis (Tables S3 and S4 and Figure S3), with no substantial difference in the results.

The decision curve analysis for simplified risk score at diagnosis is presented in Figure 4. The graph gives the expected net benefit per patient relative to no PCa mortality in any patient (Treat None). The risk prediction model is of benefit for a reasonable range of 3–25%: the curve diverges only at the threshold probability of about 3%. However, the net benefit of the model is about the same as the net benefit of Treat All below 3%.

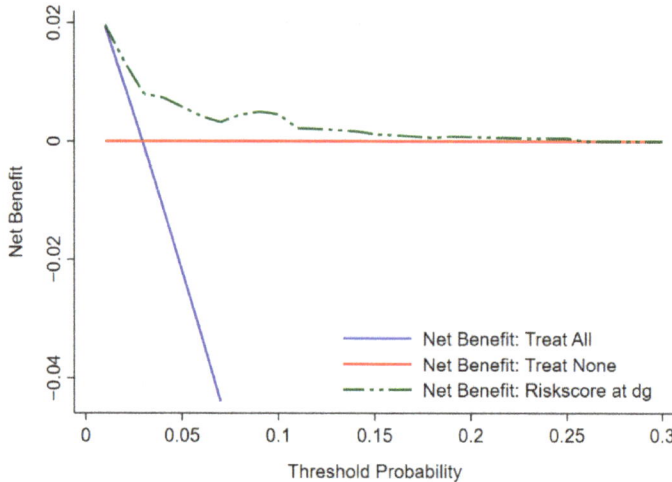

Figure 4. Decision curve analysis for the simplified risk score at diagnosis. The dotted green line is the simplified risk score (prediction model), the blue solid line assumes PCa mortality in all patients and red solid line assume no patient deaths due to PCa. Threshold probability on the x-axis is the level of diagnostic certainty above which the patient would choose to be treated.

3. Discussion

The full prognostic index with seven variables predicted PCa mortality with a performance superior to that of the EAU risk group (AUC 0.83–0.84 vs. 0.61). The robustness of the results was confirmed by sensitivity analyses including both trial arms and omitting patients with missing data.

Our model correctly predicted the 3-year survival of 99% for the patients in the lowest quintile and 97% for those in the highest quintile of the prognostic index. We divided the follow-up time into several segments due to lack of proportionality across the entire follow-up. In the second development period, 6-year survival was 99% among men in the lowest quintile, while it was 89% among men in the highest quintile.

Primary treatment predicted PCa mortality already in the early follow-up. The effect of biochemical recurrence increased with follow-up. Other factors did not show clear changes over the follow-up. Similar to earlier findings [16,17], we found no strong impact of comorbidity at baseline on PCa-specific survival. No earlier PCa survival prediction models have utilized the context of a randomized screening trial. Our approach enhances the applicability of the prognostic index to the current setting with widespread PSA testing.

A simplified risk score at diagnosis was developed using three predictors selected based on their importance and interpretability in the prognostic index model. The simplified risk score is based on a granular scale ranging from 0 to 100 with three categorical variables and can be adopted in daily clinical practice with minimal data entry.

PSA was used as a component of our prognostic index, despite being a part of the EAU risk group, because the analysis revealed that its impact was not fully captured in the EAU classification.

Our findings are mainly in line with earlier prediction models, although patient populations, outcomes and methodological approaches differ between studies. The performance of our simplified risk score tool at diagnosis was superior to that of the D'Amico risk classification and EAU risk group (AUC 0.68 vs. 0.59 and 0.61, respectively). The simplified risk score also outperformed an abridged version of the cancer of prostate risk assessment (CAPRA) risk score (AUC 0.59), though we were unable to incorporate percentage cancer in biopsy for estimating the CAPRA score in our analysis due to lack of data. Furthermore, all patients in our study were aged > 50 years at diagnosis [7].

An earlier study presented a clinical–genomic risk group classification for localized prostate cancer that showed a 10-year rate for distant metastases of 3.5% for a low-risk group, while it was 58% in a high-risk group in the training cohort, and the corresponding values for the validation cohort were 0% and 63%, respectively [9]. That risk group required extensive genomic data, restricting its applicability. Peters et al. [18] developed a prediction model for recurrent disease with three categories, which showed 60% biochemical disease-free survival and 40% composite end-point-free survival at 4 years for a low-risk group, while the corresponding figures for a high-risk group were 7% and 0%, respectively.

Decision curve analysis shows the benefit of use of the prediction model (simplified risk score at diagnosis). A net benefit was found for a reasonable range of 3–25%: the curve diverged only at the threshold probability of about 3%.

Our study had also some limitations. The patients were treated during a period spanning from the 1990s into the 2010s and treatment modalities have evolved over time. However, long follow-up is required due to the favorable prognosis to capture the full natural course of the disease and accrue a sufficient number of PCa deaths. Completeness of data was high, with the highest proportion of missing data for PSA, at 3%. However, we used imputation in the main analysis, and sensitivity analyses of complete cases yielded comparable results, suggesting that this did not affect our findings. We incorporated biochemical relapse in the prognostic index, even though in the clinical setting, it is not available at diagnosis. On the other hand, its inclusion enhances the applicability of our results in prognostic prediction after the initial phase and post-primary treatment.

4. Material and Methods

We used data from the Finnish Randomized Study of Screening for Prostate Cancer (FinRSPC). The trial protocol and main results have been described elsewhere [19]. In brief, a random sample of 8000 men aged 55–67 years were allocated to the screening arm (SA) annually in 1996–1999 and the remaining men (48,278 in total) formed the control arm that received no intervention. Men in the screening arm were invited for screening based on serum PSA. Screen-positive men (defined as those with PSA $\geq$ 4.0 ng/mL or PSA 3.0–3.9 ng/mL with free–total PSA ratio < 0.16) were referred to a local urological clinic for diagnostic examinations including transrectal ultrasound-guided biopsy. The second screening round was conducted four years later, and the final one after 8 years. Men aged > 71 years, those diagnosed with prostate cancer and men who had emigrated from the study area were no longer invited.

All men diagnosed with localized prostate cancer between randomization and the end of 2015 were included in this analysis (N = 7042). The follow-up for the primary analysis started at diagnosis and ended at death, emigration or the common closing (31 December 2015). Death from prostate cancer was the end-point in the analysis, with underlying causes of death obtained from Statistics Finland.

Information on tumor, lymph node, and metastasis (TNM) stage and Gleason score were abstracted from medical records. For previous cases, Gleason scores were revised according to the 2002 system by two pathologists. PSA at diagnosis was used for all men. Information on biochemical recurrence was obtained from laboratory databases.

Biochemical recurrence (BCR) was defined as PSA reaching at least 0.2 ng/mL in two measurements after prostatectomy, while BCR after radiotherapy was defined as a rise in PSA by at least 2.0 ng/mL above the lowest level (nadir). A modified version of the Charlson comorbidity index [20] was constructed based on hospital inpatient episodes obtained from the nationwide hospital discharge registry and categorized into no versus any comorbidity (score 0 versus 1 to 8) [21]. Prognostic risk group for PCa survival at diagnosis was classified as low, moderate and high, according to the European Association of Urology (EAU) criteria [10]. Low-risk PCa was defined as stage T1–T2a with Gleason score < 7 and PSA < 10 ng/mL; intermediate risk as T1–T2b with either Gleason 7 or PSA 10–20; high risk was stage T1–T2c with either Gleason > 7 or PSA 20–100, or T2c.

Primary treatment was retrieved from medical records and classified as radical prostatectomy, curative radiation therapy (external beam or brachytherapy), endocrine therapy (luteinizing hormone-releasing hormone agonist/antagonist, anti-androgen, or both or surgical castration), observation (watchful waiting or active surveillance) or no treatment.

4.1. Ethical Issues

Helsinki and Tampere University Hospital Ethics committees reviewed the study protocol (tracking number R10167). Cancer registry data were obtained with permission from the National Institute for Health and Welfare (Dnro THL/1601/5.05.00/2015). Written informed consent was obtained from the men participating in the screening arm.

4.2. Statistical Analysis

For the preliminary investigation, the proportional hazards assumption of the factors in the Cox regression was evaluated by graphical examination in log–log plots. These plots formed approximate parallel straight lines as required, except those for primary treatment which crossed each other. For this reason, we divided the follow-up time into three periods (0–3, 3–9 and 9–20 years) to model the effect of the full prognostic index separately during each period.

We used a same set of variables (age at diagnosis, study arm, PSA at diagnosis, EAU risk group, comorbidity index, primary treatment and biochemical recurrence) in a complementary log–log regression model, identified by a stepwise forward selection with $p = 0.10$ as the cut-off, in each of the three periods. Only statistically significant interaction terms (5% level) were included in the final model. The prognostic index (PI) was then derived as a linear combination of the variables, including interaction terms and their coefficients from the regression model. We generated prognostic indices separately for the three follow-up periods, hereafter called development periods, using regression coefficients estimated from the complementary log–log models. The probabilities of PCa death during each development period using prognostic indices were calculated.

Missing values (3.1% in PSA and 2.6% in the EAU risk group) were imputed using a multiple imputation by chained equations (MICE) algorithm, assigning multiple likely values from a predicted distribution based on association with other variables [22]. Multiple imputation creates multiple copies of the dataset, which are analyzed separately. Finally, the results were appropriately combined [23].

To avoid overfitting and overestimation of the predictive ability, we validated the results by applying them to a subsequent follow-up period for the first two development periods (i.e., prognostic index derived from the development period of 0–3 years from diagnosis to predict survival during the validation period of 3–6 years, and the index derived from the development period of 3–9 years to predict survival during years 9–15) [24]. Expected and observed probabilities and numbers of PCa deaths were calculated for each development and validation period. Expected probability of PCa death for the validation period was calculated as the inverse of the complementary log–log transformation.

Reclassification probabilities of men in the quintiles of prognostic index for the first development period (0–3 years) and validation period (3–6 years) were calculated. Moreover, we presented the distribution and the mean values of full prognostic indices by risk group for all development periods. Cumulative distribution of prognostic index values according to the survival status of the patients for all three development periods was plotted.

Receiver operating characteristic (ROC) was calculated for the test and validation periods to illustrate sensitivity and specificity. The area under the curve (AUC) was calculated to assess the discriminative power of the prediction models. We further developed a simplified risk score at diagnosis using the information at diagnosis (age at diagnosis, PSA at diagnosis and EAU risk group) for the control arm only to avoid lead-time by screening. Only three variables were selected based on their importance and interpretability in the prognostic index model. The simplified risk score was then used to calculate the predicted probability of 10-year PCa survival and was presented graphically. We further developed the decision curve analysis to determine the clinical usefulness of a simplified risk score at diagnosis by quantifying the net benefits at different threshold probabilities.

As a sensitivity analysis, we calculated the risk score at diagnosis based on data in both study arms. Furthermore, we performed a complete case analysis in the control arm by calculating the risk score and predicted 10-year PCa survival to examine the potential influence of imputation.

Analyses were performed using Stata Statistical Software version 16.0 (StataCorp, College Station, TX, USA) and IBM SPSS Statistics 23 (IBM Corp., Armonk, NY, USA).

5. Conclusion

The prognostic index accurately predicted prostate cancer survival at follow-up reaching 20 years. A simplified risk score at diagnosis using the three most relevant parameters to predict the survival at 10 years can be helpful for providing more accurate risk estimation as the basis for decision making. However, our prediction model requires further external validation.

Supplementary Materials: The following are available online at https://www.mdpi.com/2072-6694/13/3/435/s1. Table S1: Prognostic factors for prostate cancer mortality from the prediction model. Estimates from the complementary log–log regression models with their 95% confidence intervals in different time periods. Table S2: Distribution of prognostic index quintiles for the development period 0–3 years reclassified in new prognostic index quintiles for the validation period 3–6 years. Table S3: Risk score at diagnosis for PCa survival based on the complementary log–log regression model among all men with PCa diagnosis. Table S4: Risk score at diagnosis for PCa survival based on the complementary log–log regression model among men with PCa diagnosis in the control arm (complete case analysis). Figure S1: Box plots showing the distribution of the prognostic index by EAU risk groups and by period (a) 0 to 3 years, (b) 3 to 9 years and (c) 9 to 20 years. Figure S2: ROC curve for the simplified risk score at diagnosis. Figure S3: Predicted probability of PCa survival at 10-year from diagnosis among controls in a Finnish randomized screening trial (complete case analysis).

Author Contributions: Study concept: A.A., J.N. and S.N. Study design: A.A., T.L.J.T. and K.T. (Kimmo Taari). Data acquisition: J.R., J.N. and S.N. Quality control of data and algorithms: S.N., J.N., J.R. and K.T. (Kirsi Talala). Data analysis and interpretation: S.N., J.R., J.N. and A.A. Statistical analysis: S.N. and J.R. Article preparation: S.N., A.A. and J.N. Article editing: S.N., A.A., K.T. (Kirsi Talala), J.R., J.N., E.W.S. and P.K. Article review: All authors. All authors have read and agreed to the published version of the manuscript.

Funding: This study was funded in part by competitive state research funding administered by Expert Responsibility area of Tampere University Hospital, grants 9N064 and 9R002, the Finnish Academy (grant number #260931), and the Cancer Society of Finland grant to Prof. Anssi Auvinen.

Institutional Review Board Statement: The study was conducted according to the guidelines of the Declaration of Helsinki, and approved by the Ethics Committee of Helsinki and Tampere University Hospital (protocol code R10167 and date of approval 18.04.1995).

Informed Consent Statement: Informed consent was obtained from all subjects involved in the study.

Data Availability Statement: The data presented in this study are available on request from the corresponding author.

Conflicts of Interest: The authors have declared no conflict of interest.

References

1. Tosoian, J.J.; Carter, H.B.; Lepor, A.; Loeb, S. Active surveillance for prostate cancer: Current evidence and contemporary state of practice. *Nat. Rev. Urol.* **2016**, *13*, 205–215. [CrossRef]
2. Kearns, J.T.; Lin, D.W. Prediction models for prostate cancer outcomes: What is the state of the art in 2017? *Curr. Opin. Urol.* **2017**, *27*, 469–474. [CrossRef]
3. D'amico, A.V.; Whittington, R.; Malkowicz, S.B.; Schultz, D.; Blank, K.; Broderick, G.A.; Tomaszewski, J.E.; Renshaw, A.A.; Kaplan, I.; Beard, C.J.; et al. Biochemical outcome after radical prostatectomy, external beam radiation therapy, or interstitial radiation therapy for clinically localized prostate cancer. *JAMA* **1998**, *280*, 969–974. [CrossRef] [PubMed]
4. Partin, A.W.; Mangold, L.A.; Lamm, D.M.; Walsh, P.C.; Epstein, J.I.; Pearson, J.D. Contemporary update of prostate cancer staging nomograms (Partin Tables) for the new millennium. *Urology* **2001**, *58*, 843–848. [CrossRef]
5. Dess, R.T.; Suresh, K.; Zelefsky, M.J.; Freedland, S.J.; Mahal, B.A.; Cooperberg, M.R.; Davis, B.J.; Horwitz, E.M.; Terris, M.K.; Amling, C.L.; et al. Development and Validation of a Clinical Prognostic Stage Group System for Nonmetastatic Prostate Cancer Using Disease-Specific Mortality Results From the International Staging Collaboration for Cancer of the Prostate. *JAMA Oncol.* **2020**, *6*, 1912–1920. [CrossRef] [PubMed]
6. Kattan, M.W.; Eastham, J.A.; Stapleton, A.M.; Wheeler, T.M.; Scardino, P.T. A preoperative nomogram for disease recurrence following radical prostatectomy for prostate cancer. *JNCI J. Natl. Cancer Inst.* **1998**, *90*, 766–771. [CrossRef] [PubMed]
7. Cooperberg, M.R.; Pasta, D.J.; Elkin, E.P.; Litwin, M.S.; Latini, D.M.; Du Chane, J.; Carroll, P.R. The University of California, San Francisco Cancer of the Prostate Risk Assessment score: A straightforward and reliable preoperative predictor of disease recurrence after radical prostatectomy. *J. Urol.* **2005**, *173*, 1938–1942. [CrossRef]
8. Eggener, S.E.; Scardino, P.T.; Walsh, P.C.; Han, M.; Partin, A.W.; Trock, B.J.; Feng, Z.; Wood, D.P.; Eastham, J.A.; Yossepowitch, O.; et al. Predicting 15-year prostate cancer specific mortality after radical prostatectomy. *J. Urol.* **2011**, *185*, 869–875. [CrossRef]
9. Spratt, D.E.; Zhang, J.; Santiago-Jiménez, M.; Dess, R.T.; Davis, J.W.; Den, R.B.; Dicker, A.P.; Kane, C.J.; Pollack, A.; Stoyanova, R.; et al. Development and validation of a novel clinical-genomic risk group classification for localized prostate cancer. *J. Clin. Oncol.* **2017**. [CrossRef]
10. Mottet, N.; Bellmunt, J.; Bolla, M.; Briers, E.; Cumberbatch, M.G.; De Santis, M.; Fossati, N.; Gross, T.; Henry, A.M.; Joniau, S.; et al. EAU-ESTRO-SIOG guidelines on prostate cancer. Part 1: Screening, diagnosis, and local treatment with curative intent. *Eur. Urol.* **2017**, *71*, 618–629. [CrossRef]
11. Conrad, S.; Graefen, M.; Pichlmeier, U.; Henke, R.P.; Erbersdobler, A.; Hammerer, P.G.; Huland, H. Prospective validation of an algorithm with systematic sextant biopsy to predict pelvic lymph node metastasis in patients with clinically localized prostatic carcinoma. *J. Urol.* **2002**, *167*, 521–525. [PubMed]
12. Thompson, S.R.; Delaney, G.P.; Jacob, S.; Shafiq, J.; Wong, K.; Hanna, T.P.; Gabriel, G.S.; Barton, M.B. Estimation of the optimal utilisation rates of radical prostatectomy, external beam radiotherapy and brachytherapy in the treatment of prostate cancer by a review of clinical practice guidelines. *Radiother. Oncol.* **2016**, *118*, 118–121.
13. Chi, K.N.; Kheoh, T.; Ryan, C.J.; Molina, A.; Bellmunt, J.; Vogelzang, N.J.; Rathkopf, D.E.; Fizazi, K.; Kantoff, P.W.; Li, J.; et al. A prognostic index model for predicting overall survival in patients with metastatic castration-resistant prostate cancer treated with abiraterone acetate after docetaxel. *Ann. Oncol.* **2016**, *27*, 454–460. [PubMed]
14. Stephenson, A.J.; Scardino, P.T.; Eastham, J.A.; Bianco Jr, F.J.; Dotan, Z.A.; Fearn, P.A.; Kattan, M.W. Preoperative nomogram predicting the 10-year probability of prostate cancer recurrence after radical prostatectomy. *J. Natl. Cancer Inst.* **2006**, *98*, 715–717. [PubMed]
15. Steyerberg, E.W. *Clinical Prediction Models: A Practical Approach to Development, Validation, and Updating*; Springer: New York, NY, USA, 2009.
16. Rajan, P.; Sooriakumaran, P.; Nyberg, T.; Akre, O.; Carlsson, S.; Egevad, L.; Steineck, G.; Wiklund, N.P. Effect of comorbidity on prostate cancer–specific mortality: A prospective observational study. *J. Clin. Oncol.* **2017**, *35*, 3566. [PubMed]
17. Daskivich, T.J.; Fan, K.H.; Koyama, T.; Albertsen, P.C.; Goodman, M.; Hamilton, A.S.; Hoffman, R.M.; Stanford, J.L.; Stroup, A.M.; Litwin, M.S.; et al. Effect of age, tumor risk, and comorbidity on competing risks for survival in a U.S. population-based cohort of men with prostate cancer. *Ann. Intern. Med.* **2013**, *158*, 709–717.
18. Peters, M.; Kanthabalan, A.; Shah, T.T.; McCartan, N.; Moore, C.M.; Arya, M.; Moerland, M.A.; Hindley, R.G.; Emberton, M.; Ahmed, H.U. Development and internal validation of prediction models for biochemical failure and composite failure after focal salvage high intensity focused ultrasound for local radiorecurrent prostate cancer: Presentation of risk scores for individual patient prognoses. In *Urologic Oncology: Seminars and Original Investigations*; Elsevier: Amsterdam, The Netherlands, 2018; Volume 36, p. 13-e1.
19. Kilpeläinen, T.P.; Tammela, T.L.; Malila, N.; Hakama, M.; Santti, H.; Määttänen, L.; Stenman, U.H.; Kujala, P.; Auvinen, A. The Finnish prostate cancer screening trial: Analyses on the screening failures. *Int. J. Cancer* **2015**, *136*, 2437–2443.

20. Pylväläinen, J.; Talala, K.; Murtola, T.; Taari, K.; Raitanen, J.; Tammela, T.L.; Auvinen, A. Charlson Comorbidity Index Based On Hospital Episode Statistics Performs Adequately In Predicting Mortality, But Its Discriminative Ability Diminishes Over Time. *Clin. Epidemiol.* **2019**, *11*, 923.
21. Neupane, S.; Steyerberg, E.; Raitanen, J.; Talala, K.; Pylväläinen, J.; Taari, K.; Tammela, T.L.J.; Auvinen, A. Prognostic factors of prostate cancer mortality in a Finnish randomzed screening trial. *Int. J. Urol.* **2018**, *25*, 270–276. [CrossRef]
22. Van Buuren, S.; Oudshoorn, C.G.M. *Multivariate Imputation by Chained Equations: MICE V1.0 User's Manual*; TNO Report PG/VGZ/00.038; TNO Preventie en Gezondheid: Leiden, The Netherlands, 2000.
23. Sterne, J.A.; White, I.R.; Carlin, J.B.; Spratt, M.; Royston, P.; Kenward, M.G.; Wood, A.M.; Carpenter, J.R. Multiple imputation for missing data in epidemiological and clinical research: Potential and pitfalls. *BMJ* **2009**, *338*, b2393.
24. Bradburn, M.J.; Clark, T.G.; Love, S.B.; Altman, D.G. Survival Analysis Part III: Multivariate data analysis—Choosing a model and assessing its adequacy and fit. *Br. J. Cancer* **2003**, *89*, 605–611. [CrossRef] [PubMed]

Review

Global Trends of Latent Prostate Cancer in Autopsy Studies

Takahiro Kimura [1,*], Shun Sato [2], Hiroyuki Takahashi [2] and Shin Egawa [1]

1. Department of Urology, The Jikei University School of Medicine, Tokyo 105-8461, Japan; s-egpro@jikei.ac.jp
2. Department of Pathology, The Jikei University School of Medicine, Tokyo 105-8461, Japan; casinodrive@jikei.ac.jp (S.S.); hawk1bridge@gmail.com (H.T.)
* Correspondence: tkimura@jikei.ac.jp; Tel.: +81-3-3433-1111

Simple Summary: The incidence of prostate cancer (PC) is statistically biased due to the increase in prostate-specific antigen (PSA) screening and the accuracy of national cancer registration systems. However, studies on latent PC provide less biased information. This comprehensive review included studies evaluating latent PC in several countries. The prevalence of latent PC has been stable since 1950 in Western countries, but it has increased over time in Asian countries. Latent PC in Asian men has increased in prevalence and is higher in grade. This increase occurred not only due to the increase in PSA screening, but also due to increasing adoption of a Westernized lifestyle. Racial differences between Caucasian and Asian men may also explain the tumor location of latent PC. The autopsy findings in patients with latent PC included a significant proportion of high grade and stage cancers, suggesting a need to reconsider the definition of clinically insignificant PC.

Abstract: The incidence of prostate cancer (PC) has been increasing in Asian countries, where it was previously low. Although the adoption of a Westernized lifestyle is a possible explanation, the incidence is statistically biased due to the increase in prostate-specific antigen (PSA) screening and the accuracy of national cancer registration systems. Studies on latent PC provide less biased information. This review included studies evaluating latent PC in several countries after excluding studies using random or single-section evaluations and those that did not mention section thickness. The findings showed that latent PC prevalence has been stable since 1950 in Western countries, but has increased over time in Asian countries. Latent PC in Asian men has increased in both prevalence and number of high-grade cases. Racial differences between Caucasian and Asian men may explain the tumor location of latent PC. In conclusion, the recent increase in latent PC in Asian men is consistent with an increase in clinical PC. Evidence suggests that this increase is caused not only by the increase in PSA screening, but also by the adoption of a more Westernized lifestyle. Autopsy findings suggest the need to reconsider the definition of clinically insignificant PC.

Keywords: latent cancer; prostate cancer; autopsy

1. Introduction

The incidence of prostate cancer (PC) has been increasing globally in recent years. It is the second most frequently diagnosed cancer and the fifth leading cause of cancer-related deaths among men worldwide [1]. The incidence of PC in recent decades has been heavily influenced by the emergence of prostate-specific antigen (PSA) testing. The availability of PSA testing from the middle to the late 1980s led to the intensive use of the test for screening, with a subsequent rapid increase in the incidence rate in Western countries. This trend has also been growing in Asian countries, where the incidence of PC was previously low [1,2]. The cause of this increase in Asian countries is thought to be multifactorial. Although the spread of PSA screening may be a major cause, changes in lifestyle due to more Westernized diets might be another [3,4]. The accuracy of national cancer registration systems may also influence the incidence, as national cancer registration has not been developed in some Asian countries. However, PC mortality has been decreasing in many Western countries, possibly linked to earlier diagnosis due to PSA screening and improved treatment. In contrast, PC mortality is increasing in several Asian and developing countries [1]. These reports may support the influence of changes in risk factors due to more Westernized lifestyles in such countries.

Studies on latent PC provide less biased information about PC incidence compared to studies on clinical PC. Latent PC is defined as PC that is first detected in autopsy without any clinical signs of PC during the patient's lifetime. Since Mintz and Smith first reported latent PC in 13% of 100 autopsied cases in 1934 [5], many studies have been reported globally. A recent meta-analysis of 29 studies from 1948 to 2013 indicated that while the prevalence of latent PC significantly increases with age, there is no obvious time trend [6]. However, the time trend of latent PC prevalence might differ among countries. For example, a more recent study indicated that the prevalence of latent PC in Japanese men has been increasing [7]. In addition, a prospective study comparing latent PC in Asian and Caucasian men indicated that the prevalence in Asian men did not differ significantly from that in Caucasian men [8]. These results suggest that not only recent efforts for early detection, such as PSA screening, but also the change to Westernized diets and lifestyles may have influenced the increase in PC in Asian countries.

Information from studies on latent PC provides important insights from a different viewpoint. This review comprehensively discusses the results of latent PC studies in Western and Asian countries.

2. Potential Biases in Methodology in Latent PC Studies

As the methodology for latent PC studies has not yet been standardized and there are several biases between the studies, careful evaluation of their methods is required for precise interpretation. First, study populations, subject sources, and inclusion criteria differed among the studies. While most of the studies involved autopsies performed in hospitals, other studies assessed forensic autopsies. In addition, some studies have analyzed databases of institutional autopsy records or national or regional autopsy registries. However, a meta-analysis indicated that the source of subjects (population vs. hospital-based) was not significantly associated with the prevalence of latent PC [6]. Age was significantly associated with latent PC prevalence, which increased with each decade of age [6]. Thus, the inclusion and exclusion criteria of age and its distribution significantly influenced the prevalence of latent PC. Race is another major factor that affects the prevalence of latent PC. These results must be presented separately in studies that include various races. The methods of sample preparation also differed among the studies. The time elapsed from death to autopsy, step-sectioning versus random and/or single-section evaluation, and the interval between sections in step sectioning can also influence the prevalence of latent PC. The prevalence was reportedly higher in step-sectioned tissues than in randomly divided tissues [9], whereas there was no evidence that the section thickness or delay of autopsy affect PC prevalence [6]. However, information regarding the delay of autopsies is limited in most studies. The methods of diagnosis such as central review or not and use of immunohistochemical evaluation may also differ among studies, although a meta-analysis concluded that the use of immunohistochemistry was not associated with PC prevalence [6].

3. Prevalence of Latent PC in Western Countries

The most important topic in autopsy studies was the prevalence of latent PC. After the first report by Mintz and Smith in 1934, many such studies have been conducted in Western countries [5]. Studies evaluating latent PC by step-sectioning of the prostate in the US and Europe are listed in Table 1 [8,10–33]. Studies using random or single-section evaluations and those that did not mention section thickness were excluded. Cohorts of different nationalities or races are listed separately even if they were reported within the same study. The prevalence of latent PC varied from 9.6% to 58.6% between the studies. The ranges and distributions of age also varied between the studies. For example, some studies included men under 20 years of age [25,29], while another study included only men older than 70 years [15]. Age influenced the prevalence of latent PC, as it was one of the most significant factors associated with prevalence [6,34]. A recent meta-analysis of 29 studies reported an estimated mean cancer prevalence at age < 30 years of 5% (95% confidence interval (CI): 3–8%), which increased nonlinearly to 59% (95% CI: 48–71%) by age > 79 years [6]. Race is another factor that affects prevalence. Six studies in the US reported the prevalence of latent PC in Caucasian and Black men separately [15,22,23,25,26,33], with reported prevalence rates of 25.9–58.6% and 19.4–43.3%, respectively. All four studies that conducted statistical analyses on the prevalence of latent PC in Caucasian and Black men concluded that racial differences did not exist. However, these results require careful interpretation, as age distributions may differ between these races, especially in forensic studies. A recent review of 19 studies including 6,024 autopsies suggested a racial difference in latent PC prevalence between Caucasian and Black men (35.7% vs. 50.5%), but did not conduct a statistical analysis [34].

Table 1. Studies evaluating latent PC by step-sectioning of prostate in the US and Europe.

Author	Year Published	Country/Ethnicity	Duration of Study	Study Population	No. of Cases	Age	No. of Cancers	%	Pathology Section Width (mm)	Ref. No.
Moore	1935	Austria	1931–1932	Hospital	304	Range, 21–90	52	16.7	4	[10]
Andrews	1949	UK	NA	Hospital	142	Range, 40–79	17	12.0	4	[11]
Edwards	1953	Canada	1942–1945	Hospital	173	Mean, 64.1	35	16.7	4	[12]
Franks	1954	US	NA	Forensic	220	NA	69	31.4	4	[13]
Viitanen	1958	Finland	NA	Hospital	100	≥50	22	22.0	5	[14]
Halpert	1965	US/Black	NA	Hospital	30	Range, 70–79	13	43.3	4	[15]
Halpert	1965	US/Caucasian	NA	Hospital	70	Range, 70–79	41	58.6	4	[15]
Liavag	1968	Norway	NA	Hospital	340	≥40	90	26.5	4	[16]
Lundberg	1970	Sweden	1967	Hospital	292	NA	116	39.7	5	[17]
Harbitz	1973	Norway	1967–1968	Hospital	172	≥40	54	31.4	4–6	[18]
Akazaki	1973	US/men of Japanese ancestry	1969–1972	Hospital	158	≥50	46	29.1	3	[19]
Breslow	1977	Germany	NA	Hospital	145	Mean, 65	43	29.7	5	[20]
Breslow	1977	Israel	NA	Hospital	143	Mean, 65	32	22.4	5	[20]
Breslow	1977	Sweden	NA	Hospital	306	Mean, 65	123	40.2	5	[20]
Hølund	1980	Denmark	1971–1977	Hospital	223	Range, 36–94	57	25.6	3	[21]
Gulleyardo	1980	US/Black	NA	Hospital	207	NA	65	31.4	3	[22]
Gulleyardo	1980	US/Caucasian	NA	Hospital	293	NA	85	29	3	[22]
Yatani	1982	Colombia	1967–1970	Hospital	182	Mean, 64.4	NA	31.5	3	[23]
Yatani	1982	US/Black	1969–1978	Hospital	178	Mean, 63.6	NA	36.9	3	[23]
Yatani	1982	US/Caucasian	1969–1978	Hospital	253	Mean, 63.2	NA	34.6	3	[23]
Yatani	1982	US/men of Japanese ancestry	1969–1978	Hospital	417	Mean, 70.1	NA	25.6	3	[23]
Stemmermann	1992	US/men of Japanese ancestry	1970–1990	Hospital	293	Mean, 67.9	80	27.3	3	[24]
Sakr	1993	US/Black	NA	Forensic	98	10–50	19	19.4	3–4	[25]
Sakr	1993	US/Caucasian	NA	Forensic	54	10–50	14	25.9	3–4	[25]
Brawn	1996	US/Black	NA	Hospital	15	≥50	5	33.3	3	[26]
Brawn	1996	US/Caucasian	NA	Hospital	89	≥50	39	43.8	3	[26]
Sanchez-Chapado	2003	Spain	NA	Forensic	146	Mean, 48.5	27	18.5	3–4	[28]
Soos	2005	Hungary	NA	Hospital	139	18–95	54	38.8	4	[29]
Stamtiou	2007	Greece	2002–2004	Hospital	212	≥30	40	18.8	4	[30]
Haas	2007	US (92% Caucasian)	NA	Hospital	164	Median, 64	47	28.7	4	[31]
Polat	2009	Turkey	NA	Hospital	114	Mean, 55	11	9.6	4	[32]
Powell	2010	US/Black	1993–2004	Forensic	630	20–79	NA	35.1	2.5	[33]
Powell	2010	US/Caucasian	1993–2004	Forensic	426	20–79	NA	48.1	2.5	[33]
Zlotta	2013	Russia	2008–2011	Hospital	220	Mean, 62.5	82	37.3	4	[8]

NA: not available.

Figure 1 shows the prevalence of latent PC in studies of Caucasians in the US and Europe published after 1950 by year of publication. The size of each point was proportional to the number of men included in each study. The analytic linear approximation line of the datapoints indicated that the latent PC prevalence was stable over time. The spread of PSA screening programs is thought to have increased the diagnosis of insignificant PC and decreased the prevalence of latent PC. However, few studies have examined the changes in the prevalence of latent PC before and after the PSA era. A retrospective study using an autopsy record database from a single institution in the US reported that the prevalence of latent PC decreased three-fold with the widespread use of PSA screening [35]. In this study, the prevalence was 4.8% in men older than 40 years between 1955 and 1960, compared to 1.2% between 1991 and 2001. However, this study was limited by the lack of whole-mount sections to examine the prostate, which might lead to a lower prevalence compared to those in other autopsy studies using step-sectioning. However, the prevalence of latent PC in Japan has increased despite the spread of PSA screening, although the exposure rate of PSA testing in Asian countries is still lower than that in Western countries [2,7]. The trends in Asian countries are discussed in Section 4.

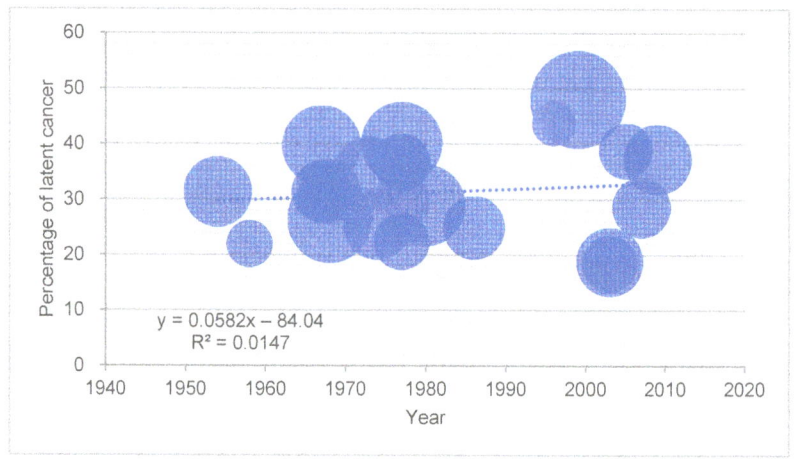

Figure 1. Prevalence of latent PC in studies of Caucasians in the US and Europe.

The prevalence of latent PC in studies of US and European Caucasians published after 1950 by year of publication. The size of each point was proportional to the number of men included in each study.

4. The prevalence of Latent PC in Asian and Other Countries

Studies investigating latent PC are fewer in Asian countries than in Western countries. In 1937, Yotsuyanagi et al. first reported a 3% prevalence of latent PC in Japanese men in a domestic journal [36]. Among the literature published in international journals, in 1961, Karube first reported a latent PC prevalence of 10.9% in Japanese men older than 40 years by step-sectioning [36]. Studies on latent PC in Asia and Africa are listed in Table 2 [7,8,19,20,23,36–44]. Studies using random or single-section evaluations that were not published in English were excluded. Some results are part of a multinational study. Most studies in Asia were from Japan [7,8,19,23,36,38–40], with the exception of two studies from Singapore [20,37] and one each from China [41], Hong Kong [20], and Iran [42]. Reports from other regions include Jamaica in Latin America and Uganda in Africa as part of a multinational study in 1977 [20]. The prevalence of latent PC in Jamaica and Uganda was 32.7% and 24.0%, respectively, which were higher than those in Asian countries in the same reports (15.0% and 14.5% in Hong Kong and Singapore, respectively). In addition, the mean age of men in Uganda was 58.3 years, which was 5 years younger than of those in other countries (64.5 in Singapore, 63.4 in Hong Kong, and 63.2 in Jamaica). Updated data for prevalence in Latin America and Africa are required.

Table 2. Studies evaluating latent PC by step-sectioning of prostate in Asian and other countries.

Author	Year Published	Country/Ethnicity	Duration of Study	Study Population	No. of Cases	Age	No. of Cancers	%	Pathology Section Width (mm)	Ref. No.
Karube	1961	Japan	1954–1959	Hospital	229	≥40	25	10.9	4–5	[36]
Lee	1972	Singapore	NA	Hospital	156	Range, 42–87	13	8.3	4	[37]
Akazaki	1973	Japan	1969–1972	Hospital	239	≥50	47	19.7	3	[19]
Bean	1973	Japan	1961–1969	Hospital	213	≥50	58	27.2	5	[38]
Breslow	1977	Hong Kong	NA	Hospital	173	Mean, 65	26	15.0	5	[20]
Breslow	1977	Jamaica	NA	Hospital	168	Mean, 65	55	32.7	5	[20]
Breslow	1977	Singapore	NA	Hospital	242	Mean, 65	35	14.5	5	[20]
Breslow	1977	Uganda	NA	Hospital	150	Mean, 65	36	24.0	5	[20]
Yatani	1982	Japan	1965–1979	Hospital	576	Mean, 67.7	NA	20.5	3	[23]
Billis	1986	Brazil	NA	Hospital	180	Range, 40–88	45	25.0	3–5	[27]
Yatani	1988	Japan	1965–1979	Hospital	576	Mean, 67.6	NA	22.5	3	[39]
Yatani	1988	Japan	1982–1986	Hospital	660	Mean, 68.7	NA	34.6	3	[39]
Takahashi	1992	Japan	NA	Hospital	29	≥90	17	58.6	3–4	[40]
Gu	1994	China	1989–1992	Hospital	381 (including 60 RCP)	NA	21	5.5	5	[41]
Zare-Mirzaie	2012	Iran	2008–2009	Hospital	149	Mean, 64.5	14	9.4	4	[42]
Zlotta	2013	Japan	2008–2011	Hospital	100	Mean, 68.5	35	35.0	4	[8]
Kimura	2016	Japan	1983–1987	Hospital	501	Mean, 63.5	104	20.8	5	[7]
Inaba	2020	Japan	2009–2017	Hospital	182	Median, 72	71	39.0	5	[43]

Figure 2 shows the prevalence of latent PC in studies of Asian countries published after 1950. Data from the study conducted by Takahashi et al. in Japan were excluded because they focused on men older than 90 years of age [40]. The size of each point is proportional to the number of men in the study. The analytic linear approximation line of the datapoints indicated that the prevalence of latent PC in Asia increased over time compared to that in the US and Europe (Figure 1). The prevalence was 8.3–27.2% from the 1960s to 1970s, 5.5–34.6% from the 1980s to the 1990s, and 9.4–39.0% after 2000. Two Japanese studies directly compared the time trends in latent PC within the same institutions. Yatani et al. compared the latent PC prevalence within the same institution between men from 1965 to 1979 and from 1982 to 1986, both in pre-PSA era periods. The prevalence increased significantly from 22.5% to 34.6% [39]. More recently, Kimura et al. compared the prevalence of latent PC between Japanese men in pre- and post-PSA eras. The prevalence in men was 20.8% in 1983–1987 and 43.3% in 2008–2013 [7]. Both studies indicated a significant increase in higher-grade and larger cancers. Yatani et al. reported a higher rate of infiltrative tumors in the cohort in 1965–1979 than in 1982–1986, at 9.8% and 17.8%, respectively [39]. Kimura et al. reported a significantly larger index cancer volume in men in 2008–2013 compared to that in 1983–1987 [7].

The increased prevalence of latent PC in Asian men is consistent with the increased prevalence of clinical PC [45]. A major explanation for the increase in latent PC in Asian countries may be lifestyle changes due to more Westernized diets. The incidence of clinical PC in US men of Japanese ancestry in 1973–1986 was between that of Caucasians in the US and Japanese men born in Japan within the same period, suggesting the influence of both genetic and lifestyle factors on PC incidence [46]. In contrast, a comparative study published in 1973 showed that the age-adjusted prevalence of latent PC did not differ significantly between Japanese men in Japan and those in Hawaii (20.5% and 26.7%, respectively). However, the age-adjusted prevalence of the proliferative type of latent PC was higher in Japanese Hawaiians than in native Japanese (19.1% and 8.7%, respectively) [19].

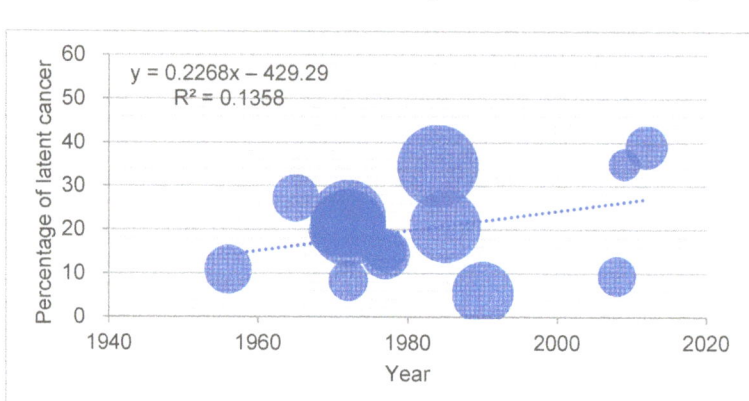

Figure 2. Prevalence of latent PC in the studies on Asian populations. Prevalence of latent PC in studies of Asian countries published after 1950 by year of publication. The size of each point was proportional to the number of men in the study.

Zlotta et al. prospectively compared the prevalence of latent PC in 100 Japanese men and 220 Russian men [8]. The prevalence was 35.0% and 37.3% in Japanese and Russian men, respectively, and did not differ significantly. However, Japanese men had a greater probability of having a PC Gleason score (GS) of 7 than Russian men after adjusting for age and prostate weight. These results suggest the increasing prevalence and grade of latent PC in Asian men over the past few decades.

5. Pathological Findings from Latent PC

Although latent PC does not cause clinical symptoms and is not generally detected during the lifetime, most studies showed that a significant proportion of latent PC had high-grade, capsular, or seminal vesicle invasion. In studies in the US and Europe, 5.43% of cancers were GS 7 or greater and 11–13% were pT3 or greater [8,27,29–31]. In contrast, in Asian studies, 35.7–51.4% of latent PC was GS 7 or higher, with proportions higher than those in Western reports, although the proportion of cancers with pT3 or greater was similar (11.5–12.7%) [7,8,42]. Consequently, these cases of latent PC included clinically significant cancer as defined by Epstein (the presence of T3 or greater and/or index tumor volume of 500 mm^3 or greater and/or GS $\geq$ 7 [47]). A prospective comparative study by Zlotta et al. reported that 29.3% and 51.4% of latent PC cases were clinically significant in Russian and Japanese men, respectively [8]. A comparative study of contemporary latent PC and historical controls in Japan reported an index cancer volume of 500 mm^3 in 9.6% of cancers in men in 1983–1987 and 25.5% in 2008–2013, a significant difference [7]. The increase in proportion of significant cancer in latent PC notwithstanding the spread of PSA screening might suggest an increase in high-grade cancer in Asian countries, especially in Japan. These results also suggest the need to reexamine the definition of clinically insignificant PC. Stamey et al. defined clinically significant PC as organ-confined tumors of <0.5 cm^3, GS 3+3 with no grade 4 or 5 [48]. However, it can also be defined as a cancer that does not affect the patient during the natural course of his lifetime. The requirements of Stamey's definition may be too stringent.

Investigating the tumor location of latent PC could improve our understanding of the origin of PC and how it grows [49]. Racial differences between Caucasian and Asian men have been suggested to affect not only the prevalence, but also the tumor location of PC. A comparative study of radical prostatectomy specimens reported that 35.5% and 0.6% of PCs originated in the transition zone (TZ) in Japanese men and US men, respectively [50].

Studies that categorized tumor location into anterior or posterior regions reported that anterior cancer was more prevalent in Asian men than in Caucasian men [50–56]. However, most studies evaluating tumor location in the prostate have analyzed only prostatectomy specimens. The tumor location of the prostatectomy specimens may overestimate the prevalence in the peripheral zone (PZ) or posterior cancer because of its higher detectability by digital rectal examination and transrectal prostate biopsy compared to in the TZ or anterior cancer. In this sense, the tumor location in latent PC may be less biased. There are limited reports regarding tumor location in latent PC. An autopsy study in Hungary including 139 men aged 18–95 years reported a latent PC prevalence of 38.8%; among the 64 tumor foci, 82.8% and 18.9% were present in the PZ and TZ, respectively [29]. Another study in the US including 164 men aged 54–73 years reported that latent PC was present in 29% of the cases, with 62% and 36% of PCs located in the posterior and anterior regions, respectively, and 77% and 16%—in the PZ and central zone in the prostate, respectively [31].

Reports on tumor locations of latent PC in Asian men are limited. A report of 149 autopsies of Iranian men over 50 years of age detected invasive adenocarcinoma in 14 (9.4%) cases, including nine cases (64%) in the posterior region, one case (7%) in the anterior region, and four cases (29%) in both lobes of the prostate [42]. A report of 182 Japanese men observed latent PC in 39.0% of cases, occurring in the TZ, PZ, or without dominance in 38.0%, 57.8%, and 4.2% of cases, respectively [43]. The tumors were located in the anterior and posterior regions in 49.3% and 40.8% of the cases, respectively. Approximately 40% of the tumors were located in the TZ and anterior region of the prostate, a rate higher than that reported in Western studies. The age distribution also differed between TZ and PZ cancers. In elderly men, cancer is more frequently diagnosed in the PZ than in the TZ [43]. This was consistent with the report by Takahashi et al. on autopsies in men over 90 years of age, which revealed that all latent PCs were localized in the PZ of the prostate [40]. An autopsy study in the US reported that most TZ cancers showed a different pathological pattern from that of PZ cancers, with lower GS and less aggressiveness [57]. However, a Japanese study reported that the pathological features did not differ between the TZ and PZ and between anterior and posterior cancers in terms of GS, tumor volume, or prevalence of clinically significant cancer; however, there is variation in the pT stage—PZ cancer has a significantly higher pT stage than TZ cancer. Several anatomical explanations have been proposed to explain this difference. For example, the TZ is separated from the surrounding area by fibromuscular tissue, whereas no such structure exists in the PZ. Moreover, the TZ contacts with the prostate capsule from the outside to the back of the PZ, and T3b cases of the TZ are few because of the anatomical position [43]. However, a prospective comparative study of Japanese and Russian autopsy cases reported similar tumor locations between cohorts, in which latent PC was located in the TZ in 25.9% and 20.7% and in the anterior region in 20.0% and 21.9% of the cases in Japanese and Russian men, respectively. Further investigation is required to determine whether there is a racial difference in tumor location and whether the location in Asian men has changed due to Western diet and lifestyle.

Few studies have investigated tumor location in the vertical direction. In their international multicenter study investigating 1327 autopsies from seven counties or regions, Breslow et al. reported latent PC in 350 cases. In the vertical direction, more tumors were present at the middle and apex levels than at the base. However, the evaluation method to describe the tumor distribution has not yet been standardized, and further studies are warranted.

Most latent PC cases represent the less aggressive forms of PC. Thus, comparing molecular markers or genomic aberrations between latent and clinical PCs is an ideal method to investigate their effects. Igawa et al. reported a significantly higher nm23-HI gene expression level in clinical PC than in normal prostatic tissues, latent PC, and clinical PC [58]. Watanabe et al. reported that *Ras* gene mutations in latent PC varied among ethnic groups and that the frequency in Japanese men was higher than that in US Black or Caucasian men [59]. Alipov et al. compared the expression of the ETS1 proto-oncogene in latent PC, benign prostatic hyperplasia, normal prostatic tissues, and clinical PC [60], reporting negative expression in benign tissues and higher levels in clinical PC than in latent PC. Maekawa et al. investigated the TMPRSS2 Met160Val polymorphism in Japanese men, including 518 men with sporadic PC, 433 healthy controls, and 154 men with latent PC [61]. The TMPRSS2 Met160Val polymorphism is a genetic risk factor for sporadic PC but not for latent PC in the Japanese population. However, molecular studies using latent PC are limited, possibly because of the limited quality and availability of latent PC specimens.

6. Limitations of Autopsy Studies

Latent PC has a unique cancer status compared to the malignancies of other origins. Although it has been investigated for a long time, several problems remain to be solved. Section 5 described variations in the methodologies used for sample preparation and diagnosis. Although step-sectioning of the whole prostate is a standard method, the duration from death to autopsy is difficult to control. A central review for diagnosis is mandatory because of inter- and intra-observer variability for the pathological diagnosis of PC [62]. A fundamental bias also exists in autopsy studies. As the subjects were men who died in the hospital, their backgrounds differed from those of healthy men. Inaba et al. reported that 106 of the 182 autopsy cases (58.2%) had been performed due to death of malignancy other than PC (unpublished data).

The available data on latent PC were provided by a limited number of countries and regions. Most studies were from North America, Western Europe, and Japan, whereas data from Africa and Latin America are limited. More importantly, the number of autopsies has been steadily declining over the past 30–40 years worldwide [63]. After the 2000s, the autopsy rate was only 7–9% in the US, compared to approximately 25–35% in the mid-1960s and 50% of all hospital deaths in the 1940s and 1950s [64,65]. In Japan, more than 40,000 autopsies were performed in 1985, but this number had gradually decreased to approximately 10,000 by 2018 [66].

One explanation for the limited number of studies evaluating molecular markers in latent PC is the low quality of specimens from autopsies. RNA and proteins were extracted during the time between death and autopsy. To overcome such limitations in autopsy studies, rapid autopsies have emerged [67]. In this new methodology, tissues are collected as soon as possible after the patient's death. Ideally, the quality of a rapid autopsy tissue can be considered comparable to the quality of a fresh surgical biopsy tissue.

7. Learning from Latent PC and Future Directions

While latent PC studies have a long history, the available evidence remains limited. Latent PC studies have revealed a larger prevalence of insignificant PC than the incidence of clinical PC. PC prevalence increases with age and more than half of both Caucasian and Asian men over 80 years of age have indolent PC. The recent increase in latent PC in Asian men is consistent with an increase in clinical PC in Asian countries. These findings suggest that this increase in clinical PC in Asian countries is due not only to the spread of PSA screening, but also to the adoption of Westernized lifestyles.

In addition, the results of autopsy studies suggest the need to reconsider the definition of clinically insignificant PC, which is thought to be an ideal candidate for active surveillance. The present definition might be too strict, as latent PC included a significant proportion of cancer cases thought to be life-threatening, such as with GS ≥ 7 and pT3 or greater. Cancer volume and the percentage of high-grade cancer cases also increased with age. However, the individuals lived without the influence of PC throughout their lives. Molecular analyses are required in latent PC studies to distinguish between indolent and life-threatening PC. Methodologies such as rapid autopsies have opened the door for new studies of latent PC.

Author Contributions: Writing—original draft preparation, T.K.; writing—review and editing, S.S., H.T., and S.E. Authorship must be limited to those who have contributed substantially to the work reported. All authors have read and agreed to the published version of the manuscript.

Funding: This research received no external funding.

Conflicts of Interest: T.K. is a paid consultant/advisor to Astellas, Bayer, Janssen, and Sanofi. S.E. is a paid consultant/advisor to Takeda, Astellas, AstraZeneca, Sanofi, Janssen, and Pfizer. The funders had no role in the study design; in the collection, analyses, or interpretation of data; in the writing of the manuscript; or in the decision to publish the results.

References

1. Bray, F.; Ferlay, J.; Soerjomataram, I.; Siegel, R.L.; Torre, L.A.; Jemal, A. Global cancer statistics 2018: Globocan estimates of incidence and mortality worldwide for 36 cancers in 185 countries. *CA Cancer J. Clin.* **2018**, *68*, 394–424. [CrossRef] [PubMed]
2. Ito, K. Prostate cancer in asian men. *Nat. Rev. Urol.* **2014**, *11*, 197–212. [CrossRef] [PubMed]
3. Yoshita, K.; Arai, Y.; Nozue, M.; Komatsu, K.; Ohnishi, H.; Saitoh, S.; Miura, K.; Group, N.D.R. Total energy intake and intake of three major nutrients by body mass index in japan: Nippon data80 and nippon data90. *J. Epidemiol.* **2010**, *20* (Suppl. S3), S515–S523. [CrossRef] [PubMed]
4. Iwasaki, M.; Mameri, C.P.; Hamada, G.S.; Tsugane, S. Cancer mortality among japanese immigrants and their descendants in the state of sao paulo, brazil, 1999–2001. *JPN J. Clin Oncol.* **2004**, *34*, 673–680. [CrossRef] [PubMed]
5. Mintz, E.R.; Smith, G.G. Autopsy findings in 100 cases of prostatic cancer. *N. Engl. J. Med.* **1934**, *211*, 479–487. [CrossRef]
6. Bell, K.J.; Del Mar, C.; Wright, G.; Dickinson, J.; Glasziou, P. Prevalence of incidental prostate cancer: A systematic review of autopsy studies. *Int. J. Cancer* **2015**, *137*, 1749–1757. [CrossRef]
7. Kimura, T.; Takahashi, H.; Okayasu, M.; Kido, M.; Inaba, H.; Kuruma, H.; Yamamoto, T.; Furusato, B.; Furusato, M.; Wada, T.; et al. Time trends in histological features of latent prostate cancer in japan. *J. Urol.* **2016**, *195*, 1415–1420. [CrossRef]
8. Zlotta, A.R.; Egawa, S.; Pushkar, D.; Govorov, A.; Kimura, T.; Kido, M.; Takahashi, H.; Kuk, C.; Kovylina, M.; Aldaoud, N.; et al. Prevalence of prostate cancer on autopsy: Cross-sectional study on unscreened caucasian and asian men. *J. Natl. Cancer Inst.* **2013**, *105*, 1050–1058. [CrossRef]
9. Rebbeck, T.R.; Haas, G.P. Temporal trends and racial disparities in global prostate cancer prevalence. *Can. J. Urol.* **2014**, *21*, 7496–7506. [PubMed]
10. Moore, R.A. The morphology of small prostatic carcinoma. *J. Urol.* **1935**, *33*, 224–234. [CrossRef]
11. Andrews, G.S. Latent carcinoma of the prostate. *J. Clin. Pathol* **1949**, *2*, 197–208. [CrossRef] [PubMed]
12. Edwards, C.N.; Steinthorsson, E.; Nicholson, D. An autopsy study of latent prostatic cancer. *Cancer* **1953**, *6*, 531–554. [CrossRef]
13. Franks, L.M. Latent carcinoma. *Ann. R Coll. Surg. Engl.* **1954**, *15*, 236–249. [PubMed]
14. Viitanen, I.; Von Hellens, A. Latent carcinoma of the prostate in finland; preliminary report. *Acta Pathol. Microbiol. Scand.* **1958**, *44*, 64–67. [CrossRef] [PubMed]
15. Halpert, B.; Schmalhorst, W.R. Carcinoma of the prostate in patients 70 to 79 years old. *Cancer* **1966**, *19*, 695–698. [CrossRef]
16. Liavag, I. The localization of prostatic carcinoma. An autopsy study. *Scand. J. Urol. Nephrol.* **1968**, *2*, 65–71. [CrossRef]
17. Lundberg, S.; Berge, T. Prostatic carcinoma. An autopsy study. *Scand. J. Urol. Nephrol.* **1970**, *4*, 93–97. [CrossRef]
18. Harbitz, T.B. Testis weight and the histology of the prostate in elderly men. An analysis in an autopsy series. *Acta Pathol. Microbiol. Scand. A* **1973**, *81*, 148–158. [CrossRef]
19. Akazaki, K.; Stemmerman, G.N. Comparative study of latent carcinoma of the prostate among japanese in Japan and Hawaii. *J. Natl. Cancer Inst.* **1973**, *50*, 1137–1144. [CrossRef]
20. Breslow, N.; Chan, C.W.; Dhom, G.; Drury, R.A.; Franks, L.M.; Gellei, B.; Lee, Y.S.; Lundberg, S.; Sparke, B.; Sternby, N.H.; et al. Latent carcinoma of prostate at autopsy in seven areas. The international agency for research on cancer, lyons, france. *Int. J. Cancer* **1977**, *20*, 680–688. [CrossRef]
21. Holund, B. Latent prostatic cancer in a consecutive autopsy series. *Scand. J. Urol. Nephrol.* **1980**, *14*, 29–35. [CrossRef] [PubMed]
22. Guileyardo, J.M.; Johnson, W.D.; Welsh, R.A.; Akazaki, K.; Correa, P. Prevalence of latent prostate carcinoma in two U.S. populations. *J. Natl. Cancer Inst.* **1980**, *65*, 311–316. [CrossRef] [PubMed]
23. Yatani, R.; Chigusa, I.; Akazaki, K.; Stemmermann, G.N.; Welsh, R.A.; Correa, P. Geographic pathology of latent prostatic carcinoma. *Int. J. Cancer* **1982**, *29*, 611–616. [CrossRef] [PubMed]
24. Stemmermann, G.N.; Nomura, A.M.; Chyou, P.H.; Yatani, R. A prospective comparison of prostate cancer at autopsy and as a clinical event: The Hawaii Japanese experience. *Cancer Epidemiol. Biomark. Prev.* **1992**, *1*, 189–193. [PubMed]
25. Sakr, W.A.; Haas, G.P.; Cassin, B.F.; Pontes, J.E.; Crissman, J.D. The frequency of carcinoma and intraepithelial neoplasia of the prostate in young male patients. *J. Urol.* **1993**, *150*, 379–385. [CrossRef]
26. Brawn, P.N.; Jay, D.W.; Foster, D.M.; Kuhl, D.; Speights, V.O.; Johnson, F.H.; Riggs, M.; Lind, M.L.; Coffield, K.S.; Weaver, B. Prostatic acid phosphatase levels (enzymatic method) from completely sectioned, clinically benign, whole prostates. *Prostate* **1996**, *28*, 295–299. [CrossRef]
27. Billis, A. Latent carcinoma and atypical lesions of prostate an autopsy study. *Urology* **1986**, *28*, 324–329. [CrossRef]
28. Sanchez-Chapado, M.; Olmedilla, G.; Cabeza, M.; Donat, E.; Ruiz, A. Prevalence of prostate cancer and prostatic intraepithelial neoplasia in caucasian mediterranean males: An autopsy study. *Prostate* **2003**, *54*, 238–247. [CrossRef] [PubMed]
29. Soos, G.; Tsakiris, I.; Szanto, J.; Turzo, C.; Haas, P.G.; Dezso, B. The prevalence of prostate carcinoma and its precursor in hungary: An autopsy study. *Eur. Urol.* **2005**, *48*, 739–744. [CrossRef]

30. Stamatiou, K.; Alevizos, A.; Perimeni, D.; Sofras, F.; Agapitos, E. Frequency of impalpable prostate adenocarcinoma and precancerous conditions in greek male population: An autopsy study. *Prostate Cancer Prostatic Dis.* **2006**, *9*, 45–49. [CrossRef] [PubMed]
31. Haas, G.P.; Delongchamps, N.B.; Jones, R.F.; Chandan, V.; Serio, A.M.; Vickers, A.J.; Jumbelic, M.; Threatte, G.; Korets, R.; Lilja, H.; et al. Needle biopsies on autopsy prostates: Sensitivity of cancer detection based on true prevalence. *J. Natl. Cancer Inst.* **2007**, *99*, 1484–1489. [CrossRef] [PubMed]
32. Polat, K.; Tüzel, E.; Aktepe, F.; Akdoğan, B.; Güler, C.; Uzun, I. Investigation of the incidence of latent prostate cancer and high-grade prostatic intraepithelial neoplasia in an autopsy series of Turkish males. *Turk. J. Urol.* **2009**, *35*, 96–100.
33. Powell, I.J.; Bock, C.H.; Ruterbusch, J.J.; Sakr, W. Evidence supports a faster growth rate and/or earlier transformation to clinically significant prostate cancer in black than in white american men, and influences racial progression and mortality disparity. *J. Urol.* **2010**, *183*, 1792–1796. [CrossRef] [PubMed]
34. Jahn, J.L.; Giovannucci, E.L.; Stampfer, M.J. The high prevalence of undiagnosed prostate cancer at autopsy: Implications for epidemiology and treatment of prostate cancer in the prostate-specific antigen-era. *Int. J. Cancer* **2015**, *137*, 2795–2802. [CrossRef]
35. Konety, B.R.; Bird, V.Y.; Deorah, S.; Dahmoush, L. Comparison of the incidence of latent prostate cancer detected at autopsy before and after the prostate specific antigen era. *J. Urol.* **2005**, *174*, 1785–1788; discussion 1788. [CrossRef] [PubMed]
36. Karube, K. Study of latent carcinoma of the prostate in the japanese based on necropsy material. *Tohoku J. Exp. Med.* **1961**, *74*, 265–285. [CrossRef] [PubMed]
37. Lee, Y.S.; Shanmugaratnam, K. Latent prostate carcinoma in Singapore Chinese. *Singap. Med. J.* **1972**, *13*, 1–6.
38. Bean, M.A.; Yatani, R.; Liu, P.I.; Fukazawa, K.; Ashley, F.W.; Fujita, S. Prostatic carcinoma at autopsy in hiroshima and nagasaki japanese. *Cancer* **1973**, *32*, 498–506. [CrossRef]
39. Yatani, R.; Shiraishi, T.; Nakakuki, K.; Kusano, I.; Takanari, H.; Hayashi, T.; Stemmermann, G.N. Trends in frequency of latent prostate carcinoma in japan from 1965–1979 to 1982–1986. *J. Natl. Cancer Inst.* **1988**, *80*, 683–687. [CrossRef]
40. Takahashi, S.; Shirai, T.; Hasegawa, R.; Imaida, K.; Ito, N. Latent prostatic carcinomas found at autopsy in men over 90 years old. *JPN J. Clin. Oncol.* **1992**, *22*, 117–121. [CrossRef]
41. Gu, F.-L.; Xia, T.-L.; Kong, X.-T. Preliminary study of the frequency of benign prostatic hyperplasia and prostatic cancer in china. *Urology* **1994**, *44*, 688–691. [CrossRef]
42. Zare-Mirzaie, A.; Balvayeh, P.; Imamhadi, M.A.; Lotfi, M. The frequency of latent prostate carcinoma in autopsies of over 50 years old males, the Iranian experience. *Med. J. Islam. Repub. Iran.* **2012**, *26*, 73–77. [PubMed]
43. Inaba, H.; Kimura, T.; Onuma, H.; Sato, S.; Kido, M.; Yamamoto, T.; Fukuda, Y.; Takahashi, H.; Egawa, S. Tumor location and pathological features of latent and incidental prostate cancer in contemporary japanese men. *J. Urol.* **2020**, *204*, 267–272. [CrossRef] [PubMed]
44. Billis, A.; Souza, C.A.F.; Piovesan, H. Histologic carcinoma of the prosate in autopsies frequency, origin, extension, grading and terminology. *Braz. J. Urol.* **2002**, *28*, 197–205.
45. Kimura, T.; Egawa, S. Epidemiology of prostate cancer in asian countries. *Int. J. Urol.* **2018**, *25*, 524–531. [CrossRef]
46. Cook, L.S.; Goldoft, M.; Schwartz, S.M.; Weiss, N.S. Incidence of adenocarcinoma of the prostate in asian immigrants to the united states and their descendants. *J. Urol.* **1999**, *161*, 152–155. [CrossRef]
47. Epstein, J.I.; Walsh, P.C.; Carmichael, M.; Brendler, C.B. Pathologic and clinical findings to predict tumor extent of nonpalpable (stage t1c) prostate cancer. *JAMA* **1994**, *271*, 368–374. [CrossRef] [PubMed]
48. Stamey, T.A.; Freiha, F.S.; McNeal, J.E.; Redwine, E.A.; Whittemore, A.S.; Schmid, H.P. Localized prostate cancer. Relationship of tumor volume to clinical significance for treatment of prostate cancer. *Cancer* **1993**, *71*, 933–938. [CrossRef]
49. Chen, Y.; Yan, W. Implications from autopsy studies of latent prostate cancer. *Nat. Rev. Urol.* **2020**, *17*, 428–429. [CrossRef] [PubMed]
50. Takahashi, H.; Epstein, J.I.; Wakui, S.; Yamamoto, T.; Furusato, B.; Zhang, M. Differences in prostate cancer grade, stage, and location in radical prostatectomy specimens from united states and japan. *Prostate* **2014**, *74*, 321–325. [CrossRef]
51. Hashine, K.; Ueno, Y.; Shinomori, K.; Ninomiya, I.; Teramoto, N.; Yamashita, N. Correlation between cancer location and oncological outcome after radical prostatectomy. *Int. J. Urol.* **2012**, *19*, 855–860. [CrossRef] [PubMed]
52. Takashima, R.; Egawa, S.; Kuwao, S.; Baba, S. Anterior distribution of stage t1c nonpalpable tumors in radical prostatectomy specimens. *Urology* **2002**, *59*, 692–697. [CrossRef]
53. Koppie, T.M.; Bianco, F.J., Jr.; Kuroiwa, K.; Reuter, V.E.; Guillonneau, B.; Eastham, J.A.; Scardino, P.T. The clinical features of anterior prostate cancers. *BJU Int.* **2006**, *98*, 1167–1171. [CrossRef] [PubMed]
54. Al-Ahmadie, H.A.; Tickoo, S.K.; Olgac, S.; Gopalan, A.; Scardino, P.T.; Reuter, V.E.; Fine, S.W. Anterior-predominant prostatic tumors: Zone of origin and pathologic outcomes at radical prostatectomy. *Am. J. Surg. Pathol.* **2008**, *32*, 229–235. [CrossRef] [PubMed]
55. Hossack, T.; Patel, M.I.; Huo, A.; Brenner, P.; Yuen, C.; Spernat, D.; Mathews, J.; Haynes, A.M.; Sutherland, R.; del Prado, W.; et al. Location and pathological characteristics of cancers in radical prostatectomy specimens identified by transperineal biopsy compared to transrectal biopsy. *J. Urol.* **2012**, *188*, 781–785. [CrossRef] [PubMed]
56. Mygatt, J.; Sesterhenn, I.; Rosner, I.; Chen, Y.; Cullen, J.; Morris-Gore, T.; Barton, J.; Dobi, A.; Srivastava, S.; McLeod, D.; et al. Anterior tumors of the prostate: Clinicopathological features and outcomes. *Prostate Cancer Prostatic Dis.* **2014**, *17*, 75–80. [CrossRef] [PubMed]

57. McNeal, J.E. Cancer volume and site of origin of adenocarcinoma in the prostate: Relationship to local and distant spread. *Hum. Pathol.* **1992**, *23*, 258–266. [CrossRef]
58. Igawa, M.; Urakami, S.; Shiina, H.; Ishibe, T.; Usui, T.; Chodak, G.W. Association of nm23 protein levels in human prostates with proliferating cell nuclear antigen expression at autopsy. *Eur. Urol.* **1996**, *30*, 383–387. [CrossRef]
59. Watanabe, M.; Shiraishi, T.; Yatani, R.; Nomura, A.M.; Stemmermann, G.N. International comparison on ras gene mutations in latent prostate carcinoma. *Int. J. Cancer* **1994**, *58*, 174–178. [CrossRef]
60. Alipov, G.; Nakayama, T.; Ito, M.; Kawai, K.; Naito, S.; Nakashima, M.; Niino, D.; Sekine, I. Overexpression of ets-1 proto-oncogene in latent and clinical prostatic carcinomas. *Histopathology* **2005**, *46*, 202–208. [CrossRef] [PubMed]
61. Maekawa, S.; Suzuki, M.; Arai, T.; Suzuki, M.; Kato, M.; Morikawa, T.; Kasuya, Y.; Kume, H.; Kitamura, T.; Homma, Y. Tmprss2 met160val polymorphism: Significant association with sporadic prostate cancer, but not with latent prostate cancer in Japanese men. *Int. J. Urol.* **2014**, *21*, 1234–1238. [CrossRef] [PubMed]
62. Allsbrook, W.C., Jr.; Mangold, K.A.; Johnson, M.H.; Lane, R.B.; Lane, C.G.; Amin, M.B.; Bostwick, D.G.; Humphrey, P.A.; Jones, E.C.; Reuter, V.E.; et al. Interobserver reproducibility of gleason grading of prostatic carcinoma: Urologic pathologists. *Hum. Pathol.* **2001**, *32*, 74–80. [CrossRef] [PubMed]
63. Dehner, L.P. The medical autopsy: Past, present, and dubious future. *Mo. Med.* **2010**, *107*, 94–100. [PubMed]
64. Nemetz, P.N.; Tanglos, E.; Sands, L.P.; Fisher, W.P., Jr.; Newman, W.P., 3rd; Burton, E.C. Attitudes toward the autopsy—An 8-state survey. *MedGenMed* **2006**, *8*, 80. [PubMed]
65. Xiao, J.; Krueger, G.R.; Buja, L.M.; Covinsky, M. The impact of declining clinical autopsy: Need for revised healthcare policy. *Am. J. Med. Sci.* **2009**, *337*, 41–46. [CrossRef] [PubMed]
66. Japanese Pathological Society. Annual of the Pathological Autopsy Cases in Japan. Available online: http://pathology.or.jp/kankoubutu/autopsy-index.html (accessed on 19 January 2021).
67. Duregon, E.; Schneider, J.; DeMarzo, A.M.; Hooper, J.E. Rapid research autopsy is a stealthy but growing contributor to cancer research. *Cancer* **2019**, *125*, 2915–2919. [CrossRef] [PubMed]

Article

Morphological and Molecular Characterization of Proliferative Inflammatory Atrophy in Canine Prostatic Samples

Giovana de Godoy Fernandes [1,†], Bruna Pedrina [1,†], Patrícia de Faria Lainetti [1], Priscila Emiko Kobayashi [1], Verônica Mollica Govoni [1], Chiara Palmieri [2], Veridiana Maria Brianezi Dignani de Moura [3], Renée Laufer-Amorim [1] and Carlos Eduardo Fonseca-Alves [1,4,*]

1. School of Veterinary Medicine and Animal Science, São Paulo State University—UNESP, Botucatu 18618-681, Brazil; giovana.godoy@unesp.br (G.d.G.F.); bruna.pedrina@unesp.br (B.P.); patricia.lainetti@unesp.br (P.d.F.L.); priscila.e.kobayashi@unesp.br (P.E.K.); veronica.m.govoni@unesp.br (V.M.G.); renee.laufer-amorim@unesp.br (R.L.-A.)
2. Gatton Campus, School of Veterinary Science, The University of Queensland, Gatton, QLD 4343, Australia; c.palmieri@uq.edu.au
3. School of Veterinary Medicine and Animal Science, Federal University of Goiás, Goiânia 18531-883, Brazil; vdmoura@ufg.br
4. Institute of Health Sciences, Paulista University-UNIP, Bauru 17048-290, Brazil
* Correspondence: carlos.e.alves@unesp.br
† These authors contributed equally to this work.

Simple Summary: Prostatic diseases are important worldwide, being the prostate cancer (PC) the most common tumor in men. Among the factors associated with PC development, the preneoplastic lesions are well-recognized. Preneoplastic lesions are cellular morphological alterations, induced by different factors and present a potential to progression for PC. In this scenario, dogs are considered spontaneous models. Dogs naturally develops prostatic hyperplasia, preneoplastic lesions and PC. Among the preneoplastic lesions, the proliferative inflammatory atrophy (PIA) develops spontaneously in dogs. PIA is an epithelial lesion induced by prostatic chronic inflammation, leading to a proliferative atrophy of the prostate gland. Thus, this study aimed to perform a full PIA morphological, phenotypical and molecular characterization in dogs. After reviewing the archives of the veterinary pathology service, it was identified 171 dogs containing PIA in the prostate gland, and among the PC cases (N = 84), it was identified PIA lesions surrounding 60.7% of PC cases. Besides that, we identified loss of genes related to the maintenance of prostatic tissue and can predispose to malignant transformation. Moreover, mutations in androgen receptor gene were identified, demonstration alteration in DNA in PIA. Overall, these results support the hypothesis that PIA can be considered a preneoplastic lesion in canine prostate.

Abstract: Proliferative inflammatory atrophy (PIA) is an atrophic lesion of the prostate gland that occurs in men and dogs and is associated with a chronic inflammatory infiltrate. In this study, we retrospectively reviewed canine prostatic samples from intact dogs, identifying 50 normal prostates, 140 cases of prostatic hyperplasia, 171 cases of PIA, 84 with prostate cancer (PC), 14 with prostatic intraepithelial neoplasia (PIN) and 10 with bacterial prostatitis. PIA samples were then selected and classified according to the human classification. The presence of PIA lesions surrounding neoplastic areas was then evaluated to establish a morphological transition from normal to preneoplastic and neoplastic tissue. In addition, the expression of PTEN, P53, MDM2 and nuclear androgen receptor (AR) were analyzed in 20 normal samples and 20 PIA lesions by immunohistochemistry and qPCR. All PIA lesions showed variable degrees of mononuclear cell infiltration around the glands and simple atrophy was the most common histopathological feature. PIA was identified between normal glands and PC in 51 (61%) out of the 84 PC samples. PIA lesions were diffusely positive for molecular weight cytokeratin (HMWC). Decreased PTEN and AR gene and protein expression was found in PIA compared to normal samples. Overall, our results strongly suggest that PIA is a frequent lesion associated with PC. Additionally, this finding corroborates the hypothesis that in dogs, as is the case in humans, PIA is a pre neoplastic lesion that has the potential to progress into PC, indicating an alternative mechanism of prostate cancer development in dogs.

Keywords: dog; comparative oncology; inflammation; prostatic atrophy; preneoplastic lesion

1. Introduction

The contribution of inflammation to prostate carcinogenesis is well-known [1,2], occurring through a combination of repeated damage to the genome and increased cell proliferation [3]. The inflammatory process in the prostate gland is associated with a morphologically atrophic epithelium, characterized by a high proliferative index and decreased expression of apoptotic markers [1,3]. De Marzo et al. [4] proposed the term proliferative inflammatory atrophy (PIA) to designate the discrete focus of glandular epithelial proliferation with the morphological appearance of simple atrophy, surrounded by variable degrees of inflammation.

In men, PIA occurs in the peripheral zone of the prostate gland, where prostate cancer (PC) is also more commonly observed [4]. A mononuclear inflammatory infiltrate is frequently associated with PIA lesions [4–6] and these inflammatory cells secrete proteases, as well as mitogenic, antiapoptotic and angiogenic factors in the prostatic microenvironment [7], which ultimately induce epithelial cell atrophy, followed by cell proliferation [5,8]. PIA can occur adjacent to high-grade prostatic intraepithelial neoplasia (HGPIN) and prostate cancer (PC), with some studies highlighting the preneoplastic significance of PIA in human PC development [9,10]. The most accepted theory suggests a progression from PIA to HGPIN and subsequently PC [10,11]. De Marzo et al. [4] previously characterized the morphology and immunophenotype of PIA, describing 34% of focal atrophic lesions surrounding high-grade intraepithelial neoplasia (PIN), and thus, hypothesizing a morphological progression from PIA to HGPIN. PIA has been described in the canine prostate [12–16], although only from a morphological perspective, without analyzing its preneoplastic potential.

Chromosomal abnormalities were previously described in canine PIA lesions. Copy number gain was detected in *CRYGS, ADIPOQ, SST* genes and copy number losses were identified in *CD38, ZNF518B, WDR1, SLC2A9* genes [16]. These results demonstrated the occurrence of chromosomal instability in canine PIA lesions and represent the first evidence of their premalignant potential. Loss of E-cadherin expression has also been reported in canine PIA [15]. Normally dividing prostatic epithelial cells can lose E-cadherin during cell division and re-express the same protein after the replication is completed [17]. Thus, all these previous studies suggest the occurrence of several genomic and protein alterations in canine PIA, that favors its potential preneoplastic lesion.

This study aimed to evaluate the immunophenotype of PIA lesions in dogs, establish a topographic relationship between normal, PIA and PC lesions, as well as to analyze TP53, MDM2, nuclear androgen receptor (AR) and PTEN protein and gene expression in PIA lesions, compared to normal prostates, in order to better characterize its preneoplastic potential in dogs.

2. Results

2.1. Animal Demographic Data

Prostate samples were collected from adult intact dogs of different ages and breeds. The group of dogs with normal prostate consisted of animals with a median age of 6 years (4–8 years), overrepresented by mixed breed dogs (30/50). Dogs with prostatic hyperplasia (PH) had a median age of 9 years (6–12 years) and this group was also overrepresented by mixed breed dogs (87/140). Among the pure breeds, the American pit bull terrier (15/140) was the most affected. The median age of dogs diagnosed with PIA was 9 years (7–14 years), with mixed breed dogs being the most affected (95/171). PC-affected dogs had a median age of 11 years (9–16 years) with the highest prevalence in mixed breed dogs (21/84). The median age of dogs with PIN and bacterial prostatitis was 9 (7–10) and 7 (5–8) years, respectively.

2.2. Morphology and Immunophenotype of PIA

All PIA lesions (N = 171) were surrounded by mononuclear cells, and specifically, by a low inflammatory infiltrate (Figure 1A) in 35.5% of samples (61/171), moderate inflammation in 42.1% of samples (72/171) (Figure 1B) and severe inflammatory infiltrate in 22.4% (38/171) (Figure 1C). Simple atrophy was the most common histopathological feature, observed in 73% of all cases (125/171). A mixed pattern was prevalent in 17.5% of cases (30/171) and post atrophic hyperplasia (PAH) in 9.5% (16/171).

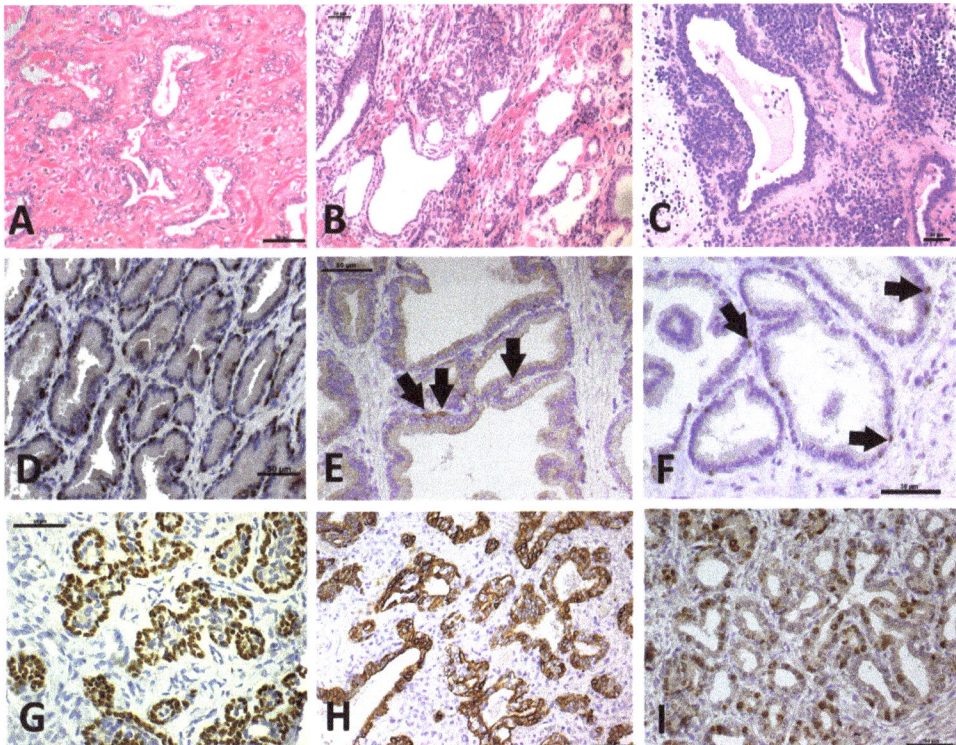

Figure 1. Hematoxylin and eosin (H&E) staining and immunohistochemistry of canine prostate with proliferative inflammatory atrophy (PIA). (**A**): Discrete mononuclear inflammatory infiltrate admixed with multifocal atrophic glands with at least two layers of epithelial cells. (**B**): Moderate mononuclear inflammatory infiltrate and areas of prostatic gland atrophy with cells showing hyperchromatic nuclei and at least two layers of epithelial cells. (**C**): Severe mononuclear inflammatory infiltrate with atrophic epithelial cells showing at least two layers of epithelial cells, hyperchromatic nuclei and evident nucleolus. (**D**): Nuclear p63 expression in a normal prostate gland. There is a discontinuous basal cell layer. (**E**): High molecular wight cytokeratin expression by normal basal cells (arrows). (**F**): Nuclear Ki67 expression by normal basal cells (arrows). There is an absence of Ki67 in luminal epithelial cells and only scatted basal cells express Ki67. (**G**): Diffuse P63 expression by atrophic cells in a PIA. (**H**): Diffuse membranous expression of high molecular weight cytokeratin in a PIA. (**I**): Nuclear positive Ki67 expression by atrophic cells in a PIA lesion. Scale bar = 50 µm.

In normal prostatic tissue, high molecular weight cytokeratin (HMWC) (Figure 1D) and P63 (Figure 1E) were expressed by basal cells, showing a discontinuous layer (100%—20/20). Interestingly, PIA lesions showed a continuous basal cell layer that was positive for P63 (Figure 1D) and HMWC (Figure 1E); with a score of four for both markers. The normal prostatic tissue had no Ki-67 expression in the epithelial luminal cells; only a few basal cells were Ki-67 positive (Figure 1F). The normal prostatic tissue had a mean of 2.2 ± 1.5 Ki-67 positive cells, while PIA lesions (Figure 1I) showed a mean of 45.1 ± 31.8 Ki-67 luminal and basal positive cells with a statistically significant difference between the two groups ($p < 0.0001$) Figure 2.

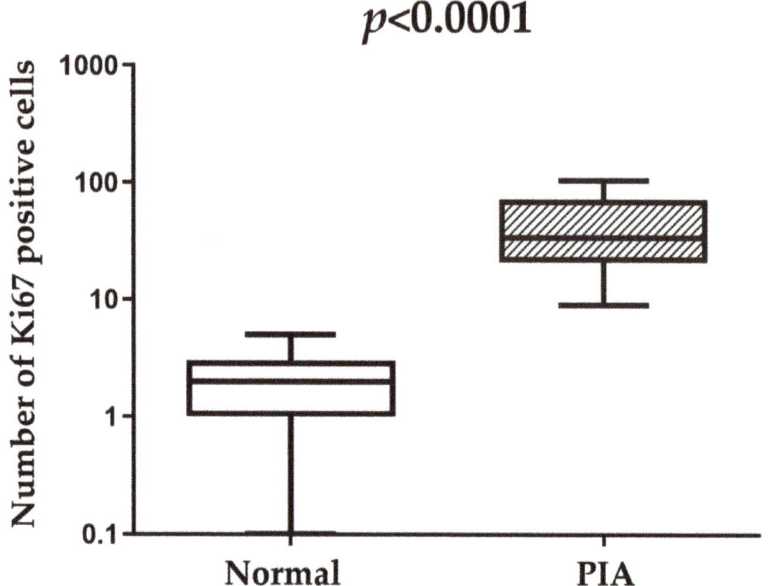

Figure 2. Ki67 protein expression in normal and proliferative inflammatory atrophy (PIA) samples. A high Ki67 index was observed in PIA compared to normal samples, indicating a high proliferative rate in PIA ($p < 0.0001$).

All samples with PIN (14/14) contained adjacent PIA lesions. No PIN foci were observed adjacent to PC lesions. A total of 61% (51/84) of the PC samples had PIA lesions close to the neoplastic areas and an evident histological transition from normal tissue (normal or hyperplastic) to PIA and PC was observed in 21.5% (11/51) of the PC samples (Figure 3).

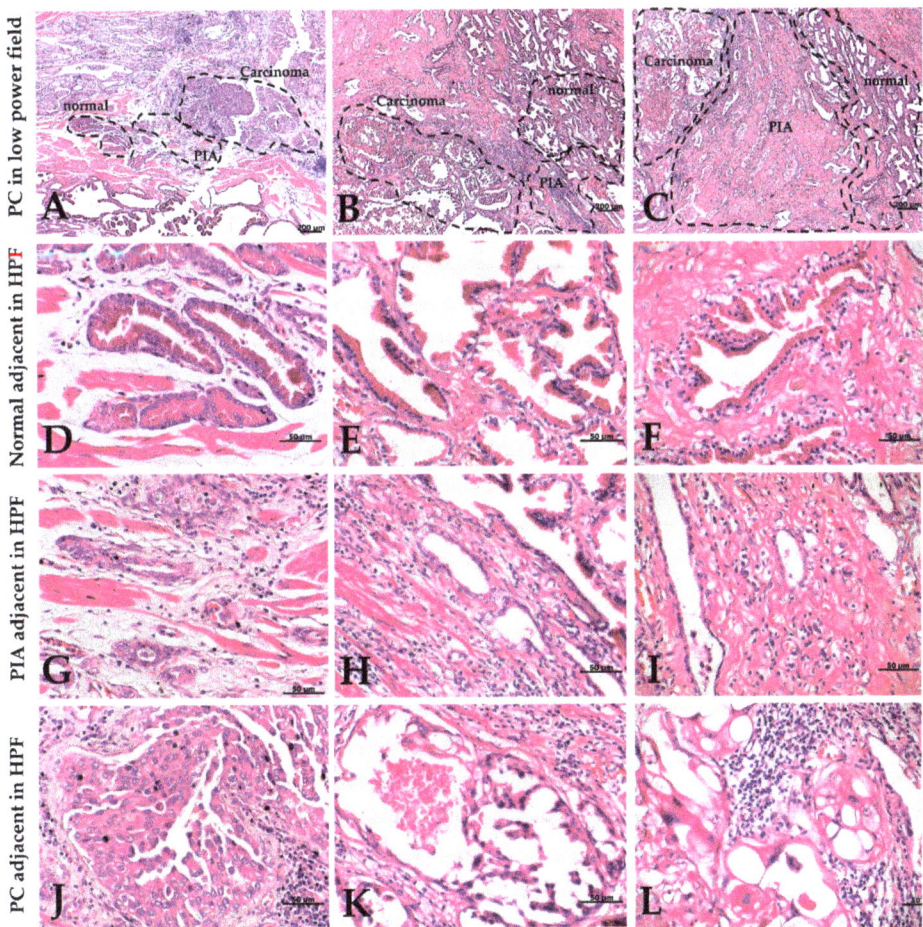

Figure 3. Hematoxylin and eosin (H&E) staining from three different canine prostate carcinomas (PC). (**A–C**): Low power field containing a carcinoma area surrounded by PIA and a normal adjacent area. A clear morphological transition between normal, proliferative inflammatory atrophy (PIA) and PC areas is evident. (**D–F**): High power field (inset of figures (**A–C**), respectively) of adjacent normal tissue. Note the prostatic gland epithelium consisting of columnar and basal epithelial cells. (**G–I**): High power field of surrounding PIA (areas outlined by the plotted areas in figures (**A–C**)). Characteristic simple atrophy with one epithelial layer and hyperchromatic nuclei. (**J–L**): High power field of a carcinoma area. (**J**): Neoplastic cells disposed in nests and moderate pleomorphism. (**K**): Cells present a pseudopapillary shape with presence of degeneration. (**L**): Ballooning neoplastic cells showing degeneration associated intense surrounding inflammation. HPF: High power field.

2.3. Immunohistochemical Features

The immunohistochemical results are presented in Table 1. Both basal and luminal cells were p53 positive, with a nuclear and cytoplasmic expression pattern (Figure 4A). The p53 score was four in nine out of 20 (45%) normal prostatic tissues and three in the remaining cases (11/20). PIA samples were assigned a score of four in 40% of cases (8/20), a score of three in 40% (8/20) and a score of two in 20% (4/20) (Figure 4B). There was no statistical difference in P53 scores between normal and PIA samples. PTEN expression was only observed in the nuclei and cytoplasm of luminal cells Figure 4C. All normal samples showed more than 75% of positive cells (score of four). Two PIA samples had a score of four, 25% (5/20) had a score of three, 25% (5/20) had a score of two and 40% (8/20) had a score of one (Figure 4D). We identified decreased staining in PIA samples compared to normal samples ($p = 0.003$). MDM2 was expressed in the nuclei of luminal and basal cells (Figure 4E) of one normal sample (5%) with a score of four, 30% (6/20) of samples with a score of three and 65% (13/20) with a score of two. Fifteen percent (3/20) of PIA samples showed a score of four, 30% (6/20) showed a score of three and 55% (11/20) showed a score of two (Figure 4F). There was no statistical difference in MDM2 expression between normal and PIA samples.

Table 1. PTEN, P53, MDM2 and AR immunohistochemical expression in canine normal prostate and PIA samples.

	Group	Score				p
		1	2	3	4	
PTEN	Normal	0% (0/20)	0% (0/20)	0% (0/20)	100% (20/20)	$p = 0.003$
	PIA	40% (8/20)	25% (5/20)	25% (5/20)	10% (2/20)	
P53	Normal	0% (0/20)	0% (0/20)	55% (11/20)	45% (9/20)	$p = 0.3568$
	PIA	0% (0/20)	20% (4/20)	40% (8/20)	40% (8/20)	
MDM2	Normal	0% (0/20)	65% (13/20)	30% (6/20)	5% (1/20)	$p = 0.5784$
	PIA	0% (0/20)	55% (11/20)	30% (6/20)	15% (3/20)	
AR	Normal	0% (0/20)	0% (0/20)	0% (0/20)	100% (20/20)	$p = 0.01$
	PIA	0% (0/20)	45% (9/20)	55% (11/20)	0% (0/20)	

Legend: Score 1 = 11–25% positive cells; score 2 = 26–50% positive cells; score 3 = 51–75% positive cells; score 4 = more than 75% positive cells.

AR was positive in luminal cells and occasionally in basal cells (Figure 4G). All normal prostate samples (20/20) showed more than 75% AR-positive cells. The expression pattern was significantly decreased in PIA samples compared to normal samples ($p = 0.01$). Fifty-five percent (11/20) of PIA samples showed a score of three and 45% (9/20) had a score of two (Figure 4H).

2.4. Gene Expression

There was no difference in *TP53* and *MDM2* transcript levels between normal and PIA samples ($p > 0.05$). A positive correlation between *TP53* and *MDM2* transcript levels in normal (R = 0.7754; $p < 0.0001$) and PIA (R = 0.6573; $p = 0.0202$) samples was observed, suggesting that in both cases the increased *TP53* transcript level is correlated with a simultaneous increased expression of *MDM2* (Supplementary Figure S1). No statistical correlations were observed between the remaining comparisons.

PTEN ($P = 0.0307$) and *AR* ($P = 0.0008$) expression was decreased in PIA compared to normal samples (Figure 5). The median relative quantification (RQ) of *AR* in normal samples was 1.8 (0.3–9), while in PIA samples it was 0.7± 0.1–1.6. The median *PTEN* RQ was 1.4 ± 0.3–6.7 and 0.7 ± 0.2–3 in normal and PIA samples, respectively.

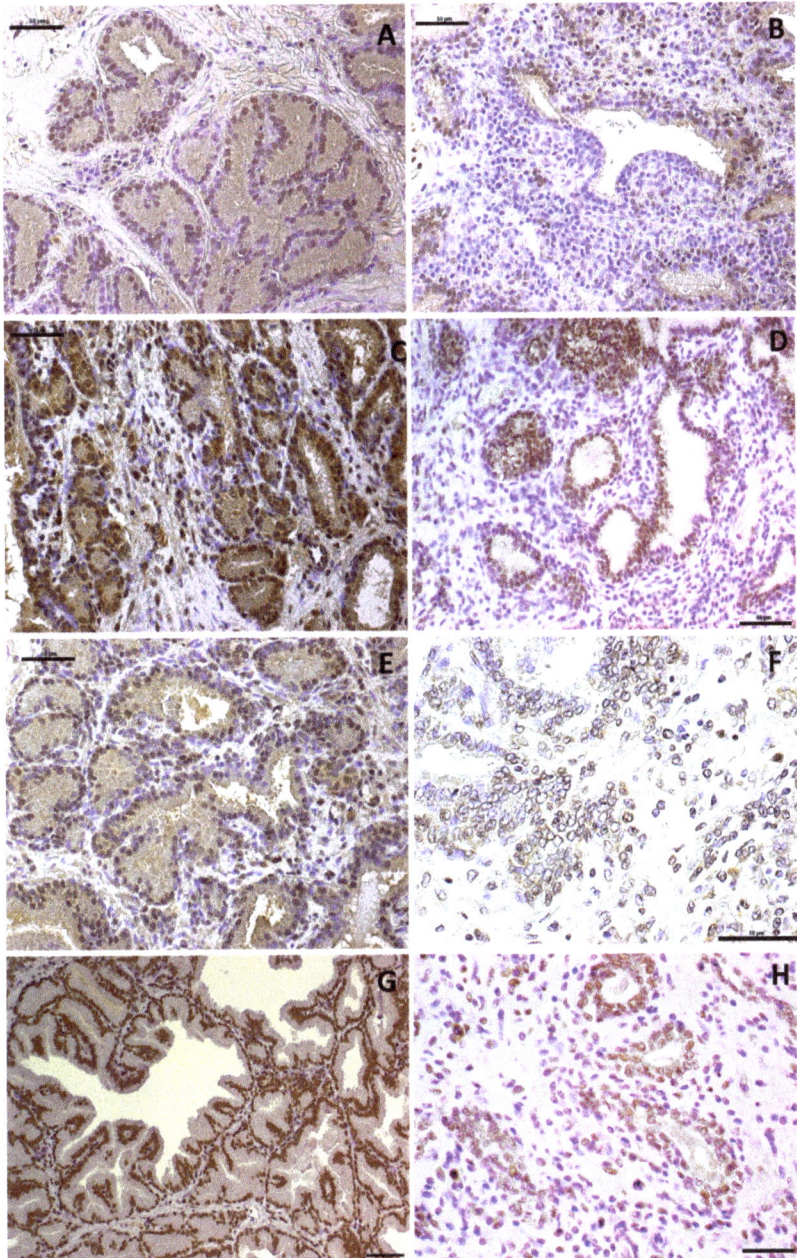

Figure 4. Immunohistochemical staining of the normal canine prostate and canine proliferative inflammatory atrophy (PIA). (**A**): PTEN expression in a canine normal sample: nuclear and cytoplasmic positive staining of luminal epithelial cells. (**B**): PTEN expression in proliferative inflammatory (PIA) samples: atrophic prostatic luminal cells with a lack of PTEN expression. (**C**): Nuclear and cytoplasmic P53 expression in a normal canine prostatic sample and (**D**): PIA lesions. (**E**,**F**): Positive cytoplasmic and nuclear MDM2 expression in normal (**E**) and PIA (**F**) samples. (**G**,**H**): nuclear androgen receptor (AR) expression in a normal prostatic sample and lack of expression in PIA samples (**H**). Counterstaining with Harris hematoxylin, DAB, 20×.

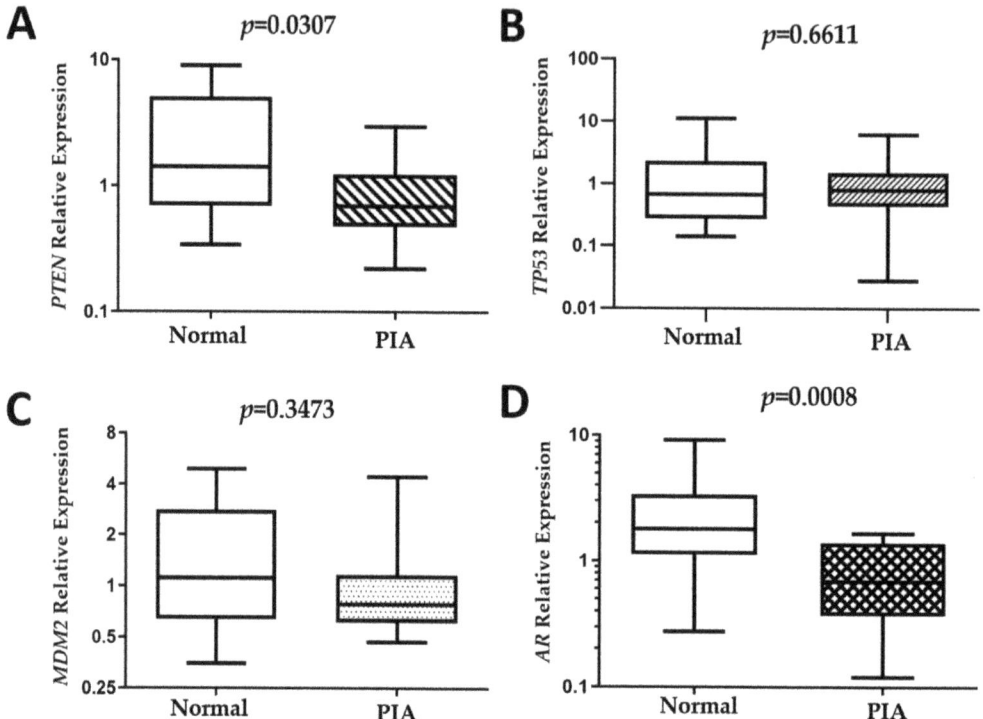

Figure 5. Evaluation of PTEN, TP53, MDM2 and AR gene expression in normal compared to proliferative inflammatory atrophy (PIA) samples. (**A**) Decreased PTEN expression was observed in PIA samples compared to normal samples (**A**). TP53 (**B**) and MDM2 (**C**) transcripts showed no statistically significant differences. AR expression decreased in PIA samples compared to normal samples (**D**).

2.5. Matrix of Multiple Correlation

Using a matrix of multiple correlation among PTEN, P53, MDM2 and AR genes and proteins, a positive strong correlation was demonstrated between *PTEN* and *MDM2* gene expression (Spearman r = 0.6) and between *AR* and *MDM2* gene expression. The matrix of multiple correlation did not show a positive correlation with the respective gene and protein expression (Figure 6).

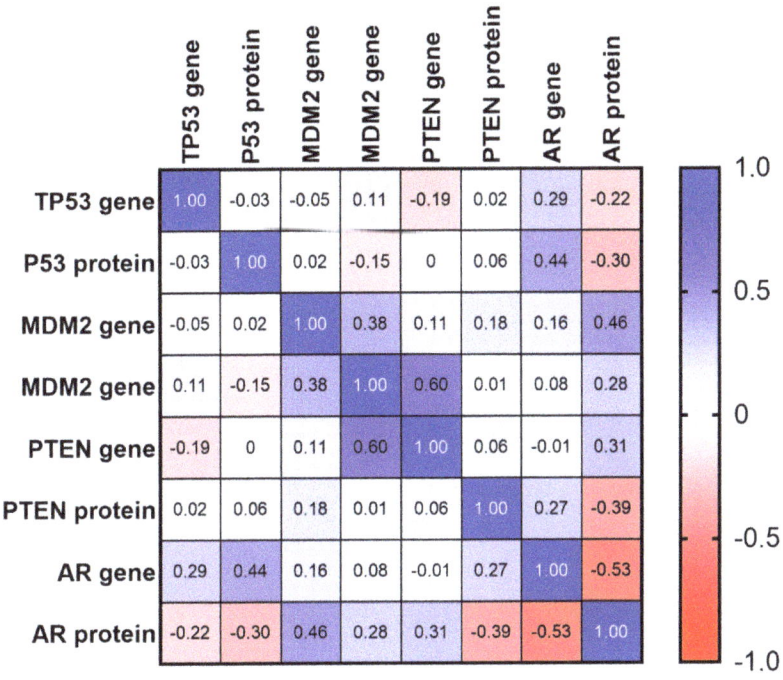

Figure 6. Matrix of multiple correlation among PTEN, TP53, MDM2 and AR gene and protein expression. The red color represents negative correlation and the blue color represents a positive correlation. The color intensity is associated with a strongest correlation. The blue color represents a positive correlation and the red color represents a negative correlation.

2.6. AR Sequencing

All samples were amplified in PCR analysis in the electrophoresis. However, after sequencing, eight out of 20 samples showed a very low sequencing quality and were excluded from the alignment analysis. In the PIA samples aligned in the investigation of *AR* mutations ($N = 12$), six out of 12 samples (50%) showed mutations, with one sample showing several nucleotide alterations (Figure 7).

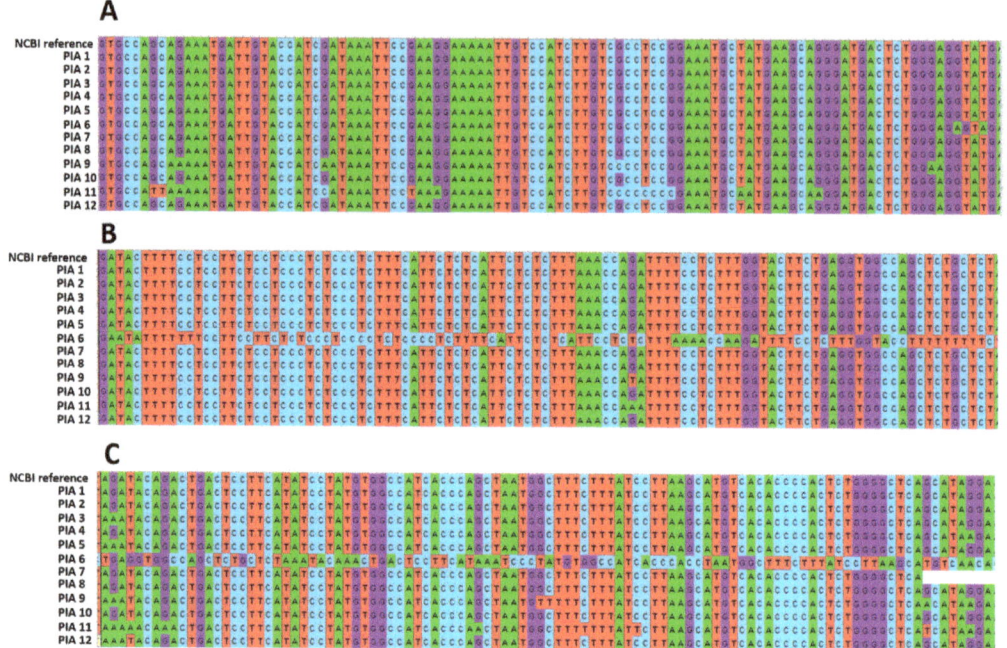

Figure 7. Androgen receptor (AR) sequencing analysis of the 12 proliferative inflammatory atrophy (PIA) samples after alignment. Each figure (**A**–**C**) represents different regions of the AR gene alignment. There are six samples containing mutations, with PIA 6 representing the samples with a higher number of nucleotide modification. (**A**,**B**): sequencing alignment showing mutations in PIA6, PIA9 and PIA 11 samples. C: Sequencing alignment evidencing mutation in the samples PIA3, PIA4, PIA5, PIA6, PIA9, PIA11.

3. Discussion

In human PC, the progression of PIA to high grade-PIN (HGPIN) and PC cancer has been described over recent years, due the role of chronic inflammation in cancer development [1–6]. In dogs, little is known in regard to the carcinogenic process and the progression of preneoplastic lesions to PC. It seems that canine PC develops from androgen-independent cells and most cases of canine PC have been found to be negative for AR [16]. Thus, alterations to the AR receptor are considered to be one of the most important steps in PC development. The previously published literature has focused on studying PIA as a preneoplastic lesion and has shown copy number variations, indicating molecular instability [16]. Although an increased number of studies have been published in recent years, the carcinogenic process of canine PC development is still unclear.

Based on this gap in the literature, our results suggest that PIA is a very common histological lesion in the prostates of intact dogs, with a prevalence of 36.5% in our dataset. On the other side, PIN was present in only 3% of selected cases (14/469). The real incidence of PIN lesions in the canine prostate is controversial. Waters and Bostwick [18,19] reported an incidence of 55–65%, while in two larger studies conducted by Aquilina et al. [20] and Madewell et al. [21], the incidence was low (less than 3%) without any histological association between PIN and PC [20,21]. Thus, PIN lesions have been most likely overestimated or not correctly diagnosed on histological sections and their role in canine prostatic carcinogenesis is still unknown.

The studied population was based on intact dogs, because in Brazil, castration of male dogs is not routinely performed due to cultural factors. Interestingly, we did not see the frequent occurrence of prostate cancer in castrated dogs. This is important in studies evaluating PIA, because castrated dogs show atrophy of the prostate during hormone deprivation and present hormonal atrophy instead of an inflammatory atypical atrophy.

PIA is a common finding in the canine prostate, although its pathogenesis, mechanisms of occurrence and role on prostatic carcinogenesis are unclear. PIA lesions surrounded canine PC areas in 61% of carcinoma samples and, in 21.5% of these cases, the morphological transition between benign tissue, PIA lesions and invasive prostate cancer was evident. Taking into account previously published articles assessing the molecular basis of canine PIA, in addition to the findings from the present study, this result may suggest a potential association between PIA and PC. The morphological transition term was previously introduced by Wang et al. [22] to associate PIA as a preneoplastic lesion in human prostatic pathology. Morphological transition refers to normal, followed by PIA areas colliding to prostatic carcinoma in histological specimens. In our samples, PIA lesions were characterized by a higher proliferative index (shown by Ki67) compared to normal prostates, underlining their high proliferative potential. Previously, our research group investigated the proliferative index of 12 PIA and 18 PC samples, demonstrating a median Ki67 expression level of 54.5 and 366, respectively [23]. Ki67 expression is higher in PIA and PC compared to normal samples. In the carcinogenic process of the human prostate, a dynamic progression from low-PIN, to HGPIN, and finally PC is widely accepted [5,24]. De Marzo et al. [4] then identified PIA lesions as a new precursor of HGPIN. Thus, the carcinogenic process of human and canine prostates might occur through different, a yet unexplored, mechanisms.

Canine PC is characterized by a heterogeneous pattern of cytokeratin expression, as attested by the consensus reached on the intermediate phenotype (luminal markers+ /basal markers+) of canine PC [12,13,25–27]. As evidenced in this study, intermediate cells are also found in greater quantities in PIA lesions, similar to the prevalent cell population of PIA in humans. It is likely that, even in dogs, PIA lesions arise from the proliferation of basal cells stimulated by the inflammatory environment. During normal prostate development and differentiation, basal cells proliferate, reducing the expression of basal cells markers (CK5, HMWC and P63) and overexpressing luminal markers (CK18), until they completely lack the basal cell markers and express luminal markers only [28]. Due to the high proliferative index of PIA lesions, demonstrated by the expression of Ki67, it is most likely that cells are constantly stimulated to divide without further developing into luminal cells. Interestingly, many canine PCs show an intermediate phenotype (concomitant expression of P63 or CK5 or CK8/18) [13,27,29]. Thus, the same cell phenotype may be involved in the development of both PIA and PC in dogs and both lesions may have a common origin in most—but not all—cases.

PTEN, P53, MDM2 and AR dysregulation in human prostate cancer has been widely described [30–32]. PTEN and P53 downregulation plays an important role in cell growth and apoptosis [33]. Furthermore, *TP53* and *MDM2* genes play a dual role during cell proliferation: *MDM2* is an oncogene associated with the proliferation of prostatic cells and *TP53* acts as an *MDM2* inhibitor [30]. During PC progression, tumor cells become androgen-independent and the AR inhibition of human prostate cells promotes cell migration and invasiveness [34].

P53 and MDM2 protein expression were similar in normal prostates and PIA, although a positive correlation between *TP53* and *MDM2* transcript levels in both cases was observed, i.e., higher *TP53* transcripts had higher *MDM2* levels. This correlation could suggest a role of *MDM2* in controlling *TP53* transcripts, as is the case in the human prostate [35]. *TP53* and *MDM2* expression have been widely studied in human cancers and *MDM2* is a negative regulator of the *TP53* transcript [35–37]. In canine PC, upregulation of *MDM2* and downregulation of *TP53* transcript levels have previously been described [38]. Recently, *TP53* copy number loss and *MDM2* gains were described in canine PC, suggesting that P53 and MDM2 could be important drivers in canine prostatic carcinogenesis [16]. Since P53 and MDM2 were not deregulated in PIA samples, and both markers have instead been described as altered in PC [16,38], P53 downregulation and MDM2 overexpression may occur only in the advanced stage of prostatic carcinogenesis, after malignant transformation.

Our results show downregulation of the AR gene and protein levels in PIA samples, as compared to normal prostate samples. The *AR* gene has a key role in prostatic development and maintenance [39]. In dogs, downregulation of the *AR* gene is common in PC development [39]. Previously, our research group demonstrated an *AR* copy number loss in canine PC [16], indicating *AR* loss as an important event in canine PC development. Since we demonstrated the downregulation of *AR*, this gene may be altered in the early stages of neoplastic development and this result supports the hypothesis of the potential progression from PIA to PC. The lack of *AR* expression in canine PC has been widely demonstrated in previous studies [16,25–27,38]. Additionally, our results have demonstrated that PIA with *AR* downregulation may be a precursor lesion to PC. Together, this information reinforces the idea of PIA progression into invasive PC in dogs. However, since we demonstrated an intermediate phenotype for PIA lesions, the AR downregulation may occur due the cell differentiation from basal cells into luminal cells. Thus, this result should be evaluated carefully.

Another gene undergoing downregulation in PIA lesions is *PTEN*. In human PC, PTEN loss is correlated with PC androgen independence [40]. PTEN downregulation is associated with decreased AR levels and these alterations are associated with activation of the anti-apoptotic pathway [40,41]. In knockout *PTEN* mice (Pten+/−), used as models of prostate carcinogenesis, the animals developed spontaneous inflammation with a high number of preneoplastic lesions and carcinomas compared to the controls [42]. Thus, *PTEN* loss is likely important during the carcinogenic process of the prostate gland [42].

Mice are the classical animal models used for studying the development of PIN and PC [43]. However, even though they may experience high-grade inflammation, PIA cannot be induced in these experimental models [42]. Thus, dogs represent a unique opportunity for comparative studies in order to evaluate the role of PIA as preneoplastic lesions in the prostate.

The high proliferative index and the downregulation of *PTEN* and *AR* in PIA lesions may reflect their preneoplastic potential. The PIA premalignant potential is strongly supported by the high mutation index identified in our sequencing analysis. A total of six out of 12 samples (50%) showed mutations, and canine PC is strongly associated with AR decreased expression (even in castrated dogs). Thus, mutations in the AR gene can lead to the progression of PIA to PC in intact dogs.

4. Materials and Methods

4.1. Animals and Experimental Design

This study was performed in accordance with the National and International Recommendations for the Care and Use of Animals (National Research Council) [44]. All procedures have been approved by the Ethics Committee on Animal Use (CEUA) of the Veterinary Teaching Hospital of São Paulo State University (CEUA/UNESP, #0208/2016). All samples were obtained from Brazilian dogs.

4.2. Tissue Selection

Four hundred and sixty nine hematoxylin and eosin (H&E) slides from different canine prostates were selected from the Veterinary Pathology archive of São Paulo State University—UNESP, between 2011 and 2019. Histological evaluation of normal, PIN, and PC samples was performed according to the human WHO Tumors of the Urinary System and Male Genital Organs guidelines [45], which were recently adapted to canine PC [14]. PIA lesions were identified based on morphological features, as described by De Marzo et al. [4]. All samples were from adult intact dogs: 50 normal prostates, 140 prostatic hyperplasia, 171 proliferative inflammatory atrophy (PIA), 84 prostate carcinoma (PC), 14 prostatic intraepithelial neoplasia (PIN) and 10 bacterial prostatitis samples were obtained. The H&E slides from the 84 PC cases were selected to evaluate PIA and PIN surrounding the neoplastic tissue, as previously described [22].

4.3. PIA Morphological Features

The intensity of inflammation was scored according to De Marzo et al. [4] with modifications. Briefly, all H&E PIA slides (N = 171) were evaluated using a numerical scale from 0 to 6, with 0 representing no inflammation, 1 and 2 mild inflammation, 3 and 4 moderate inflammation and 5 and 6 severe inflammation. The key morphological findings to identify PIA lesions in a low power field (5×) were overall hyperchromatic appearance, associated inflammatory infiltrate, loss of papillary architecture and cuboidal cell morphology. In a higher power field, key findings were acini showing at least two layers of epithelial cell, atrophic appearance of cuboidal cells with scant cytoplasm, and in some cases, the presence of evident nucleolus and mitotic figures [4,12].

Lesions were divided into simple atrophy, PAH and mixed pattern (simple atrophy and PAH), according to De Marzo et al. [4]. The presence of dilated glands was also evaluated and referred to as cystic atrophy.

4.4. Immunohistochemistry

Twenty normal prostate samples and 20 PIA paraffin blocks were further selected to investigate both gene and protein expression of specific markers as outlined below. Immunohistochemistry for HMWC, P63, Ki67, PTEN, P53, MDM2 and androgen receptor (AR) was performed with antibodies that had been previously validated to cross-react with canine tissue samples [15,38]. Slide sections were dewaxed in xylol and rehydrated in graded ethanol. For antigen retrieval, the slides were incubated with citrate buffer (pH 6.0) in a pressure cooker (Pascal®; Dako, Carpinteria, CA, USA). The slides were treated with freshly prepared 3% hydrogen peroxide in methanol for 20 min and further washed in Tris-buffered saline. The primary antibodies were diluted to 1:500 for PTEN (Bioss, Massachusets, USA -Bs-0686R), 1:50 for MDM2 (Abcam, Cambridge, UK-ab38618), 1:50 for P53 (Santa Cruz Biotecnology, sc-75366), 1:100 for AR (Abcam, Cambridge, UK-ab77557), 1:300 for HMWC (DakoCytomation, clone: 34βE12), 1:100 for P63 (DakoCytomation, clone 4A4), 1:300 for Pan-cytokeratin (Invitrogen), 1:50 for Ki-67 (DakoCytomation, clone: MIB1) with an overnight incubation at 4 °C. A polymer system (Envision, Dako, Carpinteria, CA, USA) was applied as a secondary antibody conjugated to peroxidase and 3′-diaminobenzidine tetrahydrochloride (DAB1, Dako, Carpinteria, CA, USA) was used as the chromogen, for 5 min, followed by Harris hematoxylin counterstain.

Negative controls were performed for all antibodies by omitting the primary antibody, replacing them with Tris-buffered saline and also with an iso-type matched immunoglobulin according to Hewitt et al. [46]. Normal prostate was used as a positive control for AR, p63, HMWC and PTEN. A normal lymph node was used as a positive control for p53, MDM2 and Ki67 according to the Human Protein Atlas recommendation (www.proteinatlas.org, accessed on 15 November 2020).

4.5. Immunohistochemical Score

As regards P53, MDM2, AR, PTEN, P63 and HMWC, samples were scored based on the percentage of positive cells for each antibody with the following scores: 1–10% positive cells (score 0), 11–25% positive cells (score 1); 26–50% positive cells (score 2); 51–75% positive cells (score 3); and more than 76% positive cells (score 4). The immunohistochemical results were independently interpreted by three investigators (C.E.F.A., P.E.K. and R.L.A.). The distribution (basal or luminal cells) of P63 and HMWC was also evaluated. Ten high power fields were used to establish the final IHC score. Ki-67 analysis was made by counting the number of positive cells per total of 1000 cells.

4.6. Laser-Capture Microdissection and qPCR

For mRNA extraction, microdissection was performed using a Leica AS LMD laser-capture system (Leica Microsystems, Wetslzar, Germany), according to [47] with modifications. mRNA analysis was performed with 20 normal and 20 PIA fresh frozen tissues (same animal prostates as those used for immunohistochemistry). Briefly, 10 µm tissue sections were cut in a cryostat (Leica Microsystems) and mounted on Nuclease and Human Nucleic acid free PEN-membrane slides 2.0 µm (MembraneSlides, Leica Microsystems). The slides were stained (H&E) and once air-dried, PIA areas were microdissected [47]. mRNA was extracted with RNeasy Micro Kit (Qiagen, Hilden. Germany) according to the manufacturer instructions. cDNA was synthesized in a final volume of 20 µL, and each reaction contained 1 µg of total RNA treated with DNAse I (Life Technologies, Rockville, MD, USA), 200 U of SuperScript III reverse transcriptase (Life Technologies), 4 µL of 5X SuperScript First-Strand Buffer, 1 µL of each dNTP at 10 mM (Life Technologies), 1 µL of Oligo-(dT)18 (500 ng/µL) (Life Technologies), 1 µL of random hexamers (100 ng/µL) (Life Technologies), and 1 µL of 0.1 M DTT (Life Technologies). Reverse transcription was performed for 60 min at 50 °C, and the enzyme was subsequently inactivated for 15 min at 70 °C. cDNA was stored at −80 °C.

TP53, MDM2, AR and PTEN and the endogenous genes (Table 2) were conducted in a total volume of 10 µL containing Power SYBR Green PCR Master Mix (Applied Biosystems; Foster City, CA, USA), 1 µL of cDNA (1:10) and 0.3 µM of each primer. The reactions were performed in triplicate in 384-well plates using QuantStudio 12K Flex Thermal Cycler equipment (Applied Biosystems; Foster City, CA, USA). A dissociation curve was included in all experiments to determine the PCR product specificity. Relative gene expression was quantified using the 2-ΔΔCT method [48].

Table 2. Forward and reverse primer sequence for each gene used in RT-qPCR analyses.

Gene Symbol	Location	Primer Sequences
AR	Chromosome 24	F: 5'-CGCCCCTGACCTGGTTT-3' R: 5'-GGCTGTACATCCGGGACTTG-3'
PTEN	Chromosome 26	F: 5'-CGACGGGAAGACAAGTTCATG-3' R: 5'-TCACCGCACACAGGCAAT-3'
MDM2	Chromosome 10	F: 5'-GGGCCCCTTCGTGAGAATTG-3' R: 5'-GGTGTGGCTTTTCTCAGGGATT-3'
TP53	Chromosome 5	F: 5'-GAACGCTGCTCTGACAGTAGTGA-3' R: 5'-CCCGCAAATTTCCTTCCA-3'
HPRT	Chromosome X	F: 5'-AGCTTGCTGGTGAAAAGGAC-3' R: 5'-TTATAGTCAAGGGCATATCC-3'

F: forward; R: reverse.

4.7. DNA Extraction and Sequencing

DNA extraction of 20 PIA tissues used in mRNA expression was performed using DNeasy Blood and Tissue Kits (Ambion, Life Technologies, MA, USA), according to manufacturer's instructions. The amplification of the androgen receptor (AR) DNA-binding domain was performed in a Veriti 96-Well Thermal Cycler (Thermo Fisher Scientific, MA, USA). The methodology was performed according to the previously described by Rivera-Calderon et al. [49]. Briefly, 5 µL of 10× PfuUltra II, 0.5 µL of 100 µM dNTP, 1 µL of each oligonucleotide in 10 µM of each primer, 1 µL of genomic DNA (100 ng/µL) and 1 µL PfuUltra II fusion HS DNA polymerase (Agilent, CA, USA). The primers used were previously published by Lai et al. [50].

After of the amplification, the AR DNA-binding domain was analyzed in 2% agarose gel stained with Neotaq Brilliant Plus DNA Stain (Neobio, Brazil). Then, the PCR products were cut from the agarose gel were and the DNA sequencing was evaluated with a BigDye™ Terminator v 3.1 Cycle Sequencing Kit version 3.1, according to the manufacturer's instructions (Applied Biosystems, Thermo Fisher Scientific, MA, USA). The RT-amplicons were directly sequenced at ABI 3500 (Applied Biosystems, Thermo Fisher Scientific, MA, USA).

The resulting nucleotide sequences were compared to data of the NBCI (National Center for Biotechnology Information, Bethesda, MD, USA) (gi:6578766).

Sequences analysis was performed in the MEGA Software (www.megasoftware.net, accessed on 17 February 2021) [51], a FASTA analysis was performed, and the sequences aligned with the reference gene from NCBI (gi:6578766), according to the previous literature description [50]. Then, sequences with low quality were excluded.

4.8. Data Analysis

Statistical analyses were performed using GraphPad Prism v.5.0 (GraphPad Software Inc., La Jolla, CA, USA). Kruskal–Wallis or Mann–Whitney U tests was applied to compare TP53, MDM2, AR and PTEN transcription levels between normal and PIA samples. The chi-square exact test was used to evaluate difference in the immunohistochemical expression between normal and PIA samples.

4.9. Data Availability

The authors confirm that the data supporting the findings of this study are available within the article and/or its Supplementary Materials.

5. Conclusions

Canine PIA is a common lesion in the dog prostate and shares many morphological similarities with human PIA. Our results strongly suggest that PIA is a potential pre neoplastic lesion and is associated with the progression to PC.

Supplementary Materials: Supplementary materials can be found at https://www.mdpi.com/article/10.3390/cancers13081887/s1. Figure S1. Canine prostate gland with atrophic lesion. The slides were stained with hematoxylin and eosin staining and tissue area with inflammatory atrophic gland were selected (A). Then, lase microdissection was performed to obtain only the epithelial portion of these atrophic glands (B). Hematoxylin and eosin staining, 200×.

Author Contributions: Conceptualization, G.d.G.F. and B.P.; methodology, G.d.G.F., B.P., P.E.K., P.d.F.L., C.E.F.-A.; software, V.M.G. and C.E.F.-A.; validation, C.E.F.-A. and R.L.-A.; formal analysis, G.d.G.F., B.P., P.E.K., V.M.B.D.d.M., C.P., R.L.-A., C.E.F.-A.; investigation, G.d.G.F., B.P., P.E.K. and C.E.F.-A.; resources, C.E.F.-A. and R.L.-A.; data curation, P.d.F.L. and C.E.F.-A.; writing—original draft preparation, G.d.G.F., B.P. and C.E.F.-A.; writing—review and editing, G.d.G.F., B.P., P.E.K., P.d.F.L., V.M.B.D.d.M., V.M.G., C.P., R.L.-A and C.E.F.-A.; visualization, G.d.G.F., B.P., P.E.K., P.d.F.L., V.M.B.D.d.M., V.M.G., C.P., R.L.-A., C.E.F.-A.; supervision, C.E.F.-A.; project administration, C.E.F.-A. and R.L.-A.; funding acquisition, C.E.F.-A. All authors have read and agreed to the published version of the manuscript.

Funding: This research was funded by Sao Paulo Research Foundation (FAPESP), grants numbers #2019/00497-9 and #2012/18426-1.

Institutional Review Board Statement: The study was conducted according to the guidelines of the Declaration of Helsinki, and approved by the Ethics Committee of the Veterinary Teaching Hospital of São Paulo State University (CEUA/UNESP, #0208/2016).

Informed Consent Statement: Not applicable.

Data Availability Statement: The data presented in this study are available in Supplementary Materials.

Acknowledgments: The authors would like to thank Marcio Carvalho for his help during q-PCR experiments. The content of this manuscript has been published in part of the thesis of Carlos Eduardo Fonseca Alves.

Conflicts of Interest: The authors declare no conflict of interest. The funders had no role in the design of the study; in the collection, analyses, or interpretation of data; in the writing of the manuscript, or in the decision to publish the results.

References

1. Palapattu, G.S.; Sutcliffe, S.; Bastian, P.J.; Platz, E.A.; De Marzo, A.M.; Isaacs, W.B.; Nelson, W.G. Prostate carcinogenesis and inflammation: Emerging insights. *Carcinogenesis* 2005, 26, 1170–1181. [CrossRef]
2. de Bono, J.S.; Guo, C.; Gurel, B.; De Marzo, A.M.; Sfanos, K.S.; Mani, R.S.; Gil, J.; Drake, C.G.; Alimonti, A. Prostate carcinogenesis: Inflammatory storms. *Nat. Rev. Cancer* 2020, 20, 455–469. [CrossRef]
3. Montironi, R.; Mazzucchelli, R.; Lopez-Beltran, A.; Cheng, L.; Scarpelli, M. Mechanisms of disease: High-grade prostatic intraepithelial neoplasia and other proposed preneoplastic lesions in the prostate. *Nat. Clin. Pract. Urol.* 2007, 4, 321–332. [CrossRef]
4. De Marzo, A.M.; Marchi, V.L.; Epstein, J.I.; Nelson, W.G. Proliferative inflammatory atrophy of the prostate: Implications for prostatic carcinogenesis. *Am. J. Pathol.* 1999, 155, 1985–1992. [CrossRef]
5. De Marzo, A.M.; Platz, E.A.; Epstein, J.I.; Ali, T.; Billis, A.; Chan, T.Y.; Cheng, L.; Datta, M.; Egevad, L.; Ertoy-Baydar, D.; et al. A working group classification of focal prostate atrophy lesions. *Am. J. Surg Pathol.* 2006, 30, 1281–1291. [CrossRef] [PubMed]
6. Gao, Y.; Wei, L.; Wang, C.; Huang, Y.; Li, W.; Li, T.; Mo, C.; Qin, H.; Zhong, X.; Wang, Y.; et al. Chronic prostatitis alters the prostatic microenvironment and accelerates preneoplastic lesions in C57BL/6 mice. *Biol. Res.* 2019, 52, 30. [CrossRef] [PubMed]
7. Wang, W.; Bergh, A.; Damber, J.E. Cyclooxygenase-2 expression correlates with local chronic inflammation and tumor neovascularization in human prostate cancer. *Clin. Cancer Res.* 2005, 11, 3250–3256. [CrossRef] [PubMed]
8. Ashok, A.; Keener, R.; Rubenstein, M.; Stookey, S.; Bajpai, S.; Hicks, J.; Alme, A.K.; Drake, C.G.; Zheng, Q.; Trabzonlu, L.; et al. Consequences of interleukin 1β-triggered chronic inflammation in the mouse prostate gland: Altered architecture associated with prolonged CD4. *Prostate* 2019, 79, 732–745. [CrossRef]
9. Sugar, L.M. Inflammation and prostate cancer. *Can. J. Urol.* 2006, 13 (Suppl. 1), 46–47.
10. Karaivanov, M.; Todorova, K.; Kuzmanov, A.; Hayrabedyan, S. Quantitative immunohistochemical detection of the molecular expression patterns in proliferative inflammatory atrophy. *J. Mol. Histol.* 2007, 38, 1–11. [CrossRef]
11. De Marzo, A.M.; Platz, E.A.; Sutcliffe, S.; Xu, J.; Grönberg, H.; Drake, C.G.; Nakai, Y.; Isaacs, W.B.; Nelson, W.G. Inflammation in prostate carcinogenesis. *Nat. Rev. Cancer* 2007, 7, 256–269. [CrossRef]
12. Palmieri, C.; Story, M.; Lean, F.Z.X.; Akter, S.H.; Grieco, V.; De Marzo, A.M. Diagnostic Utility of Cytokeratin-5 for the Identification of Proliferative Inflammatory Atrophy in the Canine Prostate. *J. Comp. Pathol.* 2018, 158, 1–5. [CrossRef]
13. Fonseca-Alves, C.E.; Rodrigues, M.M.; de Moura, V.M.; Rogatto, S.R.; Laufer-Amorim, R. Alterations of C-MYC, NKX3.1, and E-cadherin expression in canine prostate carcinogenesis. *Microsc. Res. Tech.* 2013, 76, 1250–1256. [CrossRef]
14. Palmieri, C.; Lean, F.Z.; Akter, S.H.; Romussi, S.; Grieco, V. A retrospective analysis of 111 canine prostatic samples: Histopathological findings and classification. *Res. Vet. Sci.* 2014, 97, 568–573. [CrossRef] [PubMed]
15. Fonseca-Alves, C.E.; Kobayashi, P.E.; Rivera-Calderón, L.G.; Laufer-Amorim, R. Evidence of epithelial-mesenchymal transition in canine prostate cancer metastasis. *Res. Vet. Sci.* 2015, 100, 176–181. [CrossRef] [PubMed]
16. Laufer-Amorim, R.; Fonseca-Alves, C.E.; Villacis, R.A.R.; Linde, S.A.D.; Carvalho, M.; Larsen, S.J.; Marchi, F.A.; Rogatto, S.R. Comprehensive Genomic Profiling of Androgen-Receptor-Negative Canine Prostate Cancer. *Int. J. Mol. Sci.* 2019, 20, 1555. [CrossRef]
17. Fonseca-Alves, C.D.; Busso, A.F.; Silveira, S.M.; Rogatto, S.R.; Laufer-Amorim, R. Genomic gains in prostatic carcinoma and proliferative inflammatory atrophy in dogs. *Cancer Res.* 2012, 72, 5260. [CrossRef]
18. Waters, D.J.; Bostwick, D.G. Prostatic intraepithelial neoplasia occurs spontaneously in the canine prostate. *J. Urol.* 1997, 157, 713–716. [CrossRef]
19. Waters, D.J.; Bostwick, D.G. The canine prostate is a spontaneous model of intraepithelial neoplasia and prostate cancer progression. *Anticancer Res.* 1997, 17, 1467–1470. [PubMed]
20. Aquilina, J.W.; McKinney, L.; Pacelli, A.; Richman, L.K.; Waters, D.J.; Thompson, I.; Burghardt, W.F.; Bostwick, D.G. High grade prostatic intraepithelial neoplasia in military working dogs with and without prostate cancer. *Prostate* 1998, 36, 189–193. [CrossRef]
21. Madewell, B.R.; Gandour-Edwards, R.; DeVere White, R.W. Canine prostatic intraepithelial neoplasia: Is the comparative model relevant? *Prostate* 2004, 58, 314–317. [CrossRef]
22. Wang, W.; Bergh, A.; Damber, J.E. Morphological transition of proliferative inflammatory atrophy to high-grade intraepithelial neoplasia and cancer in human prostate. *Prostate* 2009, 69, 1378–1386. [CrossRef] [PubMed]
23. Fonseca-Alves, C.E.; Kobayashi, P.E.; Palmieri, C.; Laufer-Amorim, R. Investigation of c-KIT and Ki67 expression in normal, preneoplastic and neoplastic canine prostate. *BMC Vet. Res.* 2017, 13, 380. [CrossRef] [PubMed]
24. Woenckhaus, J.; Fenic, I. Proliferative inflammatory atrophy: A background lesion of prostate cancer? *Andrologia* 2008, 40, 134–137. [CrossRef] [PubMed]
25. Leav, I.; Schelling, K.H.; Adams, J.Y.; Merk, F.B.; Alroy, J. Role of canine basal cells in prostatic post natal development, induction of hyperplasia, sex hormone-stimulated growth; and the ductal origin of carcinoma. *Prostate* 2001, 47, 149–163. [CrossRef]
26. Romanucci, M.; Frattone, L.; Ciccarelli, A.; Bongiovanni, L.; Malatesta, D.; Benazzi, C.; Brachelente, C.; Della Salda, L. Immunohistochemical expression of heat shock proteins, p63 and androgen receptor in benign prostatic hyperplasia and prostatic carcinoma in the dog. *Vet. Comp. Oncol.* 2016, 14, 337–349. [CrossRef]
27. Akter, S.H.; Lean, F.Z.; Lu, J.; Grieco, V.; Palmieri, C. Different Growth Patterns of Canine Prostatic Carcinoma Suggests Different Models of Tumor-Initiating Cells. *Vet. Pathol.* 2015, 52, 1027–1033. [CrossRef]

28. Leis-Filho, A.F.; Fonseca-Alves, C.E. Anatomy, histology, and physiology of the canine prostate gland. In *Veterinary Anatomy and Physiology*; IntechOpen: London, UK, 2018; pp. 47–67. [CrossRef]
29. Fonseca-Alves, C.E.; Kobayashi, P.E.; Rivera Calderón, L.G.; Felisbino, S.L.; Rinaldi, J.C.; Drigo, S.A.; Rogatto, S.R.; Laufer-Amorim, R. Immunohistochemical panel to characterize canine prostate carcinomas according to aberrant p63 expression. *PLoS ONE* **2018**, *13*, e0199173. [CrossRef]
30. Gupta, A.; Behl, T.; Heer, H.R.; Deshmukh, R.; Sharma, P.L. Mdm2-P53 Interaction Inhibitor with Cisplatin Enhances Apoptosis in Colon and Prostate Cancer Cells In-Vitro. *Asian Pac. J. Cancer Prev.* **2019**, *20*, 3341–3351. [CrossRef]
31. McClurg, U.L.; Chit, N.C.T.H.; Azizyan, M.; Edwards, J.; Nabbi, A.; Riabowol, K.T.; Nakjang, S.; McCracken, S.R.; Robson, C.N. Molecular mechanism of the TP53-MDM2-AR-AKT signalling network regulation by USP12. *Oncogene* **2018**, *37*, 4679–4691. [CrossRef]
32. Chopra, H.; Khan, Z.; Contreras, J.; Wang, H.; Sedrak, A.; Zhu, Y. Activation of p53 and destabilization of androgen receptor by combinatorial inhibition of MDM2 and MDMX in prostate cancer cells. *Oncotarget* **2018**, *9*, 6270–6281. [CrossRef] [PubMed]
33. Bouali, S.; Chrétien, A.S.; Ramacci, C.; Rouyer, M.; Marchal, S.; Galenne, T.; Juin, P.; Becuwe, P.; Merlin, J.L. P53 and PTEN expression contribute to the inhibition of EGFR downstream signaling pathway by cetuximab. *Cancer Gene Ther.* **2009**, *16*, 498–507. [CrossRef]
34. Yang, W.; Wang, K.; Ma, J.; Hui, K.; Lv, W.; Ma, Z.; Huan, M.; Luo, L.; Wang, X.; Li, L.; et al. Inhibition of Androgen Receptor Signaling Promotes Prostate Cancer Cell Migration via Upregulation of Annexin A1 Expression. *Arch. Med. Res.* **2020**. [CrossRef]
35. Nakano, M.; Taura, Y.; Inoue, M. Protein expression of Mdm2 and p53 in hyperplastic and neoplastic lesions of the canine circumanal gland. *J. Comp. Pathol.* **2005**, *132*, 27–32. [CrossRef]
36. Mayo, L.D.; Donner, D.B. The PTEN, Mdm2, p53 tumor suppressor-oncoprotein network. *Trends Biochem. Sci.* **2002**, *27*, 462–467. [CrossRef]
37. Pant, V.; Lozano, G. Dissecting the p53-Mdm2 feedback loop in vivo: Uncoupling the role in p53 stability and activity. *Oncotarget* **2014**, *5*, 1149–1156. [CrossRef]
38. Rivera-Calderón, L.G.; Fonseca-Alves, C.E.; Kobayashi, P.E.; Carvalho, M.; Drigo, S.A.; de Oliveira Vasconcelos, R.; Laufer-Amorim, R. Alterations in PTEN, MDM2, TP53 and AR protein and gene expression are associated with canine prostate carcinogenesis. *Res. Vet. Sci.* **2016**, *106*, 56–61. [CrossRef] [PubMed]
39. Edwards, J.; Krishna, N.S.; Grigor, K.M.; Bartlett, J.M. Androgen receptor gene amplification and protein expression in hormone refractory prostate cancer. *Br. J. Cancer* **2003**, *89*, 552–556. [CrossRef] [PubMed]
40. Sun, C.; Dobi, A.; Mohamed, A.; Li, H.; Thangapazham, R.L.; Furusato, B.; Shaheduzzaman, S.; Tan, S.H.; Vaidyanathan, G.; Whitman, E.; et al. TMPRSS2-ERG fusion, a common genomic alteration in prostate cancer activates C-MYC and abrogates prostate epithelial differentiation. *Oncogene* **2008**, *27*, 5348–5353. [CrossRef]
41. Attard, G.; Swennenhuis, J.F.; Olmos, D.; Reid, A.H.; Vickers, E.; A'Hern, R.; Levink, R.; Coumans, F.; Moreira, J.; Riisnaes, R.; et al. Characterization of ERG, AR and PTEN gene status in circulating tumor cells from patients with castration-resistant prostate cancer. *Cancer Res.* **2009**, *69*, 2912–2918. [CrossRef] [PubMed]
42. Burcham, P.C.; Raso, A.; Henry, P.J. Airborne acrolein induces keratin-8 (Ser-73) hyperphosphorylation and intermediate filament ubiquitination in bronchiolar lung cell monolayers. *Toxicology* **2014**, *319*, 44–52. [CrossRef] [PubMed]
43. Ittmann, M.; Huang, J.; Radaelli, E.; Martin, P.; Signoretti, S.; Sullivan, R.; Simons, B.W.; Ward, J.M.; Robinson, B.D.; Chu, G.C.; et al. Animal models of human prostate cancer: The consensus report of the New York meeting of the Mouse Models of Human Cancers Consortium Prostate Pathology Committee. *Cancer Res.* **2013**, *73*, 2718–2736. [CrossRef] [PubMed]
44. Committee for the Update of the Guide for the Care and Use of Laboratory Animals. *Guide for the Care and Use of Laboratory Animals*; The National Academies Press: Washington, DC, USA, 2011.
45. Eble, J.N.; Sauter, G.; Epstein, J.I.; Sesterhenn, I.A. Pathology and Genetics of Tumours of the Urinary System and Male Genital Organs. In *Classification of Tumours*; WHO, Ed.; WHO: Geneva, Switzerland, 2004; pp. 255–257.
46. Hewitt, S.M.; Baskin, D.G.; Frevert, C.W.; Stahl, W.L.; Rosa-Molinar, E. Controls for immunohistochemistry: The Histochemical Society's standards of practice for validation of immunohistochemical assays. *J. Histochem. Cytochem.* **2014**, *62*, 693–697. [CrossRef]
47. Ito, S.; Ohga, T.; Saeki, H.; Nakamura, T.; Watanabe, M.; Tanaka, S.; Kakeji, Y.; Maehara, Y. p53 mutation profiling of multiple esophageal carcinoma using laser capture microdissection to demonstrate field carcinogenesis. *Int. J. Cancer* **2005**, *113*, 22–28. [CrossRef]
48. Livak, K.J.; Schmittgen, T.D. Analysis of relative gene expression data using real-time quantitative PCR and the 2(-Delta Delta C(T)) Method. *Methods* **2001**, *25*, 402–408. [CrossRef] [PubMed]
49. Rivera-Calderón, L.G.; Fonseca-Alves, C.E.; Kobayashi, P.E.; Carvalho, M.; Vasconcelos, R.O.; Laufer-Amorim, R. p-mTOR, p-4EBP-1 and eIF4E expression in canine prostatic carcinoma. *Res. Vet. Sci.* **2019**, *122*, 86–92. [CrossRef]
50. Lai, C.L.; van den Ham, R.; Mol, J.; Teske, E. Immunostaining of the androgen receptor and sequence analysis of its DNA-binding domain in canine prostate cancer. *Vet. J.* **2009**, *181*, 256–260. [CrossRef]
51. Kumar, U.; Stecher, G.; Li, M.; Knyaz, C.; Tamura, K. MEGA X: Molecular Evolutionary Genetics Analysis across computing platforms. *Mol. Biol. Evol.* **2018**, *35*, 1547–1549. [CrossRef]

Review

Prognostic and Predictive Factors in Advanced Urothelial Carcinoma Treated with Immune Checkpoint Inhibitors: A Review of the Current Evidence

Sara Elena Rebuzzi [1,2,*,†], Giuseppe Luigi Banna [3,†], Veronica Murianni [4], Alessandra Damassi [5], Emilio Francesco Giunta [6], Filippo Fraggetta [7], Ugo De Giorgi [8], Richard Cathomas [9], Pasquale Rescigno [10], Matteo Brunelli [11] and Giuseppe Fornarini [4]

1. Medical Oncology, Ospedale San Paolo, 17100 Savona, Italy
2. Department of Internal Medicine and Medical Specialties (Di.M.I.), University of Genova, 16132 Genova, Italy
3. Candiolo Cancer Institute, FPO-IRCCS, 10060 Candiolo, Italy; gbanna@yahoo.com
4. Medical Oncology Unit 1, IRCCS Ospedale Policlinico San Martino, 16132 Genova, Italy; murianni.veronica@gmail.com (V.M.); giuseppe.fornarini@hsanmartino.it (G.F.)
5. Academic Unit of Medical Oncology, IRCCS Ospedale Policlinico San Martino, 16132 Genova, Italy; alessandra.damassi@gmail.com
6. Department of Precision Medicine, Università Degli Studi della Campania Luigi Vanvitelli, 80131 Naples, Italy; emiliofrancescogiunta@gmail.com
7. Pathology Department, Ospedale "Gravina", 95041 Caltagirone, Italy; filippofra@hotmail.com
8. Department of Medical Oncology, IRCCS Istituto Romagnolo per lo Studio dei Tumori (IRST) "Dino Amadori", 47014 Meldola, Italy; ugo.degiorgi@irst.emr.it
9. Division of Oncology/Hematology, Kantonsspital Graubünden, 7000 Chur, Switzerland; richard.cathomas@ksgr.ch
10. Interdisciplinary Group for Translational Research and Clinical Trials, Urogenital Cancers GIRT-Uro, Candiolo Cancer Institute, FPO-IRCCS, Candiolo, 10060 Turin, Italy; pasquale.rescigno@ircc.it
11. Department of Diagnostics and Public Health, Section of Pathology, University and Hospital Trust of Verona, 37134 Verona, Italy; matteo.brunelli@univr.it
* Correspondence: saraelena89@hotmail.com
† These authors equally contributed as first co-authors.

Simple Summary: Despite immune checkpoint inhibitors' (ICIs) improved overall survival in urothelial carcinoma patients, only a minority of them benefit from immunotherapy. Therefore, there is an unmet clinical need to identify biomarkers which are useful to select the patients who are most likely to respond to ICIs. This review describes the prognostic and predictive role, and potential clinical applicability, of patient- and tumour-related factors. These factors include new molecular classes, tumour mutational burden, mutational signatures, circulating tumour DNA, programmed death-ligand 1, inflammatory indices and clinical characteristics. This summary may help clinicians to assess patients who are considered for ICI treatment, and may drive further prospective research on these biomarkers.

Abstract: In recent years, the treatment landscape of urothelial carcinoma has significantly changed due to the introduction of immune checkpoint inhibitors (ICIs), which are the standard of care for second-line treatment and first-line platinum-ineligible patients with advanced disease. Despite the overall survival improvement, only a minority of patients benefit from this immunotherapy. Therefore, there is an unmet need to identify prognostic and predictive biomarkers or models to select patients who will benefit from ICIs, especially in view of novel therapeutic agents. This review describes the prognostic and predictive role, and clinical readiness, of clinical and tumour factors, including new molecular classes, tumour mutational burden, mutational signatures, circulating tumour DNA, programmed death-ligand 1, inflammatory indices and clinical characteristics for patients with urothelial cancer treated with ICIs. A classification of these factors according to the levels of evidence and grades of recommendation currently indicates both a prognostic and predictive value for ctDNA and a prognostic relevance only for concomitant medications and patients' characteristics.

Keywords: advanced urothelial carcinoma; immune checkpoint inhibitor; immunotherapy; prognostic; biomarkers; tumour mutational board; genomic signatures; ctDNA; PD-L1; inflammatory indices

1. Introduction

Worldwide, urothelial carcinoma (UC) represents the seventh most common cancer and the ninth most deadly tumour, with about 212,000 related deaths [1].

For a long time, the only effective treatment of metastatic UC (mUC) was platinum-based chemotherapy, which is still the standard of care in the first-line setting [2]. In recent years, the treatment landscape of mUC has been changed profoundly by the introduction of immune-checkpoint inhibitors (ICIs) [3] and, even more recently, antibody–drug conjugates (anti-nectin 4 and anti-Trop2) [4] and FGFR inhibitors [5].

Since 2016, the U.S. Food and Drug Administration (FDA) has approved two monoclonal antibodies targeting PD-1 (nivolumab and pembrolizumab) and three antibodies targeting PD-L1 (atezolizumab, avelumab, and durvalumab) for mUC [6].

The introduction of these new drugs urges the identification of potential biomarkers which are able to select the patients most likely to respond to immunotherapy. Promising prognostic and predictive factors in patients with mUC treated with ICIs include clinical features, new tumour molecular classes, the tumour mutational burden (TMB), mutational signatures, circulating tumour DNA (ctDNA) and programmed death-ligand 1 (PD-L1). Despite the increasing number of biomarkers under investigation, these factors still need validation for their application in clinical practice.

In this review, we summarize the landscape of clinical, molecular and genomic determinants of the prognosis and response to ICIs in patients with mUC. A classification of these factors by their prognostic and predictive value according to levels of evidence and grades of recommendation [7] based on the available evidence has been attempted.

2. Molecular Factors

2.1. Molecular Classes

Muscle-invasive bladder cancer (MIBC) is a heterogeneous disease characterized by genomic instability and a high mutation rate [8]. In this scenario, transcriptome profiling may be helpful with the classification of UC into molecular subtypes in order to stratify the prognosis more precisely and drive more effective therapeutic choices. Indeed, the assumption for a molecular tumour classification is the understanding of cancer biology by identifying the specific genomic alterations of which the molecular subtypes are enriched, and which could be clinically significant as prognostic and druggable [9].

Several molecular classifications have been attempted for UC, advancing our knowledge about its biology [9]. The response to chemotherapy and immunotherapy may be enriched in specific MIBC subtypes [10–13]. However, the diversity of their subtype sets, so far, has impeded their clinical application.

Based on the transcriptomic profiles of 1750 MIBCs from 16 published datasets, two additional cohorts, and a network-based analysis of six independent MIBC classification systems, a consensus set of six molecular classes has been identified by a single-sample transcriptomic classifier [9]. The six molecular classes included: the basal/squamous class (Ba/Sq), accounting for 35% of all MIBC; the luminal/papillary class (LumP), accounting for 24%; the luminal unstable class (LumU), accounting for 15%; the stroma-rich class, accounting for 15%; the luminal non-specified class (LumNS), accounting for 8%; and the neuroendocrine (NE)-like class, accounting for 3% [9].

An association of some of these classes with The Cancer Genome Atlas (TCGA) PanCancer clusters has been observed between the Ba/Sq and the Squamous cell carcinoma (C27: Pan-SCC pan-cancer cluster) ($p < 0.001$), and the stroma-rich class with the stroma-driven class (C20: Mixed stromal/immune) ($p < 0.001$) [9].

The six molecular classes differ by their underlying oncogenic mechanisms, infiltration by immune and stromal cells, histological and clinical characteristics, and survival outcomes.

Regarding their prognostic value, based on 872 patients, and taking the LumP class as the reference for class-based survival, the LumNS (hazard ratio [HR] 1.07, 95% confidence interval [CI] 0.63–1.82) and stroma-rich classes (HR 0.98, 95% CI 0.65–1.49) showed a similar prognosis to the LumP class; the LumU class did not have significantly inferior overall survival (OS) (HR 1.49, 95% CI 0.93–2.39), the Ba/Sq class had a significantly poorer prognosis (HR 1.83, 95% CI 1.30–2.58, $p < 0.001$), and the NE-like class had the worst prognosis (HR 2.34, 95% CI 1.09–5.05, $p < 0.03$) [9].

In addition to common urothelial differentiation signatures, like the PPARG/GATA3/FOXA1-related Lund urothelial ones, the luminal classes may selectively bear gene alterations which are potentially relevant as drug targets, such as the FGFR3 in the LumP, and the TP53, high TMB and ERCC2—which are related to higher cell cycle activity and genomic instability—in the LumU. Similarly, the Ba/Sq was enriched in cytotoxic lymphocytes (CTL) and NK, alongside EGFR mutations, the Stroma-rich in T- and B-Cells, and the NE-like in TP53 and RB1 alterations [9].

Concerning treatment responses, an enrichment in responders to atezolizumab was observed among LumNS, LumU and NE-like [9]. However, previous discordant results from the TCGA and different ICIs [11,14] indicate the need for prospective validation.

The molecular classification might be used for prognostication to assist treatment evaluation, and, at the same time, for a more productive collection of clinical information. However, prospective validation is warranted because the use of such classification is only supported by retrospective clinical data lacking the complete patients' treatment history. Furthermore, it might represent a robust framework enabling the testing and validation of predictive biomarkers in future prospective clinical trials, including basket trials, based on similarities with other cancer molecular subtypes according to the PanCancer Atlas, such as the Ba/Sq and Lum.

On the other hand, an integrative multi-omics analysis has recently been performed to better characterize Non-Invasive Muscle Bladder Cancer (NIMBC) [15]. In this study, the authors identified four molecular classes reflecting the behaviour and aggressiveness of UC, driving the implementation of biomarkers with predictive and prognostic value [15].

The genomic landscape of NMIBC showed complex genomic patterns, with activating mutations in FGFR3 and PIK3CA, such as chromosome 9 deletions in early disease [16,17]. The NMIBCs were also subdivided into different progression risk groups based on mutations in FGFR3, the methylation of GATA2 and copy number alterations (CNAs). All of these data could provide new molecular therapeutic targets [16,17].

2.2. Tumour Mutational Burden (TMB)

TMB can be defined as the total number of non-synonymous mutations per coding area of a tumour genome [18]. These mutations can be transcribed and translated to generate neoantigens displayed on the cell surface; T-lymphocytes can recognise some of these neoantigens and promote the apoptosis of tumour cells [18]. Tumours with a high mutational load are more likely to express neoantigens, and to induce a strong immune reaction [19]. Several studies have demonstrated an association between a high TMB and the response to immunotherapy in different locally advanced or metastatic solid tumour types [20].

In June 2020, the FDA approved pembrolizumab for the treatment of adult and paediatric patients with unresectable or metastatic solid cancers and a high TMB [>10 mutations/megabase (mut/Mb)] [21]. This approval was based on the efficacy data from the phase II KEYNOTE-158 trial, which demonstrated an association between a high TMB and the tumour response to immunotherapy, with a durable response (>2 years) rarely being observed in heavily pre-treated metastatic cancers [22].

UC, along with melanoma and lung cancer, is characterized by a high number of somatic mutations and, therefore, a significant genomic instability [8]. The exploratory analyses of the phase II IMvigor210 trial included the quantification of the mutational load and a correlation with the clinical outcomes [23]. As expected, the TMB was significantly higher in patients who responded to atezolizumab than in non-responders. Moreover, patients with a higher mutational load had significantly longer overall survival than patients with a lower load [23].

However, the TMB status alone was not able to stratify the patients according to the survival benefit. In this regard, exploratory analyses of the JAVELIN Bladder 100 trial demonstrated that some DNA mutational signatures, involving certain base pair alterations, were associated with a survival benefit from avelumab, while other mutations were not [24]. This suggests that the type and the location of mutations may influence the predictive role of TMB assessment more accurately than the degree of the mutational load.

The predictive value of TMB for the pathological response to immunotherapy has also been explored in the neoadjuvant setting [25,26]. In patients with MIBC, a high TMB was generally observed in responders versus non-responders to neoadjuvant pembrolizumab, irrespective of the histological subtypes [27].

The predictive role of combining TMB with PD-L1 was also investigated. Patients with high TMB and PD-L1-positive tumours were more likely to derive a survival benefit from avelumab maintenance therapy [24]. Similar findings were observed in the randomised phase III IMvigor130 study [28]. The association of high PD-L-1 (>5% of immune cells) and a high TMB (>10 mut/MB) was associated with improved OS in the atezolizumab monotherapy arm, as compared to the chemotherapy arm [28]. Nevertheless, similar outcomes were not seen with the combination of atezolizumab and chemotherapy compared to the chemotherapy alone, suggesting a potentially distinct biology driving the benefit from atezolizumab and its combination with chemotherapy, and eventually highlighting the predictive inconsistency of this biomarker combination [28].

As for other tumour types, many factors hinder the clinical application of TMB as a biomarker, including the variability and the lack of a validated cut-off, a clear prognostic value, the differences related to the sequencing platforms used for its assessment, and the high scoring failure rate due to the quantity and quality of tumour tissue analysed [29]. As an example for the mUC, in the IMvigor210 study, the cut-off of the TMB varied between the two different cohorts of cisplatin-ineligible mUC treated with first-line atezolizumab and the platinum-treated patients [23]. Moreover, a high tumour mutational load may not be a specific biomarker for immunotherapy, as it has also been associated with the tumour response to neoadjuvant chemotherapy [30,31].

2.3. Mulecular Signatures

The increasing interest in tumour molecular features and new omics technologies has led to the discovery of different molecular signatures. These involve genes, messenger ribonucleic acids (mRNAs) and proteins which are studied as biomarkers to better predict clinical outcomes, or to better understand the cancerogenesis process [32,33].

Furthermore, because the TMB and/or the PD-L1 expression did not precisely identify patients who were more likely to derive benefit from immunotherapy [34,35], the assessment of tumour-related or immune-related gene signatures was actively investigated.

The DNA and RNA sequencing analyses of the IMvigor130 trial investigated the apolipoprotein B editing catalytic polypeptide (APOBEC) signature [28]. The APOBEC enzymes are a family of cytidine deaminases involved in the DNA repair processes, and are responsible for a mutation signature (TCW > T/G) frequently observed in MIBCs [36,37]. Patients with a high APOBEC mutational signature had a longer survival, whether in the atezolizumab monotherapy or the combination arm with chemotherapy compared to chemotherapy alone, whereas a high TGFβ signature predicted a worse OS with the immunotherapy alone [28].

RNA-based immune-gene expression profiling has the advantage of providing information from many cancer cells and immune cells, identifying more accurately the inflammatory status. The Checkmate-275 study investigated an interferon-gamma (IFN-γ) expression signature, and found a significant correlation with the response to nivolumab in the metastatic setting [38].

At the ESMO 2020 meeting, an exploratory analysis of the Javelin Bladder 100 study on tumour biomarkers was presented. While neither PD-L1 and TMB alone nor in combination predicted the response to immunotherapy and survival benefit, the expression of the immune-related genes of both the innate and adaptive immune system (i.e., CD8, IFNG, LAG3, TIGIT and CXCL9) and the number of alleles encoding high-affinity Fc gamma receptor variants predicted the survival benefit from avelumab first-line maintenance [24]. The application of the T-cell-inflamed and JAVELIN-Immuno signatures seemed to correlate with the treatment response with HRs of 0.55 and 0.49, respectively. Furthermore, the JAVELIN-Immuno high signature showed enrichment in some signalling pathways (Notch, Hedgehog, TGFbeta) which were likely to have an increased antitumour response to immunotherapy [24].

In the PURE-01 study, investigating the efficacy of pembrolizumab as a neoadjuvant therapy, a T-cell-inflamed signature was able to identify those patients who achieved pathological T0 downstaging. In a subsequent analysis, other RNA-based immune signatures were evaluated for their association with pCR and a high Immune190 signature, as well as hallmark signatures for interferon gamma and interferon-alpha, which were significantly associated with pCR and progression-free survival (PFS) after pembrolizumab in the PURE01 study [27]. Similarly, a correlation between the tumour response to atezolizumab and the transcriptional signature of eight genes (IFNG, CXCL9, CD8A, GZMA, GZMB, CXCL10, PRF1 and TBX21) representing interferon signalling, and the presence of CD8+ effector T cells, namely the tGE8, was reported within the ABACUS neoadjuvant study [39].

Moreover, a strong immune-mediated adaptive resistance was observed in non-responding tumours, suggesting the investigation of ICI combinations to counteract the expression of negative regulators of the immune response [40].

The BISCAY study combined durvalumab (anti-PD-L1) with different targeted therapies depending on tumour gene alterations determined by NGS [41]. Despite the overall negative results, this study suggested a negative predictive role for the FGFR expression with regards to immunotherapy. In fact, FGFR-mutated tumours did not have a high expression of immune-active T-cell signatures, and the addition of durvalumab did not result in enhanced activity from the targeting agent only [9,41].

Although the above data look promising, there is a need to standardize molecular assays and find a molecular panel which is applicable to daily clinical practice.

2.4. ctDNA

The sequencing of the tumour fraction of the cell-free DNA (ctDNA) is an emerging and sensitive method to detect residual disease, anticipate relapse, and monitor the therapeutic efficacy in patients with several cancer types [42,43]. Furthermore, it provides a basis for clinical studies evaluating early therapeutic interventions [42,43].

In UCs, proof-of-concept data documented that ctDNA is detectable in plasma and urine, and could be a prognostic factor [44,45].

Vandekerkhove et al. demonstrated that ctDNA profiling may also identify putative biomarkers for therapy response in bladder cancer (such as FGFR3, ERCC2, ERBB2 and TMB), and represents a cost-effective and minimally invasive method for their identification and patient stratification [46]. Furthermore, the serial monitoring of ctDNA can optimize the use of therapies, as ctDNA fractions are expected to decrease in patients responding to treatment [46].

In a longitudinal analysis, the presence of ctDNA—assessed by whole-exome sequencing (WES) in 68 patients with localised advanced bladder cancer at diagnosis—during chemotherapy, before and after cystectomy and during surveillance, resulted in being highly prognostic at diagnosis (HR 29.1) [47]. The presence of ctDNA identified all of the patients with metastatic relapse (with 100% sensitivity and 98% specificity), with a median of 96 days of diagnostic anticipation compared to the radiographic imaging [47]. Furthermore, the dynamics of ctDNA during chemotherapy, for those patients with positive ctDNA before or during the treatment, were associated with disease recurrence. It is worth noting that pathological downstaging, including related mutational signatures, was not associated with disease recurrence [47].

A post-surgical ctDNA detection was also associated with a higher risk for recurrence and death in the randomised phase III IMvigor010 study with adjuvant atezolizumab versus observation following cystectomy for patients with MIBC [48]. Moreover, the presence of ctDNA was predictive, as it was able to identify the patients who were likely to benefit from adjuvant atezolizumab. The IMvigor010 study did not meet his primary endpoint of disease-free survival (DFS) in the overall population of 809 enrolled patients [48]. The ctDNA was assessed by WES at a median of 11 weeks post-cystectomy in 581 patients with evaluable samples, comprising 72% of the intent-to-treat population of the IMvigor010 study. While the detection of ctDNA (in 214 patients, 37%) was associated with worse DFS and OS compared to the absence of ctDNA (in 367 patients, 63%), a significant difference in DFS (HR 0.58, 95% CI 0.43–0.79) and OS (0.59, 95% CI 0.41–0.86, $p = 0.0059$) was observed in favour of the patients with detectable ctDNA treated with atezolizumab compared to the observational arm. Furthermore, in patients with the presence of ctDNA, the rate of ctDNA clearance from cycle 1 to cycle 3 was significantly higher in the atezolizumab arm versus the observational one, with ctDNA clearance rates of 18.2% and 3.8%, respectively ($p = 0.0048$) [48].

2.5. Programmed Death Ligand-1 (PD-L1)

In the first-line treatment of mUC, a high PD-L1 combined positive score (CPS) of $\geq 10\%$, defined as the percentage of tumour cells (TC) and immune cells (IC)/the total tumour cells given by the 22C3 Dako assay, was associated with a prolonged median OS (mOS) in patients ineligible for cisplatin treatment with first-line pembrolizumab in the single-arm phase II KEYNOTE 052 study. The mOS was 18.3 versus 9 months in patients with PD-L1 CPS $\geq$ 10% and <10%, respectively, and it was 11.3 months in the overall population [49]. As the KEYNOTE 052 was not a controlled phase III study, a favourable prognostic value only for the high PD-L1 expression could not be rule out. Indeed, in the following KEYNOTE 361, IMvigor130 and Danube phase III randomised trials, the anti-PD1/PD-L1 agents either alone or in combination with chemotherapy were not superior to the chemotherapy according to OS in patients with high PD-L1 tumours, as well as in the overall population [50–52]. Only in the Danube trial did the combination of the anti-PD-L1 inhibitor durvalumab and anti-CTLA4 tremelimumab show a superior OS compared to chemotherapy in patients with high PD-L1. Therefore, the predictive value of high PD-L1 remains uncertain [52]. In contrast, a decreased survival with first-line pembrolizumab or atezolizumab compared to platinum-based chemotherapy was reported by the KEYNOTE 361 and IMvigor130 phase III trials for patients with low PD-L1 tumour expression, suggesting a negative predictive value for low PD-L1 expression [50,51]. A low PD-L1 expression was defined in the KEYNOTE 361 and IMvigor130 studies as a CPS < 10% or positive IC < 5% by the immunohistochemistry (IHC) 22C3 pharmDx and SP142 Ventana assays, respectively [50,51].

In the maintenance setting of mUC, following first-line chemotherapy, the Javelin-100 phase III trial showed an advantage in OS from the anti-PD-L1 avelumab compared to the best supportive care irrespective of the positive PD-L1 expression, which was assessed by a SP263 Ventana assay and classified as positive if at least one of the following three criteria were met: at least 25% TC staining for PD-L1, at least 25% IC staining for PD-L1 if more than 1% of the tumour area contained IC, or 100% ICs staining for PD-L1 if no more than 1% of the tumour area contained ICs [53].

In the adjuvant setting of UC, the IMvigor 010 with the anti-PD-L1 atezolizumab did not show a significant advantage in DFS compared to observation either in patients with positive PD-L1 tumours IC according to the SP142 Ventana assay, or in the overall population [54], while the Checkmate274 showed a significant DFS advantage in favour of nivolumab versus observation in the intention-to-treat population with a better HR in patients with a PD-L1 expression $\geq 1\%$ (HR 0.55 vs. 0.70 for the overall population), as assessed on TC by a PD-L1 IHC 28-8 pharmDx assay [55].

In the second-line of mUC, in the IMvigor 211 phase III study with anti-PD-L1 atezolizumab compared to chemotherapy, the OS was not significantly different in patients with a high PD-L1 defined by IC 2/3 (or $\geq 5\%/10\%$ of PD-L1 positive IC according to a SP142 Ventana assay) [56]. In contrast, in the KEYNOTE 045 study, a significant difference in OS was observed in favour of the anti-PD-1 pembrolizumab versus chemotherapy in patients with high PD-L1 tumours defined as CPS according the 22C3 Dako assay [57]. Although possible differences in efficacy between the anti-PD1 and anti-PD-L1 agents cannot completely be ruled out [58], plausible explanations may rely on the different tests and scores used to assess the PD-L1 expression, as well as on the tumoural site and timing of the biopsy in respect to the disease history. In this regard, it is noteworthy that the mOS duration in the high PD-L1 subgroup of patients differed between those two trials despite the similar populations enrolled and the mOS observed in the overall population of the chemotherapy control arm (of 8.4 and 7.3 months, for the IMvigor211 and the Keynote-045 study, respectively) [56,57]. In the KEYNOTE 045 study, the mOS of patients with high CPS ($\geq 10\%$) treated with pembrolizumab was of 8.0 months, which was significantly longer than 5.2 months with chemotherapy ($p = 0.0048$) [57]; on the other hand, in the IMvigor211 study, the patients with high IC 2/3 had an mOS of 11.1 months with atezolizumab, which was not significantly longer than 10.6 months with chemotherapy ($p = 0.41$) [56]. The OS difference seen in the chemotherapy control arms of these two studies, with a different definition of high PD-L1 tumours, might suggest a positive prognostic value which is marginally predictive for the high PD-L1 by the IC 2/3, and a negative prognostic—although predictive—value by the CPS.

In the neoadjuvant setting, only phase II studies are currently available. However, the Pure-01 study demonstrated the dynamic characteristic of the PD-L1 by showing its significantly increased expression as CPS following three administrations of the anti-PD1 pembrolizumab [40].

In conclusion, the current evidence on the prognostic and predictive value of PD-L1 expression is limited by the different diagnostic assays used in the clinical trials for each anti-PD1 or anti-PD-L1 agent, and remains inconclusive. PD-L1 may have a prognostic role, the value of which may also change depending on the type of tumour cells scored (i.e., negative if CPS, or positive if IC), and may be predictive only for the low expression. The molecular predictive and prognostic factors in UCs are summarized in Table 1.

Table 1. Molecular factors and their prognostic and predictive value in UC patients, with a particular focus on ICIs.

Category	Description	Prognostic and Predictive Value in UC	Notes
Molecular classes	6 molecular (transcriptomic) classes based on a consensus of MIBC: Squamous (Ba/Sq)-35%; luminal/papillary (LumP), 24%; luminal unstable (LumU), 15%; stroma-rich, 15%; luminal non-specified (LumNS), 8%; neuroendocrine (NE)-like, 3% [9].	Ba/Sq is associated with shorter OS, NE-like is associated with worse prognosis (LumP as reference) [9]. LUmNS, LumU, NE-like are more responsive to ICIs [9].	Need for prospective validation

Table 1. Cont.

Category	Description	Prognostic and Predictive Value in UC	Notes
TMB	Total number of non-synonymous mutations per coding area of a tumour genome. UC is characterized by high values of TMB compared to other tumours [23].	High TMB predicts OS benefit from avelumab maintenance therapy [23] and improved OS with atezolizumab when compared to CT [28]. TMB was higher in responder to neoadjuvant pembrolizumab [27].	Issues: variability, lack of a validated cut-off, differences related to the sequencing platforms
Molecular signatures	Study of involved genes (DNA sequencing), messenger ribonucleic acids (mRNAs) (RNA sequencing), and proteins (transcriptome) in tumor samples.	APOBEC mutational signature predicts OS benefit with atezolizumab ± CT compared to CT alone [28]. TGF-β signature predicts worse OS with ICIs [28]. IFN-γ signature correlates with response to nivolumab [38]. JAVELIN -Immuno high signature correlates with increased responses to ICIs [24]. Transcriptional signature of eight genes (IFNG, CXCL9, CD8A, GZMA, GZMB, CXCL10, PRF1 and TBX21) correlates with response to atezolizumab [39]. FGFR mutations predict low response to durvalumab [41].	Need to standardize molecular assays and find a molecular panel applicable to daily clinical practice
ctDNA	Quantitative and qualitative analysis of circulating tumoral DNA detected on blood samples.	ctDNA detection is associated with worse prognosis in early stages and could identify metastatic relapses before imaging [47]. ctDNA detection is predictive for adjuvant atezolizumab (both DFS and OS) [48]. ctDNA profiling may be predictive for response to specific therapies [46].	
PD-L1	Expression of the ligand of PD1 receptor has been widely studied as predictive biomarker of response to anti-PD1 and anti-PD-L1 therapies across human cancers.	Possibly predictive for anti-PD1 and anti-PD-L1 agents used in UC patients as adjuvant [54], first-line [49], maintenance after first-line [53] and second-line therapy [56].	Issues: different diagnostic assays used in clinical trials for each anti-PD1 or anti-PD-L1 agent, discordant efficacy of these agents between studies

Abbreviations: UC, urothelial carcinoma; ICIs, immune checkpoint inhibitors; TMB, tumor mutational burden; OS, overall survival; CT, chemotherapy; ctDNA, circulating tumor DNA.

3. Clinical Factors

3.1. Patient's Characteristics

Several clinical factors have been reported to be prognostic in patients with mUC treated with chemotherapy. The most important ones include performance status (PS), metastatic sites, haemoglobin levels, and time from prior chemotherapy. Bajorin et al. showed that, in untreated patients, a poor Karnofsky Performance Status (KPS) < 80% and the presence of visceral (liver, lung, and bone) metastases were independent prognostic factors associated with a worse OS [59]. Bellmunt et al. reported that a poor Eastern Cooperative Oncology Group (ECOG) PS ≥ 2, low haemoglobin levels (<10 g/dL) and the presence of liver metastasis were independent poor prognostic factors in patients who failed platinum-based chemotherapy [60]. Moreover, a shorter time from prior chemotherapy

(<3 months) has been reported to enhance the prognostic classification of patients receiving second-line therapy [61].

The survival benefit of immunotherapy in both untreated and pretreated patients with mUC has been observed regardless of all of those clinical factors, confirming their prognostic but not predictive value, in many retrospective analyses [62–70] and subgroup and post-hoc analyses of prospective randomised trials [71,72].

3.2. Concomitant Medications

Several retrospective analyses and meta-analyses have recently been reported in the literature about the impact of concomitant medications on the clinical outcomes of patients with cancer treated with ICIs [73–78]. These studies mainly included patients with non-small cell lung cancer (NSCLC), melanoma or renal cell carcinoma (RCC). Commonly used drugs in clinical practice—such as corticosteroids, antibiotics and proton pump inhibitors (PPIs)—have been reported to negatively affect the activity of immunotherapy through immune-modulatory effects. In fact, these drugs may induce a detrimental effect on the immune system and gut microbiota, which is a well-known regulator of immune homeostasis [79]. A drug-based prognostic score developed by Buti et al.—including the cumulative exposure to high-dose steroid therapy (a dose of ≥ 10 mg prednisone-equivalents per day), antibiotics and PPIs—can help to stratify patients treated with ICIs in routine practice and clinical trials [73].

The post-hoc analyses of the single-arm phase II IMvigor210 trial and the randomised phase III IMvigor211 trial confirmed the negative predictive role of PPI and antibiotic use in patients with mUC treated with ICI, but not in the case of treatment with chemotherapy [80,81].

3.3. Inflammatory Indices

An inflammatory condition in patients with cancer has been associated with worse outcomes and a lower therapeutic response across different tumour types [82]. Inflammatory indices from peripheral blood have been investigated as potential biomarkers in different tumours, settings and therapies [83–85], including patients with UC treated with surgery or chemotherapy [64,86–88].

As far as ICIs are concerned, inflammatory indices have been mostly studied in patients with advanced NSCLC and melanoma [89,90], while fewer and smaller studies have been conducted in patients treated with ICIs for genitourinary tumours, including UC [62,63,65,69,70,91–97]. The inflammatory indices most commonly studied in patients with UC treated with ICIs include the neutrophil-to-lymphocyte ratio (NLR), baseline platelet-to-lymphocyte ratio (PLR), lymphocyte-to-monocyte ratio (LMR), lactate dehydrogenase (LDH), C-reactive protein (CRP) and albumin. High levels of NLR, PLR, LDH and CRP, and low levels of albumin have been correlated with worse survival and efficacy outcomes, either as single parameters or within combined prognostic scores [62,63,65,69,70,92–98].

3.4. Combined Prognostic Tools

Inflammatory indices from peripheral blood have been investigated in combination with other clinical prognostic factors within prognostic models for risk-stratification in several cancer types treated with ICIs, especially the NSCLC [91,98–100]. The interest in prognostic models has also recently been increasing for genitourinary tumours, including RCC [101–103] and UC [63,65,68,95–97,104].

In patients with mUC treated with ICIs, the most frequently included factors in prognostic models are inflammatory indices (such as NLR, C-reactive protein and albumin) and pretreatment clinical parameters (i.e., the PS and metastatic site) [63,65,68,95–97,104–106]. The molecular factors included in the prognostic scores were PD-L1 expression and genomic parameters (e.g., single-nucleotide variants) [96,104].

Most of these analyses are derived from retrospective studies, except for the two prognostic scores developed by Fornarini et al. and Sonpavde et al., and the machine learning (ML) model developed by Abuhelwa et al. [104–106]. The two prognostic models were developed by the retrospective analysis of the phase IIIb SAUL trial [104] and the two phase I/II trials [105], both evaluating second-line ICI monotherapy. The ML model was built using the atezolizumab cohort of the IMvigor210 trial as the training cohort, and the IMvigor211 trial population as the external validation [106].

Most of these prognostic models (Table 2) need external validation with prospective studies before being incorporated into clinical practice; only one of them has shown a predictive value for immunotherapy but not for chemotherapy in a retrospective analysis [96].

Table 2. Clinical factors and their prognostic and predictive value in UC patients, with a particular focus on ICIs.

Category	Description	Prognostic and Predictive Value in UC
Patient's characteristics	Performance status (PS), metastatic sites, haemoglobin levels are prognostic factors in human cancer.	Karnofsky PS < 80%, ECOG PS ≥ 2, low haemoglobin levels and the presence of visceral metastases are associated with worse OS in mUC patients treated with CT [59,61]. Prognostic but not predictive in mUC patients treated with ICIs [62–72].
Concomitant medications	Commonly used drugs in clinical practice may affect clinical outcomes of cancer patients treated with ICIs.	Use of antibiotics and PPIs have a negative predictive role in mUC patients treated with ICIs but not in those treated with CT [80,81].
Inflammatory indices	Inflammatory indices, like NLR, CMP and albumin, have been investigated in several human cancer types as prognostic tools, especially regarding ICIs.	High levels of NLR, PLR, LDH, CRP and low levels of albumin have been correlated with worse survival and efficacy outcomes in mUC patients [62–64,70,92–98].
Combined prognostic tools	Combinations of inflammatory indices and clinical factors have been investigated in several human cancer types as prognostic tools, especially regarding ICIs.	Two prognostic models were developed in second-line ICIs mono-therapy [104,105]. A machine learning model was built from IMvigor210 atezolizumab arm and validated in IMvigor211 atezolizumab arm [106].

Abbreviations: UC, urothelial carcinoma; mUC, metastatic urothelial carcinoma; ICIs, immune checkpoint inhibitors; OS, overall survival; CT, chemotherapy; PPIs, proton pump inhibitors; NLR, neutrophil-to-lymphocyte ratio; PLR, platelet-to-lymphocyte ratio; LMR, lymphocyte-to-monocyte ratio; LDH, lactate dehydrogenase; CRP, C-reactive protein (CRP).

4. Radiomics

Due to the growing need to identify useful biomarkers to select the patients who are most likely to benefit from ICIs, the quantitative analysis of imaging features by artificial intelligence algorithms, namely radiomics, has recently been investigated as a possible surrogate marker to predict the outcome of patients treated with immunotherapy [107,108]. Radiomics has been reported as a promising approach to predict the response and survival outcomes in patients with NSCLC and melanoma receiving ICIs [109,110]. Three retrospective analyses investigated radiomics from baseline contrast-enhanced computed tomography (CT) images in patients with mUC receiving anti-PDL1 or anti-PD1 monotherapy [111–113].

Artificial intelligence algorithms may also allow us to combine the information obtained by the image features with other clinical and laboratory prognostic factors, thus increasing the diagnostic accuracy of predictive models [112,113].

Artificial intelligence and deep machine learning-based models may provide helpful decision tools to clinicians for the selection of patients for ICIs in the near future. However, none of these radiomics-based models have been assessed in patients with UC treated with chemotherapy yet; further studies on larger populations and external validation are still needed to confirm their predictive value.

5. Conclusions

In the current era, in which ICIs have been tested for several tumour types, it is crucial to identify prognostic and predictive biomarkers or models which are able to select UC patients who will benefit from ICIs, especially as novel therapeutic agents could soon become available in clinical practice for the treatment of UC. In addition, adjuvant ICIs have recently shown a benefit in terms of disease-free survival in patients with radically resected muscle-invasive urothelial carcinoma [112]. With more follow-up, a survival benefit is awaited, opening new methods for biomarker identification even in this setting.

By reviewing the available evidence and attempting a classification of clinical and tumour factors according to levels of evidence and grades of recommendation (see Table 3), both a prognostic and a predictive value is currently suggested for ctDNA, while only a prognostic relevance seems to apply to concomitant medications and patient's characteristics.

Table 3. Prognostic and predictive values of tumour and clinical factors for their impact on the response and survival outcomes in mUC treated with ICIs.

Variable Parameter	Prognostic			Predictive		
	Clinical Value	Strenght of Evidence	Outcome Variable	Clinical Value	Strenght of Evidence	Outcome Variable
Tumour molecular factors						
Molecular classes for MIBC	🟡	IV, B	PFS, OS	🟡	IV, B	PFS, OS
TMB	🟡	IV, C	ORR, PFS, OS	🟡	I, C	ORR, PFS, OS
Mutational signatures [1]	🟡	IV, C	ORR, PFS, OS	🟡	I, C	ORR, PFS, OS
ctDNA	🟢	II, A	DFS, OS	🟢	II, A	DFS, OS
PD-L1	🟡	I, C	DFS, PFS, OS	🟡	I, B [2]	DFS, PFS, OS
Clinical factors						
Patient's characteristics [3]	🟢	I, A	ORR, PFS, OS	🔴	I, A	ORR, PFS, OS
Concomitant medications [4]	🟢	I, A	ORR, PFS, OS	🟡	I, B	ORR, PFS, OS
Inflammatory indices [5]	🟡	IV, A	ORR, PFS, OS	🔴	IV, A	ORR, PFS, OS
Combined tools [6]	🟡	III, B	ORR, PFS, OS	🟡	IV, C	ORR, PFS, OS
Radiomics						
Radiomics-based models [7]	🟡	IV, B	ORR, PFS, OS	🔴	NI	NI

PFS, progress-free survival; OS, overall survival; ORR, overall response rate; DFS, disease-free survival; NI, not investigated. [1] DNA and RNA signatures such as the APOBEC, TGFbeta, IFN-γ and IFN-alpha; immune-related signatures such as the JAVELIN-Immuno or Immune190. [2] Evidence was limited due to its low expression. [3] Poor PS, visceral metastases, mainly liver metastases, low hemoglobin levels, and a shorter time from prior chemotherapy (<3 months). [4] High-dose steroid therapy, antibiotics and PPIs. [5] Inflammatory indices from peripheral blood, such as ratios of immune system cells (e.g., NLR) or LDH. [6] Combination of inflammatory indices and clinical parameters. [7] Radiomic features with/without clinical factors. 🟢 Green circle: clinically useful. 🟡 Yellow circle: uncertain clinical usefulness. 🔴 Red circle: not clinically useful. The strength of evidence was adapted from the Infectious Diseases Society of America-United States Public Health Service Grading System [7]. Levels of evidence: I, evidence from at least one large randomised, controlled trial of good methodological quality (low potential for bias) or meta-analyses of well-conducted randomised trials without heterogeneity; II, small randomised trials or large randomised trials with a suspicion of bias (lower methodological quality), or meta-analyses of such trials or of trials with demonstrated heterogeneity; III, prospective cohort studies; IV, retrospective cohort studies or case-control studies; V, studies without a control group, case reports, or expert opinions. Grades of recommendation: A, strong evidence for efficacy with a substantial clinical benefit, strongly recommended; B, strong or moderate evidence for efficacy, but with a limited clinical benefit, generally recommended; C, insufficient evidence for efficacy or benefit does not outweigh the risk or the disadvantages (adverse events, costs, etc.), optional; D, moderate evidence against efficacy or for adverse outcomes, generally not recommended; E, strong evidence against efficacy or for adverse outcomes, never recommended.

We think this information may be helpful to clinically assess mUCs patients who are considered for treatment with ICIs, and could also drive further prospective research on these biomarkers, either as single factors or within combined prognostic models implemented by artificial intelligence algorithms.

Author Contributions: Conceptualization, S.E.R., G.L.B. and G.F.; methodology, S.E.R. and G.L.B.; writing—original draft preparation, S.E.R., G.L.B., V.M., A.D., F.F. and G.F.; writing—review and editing, S.E.R., G.L.B., E.F.G., U.D.G., R.C., P.R. and M.B.; supervision, S.E.R., G.L.B. and G.F. All authors have read and agreed to the published version of the manuscript.

Funding: This research received no external funding.

Acknowledgments: S.E.R. and G.F. would like to thank the Italian Ministry of Health (Ricerca Corrente 2018–2021 grants), which financially supports their current research focused on the identification of prognostic and predictive markers for patients with genitourinary tumours.

Conflicts of Interest: The authors declare no conflict of interest for this manuscript.

References

1. Sung, H.; Ferlay, J.; Siegel, R.L.; Laversanne, M.; Soerjomataram, I.; Jemal, A.; Bray, F. Global Cancer Statistics 2020: GLOBOCAN Estimates of Incidence and Mortality Worldwide for 36 Cancers in 185 Countries. *CA Cancer J. Clin.* **2021**, *71*, 209–249. [CrossRef]
2. Mori, K.; Pradere, B.; Moschini, M.; Mostafaei, H.; Laukhtina, E.; Schuettfort, V.M.; Sari Motlagh, R.; Soria, F.; Teoh, J.Y.C.; Egawa, S.; et al. First-Line Immune-Checkpoint Inhibitor Combination Therapy for Chemotherapy-Eligible Patients with Metastatic Urothelial Carcinoma: A Systematic Review and Meta-Analysis. *Eur. J. Cancer* **2021**, *151*, 35–48. [CrossRef] [PubMed]
3. Lopez-Beltran, A.; Cimadamore, A.; Blanca, A.; Massari, F.; Vau, N.; Scarpelli, M.; Cheng, L.; Montironi, R. Immune Checkpoint Inhibitors for the Treatment of Bladder Cancer. *Cancers* **2021**, *13*, 131. [CrossRef] [PubMed]
4. Lattanzi, M.; Rosenberg, J.E. The Emerging Role of Antibody-Drug Conjugates in Urothelial Carcinoma. *Expert Rev. Anticancer Ther.* **2020**, *20*, 551–561. [CrossRef]
5. Loriot, Y.; Necchi, A.; Park, S.H.; Garcia-Donas, J.; Huddart, R.; Burgess, E.; Fleming, M.; Rezazadeh, A.; Mellado, B.; Varlamov, S.; et al. Erdafitinib in Locally Advanced or Metastatic Urothelial Carcinoma. *N. Engl. J. Med.* **2019**, *381*, 338–348. [CrossRef] [PubMed]
6. Patel, A.; Bisno, D.I.; Patel, H.V.; Ghodoussipour, S.; Saraiya, B.; Mayer, T.; Singer, E.A. Immune Checkpoint Inhibitors in the Management of Urothelial Carcinoma. *J. Cancer Immunol.* **2021**, *3*, 115–136. [CrossRef]
7. Dykewicz, C.A.; Centers for Disease Control and Prevention (US). Infectious Diseases Society of America, American Society of Blood and Marrow Transplantation Summary of the Guidelines for Preventing Opportunistic Infections among Hematopoietic Stem Cell Transplant Recipients. *Clin. Infect. Dis.* **2001**, *33*, 139–144. [CrossRef] [PubMed]
8. Alexandrov, L.B.; Nik-Zainal, S.; Wedge, D.C.; Aparicio, S.A.J.R.; Behjati, S.; Biankin, A.V.; Bignell, G.R.; Bolli, N.; Borg, A.; Børresen-Dale, A.-L.; et al. Signatures of Mutational Processes in Human Cancer. *Nature* **2013**, *500*, 415–421. [CrossRef]
9. Kamoun, A.; de Reyniès, A.; Allory, Y.; Sjödahl, G.; Robertson, A.G.; Seiler, R.; Hoadley, K.A.; Groeneveld, C.S.; Al-Ahmadie, H.; Choi, W.; et al. A Consensus Molecular Classification of Muscle-Invasive Bladder Cancer. *Eur. Urol.* **2020**, *77*, 420–433. [CrossRef] [PubMed]
10. Choi, W.; Porten, S.; Kim, S.; Willis, D.; Plimack, E.R.; Hoffman-Censits, J.; Roth, B.; Cheng, T.; Tran, M.; Lee, I.-L.; et al. Identification of Distinct Basal and Luminal Subtypes of Muscle-Invasive Bladder Cancer with Different Sensitivities to Frontline Chemotherapy. *Cancer Cell* **2014**, *25*, 152–165. [CrossRef] [PubMed]
11. Rosenberg, J.E.; Hoffman-Censits, J.; Powles, T.; van der Heijden, M.S.; Balar, A.V.; Necchi, A.; Dawson, N.; O'Donnell, P.H.; Balmanoukian, A.; Loriot, Y.; et al. Atezolizumab in Patients with Locally Advanced and Metastatic Urothelial Carcinoma Who Have Progressed Following Treatment with Platinum-Based Chemotherapy: A Single-Arm, Multicentre, Phase 2 Trial. *Lancet* **2016**, *387*, 1909–1920. [CrossRef]
12. Seiler, R.; Ashab, H.A.D.; Erho, N.; van Rhijn, B.W.G.; Winters, B.; Douglas, J.; Van Kessel, K.E.; Fransen van de Putte, E.E.; Sommerlad, M.; Wang, N.Q.; et al. Impact of Molecular Subtypes in Muscle-Invasive Bladder Cancer on Predicting Response and Survival after Neoadjuvant Chemotherapy. *Eur. Urol.* **2017**, *72*, 544–554. [CrossRef] [PubMed]
13. Mariathasan, S.; Turley, S.J.; Nickles, D.; Castiglioni, A.; Yuen, K.; Wang, Y.; Kadel, E.E.; Koeppen, H.; Astarita, J.L.; Cubas, R.; et al. TGFβ Attenuates Tumour Response to PD-L1 Blockade by Contributing to Exclusion of T Cells. *Nature* **2018**, *554*, 544–548. [CrossRef]
14. Galsky, M.D.; Retz, M.; Siefker-Radtke, A.O.; Baron, A.; Necchi, A.; Bedke, J.; Plimack, E.R.; Vaena, D.; Grimm, M.-O.; Bracarda, S.; et al. Efficacy and Safety of Nivolumab Monotherapy in Patients with Metastatic Urothelial Cancer (MUC) Who Have Received Prior Treatment: Results from the Phase II CheckMate 275 Study. *Ann. Oncol.* **2016**, *27*, vi567. [CrossRef]
15. Lindskrog, S.V.; Prip, F.; Lamy, P.; Taber, A.; Groeneveld, C.S.; Birkenkamp-Demtröder, K.; Jensen, J.B.; Strandgaard, T.; Nordentoft, I.; Christensen, E.; et al. An Integrated Multi-Omics Analysis Identifies Prognostic Molecular Subtypes of Non-Muscle-Invasive Bladder Cancer. *Nat. Commun.* **2021**, *12*, 2301. [CrossRef]
16. van Kessel, K.E.M.; van der Keur, K.A.; Dyrskjøt, L.; Algaba, F.; Welvaart, N.Y.C.; Beukers, W.; Segersten, U.; Keck, B.; Maurer, T.; Simic, T.; et al. Molecular Markers Increase Precision of the European Association of Urology Non-Muscle-Invasive Bladder Cancer Progression Risk Groups. *Clin. Cancer Res.* **2018**, *24*, 1586–1593. [CrossRef]
17. Hurst, C.D.; Platt, F.M.; Taylor, C.F.; Knowles, M.A. Novel Tumor Subgroups of Urothelial Carcinoma of the Bladder Defined by Integrated Genomic Analysis. *Clin. Cancer Res.* **2012**, *18*, 5865–5877. [CrossRef]

18. Meléndez, B.; Van Campenhout, C.; Rorive, S.; Remmelink, M.; Salmon, I.; D'Haene, N. Methods of Measurement for Tumor Mutational Burden in Tumor Tissue. *Transl. Lung Cancer Res.* **2018**, *7*, 661–667. [CrossRef] [PubMed]
19. Chan, T.A.; Yarchoan, M.; Jaffee, E.; Swanton, C.; Quezada, S.A.; Stenzinger, A.; Peters, S. Development of Tumor Mutation Burden as an Immunotherapy Biomarker: Utility for the Oncology Clinic. *Ann. Oncol.* **2019**, *30*, 44–56. [CrossRef]
20. Klempner, S.J.; Fabrizio, D.; Bane, S.; Reinhart, M.; Peoples, T.; Ali, S.M.; Sokol, E.S.; Frampton, G.; Schrock, A.B.; Anhorn, R.; et al. Tumor Mutational Burden as a Predictive Biomarker for Response to Immune Checkpoint Inhibitors: A Review of Current Evidence. *Oncologist* **2020**, *25*, e147–e159. [CrossRef]
21. Subbiah, V.; Solit, D.B.; Chan, T.A.; Kurzrock, R. The FDA Approval of Pembrolizumab for Adult and Pediatric Patients with Tumor Mutational Burden (TMB) ≥10: A Decision Centered on Empowering Patients and Their Physicians. *Ann. Oncol.* **2020**, *31*, 1115–1118. [CrossRef] [PubMed]
22. Marabelle, A.; Fakih, M.; Lopez, J.; Shah, M.; Shapira-Frommer, R.; Nakagawa, K.; Chung, H.C.; Kindler, H.L.; Lopez-Martin, J.A.; Miller, W.H.; et al. Association of Tumour Mutational Burden with Outcomes in Patients with Advanced Solid Tumours Treated with Pembrolizumab: Prospective Biomarker Analysis of the Multicohort, Open-Label, Phase 2 KEYNOTE-158 Study. *Lancet Oncol.* **2020**, *21*, 1353–1365. [CrossRef]
23. Balar, A.V.; Galsky, M.D.; Rosenberg, J.E.; Powles, T.; Petrylak, D.P.; Bellmunt, J.; Loriot, Y.; Necchi, A.; Hoffman-Censits, J.; Perez-Gracia, J.L.; et al. Atezolizumab as First-Line Treatment in Cisplatin-Ineligible Patients with Locally Advanced and Metastatic Urothelial Carcinoma: A Single-Arm, Multicentre, Phase 2 Trial. *Lancet* **2017**, *389*, 67–76. [CrossRef]
24. Powles, T.; Park, S.H.; Voog, E.; Caserta, C.; Valderrama, B.P.; Gurney, H.; Kalofonos, H.; Radulović, S.; Demey, W.; Ullén, A.; et al. Avelumab Maintenance Therapy for Advanced or Metastatic Urothelial Carcinoma. *N. Engl. J. Med.* **2020**, *383*, 1218–1230. [CrossRef]
25. Karn, T.; Denkert, C.; Weber, K.E.; Holtrich, U.; Hanusch, C.; Sinn, B.V.; Higgs, B.W.; Jank, P.; Sinn, H.P.; Huober, J.; et al. Tumor Mutational Burden and Immune Infiltration as Independent Predictors of Response to Neoadjuvant Immune Checkpoint Inhibition in Early TNBC in GeparNuevo. *Ann. Oncol.* **2020**, *31*, 1216–1222. [CrossRef]
26. Rozeman, E.A.; Hoefsmit, E.P.; Reijers, I.L.M.; Saw, R.P.M.; Versluis, J.M.; Krijgsman, O.; Dimitriadis, P.; Sikorska, K.; van de Wiel, B.A.; Eriksson, H.; et al. Survival and Biomarker Analyses from the OpACIN-Neo and OpACIN Neoadjuvant Immunotherapy Trials in Stage III Melanoma. *Nat. Med.* **2021**, *27*, 256–263. [CrossRef]
27. Necchi, A.; Raggi, D.; Gallina, A.; Madison, R.; Colecchia, M.; Lucianò, R.; Montironi, R.; Giannatempo, P.; Farè, E.; Pederzoli, F.; et al. Updated Results of PURE-01 with Preliminary Activity of Neoadjuvant Pembrolizumab in Patients with Muscle-Invasive Bladder Carcinoma with Variant Histologies. *Eur. Urol.* **2020**, *77*, 439–446. [CrossRef]
28. Galsky, M.D.; Banchereau, R.; Hamidi, H.R.; Leng, N.; Harris, W.; O'Donnell, P.H.; Kadel, E.E.; Yuen, K.C.Y.; Jin, D.; Koeppen, H.; et al. Tumor, Immune, and Stromal Characteristics Associated with Clinical Outcomes with Atezolizumab (Atezo) + Platinum-Based Chemotherapy (PBC) or Atezo Monotherapy (Mono) versus PBC in Metastatic Urothelial Cancer (MUC) from the Phase III IMvigor130 Study. *JCO* **2020**, *38*, 5011. [CrossRef]
29. Addeo, A.; Banna, G.L.; Weiss, G.J. Tumor Mutation Burden-From Hopes to Doubts. *JAMA Oncol.* **2019**, *5*, 934–935. [CrossRef] [PubMed]
30. Allen, E.M.V.; Mouw, K.W.; Kim, P.; Iyer, G.; Wagle, N.; Al-Ahmadie, H.; Zhu, C.; Ostrovnaya, I.; Kryukov, G.V.; O'Connor, K.W.; et al. Somatic ERCC2 Mutations Correlate with Cisplatin Sensitivity in Muscle-Invasive Urothelial Carcinoma. *Cancer Discov.* **2014**, *4*, 1140–1153. [CrossRef] [PubMed]
31. Plimack, E.R.; Dunbrack, R.L.; Brennan, T.A.; Andrake, M.D.; Zhou, Y.; Serebriiskii, I.G.; Slifker, M.; Alpaugh, K.; Dulaimi, E.; Palma, N.; et al. Defects in DNA Repair Genes Predict Response to Neoadjuvant Cisplatin-Based Chemotherapy in Muscle-Invasive Bladder Cancer. *Eur. Urol.* **2015**, *68*, 959–967. [CrossRef]
32. Nilsson, R.; Björkegren, J.; Tegnér, J. On Reliable Discovery of Molecular Signatures. *BMC Bioinform.* **2009**, *10*, 38. [CrossRef] [PubMed]
33. Sung, J.; Wang, Y.; Chandrasekaran, S.; Witten, D.M.; Price, N.D. Molecular Signatures from Omics Data: From Chaos to Consensus. *Biotechnol. J.* **2012**, *7*, 946–957. [CrossRef]
34. Sharma, P.; Siddiqui, B.A.; Anandhan, S.; Yadav, S.S.; Subudhi, S.K.; Gao, J.; Goswami, S.; Allison, J.P. The Next Decade of Immune Checkpoint Therapy. *Cancer Discov.* **2021**, *11*, 838–857. [CrossRef]
35. Lopez-Beltran, A.; López-Rios, F.; Montironi, R.; Wildsmith, S.; Eckstein, M. Immune Checkpoint Inhibitors in Urothelial Carcinoma: Recommendations for Practical Approaches to PD-L1 and Other Potential Predictive Biomarker Testing. *Cancers* **2021**, *13*, 1424. [CrossRef]
36. Mullane, S.A.; Werner, L.; Rosenberg, J.; Signoretti, S.; Callea, M.; Choueiri, T.K.; Freeman, G.J.; Bellmunt, J. Correlation of Apobec Mrna Expression with Overall Survival and Pd-L1 Expression in Urothelial Carcinoma. *Sci. Rep.* **2016**, *6*, 27702. [CrossRef]
37. Glaser, A.P.; Fantini, D.; Wang, Y.; Yu, Y.; Rimar, K.J.; Podojil, J.R.; Miller, S.D.; Meeks, J.J. APOBEC-Mediated Mutagenesis in Urothelial Carcinoma Is Associated with Improved Survival, Mutations in DNA Damage Response Genes, and Immune Response. *Oncotarget* **2018**, *9*, 4537–4548. [CrossRef]
38. Tu, M.M.; Ng, T.L.; De Jong, F.C.; Zuiverloon, T.C.M.; Fazzari, F.G.T.; Theodorescu, D. Molecular Biomarkers of Response to PD-1/PD-L1 Immune Checkpoint Blockade in Advanced Bladder Cancer. *Bladder Cancer* **2019**, *5*, 131–145. [CrossRef] [PubMed]

39. Powles, T.; Kockx, M.; Rodriguez-Vida, A.; Duran, I.; Crabb, S.J.; Van Der Heijden, M.S.; Szabados, B.; Pous, A.F.; Gravis, G.; Herranz, U.A.; et al. Clinical Efficacy and Biomarker Analysis of Neoadjuvant Atezolizumab in Operable Urothelial Carcinoma in the ABACUS Trial. *Nat. Med.* **2019**, *25*, 1706–1714. [CrossRef] [PubMed]
40. Necchi, A.; Anichini, A.; Raggi, D.; Briganti, A.; Massa, S.; Lucianò, R.; Colecchia, M.; Giannatempo, P.; Mortarini, R.; Bianchi, M.; et al. Pembrolizumab as Neoadjuvant Therapy before Radical Cystectomy in Patients With Muscle-Invasive Urothelial Bladder Carcinoma (PURE-01): An Open-Label, Single-Arm, Phase II Study. *J. Clin. Oncol.* **2018**, *36*, 3353–3360. [CrossRef]
41. Powles, T.; Carroll, D.; Chowdhury, S.; Gravis, G.; Joly, F.; Carles, J.; Fléchon, A.; Maroto, P.; Petrylak, D.; Rolland, F.; et al. An Adaptive, Biomarker-Directed Platform Study of Durvalumab in Combination with Targeted Therapies in Advanced Urothelial Cancer. *Nat. Med.* **2021**, *27*, 793–801. [CrossRef]
42. Cabel, L.; Proudhon, C.; Romano, E.; Girard, N.; Lantz, O.; Stern, M.-H.; Pierga, J.-Y.; Bidard, F.-C. Clinical Potential of Circulating Tumour DNA in Patients Receiving Anticancer Immunotherapy. *Nat. Rev. Clin. Oncol.* **2018**, *15*, 639–650. [CrossRef]
43. Corcoran, R.B.; Chabner, B.A. Application of Cell-Free DNA Analysis to Cancer Treatment. *N. Engl. J. Med.* **2018**, *379*, 1754–1765. [CrossRef] [PubMed]
44. Birkenkamp-Demtröder, K.; Christensen, E.; Nordentoft, I.; Knudsen, M.; Taber, A.; Høyer, S.; Lamy, P.; Agerbæk, M.; Jensen, J.B.; Dyrskjøt, L. Monitoring Treatment Response and Metastatic Relapse in Advanced Bladder Cancer by Liquid Biopsy Analysis. *Eur. Urol.* **2018**, *73*, 535–540. [CrossRef]
45. Patel, K.M.; van der Vos, K.E.; Smith, C.G.; Mouliere, F.; Tsui, D.; Morris, J.; Chandrananda, D.; Marass, F.; van den Broek, D.; Neal, D.E.; et al. Association of Plasma and Urinary Mutant DNA with Clinical Outcomes in Muscle Invasive Bladder Cancer. *Sci. Rep.* **2017**, *7*, 5554. [CrossRef]
46. Vandekerkhove, G.; Lavoie, J.-M.; Annala, M.; Murtha, A.J.; Sundahl, N.; Walz, S.; Sano, T.; Taavitsainen, S.; Ritch, E.; Fazli, L.; et al. Plasma CtDNA Is a Tumor Tissue Surrogate and Enables Clinical-Genomic Stratification of Metastatic Bladder Cancer. *Nat. Commun.* **2021**, *12*, 184. [CrossRef]
47. Christensen, E.; Birkenkamp-Demtröder, K.; Sethi, H.; Shchegrova, S.; Salari, R.; Nordentoft, I.; Wu, H.-T.; Knudsen, M.; Lamy, P.; Lindskrog, S.V.; et al. Early Detection of Metastatic Relapse and Monitoring of Therapeutic Efficacy by Ultra-Deep Sequencing of Plasma Cell-Free DNA in Patients with Urothelial Bladder Carcinoma. *J. Clin. Oncol.* **2019**, *37*, 1547–1557. [CrossRef]
48. Powles, T.B.; Assaf, Z.J.; Davarpanah, N.; Hussain, M.; Oudard, S.; Gschwend, J.E.; Albers, P.; Castellano, D.; Nishiyama, H.; Daneshmand, S.; et al. 1O Clinical Outcomes in Post-Operative CtDNA-Positive Muscle-Invasive Urothelial Carcinoma (MIUC) Patients after Atezolizumab Adjuvant Therapy. *Ann. Oncol.* **2020**, *31*, S1417. [CrossRef]
49. O'Donnell, P.H.; Balar, A.V.; Vuky, J.; Castellano, D.; Bellmunt, J.; Powles, T.; Bajorin, D.F.; Grivas, P.; Hahn, N.M.; Plimack, E.R.; et al. First-Line Pembrolizumab (Pembro) in Cisplatin-Ineligible Patients with Advanced Urothelial Cancer (UC): Response and Survival Results up to Five Years from the KEYNOTE-052 Phase 2 Study. *JCO* **2021**, *39*, 4508. [CrossRef]
50. Alva, A.; Csőszi, T.; Ozguroglu, M.; Matsubara, N.; Geczi, L.; Cheng, S.Y.-S.; Fradet, Y.; Oudard, S.; Vulsteke, C.; Barrera, R.M.; et al. LBA23 Pembrolizumab (P) Combined with Chemotherapy (C) vs C Alone as First-Line (1L) Therapy for Advanced Urothelial Carcinoma (UC): KEYNOTE-361. *Ann. Oncol.* **2020**, *31*, S1155. [CrossRef]
51. Galsky, M.D.; Arija, J.Á.A.; Bamias, A.; Davis, I.D.; De Santis, M.; Kikuchi, E.; Garcia-Del-Muro, X.; De Giorgi, U.; Mencinger, M.; Izumi, K.; et al. Atezolizumab with or without Chemotherapy in Metastatic Urothelial Cancer (IMvigor130): A Multicentre, Randomised, Placebo-Controlled Phase 3 Trial. *Lancet* **2020**, *395*, 1547–1557. [CrossRef]
52. Powles, T.; van der Heijden, M.S.; Castellano, D.; Galsky, M.D.; Loriot, Y.; Petrylak, D.P.; Ogawa, O.; Park, S.H.; Lee, J.-L.; De Giorgi, U.; et al. Durvalumab Alone and Durvalumab plus Tremelimumab versus Chemotherapy in Previously Untreated Patients with Unresectable, Locally Advanced or Metastatic Urothelial Carcinoma (DANUBE): A Randomised, Open-Label, Multicentre, Phase 3 Trial. *Lancet Oncol.* **2020**, *21*, 1574–1588. [CrossRef]
53. Powles, T.B.; Loriot, Y.; Bellmunt, J.; Sternberg, C.N.; Sridhar, S.; Petrylak, D.P.; Tambaro, R.; Dourthe, L.M.; Alvarez-Fernandez, C.; Aarts, M.; et al. 699O Avelumab First-Line (1L) Maintenance + Best Supportive Care (BSC) vs. BSC Alone for Advanced Urothelial Carcinoma (UC): Association between Clinical Outcomes and Exploratory Biomarkers. *Ann. Oncol.* **2020**, *31*, S552–S553. [CrossRef]
54. Hussain, M.H.A.; Powles, T.; Albers, P.; Castellano, D.; Daneshmand, S.; Gschwend, J.; Nishiyama, H.; Oudard, S.; Tayama, D.; Davarpanah, N.N.; et al. IMvigor010: Primary Analysis from a Phase III Randomized Study of Adjuvant Atezolizumab (Atezo) versus Observation (Obs) in High-Risk Muscle-Invasive Urothelial Carcinoma (MIUC). *JCO* **2020**, *38*, 5000. [CrossRef]
55. Bajorin, D.F.; Witjes, J.A.; Gschwend, J.E.; Schenker, M.; Valderrama, B.P.; Tomita, Y.; Bamias, A.; Lebret, T.; Shariat, S.F.; Park, S.H.; et al. Adjuvant Nivolumab versus Placebo in Muscle-Invasive Urothelial Carcinoma. *N. Engl. J. Med.* **2021**, *384*, 2102–2114. [CrossRef] [PubMed]
56. Powles, T.; Durán, I.; van der Heijden, M.S.; Loriot, Y.; Vogelzang, N.J.; De Giorgi, U.; Oudard, S.; Retz, M.M.; Castellano, D.; Bamias, A.; et al. Atezolizumab versus Chemotherapy in Patients with Platinum-Treated Locally Advanced or Metastatic Urothelial Carcinoma (IMvigor211): A Multicentre, Open-Label, Phase 3 Randomised Controlled Trial. *Lancet* **2018**, *391*, 748–757. [CrossRef]
57. Bellmunt, J.; de Wit, R.; Vaughn, D.J.; Fradet, Y.; Lee, J.-L.; Fong, L.; Vogelzang, N.J.; Climent, M.A.; Petrylak, D.P.; Choueiri, T.K.; et al. Pembrolizumab as Second-Line Therapy for Advanced Urothelial Carcinoma. *N. Engl. J. Med.* **2017**, *376*, 1015–1026. [CrossRef]

68. Banna, G.L.; Cantale, O.; Bersanelli, M.; Del Re, M.; Friedlaender, A.; Cortellini, A.; Addeo, A. Are Anti-PD1 and Anti-PD-L1 Alike? The Non-Small-Cell Lung Cancer Paradigm. *Oncol. Rev.* **2020**, *14*, 490. [CrossRef]
69. Bajorin, D.F.; Dodd, P.M.; Mazumdar, M.; Fazzari, M.; McCaffrey, J.A.; Scher, H.I.; Herr, H.; Higgins, G.; Boyle, M.G. Long-Term Survival in Metastatic Transitional-Cell Carcinoma and Prognostic Factors Predicting Outcome of Therapy. *J. Clin. Oncol.* **1999**, *17*, 3173–3181. [CrossRef]
70. Bellmunt, J.; Choueiri, T.K.; Fougeray, R.; Schutz, F.A.B.; Salhi, Y.; Winquist, E.; Culine, S.; von der Maase, H.; Vaughn, D.J.; Rosenberg, J.E. Prognostic Factors in Patients with Advanced Transitional Cell Carcinoma of the Urothelial Tract Experiencing Treatment Failure with Platinum-Containing Regimens. *J. Clin. Oncol.* **2010**, *28*, 1850–1855. [CrossRef]
71. Sonpavde, G.; Pond, G.R.; Fougeray, R.; Choueiri, T.K.; Qu, A.Q.; Vaughn, D.J.; Niegisch, G.; Albers, P.; James, N.D.; Wong, Y.-N.; et al. Time from Prior Chemotherapy Enhances Prognostic Risk Grouping in the Second-Line Setting of Advanced Urothelial Carcinoma: A Retrospective Analysis of Pooled, Prospective Phase 2 Trials. *Eur. Urol.* **2013**, *63*, 717–723. [CrossRef]
72. Matsumoto, R.; Abe, T.; Ishizaki, J.; Kikuchi, H.; Harabayashi, T.; Minami, K.; Sazawa, A.; Mochizuki, T.; Akino, T.; Murakumo, M.; et al. Outcome and Prognostic Factors in Metastatic Urothelial Carcinoma Patients Receiving Second-Line Chemotherapy: An Analysis of Real-World Clinical Practice Data in Japan. *Jpn. J. Clin. Oncol.* **2018**, *48*, 771–776. [CrossRef]
73. Shabto, J.M.; Martini, D.J.; Liu, Y.; Ravindranathan, D.; Brown, J.; Hitron, E.E.; Russler, G.A.; Caulfield, S.; Kissick, H.; Alemozaffar, M.; et al. Novel Risk Group Stratification for Metastatic Urothelial Cancer Patients Treated with Immune Checkpoint Inhibitors. *Cancer Med.* **2020**, *9*, 2752–2760. [CrossRef]
74. Suh, J.; Jung, J.H.; Jeong, C.W.; Kwak, C.; Kim, H.H.; Ku, J.H. Clinical Significance of Pre-Treated Neutrophil-Lymphocyte Ratio in the Management of Urothelial Carcinoma: A Systemic Review and Meta-Analysis. *Front. Oncol.* **2019**, *9*, 1365. [CrossRef] [PubMed]
75. Kobayashi, K.; Suzuki, K.; Hiraide, M.; Aoyama, T.; Yokokawa, T.; Shikibu, S.; Hashimoto, K.; Iikura, Y.; Sato, H.; Sugiyama, E.; et al. Association of Immune-Related Adverse Events with Pembrolizumab Efficacy in the Treatment of Advanced Urothelial Carcinoma. *Oncology* **2020**, *98*, 237–242. [CrossRef]
76. Khaki, A.R.; Li, A.; Diamantopoulos, L.N.; Bilen, M.A.; Santos, V.; Esther, J.; Morales-Barrera, R.; Devitt, M.; Nelson, A.; Hoimes, C.J.; et al. Impact of Performance Status on Treatment Outcomes: A Real-World Study of Advanced Urothelial Cancer Treated with Immune Checkpoint Inhibitors. *Cancer* **2020**, *126*, 1208–1216. [CrossRef]
77. Furubayashi, N.; Negishi, T.; Miura, A.; Nakamura, N.; Nakamura, M. Organ-Specific Therapeutic Effect of Paclitaxel and Carboplatin Chemotherapy After Platinum-Based Chemotherapy and Pembrolizumab for Metastatic Urothelial Carcinoma. *Res. Rep. Urol.* **2020**, *12*, 455–461. [CrossRef]
78. Ruiz-Bañobre, J.; Molina-Díaz, A.; Fernández-Calvo, O.; Fernández-Núñez, N.; Medina-Colmenero, A.; Santomé, L.; Lázaro-Quintela, M.; Mateos-González, M.; García-Cid, N.; López-López, R.; et al. Rethinking Prognostic Factors in Locally Advanced or Metastatic Urothelial Carcinoma in the Immune Checkpoint Blockade Era: A Multicenter Retrospective Study. *ESMO Open* **2021**, *6*, 100090. [CrossRef] [PubMed]
79. Fujiwara, M.; Yuasa, T.; Urasaki, T.; Komai, Y.; Fujiwara, R.; Numao, N.; Yamamoto, S.; Yonese, J. Effectiveness and Safety Profile of Pembrolizumab for Metastatic Urothelial Cancer: A Retrospective Single-Center Analysis in Japan. *Cancer Rep.* **2021**, e1398. [CrossRef]
80. Tamura, D.; Jinnouchi, N.; Abe, M.; Ikarashi, D.; Matsuura, T.; Kato, R.; Maekawa, S.; Kato, Y.; Kanehira, M.; Takata, R.; et al. Prognostic Outcomes and Safety in Patients Treated with Pembrolizumab for Advanced Urothelial Carcinoma: Experience in Real-World Clinical Practice. *Int. J. Clin. Oncol.* **2020**, *25*, 899–905. [CrossRef] [PubMed]
81. Necchi, A.; Fradet, Y.; Bellmunt, J.; de Wit, R.; Lee, J.; Fong, L.; Vozelgang, N.J.; Climent, M.A.; Petrylak, D.P.; Choueiri, T.K.; et al. 3598-Three-Year Follow-Up From the Phase 3 KEYNOTE-045 Trial: Pembrolizumab (Pembro) Versus Investigator's Choice (Paclitaxel, Docetaxel, or Vinflunine) in Recurrent, Advanced Urothelial Cancer (UC). *Ann. Oncol.* **2019**, *30* (Suppl. 5), V356–V402. [CrossRef]
82. Grivas, P.; Park, S.H.; Voog, E.; Caserta, C.; Valderrama, B.P.; Gurney, H.; Kalofonos, H.; Radulovic, S.; Demey, W.; Ullén, A.; et al. 704MO Avelumab First-Line (1L) Maintenance + Best Supportive Care (BSC) vs BSC Alone with 1L Chemotherapy (CTx) for Advanced Urothelial Carcinoma (UC): Subgroup Analyses from JAVELIN Bladder 100. *Ann. Oncol.* **2020**, *31*, S555–S556. [CrossRef]
83. Buti, S.; Bersanelli, M.; Perrone, F.; Bracarda, S.; Di Maio, M.; Giusti, R.; Nigro, O.; Cortinovis, D.L.; Aerts, J.G.J.V.; Guaitoli, G.; et al. Predictive Ability of a Drug-Based Score in Patients with Advanced Non-Small-Cell Lung Cancer Receiving First-Line Immunotherapy. *Eur. J. Cancer* **2021**, *150*, 224–231. [CrossRef] [PubMed]
84. Cortellini, A.; Tucci, M.; Adamo, V.; Stucci, L.S.; Russo, A.; Tanda, E.T.; Spagnolo, F.; Rastelli, F.; Bisonni, R.; Santini, D.; et al. Integrated Analysis of Concomitant Medications and Oncological Outcomes from PD-1/PD-L1 Checkpoint Inhibitors in Clinical Practice. *J. Immunother. Cancer* **2020**, *8*, e001361. [CrossRef] [PubMed]
85. Petrelli, F.; Iaculli, A.; Signorelli, D.; Ghidini, A.; Dottorini, L.; Perego, G.; Ghidini, M.; Zaniboni, A.; Gori, S.; Inno, A. Survival of Patients Treated with Antibiotics and Immunotherapy for Cancer: A Systematic Review and Meta-Analysis. *J. Clin. Med.* **2020**, *9*, 1458. [CrossRef]

76. Rossi, G.; Pezzuto, A.; Sini, C.; Tuzi, A.; Citarella, F.; McCusker, M.G.; Nigro, O.; Tanda, E.; Russo, A. Concomitant Medications during Immune Checkpoint Blockage in Cancer Patients: Novel Insights in This Emerging Clinical Scenario. *Crit. Rev. Oncol. Hematol.* **2019**, *142*, 26–34. [CrossRef]
77. Li, C.; Xia, Z.; Li, A.; Meng, J. The Effect of Proton Pump Inhibitor Uses on Outcomes for Cancer Patients Treated with Immune Checkpoint Inhibitors: A Meta-Analysis. *Ann. Transl. Med.* **2020**, *8*, 1655. [CrossRef]
78. Xu, H.; He, A.; Liu, A.; Tong, W.; Cao, D. Evaluation of the Prognostic Role of Platelet-Lymphocyte Ratio in Cancer Patients Treated with Immune Checkpoint Inhibitors: A Systematic Review and Meta-Analysis. *Int. Immunopharmacol.* **2019**, *77*, 105957. [CrossRef] [PubMed]
79. Lee, K.A.; Shaw, H.M.; Bataille, V.; Nathan, P.; Spector, T.D. Role of the Gut Microbiome for Cancer Patients Receiving Immunotherapy: Dietary and Treatment Implications. *Eur. J. Cancer* **2020**, *138*, 149–155. [CrossRef]
80. Hopkins, A.M.; Kichenadasse, G.; Karapetis, C.S.; Rowland, A.; Sorich, M.J. Concomitant Proton Pump Inhibitor Use and Survival in Urothelial Carcinoma Treated with Atezolizumab. *Clin. Cancer Res.* **2020**, *26*, 5487–5493. [CrossRef]
81. Hopkins, A.M.; Kichenadasse, G.; Karapetis, C.S.; Rowland, A.; Sorich, M.J. Concomitant Antibiotic Use and Survival in Urothelial Carcinoma Treated with Atezolizumab. *Eur. Urol.* **2020**, *78*, 540–543. [CrossRef]
82. Hanahan, D.; Weinberg, R.A. Hallmarks of Cancer: The next Generation. *Cell* **2011**, *144*, 646–674. [CrossRef]
83. Templeton, A.J.; McNamara, M.G.; Šeruga, B.; Vera-Badillo, F.E.; Aneja, P.; Ocaña, A.; Leibowitz-Amit, R.; Sonpavde, G.; Knox, J.J.; Tran, B.; et al. Prognostic Role of Neutrophil-to-Lymphocyte Ratio in Solid Tumors: A Systematic Review and Meta-Analysis. *J. Natl. Cancer Inst.* **2014**, *106*, dju124. [CrossRef]
84. Templeton, A.J.; Ace, O.; McNamara, M.G.; Al-Mubarak, M.; Vera-Badillo, F.E.; Hermanns, T.; Seruga, B.; Ocaña, A.; Tannock, I.F.; Amir, E. Prognostic Role of Platelet to Lymphocyte Ratio in Solid Tumors: A Systematic Review and Meta-Analysis. *Cancer Epidemiol. Biomark. Prev.* **2014**, *23*, 1204–1212. [CrossRef] [PubMed]
85. Kumarasamy, C.; Sabarimurugan, S.; Madurantakam, R.M.; Lakhotiya, K.; Samiappan, S.; Baxi, S.; Nachimuthu, R.; Gothandam, K.M.; Jayaraj, R. Prognostic Significance of Blood Inflammatory Biomarkers NLR, PLR, and LMR in Cancer-A Protocol for Systematic Review and Meta-Analysis. *Medicine* **2019**, *98*, e14834. [CrossRef] [PubMed]
86. Rossi, L.; Santoni, M.; Crabb, S.J.; Scarpi, E.; Burattini, L.; Chau, C.; Bianchi, E.; Savini, A.; Burgio, S.L.; Conti, A.; et al. High Neutrophil-to-Lymphocyte Ratio Persistent during First-Line Chemotherapy Predicts Poor Clinical Outcome in Patients with Advanced Urothelial Cancer. *Ann. Surg. Oncol.* **2015**, *22*, 1377–1384. [CrossRef] [PubMed]
87. Yuk, H.D.; Ku, J.H. Role of Systemic Inflammatory Response Markers in Urothelial Carcinoma. *Front. Oncol.* **2020**, *10*, 1473. [CrossRef]
88. Wu, S.; Zhao, X.; Wang, Y.; Zhong, Z.; Zhang, L.; Cao, J.; Ai, K.; Xu, R. Pretreatment Neutrophil-Lymphocyte Ratio as a Predictor in Bladder Cancer and Metastatic or Unresectable Urothelial Carcinoma Patients: A Pooled Analysis of Comparative Studies. *Cell. Physiol. Biochem.* **2018**, *46*, 1352–1364. [CrossRef]
89. Xu, H.; Xu, X.; Wang, H.; Ge, W.; Cao, D. The Association between Antibiotics Use and Outcome of Cancer Patients Treated with Immune Checkpoint Inhibitors: A Systematic Review and Meta-Analysis. *Crit. Rev. Oncol. Hematol.* **2020**, *149*, 102909. [CrossRef] [PubMed]
90. Sacdalan, D.B.; Lucero, J.A.; Sacdalan, D.L. Prognostic Utility of Baseline Neutrophil-to-Lymphocyte Ratio in Patients Receiving Immune Checkpoint Inhibitors: A Review and Meta-Analysis. *Onco Targets Ther.* **2018**, *11*, 955–965. [CrossRef]
91. Banna, G.L.; Signorelli, D.; Metro, G.; Galetta, D.; De Toma, A.; Cantale, O.; Banini, M.; Friedlaender, A.; Pizzutillo, P.; Garassino, M.C.; et al. Neutrophil-to-Lymphocyte Ratio in Combination with PD-L1 or Lactate Dehydrogenase as Biomarkers for High PD-L1 Non-Small Cell Lung Cancer Treated with First-Line Pembrolizumab. *Transl. Lung Cancer Res.* **2020**, *9*, 1533–1542. [CrossRef] [PubMed]
92. Ogihara, K.; Kikuchi, E.; Shigeta, K.; Okabe, T.; Hattori, S.; Yamashita, R.; Yoshimine, S.; Shirotake, S.; Nakazawa, R.; Matsumoto, K.; et al. The Pretreatment Neutrophil-to-Lymphocyte Ratio Is a Novel Biomarker for Predicting Clinical Responses to Pembrolizumab in Platinum-Resistant Metastatic Urothelial Carcinoma Patients. *Urol. Oncol.* **2020**, *38*, 602.e1–602.e10. [CrossRef] [PubMed]
93. Shimizu, T.; Miyake, M.; Hori, S.; Ichikawa, K.; Omori, C.; Iemura, Y.; Owari, T.; Itami, Y.; Nakai, Y.; Anai, S.; et al. Clinical Impact of Sarcopenia and Inflammatory/Nutritional Markers in Patients with Unresectable Metastatic Urothelial Carcinoma Treated with Pembrolizumab. *Diagnostics* **2020**, *10*, 310. [CrossRef] [PubMed]
94. Brown, J.T.; Liu, Y.; Shabto, J.M.; Martini, D.J.; Ravindranathan, D.; Hitron, E.E.; Russler, G.A.; Caulfield, S.; Yantorni, L.B.; Joshi, S.S.; et al. Baseline Modified Glasgow Prognostic Score Associated with Survival in Metastatic Urothelial Carcinoma Treated with Immune Checkpoint Inhibitors. *Oncologist* **2021**, *26*, 397–405. [CrossRef] [PubMed]
95. Yamamoto, Y.; Yatsuda, J.; Shimokawa, M.; Fuji, N.; Aoki, A.; Sakano, S.; Yamamoto, M.; Suga, A.; Tei, Y.; Yoshihiro, S.; et al. Prognostic Value of Pre-Treatment Risk Stratification and Post-Treatment Neutrophil/Lymphocyte Ratio Change for Pembrolizumab in Patients with Advanced Urothelial Carcinoma. *Int. J. Clin. Oncol.* **2021**, *26*, 169–177. [CrossRef]
96. Nassar, A.H.; Mouw, K.W.; Jegede, O.; Shinagare, A.B.; Kim, J.; Liu, C.-J.; Pomerantz, M.; Harshman, L.C.; Van Allen, E.M.; Wei, X.X.; et al. A Model Combining Clinical and Genomic Factors to Predict Response to PD-1/PD-L1 Blockade in Advanced Urothelial Carcinoma. *Br. J. Cancer* **2020**, *122*, 555–563. [CrossRef]

97. Khaki, A.R.; Diamantopoulos, L.N.; Li, A.; Devitt, M.E.; Drakaki, A.; Shreck, E.; Joshi, M.; Velho, P.I.; Alonso, L.; Nelson, A.A.; et al. Outcomes of Patients (Pts) with Metastatic Urothelial Cancer (MUC) and Poor Performance Status (PS) Receiving Anti-PD(L)1 Agents. *JCO* **2019**, *37*, 4525. [CrossRef]
98. Banna, G.L.; Di Quattro, R.; Malatino, L.; Fornarini, G.; Addeo, A.; Maruzzo, M.; Urzia, V.; Rundo, F.; Lipari, H.; De Giorgi, U.; et al. Neutrophil-to-Lymphocyte Ratio and Lactate Dehydrogenase as Biomarkers for Urothelial Cancer Treated with Immunotherapy. *Clin. Transl. Oncol.* **2020**, *22*, 2130–2135. [CrossRef]
99. Prelaj, A.; Rebuzzi, S.E.; Pizzutilo, P.; Bilancia, M.; Montrone, M.; Pesola, F.; Longo, V.; Del Bene, G.; Lapadula, V.; Cassano, F.; et al. EPSILoN: A Prognostic Score Using Clinical and Blood Biomarkers in Advanced Non-Small-Cell Lung Cancer Treated With Immunotherapy. *Clin. Lung Cancer* **2020**, *21*, 365–377.e5. [CrossRef]
100. Mezquita, L.; Auclin, E.; Ferrara, R.; Charrier, M.; Remon, J.; Planchard, D.; Ponce, S.; Ares, L.P.; Leroy, L.; Audigier-Valette, C.; et al. Association of the Lung Immune Prognostic Index with Immune Checkpoint Inhibitor Outcomes in Patients with Advanced Non-Small Cell Lung Cancer. *JAMA Oncol.* **2018**, *4*, 351–357. [CrossRef]
101. Chrom, P.; Zolnierek, J.; Bodnar, L.; Stec, R.; Szczylik, C. External Validation of the Systemic Immune-Inflammation Index as a Prognostic Factor in Metastatic Renal Cell Carcinoma and Its Implementation within the International Metastatic Renal Cell Carcinoma Database Consortium Model. *Int. J. Clin. Oncol.* **2019**, *24*, 526–532. [CrossRef] [PubMed]
102. Martini, D.J.; Liu, Y.; Shabto, J.M.; Carthon, B.C.; Hitron, E.E.; Russler, G.A.; Caulfield, S.; Kissick, H.T.; Harris, W.B.; Kucuk, O.; et al. Novel Risk Scoring System for Patients with Metastatic Renal Cell Carcinoma Treated with Immune Checkpoint Inhibitors. *Oncologist* **2020**, *25*, e484–e491. [CrossRef] [PubMed]
103. Rebuzzi, S.E.; Signori, A.; Banna, G.L.; Maruzzo, M.; De Giorgi, U.; Pedrazzoli, P.; Sbrana, A.; Zucali, P.A.; Masini, C.; Naglieri, E.; et al. Inflammatory Indices and Clinical Factors in Metastatic Renal Cell Carcinoma Patients Treated with Nivolumab: The Development of a Novel Prognostic Score (Meet-URO 15 Study). *Ther. Adv. Med. Oncol.* **2021**, *13*, 17588359211019642. [CrossRef] [PubMed]
104. Fornarini, G.; Rebuzzi, S.E.; Banna, G.L.; Calabrò, F.; Scandurra, G.; De Giorgi, U.; Masini, C.; Baldessari, C.; Naglieri, E.; Caserta, C.; et al. Immune-Inflammatory Biomarkers as Prognostic Factors for Immunotherapy in Pretreated Advanced Urinary Tract Cancer Patients: An Analysis of the Italian SAUL Cohort. *ESMO Open* **2021**, *6*, 100118. [CrossRef]
105. Sonpavde, G.; Manitz, J.; Gao, C.; Tayama, D.; Kaiser, C.; Hennessy, D.; Makari, D.; Gupta, A.; Abdullah, S.E.; Niegisch, G.; et al. Five-Factor Prognostic Model for Survival of Post-Platinum Patients with Metastatic Urothelial Carcinoma Receiving PD-L1 Inhibitors. *J. Urol.* **2020**, *204*, 1173–1179. [CrossRef] [PubMed]
106. Abuhelwa, A.Y.; Kichenadasse, G.; McKinnon, R.A.; Rowland, A.; Hopkins, A.M.; Sorich, M.J. Machine Learning for Prediction of Survival Outcomes with Immune-Checkpoint Inhibitors in Urothelial Cancer. *Cancers* **2021**, *13*, 2001. [CrossRef]
107. Trebeschi, S.; Drago, S.G.; Birkbak, N.J.; Kurilova, I.; Călin, A.M.; Delli Pizzi, A.; Lalezari, F.; Lambregts, D.M.J.; Rohaan, M.W.; Parmar, C.; et al. Predicting Response to Cancer Immunotherapy Using Noninvasive Radiomic Biomarkers. *Ann. Oncol.* **2019**, *30*, 998–1004. [CrossRef]
108. Banna, G.L.; Olivier, T.; Rundo, F.; Malapelle, U.; Fraggetta, F.; Libra, M.; Addeo, A. The Promise of Digital Biopsy for the Prediction of Tumor Molecular Features and Clinical Outcomes Associated With Immunotherapy. *Front. Med.* **2019**, *6*, 172. [CrossRef] [PubMed]
109. Wang, J.H.; Wahid, K.A.; van Dijk, L.V.; Farahani, K.; Thompson, R.F.; Fuller, C.D. Radiomic Biomarkers of Tumor Immune Biology and Immunotherapy Response. *Clin. Transl. Radiat. Oncol.* **2021**, *28*, 97–115. [CrossRef] [PubMed]
110. Zhang, C.; de Fonseca, L.; Shi, Z.; Zhu, C.; Dekker, A.; Bermejo, I.; Wee, L. Systematic Review of Radiomic Biomarkers for Predicting Immune Checkpoint Inhibitor Treatment Outcomes. *Methods* **2021**, *188*, 61–72. [CrossRef]
111. Trebeschi, S.; Bodalal, Z.; Boellaard, T.N.; Tareco Bucho, T.M.; Drago, S.G.; Kurilova, I.; Calin-Vainak, A.M.; Delli Pizzi, A.; Muller, M.; Hummelink, K.; et al. Prognostic Value of Deep Learning-Mediated Treatment Monitoring in Lung Cancer Patients Receiving Immunotherapy. *Front. Oncol.* **2021**, *11*, 609054. [CrossRef] [PubMed]
112. Park, K.J.; Lee, J.-L.; Yoon, S.-K.; Heo, C.; Park, B.W.; Kim, J.K. Radiomics-Based Prediction Model for Outcomes of PD-1/PD-L1 Immunotherapy in Metastatic Urothelial Carcinoma. *Eur. Radiol.* **2020**, *30*, 5392–5403. [CrossRef] [PubMed]
113. Rundo, F.; Bersanelli, M.; Urzia, V.; Friedlaender, A.; Cantale, O.; Calcara, G.; Addeo, A.; Banna, G.L. Three-Dimensional Deep Noninvasive Radiomics for the Prediction of Disease Control in Patients With Metastatic Urothelial Carcinoma Treated With Immunotherapy. *Clin. Genitourin. Cancer* **2021**, *19*, 396–404. [CrossRef] [PubMed]

Review

Review of Experimental Studies to Improve Radiotherapy Response in Bladder Cancer: Comments and Perspectives

Linda Silina [1,2,*], Fatlinda Maksut [2], Isabelle Bernard-Pierrot [1], François Radvanyi [1], Gilles Créhange [3], Frédérique Mégnin-Chanet [2] and Pierre Verrelle [2,3,4,*]

[1] French League Against Cancer Team, CNRS UMR144, Curie Institute and PSL Research University, 75005 Paris, France; isabelle.bernard-pierrot@curie.fr (I.B.-P.); francois.radvanyi@curie.fr (F.R.)
[2] CNRS UMR 9187, INSERM U1196, Curie Institute, PSL Research University and Paris-Saclay University, Rue H. Becquerel, 91405 Orsay, France; fatlinda.maksut@curie.fr (F.M.); frederique.megnin@curie.fr (F.M.-C.)
[3] Radiation Oncology Department, Curie Institute, 75005 Paris, France; gilles.crehange@curie.fr
[4] Clermont Auvergne University, 63000 Clermont-Ferrand, France
* Correspondence: linda.silina@curie.fr (L.S.); pierre.verrelle@curie.fr (P.V.)

Simple Summary: Bladder cancer is a major global health problem. Bladder removal surgery is the standard treatment for muscle-invasive bladder cancer (25% of all bladder cancer), but this treatment negatively affects the quality of life, especially for elderly and frail patients. Tumour resection followed by combination of radiotherapy and chemotherapy has emerged as a promising bladder preserving strategy. However, this strategy is unable to avoid radiation-related bladder side effects. Therefore, it is of great interest to discover novel strategies radiosensitising tumours while sparing normal bladder tissue. In this review, we analysed the experimental studies of radiosensitising strategies in bladder cancer and provided suggestions to improve forthcoming studies.

Abstract: Bladder cancer is among the top ten most common cancer types in the world. Around 25% of all cases are muscle-invasive bladder cancer, for which the gold standard treatment in the absence of metastasis is the cystectomy. In recent years, trimodality treatment associating maximal transurethral resection and radiotherapy combined with concurrent chemotherapy is increasingly used as an organ-preserving alternative. However, the use of this treatment is still limited by the lack of biomarkers predicting tumour response and by a lack of targeted radiosensitising drugs that can improve the therapeutic index, especially by limiting side effects such as bladder fibrosis. In order to improve the bladder-preserving treatment, experimental studies addressing these main issues ought to be considered (both in vitro and in vivo studies). Following the Preferred Reporting Items for Systematic Reviews and Meta-Analyses (PRISMA) guidelines for systematic reviews, we conducted a literature search in PubMed on experimental studies investigating how to improve bladder cancer radiotherapy with different radiosensitising agents using a comprehensive search string. We made comments on experimental model selection, experimental design and results, formulating the gaps of knowledge still existing: such as the lack of reliable predictive biomarkers of tumour response to chemoradiation according to the molecular tumour subtype and lack of efficient radiosensitising agents specifically targeting bladder tumour cells. We provided guidance to improve forthcoming studies, such as taking into account molecular characteristics of the preclinical models and highlighted the value of using patient-derived xenografts as well as syngeneic models. Finally, this review could be a useful tool to set up new radiation-based combined treatments with an improved therapeutic index that is needed for bladder preservation.

Keywords: bladder cancer; radiotherapy; radiosensitisation; molecular subtypes; preclinical studies; bladder cancer cell lines

1. Introduction

While radical cystectomy has taken the central place in the treatment of muscle-invasive bladder cancer (MIBC) in recent decades, radiation-based treatments have also been investigated. Radiotherapy (RT) alone with curative intent for MIBC was extensively used in the 1950s through the 1980s. From 1981 to 1985, the addition of concurrent chemotherapy to RT was investigated. The National Bladder Cancer Group first used cisplatin as a radiosensitiser for MIBC patients who were ineligible for cystectomy and observed high complete response and survival rates, which consequently encouraged further studies [1] (see Table S1).

Housset and colleagues first reported promising findings using 5-fluorouracil (5-FU) + cisplatin combination as a radiosensitiser in MIBC [2]. Following further studies, it became evident that the concurrent chemoradiotherapy (CCRT) improves locoregional disease control in MIBC as compared to RT alone [3–5]. However, despite the existing volume of research, there remains no standard procedure of CCRT regimen. Although different chemotherapy (CT) agents have been investigated, most evidence exists for cisplatin [3] or mitomycin C + 5-FU, [4] and more recently for gemcitabine [6]. In addition, other approaches have been explored such as the use of nicotinamide and carbogen to fight hypoxia-related radioresistance [7]. Mitomycin C + 5-FU is a very effective radiosensitising combination that has improved clinical outcomes in head and neck and anal cancers [8–10]. Although cisplatin + 5-FU is widely delivered, mitomycin C + 5-FU is also a common combination particularly for frailer and elderly bladder cancer (BCa) patients, given the absence of nephrotoxicity when compared to platinum drugs [4,11].

With the advances of the cystectomy techniques, radical cystectomy with pelvic lymphadenectomy and cisplatin-based CT has become the gold standard treatment for patients with MIBC. RT can be considered as an adjuvant therapy following radical cystectomy in patients with pathological high-risk of loco regional relapse (i.e., pT3-4, positive nodes, positive surgical margins), but the pelvic toxicity remains significant despite the advances in RT such as intensity-modulated radiation therapy. This management approach is supported by numerous renowned organisations, such as the National Comprehensive Cancer Network in the United States [12], as well as by the European Association of Urology [13]. In fact, the latter has made strong recommendation to use cisplatin based neoadjuvant CT before radical cystectomy for treating MIBC (T2-T4aN0M0) and high-risk non-muscle invasive bladder cancer. Level 1a evidence supports that neoadjuvant cisplatin-based CT increases survival at 6 years by 8% [13,14]. Although post-radical cystectomy history can be associated with increased risk of infection, extensive bleeding, affected sexual function and quality of life, it achieves locoregional control and results in 60% of the overall 5-year survival [15,16]. The absence of prospective randomised studies has impeded comparison of radical cystectomy versus other forms of therapy [17]. The treatment choice for MIBC between radical cystectomy versus bladder preservation largely depends on the specialist expertise in the treatment centre and often varies among countries.

Although RT has been used in bladder cancer (BCa) treatment since the 1950s, there is a relatively low number of experimental studies on the topic of radiosensitisers in BCa compared to other cancer types. The first study identified was in the 1979 [18]. Altogether, we identified 85 studies investigating RT in BCa experimental models published between 1979 and October 2020.

2. Radiotherapy as Part of Bladder Preserving Treatment in Clinics

In the last decade, trimodality treatment consisting of maximal transurethral resection of bladder tumour (TURBT) coupled with CT has emerged as a bladder sparing treatment either driven by patients' choice or due to the patients' ineligibility for radical cystectomy. In most of the CCRT protocols, including the pioneering study of Housset et al. [2], following cystoscopic evaluation of the initial CCRT response, good responders complete the CCRT schedule. Bladder preservation outcomes heavily depend on tumour response to CCRT (reviewed by [19]) and in the case of poor response, radical cystectomy is planned [2].

A standard RT schedule consists of external beam radiation therapy (EBRT) to the bladder and limited pelvic lymph nodes with an initial dose of 40–46 Gy, with a boost to the whole bladder of 14–20 Gy, with a total dose of 60–66 Gy [20] with conventional fractionation. Partial bladder irradiation remains controversial [21] as well as a tumour dose escalation which is still investigational [22]. Moderate hypofractionation is a well-tolerated option for frailer and elderly patients, even in combination with CT [23]. In addition, RT has been recently successfully combined with several immunotherapy agents in clinical trials for metastatic BCa [24].

3. Limitations of Use of RT in MIBC in Clinics

There are three main limitations of using CCRT in MIBC. Firstly, there is a significant risk of pelvic recurrence (25–50%) [17]. Secondly, CCRT treatment may create damage to the bladder wall resulting in undesirable toxicity. Late toxicity is characterised by replication of the injured vascular endothelial cells and connective tissue, but failure of regeneration, and may result in fibrosis, which can lead to the need of ultimate cystectomy [25]. Presently, the underlying reasons of including whole bladder in the clinical target volume (CTV) are that the irradiation field is difficult to be adjusted to concentrate on the bulk tumour and due to the high risk of spread of bladder tumours within the urothelium layers. It is proving problematic to reliably and accurately define the CTV exact position for the RT delivery as the bladder volume is continuously changing with the level of urine and post-void residual volume [26]. Nevertheless, it is important to emphasise that modern radiotherapy techniques such as image-guided radiation therapy, intensity-modulated radiation therapy and volumetric-modulated arc therapy have significantly advanced and improved the sparing of pelvic organs, especially small intestines, while better targeting delivery to the bladder (reviewed by [23]).

Thirdly, at the present time there is a lack of validated biomarkers predicting tumour response to CCRT [27–29]. Several candidates from the DNA damage response (DDR) pathways have been investigated [30–32]. Unfortunately, even the most promising biomarkers, such as double strand break repair nuclease MRE11 (MRE11), have failed to generate reproducible data. In the latest multicentre collaborative effort to validate MRE11 as a biomarker, the immunohistochemistry scoring results varied considerably and failed to attain a reliable dataset [33]. Finally, a complete or near complete response assessed by cystoscopy after 4–5 weeks induction phase remains as the only reliable predictor of treatment outcome [34,35]

4. MIBC Molecular Subtypes as Biomarkers for CCRT Response

4.1. BCa Tumour Subtypes

MIBC can be classified into molecular subtypes by transcriptome profiling, thus allowing patient stratification to consider different therapeutic options. However, MIBC subtyping is still not included in routine clinical practice due to several classifications existing simultaneously in the last decade.In 2020, a consensus on MIBC subtypes has been reached [36], giving hope for a rapid translation into clinics. Furthermore, a single-sample classifier has been established enabling to assign a consensus class label to a tumour sample's transcriptome. There are six biologically relevant consensus molecular classes, namely, luminal papillary, luminal non-specified, luminal unstable, stroma-rich, basal/squamous and neuroendocrine-like [36]. Among the luminal subtypes, the most represented is the luminal papillary subtype (24% of MIBC), the other two luminal subtypes representing 15% (for the luminal unstable subtype) and 8% (for the luminal non-specified subtype) of MIBC (Figure 1a). The other most frequent subtype is the basal/squamous subtype representing 33% of MIBC. Further, 15% of MIBC represent stroma-rich and 3% neuroendocrine-like subtype (Figure 1a) [36].

Several retrospective studies have highlighted the clinical significance of molecular stratification of MIBC suggesting that responses to treatment could be predicted by tumour subtyping [37–40]. However, there was a lack of association between pathological responses and overall survival for the patients having basal tumours. Prospective validation in a larger cohort is required to address this issue. In addition, our group showed that basal tumours are sensitive to epidermal growth factor receptor (EGFR) inhibition in vitro and in preclinical models [41]. Sensitivity to RT of the subtypes remains yet to be investigated, but it has been suggested to be increased in two subtypes: neuroendocrine-like and luminal unstable, which show elevated cell cycle activity and low hypoxia signals [42,43]. At the present time no significant difference has been found in local relapse-free survival between bladder tumour subtypes in MIBC patients treated by TURBT followed by CCRT [44] or RT alone [45].

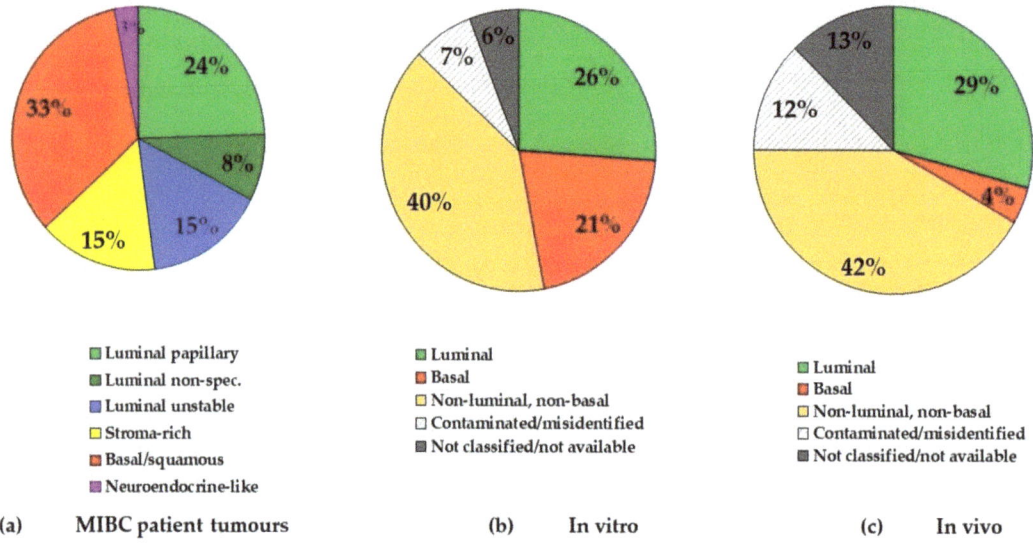

Figure 1. Comparison of muscle-invasive bladder cancer (MIBC) subtype frequency in patient tumours and human bladder cancer (BCa) cell lines used in radiotherapy (RT) experimental studies in vitro and preclinical studies in vivo. (a) MIBC tumour subtypes (classified by [36]); (b) cell line subtypes used in experimental studies of RT in BCa in vitro (classified by [46]); (c) cell line subtypes used in preclinical studies of radiotherapy in BCa in vivo (mice xenografts) (classified by [46]).

4.2. BCa Cell Line Molecular Subtypes

There is no agreement yet on the molecular classification of BCa cell lines, in particular with regard to the new consensus classification [36]. Our group made a first dichotomy between basal and non-basal cell lines [41]. Then, Earl et al. assigned the subtypes to a series of 40 BCa cell lines [47]. Our group has made more stringent classification [46] For example, five cell lines discussed in this review were identified as "basal" by Earl et al., while classified as "non-luminal, non-basal" by Shi et al. [46]. These non-luminal, non-basal cell lines expressed epithelium to mesenchyme transition markers and do not express E-cadherin. These cells could represent sarcomatoid tumours which is a rare entity in vivo. This phenotype could be present in the initial tumours or acquired in vitro. In vivo, the sarcomatoid tumours are classified mainly in the basal subtype probably due to commonalities in their stroma. However, data from experimental studies on radiosensitisation of relevant models representing different molecular subtypes are sparse. It is worth noting that none of the studies included in this review has discussed the relevance of a molecular subtype of the chosen experimental models. It is in part due to the fact that the consensus was only recently established and that there was an absence of classification of the cell lines until recently.

5. BCa Experimental Models in RT Studies

The experimental models and study design should utilise the information available regarding the subtypes of the cell lines and also revisit the information available regarding the patient and the original tumour from which the cell line has been established. BCa experimental models and their molecular features have been recently reviewed by Zuiverloon et al. [48] and rodent models with relevant molecular subtypes have been described by Ruan et al. [49]. Here, we critically discuss the BCa experimental and preclinical models used in association with RT treatment.

5.1. BCa Cell Lines
5.1.1. Molecular Subtypes

Selection of BCa cell lines with different mutation status and from different subtypes would better reflect the heterogeneity of MIBC cancer patients. We identified 29 different BCa cell lines used in the experimental studies of RT. We assigned molecular subtypes to the human BCa using a classification recently proposed by our team [46] (Table 1, additional information regarding the mutational status and origin are available in Table S2). We found that from all the BCa cell lines used in RT studies, 26% are of luminal subtype, which is in contrast to 47% of human tumours considered having luminal features (Figure 1a,b). In total, 21% of cell lines used were of basal subtype (contrary to 33% of human tumours) and the largest part (40%) of BCa cell lines was classified as neither luminal nor basal (Figure 1b). These cells, which express epithelial to mesenchyme transition markers and do not express E-cadherin, are rarely found in vivo. They could represent sarcomatoid tumours or the transient state of tumour cells representing only a fraction of tumours. This transient state could be of importance during the invasion or the metastatic process.

Table 1. BCa cell lines used in RT studies.

	Cell Line	Cellosaurus Accession No. [50]	Molecular Subtype [46]
Human	RT112	CVCL_1670	luminal
	SW780	CVCL_1728	
	UMUC5	CVCL_2750	
	UMUC9	CVCL_2753	
	RT4 [1]	CVCL_0036	
	5637	CVCL_0126	basal
	647V	CVCL_1049	
	HT1197	CVCL_1291	
	HT1376	CVCL_1292	
	KU19-19	CVCL_1344	
	UMUC6	CVCL_2751	
	VMCUB1	CVCL_1786	
	253J B-V	CVCL_7937	non-luminal, non-basal
	639-V	CVCL_1048	
	J82	CVCL_0359	
	KK47	CVCL_8253	
	T24	CVCL_0554	
	TCC-SUP	CVCL_1738	
	UMUC3	CVCL_1783	
	CAL29	CVCL_1808	n/c
	NTUB1	CVCL_RW29	n/a
	OBR	n/a	
	SW-800	CVCL_A684	
	UCRU-BL13	CVCL_M873	
	UCRU-BL17	CVCL_M007	
	UCRU-BL28	CVCL_4904	
Mouse	MB49	CVCL_7076	basal (mouse) [2]
	MB49-I	CVCL_VL62	
	MBT2	CVCL_4660	

Mean values (n = 3 repetitions) preceded by one common letter (a, b) were not significantly different (p < 5%). Authors should discuss the results and how they can be interpreted from the perspective of previous studies and of the working hypotheses.

Regrettably, we found that 7% of all in vitro studies and 12% of all in vivo studies have used cell lines that have been identified as cross-contaminated or misidentified (Figure 1b,c, Tables S3 and S4).

5.1.2. Gender

Given the difference incidence rates between men and women in BCa, it is important to exclude potential gender bias in the study design and include cell lines from both sexes. BCa is significantly more frequent in males, while the majority of studies (54 of 85 studies identified or 64%) have used RT112 and T24 cell lines which are of female origin (Table S2). Androgen receptor (AR) signalling could be implicated in the gender disparity of BCa but it remains to be confirmed [51,52]. Furthermore, AR signalling has been recently shown to reduce radiosensitivity [53].

5.1.3. Intrinsic Radiosensitivity

There is a lack of certainty when it comes to the intrinsic radiosensitivity of BCa cell lines as many have been used only in a single study and therefore obtained radiation–response curves have never been reproduced (Table S2, No. of studies). On the other hand, the wide range of different cell lines for the use of radiosensitisation studies offers opportunity to examine many potential factors influencing radiation response. Only a single dataset comparing intrinsic radiosensitivities of 19 BCa cell lines exists by Yard et al. [54]. Yard et al. used data derived from a single experimental platform and performed analysis using a rigorous statistical methodology. They studied genetic determinants influencing tumour response to DNA damage and influencing tumour survival as assessed by colony forming assays. The radiosensitivity was described by the area under the curve (AUC) and scored from 0 (completely sensitive) to 7 (completely resistant) (Table 2). There was a high heterogeneity among the BCa cell lines (radiosensitivity varying from 1.883 (radiosensitive) to 5.228 (radioresistant)) [54] (Table 2). It is interesting to note that by associating with the molecular subtype, the three most radioresistant cell lines are of basal subtype (AUC ≥ 4.4). However, there were also two basal cell lines reported to be very radiosensitive (AUC < 2.5). Luminal cell lines included in this study had a lower variation of the AUC (2.935–4.126) all being moderately radiosensitive/moderately radioresistant. However, this data set included only five luminal cell lines so more studies having stringent measurements of radiosensitivity are desirable.

Table 2. Intrinsic radiosensitivity of a panel of BCa cell lines reported by Yard et al., 2016 [54].

Cell Line	AUC [1] [54]	Molecular Subtype [46]
HT1376	5.228	basal
HT1197	4.449	basal
VMCUB1	4.412	basal
KMBC2	4.126	luminal
TCCSUP	3.539	non-luminal/non-basal
KU1919	3.503	basal
647V	3.374	basal
BC3C	3.362	basal
UMUC1	3.346	luminal
SW1710	3.309	n/c
UMUC3	3.231	luminal
J82	3.198	non-luminal/non-basal
RT112	3.038	luminal
RT4 [2]	2.935	luminal
UBLC1	2.914	n/c
JMSU1	2.792	non-luminal/non-basal
5637	2.473	basal
T24	2.366	non-luminal/non-basal
SCABER	1.883	basal

[1] AUC: Area under the curve; [2] RT4 cell line originates from re-occurring human transitional cell papilloma. Abbreviations: n/c, not categorised (not coherent classification depending on the dataset used).

5.2. BCa Xenografts

The majority of data generated from the RT studies in BCa are currently from BCa xenografts (Table 3); fewer studies have used syngeneic mouse models (Table 4), while none have used patient-derived xenografts (PDX). RT studies using 3D-BCa xenografts have the advantage to gather more clinically-relevant information when compared to in vitro 2D-models. For example, radiosensitivity can be studied in 3D considering hypoxic areas, which have been previously identified as potential cause of therapy failure in BCa. Indeed, a study by Williams et al. found that the hypoxia-targeting prodrug AQ4N efficiently sensitised luminal BCa xenografts to cisplatin-based CCRT (Table 5). However, some interactions such as immune infiltration between tumour and its microenvironment can be limited by the species barrier.

We found that seven (29%) BCa xenografts used were of luminal subtype (Table 3, Figure 1c), while only one study (4%) used the cell line of the basal subtype (Table 3, Figure 1c). Clearly, the BCa xenografts of the basal subtype have been less studied in the RT context. We analysed the in vivo study design whereby different radiation schedules were employed (single larger radiation dose versus fractionated dose delivery schedule) (Table 3). Regrettably, three in vivo studies have used cell lines reported as cross-contaminated (Table S4).

5.3. Syngeneic Models

Evidently, syngeneic mouse models are a suitable choice to test immunotherapy agents in combination with radiation treatment. Furthermore, studies of interactions of tumour cells with endothelial and fibroblastic syngeneic cells are more relevant than in a xenogeneic model. We identified seven studies using syngeneic mouse models (Table 4). The three cell lines used in these studies were all chemically induced (detailed in [48]) and resemble the human basal/squamous subtype [41,49]. We noted that all of these models were heterotopic, excluding the assessment of treatment-induced bladder toxicity. Results of most of the studies are discussed further.

Table 3. Overview of human BCa cell line xenograft models used in radiosensitisation studies.

Subtype [46]	Cell Line	IR Regimen	Radiosensitising Agent	Class	Nude Mice Genetic Background (Gender)	Initial Tumour Size (mm³) [1]	Study Follow-Up (Days) [2]	Ref.
luminal	RT112	4 × 5 Gy	Panobinostat (vs. gemcitabine)	HDAC inhibitor	(Unknown strain) (F)	100	10–60	[55]
		2 × 5 Gy	AQ4N (banoxantrone) (vs. cisplatin)	DNA intercalator and Topoisomerase II inhibitor	CBA (F)	240–280	10–60	[56]
		1 × 6 Gy	Romidepsin	HDAC inhibitor	CD1 (F)	50	25	[57]
		1 × 6 Gy	Low-/soluble high-/insoluble high- and mixed high-fibre diets	Diet	CD1 (F)	50	42	[58]
	RT4	1 × 5 or1 × 15 Gy	Photofrin II	Photosensitiser	(Unknown strain) (F)	2.6–3.0	15	[59]
		1 × 2 Gy	Caffeine	DNA Damage Response inhibitor	BALB/c (M)	30–75	0 [3]	[60]
	SW780	2 × 5 Gy	siTUG1	siRNA	(Unknown strain) (M)	100	21	[61]
basal	5637	2 × 2 Gy	Sulfoquinovosylacylpropanediol	Synthetic sulfoglycolipid	BALB/c Slc (M)	100–300	33	[62]
		1 × n/a Gy	shRNF8	shRNA	BALB/c (M)	100–150	30	[63]
		1 × 6 Gy	Chloroquine	Other	BALB/c (F)	~200	25	[64]
		1 × 6 Gy	Nanoparticles (chloroquine conjugated)	Nanoparticles	(Unknown strain) (n/a)	150	16	[65]
Non luminal/non basal		1 × 6 Gy	LY294002	TKI	Ncr-nu/n (F)	300–400	40	[66]
		1 × 6 Gy	FTI-276 or L744832	Farnesyltransferase inhibitors	Ncr-nu/n (n/a)	58	80	[67]
		2 × 3 Gy	shHMGB1	shRNA	(Unknown strain) (F)	n/a	21	[68]
	UMUC3	1 × 12 Gy	17-AAG or 17-DMAG/Trastuzumab/LY294002	Hsp90 inhibitors/ monoclonal antibody/TKI	BALB/c (M)	1000	12	[69]
		2 × 2 Gy	Flutamide/shAR	Antiandrogen/shRNA	NOD-SCID (M)	30	12	[53]
n/a	J82	1 × 5 Gy	Gefitinib ("Tressa", ZD1839)	TKI	BALB/c (n/a)	100	n/a	[70]
	KK47	1 × 4 Gy	Ad-RSV-CD+5-FC	A recombinant adenovirus vector	BALB/c (n/a)	n/a	n/a	[71]

[1] The initial size of the tumour is defined as the size of the tumour at the start of the RT or combination treatment (Day 1). [2] The minimum follow-up for the non-treated control was used to compare the growth of the xenografts. [3] In this study, the mice were sacrificed immediately after the treatment delivery. Abbreviations: AR, androgen receptor; HDAC, histone deacetylase; HMGB1, high mobility group box 1; Hsp90, heat shock protein 90; n/a, information not available; RNF8, ring finger protein 8; TKI, tyrosine kinase inhibitor; TUG1, taurine upregulated gene.

Table 4. Overview of mouse BCa syngeneic models used in radiosensitisation studies.

Cell Line	IR Regimen	Radiosensitising Agent	Class	Mouse Background (Gender)	Initial Tumour Size (mm^3) [1]	Study Follow-Up (days) [2]	Ref.
MB49	1 × 12 Gy	PD-L1 blocking antibody	Immunotherapy	C57BL/6 (F)	500	27	[72]
MB49	2 × 5 Gy	Glycyrrhizin	HMGB1 inhibitor	C57BL/6 (M)	Once palpable	7	[73]
MB49, MB49-I	6 × 3 Gy	Silybin (Sb)	Flavonoid	C57BL/6J (n/a)	50	30	[74]
MB49-I	6 × 3 Gy	Bacillus Calmette-Guérin (BCG)	Immunotherapy	C57BL/6J (n/a)	50	21	[75]
MBT-2	1 × 15 Gy	Lapatinib	TKI	C3H/HeN (F)	162	21	[76]
MBT-2	1 × 15 Gy	Afatinib	TKI	C3H/HeN (F)	162	21	[77]
MBT-2	5 × 4 Gy	Cisplatin, doxorubicin hydrochloride (adriamycin), cyclophosphamide	CT	CsH/Hej (n/a)	6	60	[18]

[1] The initial size of the tumour is defined as the size of the tumour at the start of the RT or combination treatment (Day 1). [2] The minimum follow-up for the non-treated control was used to compare the growth of the xenografts. Abbreviations: CT, chemotherapy; HMGB1, high mobility group box 1; PD-L1, programmed death ligand 1; TKI: tyrosine kinase inhibitor.

6. Use of CT Agents in Combination with RT in Experimental Studies for BCa Treatment

6.1. Cisplatin

Cisplatin is currently the most widely used radiosensitising agent in MIBC, supported by a randomised clinical trial [3]. According to the NCCN Clinical Practice Guidelines in Oncology, the most comprehensive guidelines for treatment of oncological patients in the United States, the recommended radiosensitising regimen for locally advanced or metastatic BCa is a combination of cisplatin and gemcitabine [78]. However, it should be emphasised that this combination has not been investigated in our identified experimental and preclinical studies.

Table 5. Overview of preclinical studies using Cisplatin in BCa in vivo in combination with RT.

	Yoshida et al., 2011 [69]	Williams et al., 2009 [56]	Kyriazis et al., 1986 [79]	Weldon et al., 1979 [18]
Cell lines	UMUC3 (non-luminal, non-basal)	RT112 (luminal)	SW-800 (not classified)	MBT-2
Source/ Dose rate (Gy/min)	X-rays (225 V)/ 0.83 Gy/min	X-rays (230 kV)/ 2 Gy/min	X-rays (250 kV)/ 1.23 Gy/min	X-rays (250 kV)/ n/a
IR dose and fractionation	5 × 2 Gy	5 × 2 Gy	1 × 10 Gy	5 × 4 Gy
Cisplatin dose	3 mg/kg (administered once)	2 mg/kg (administered once)	5 mg/kg once on each specified day before or after radiation	3 mg/kg once a week (3 weeks)
Treatment arms	Hsp90 inhibitors (17-AAG or 17-DMAG) Trastuzumab, LY294002	AQ4N (banoxantrone)	-	Doxorubicin hydrochloride (Adriamycin), cyclophosphamide
Normal tissue toxicity	Yes (NHU in vitro)	-	-	-

In our literature analysis, we found only two in vitro and four in vivo (preclinical) studies investigating cisplatin and RT in BCa (Table 5). Weldon et al., for the first time, used cisplatin in combination with RT in a murine BCa model and compared different administration schedules and included also two other drugs for comparison [18]. They found that the concomitant administration of cisplatin and RT was toxic, but when cisplatin was used as adjuvant therapy after completion of RT, synergistic effect was produced, but another CT drug cyclophosphamide was more effective in terms of growth delay [18]. Further, a study by Kyriazis et al. observed the most synergistic effect when cisplatin was given on days 3 and 6 post-radiation using human BCa xenograft model (SW-800, not classified cell line) [79]. In an in vitro study, Bedford et al. demonstrated that radiation-resistant cell lines are more sensitive to cisplatin and radiation compared to wild-type human BCa cell lines [80]. Kawashima and colleagues investigated whether CCRT response can be predicted using expression of excision repair cross-complementing group 1 (ERCC1) and found that its downregulation improved the effect of CCRT, but not of cisplatin alone in vitro [31]. Two further studies have used mouse BCa xenografts. Yoshida and colleagues investigate the prospect to improve CCRT response by using heat shock protein (HSp90) inhibitors in non-luminal, non-basal mouse BCa xenograft (UMUC3) while evaluating the effect on normal human urothelial (NHU) cells in vitro [69]. They found greater BCa sensitisation to cisplatin-based CCRT after low-dose Hsp90 inhibitor treatment than with the combination of trastuzumab (HER2 blocking antibody) or LY294002 (PI3K inhibitor). A few sensitising effects of NHU to CCRT were found [69]. Furthermore, Williams et al. found that Hypoxia-Activated ProDrug AQ4N increased the efficacy of RT alone and cisplatin-based CCRT in vivo [81].

6.2. Gemcitabine

Results from eight phase I-II trials concluded that there is strong evidence that CCRT regimens with concurrent gemcitabine are feasible and well tolerated in BCa [82]. Prospective randomised controlled trials are ongoing to definitively assess the efficacy of gemcitabine-based CCRT for MIBC. From the experimental studies identified in our literature search, two have made use of gemcitabine as a radiosensitiser in vitro, while two have used preclinical mouse models in vivo (Table 6).

There have been some conflicting results among the early studies of the use of gemcitabine in BCa in vitro models. In 2003, Fechner and colleagues showed no effect of radiosensitivity of gemcitabine in four different BCa cell lines (RT112, RT4, T24 and TCC-SUP) with differing p53 status [83]. In contrast, Pauwels et al. demonstrated correlation between gemcitabine-induced S phase block resulting and sensitization to RT of a BCa cell line ECV304 and cell lines from other cancers in vitro [84]. It is worth noting that the cell line used has been recognised as being contaminated by another BCa cell line (T24, non-luminal, non-basal, Table S3. These studies did not use colony forming assay to investigate radiosensitisation, which is considered the gold standard for assessing RT efficiency. Another study used colony forming assay to compare gemcitabine radiosensitising effect in related bladder cancer cell line MGHU1 and its radiosensitive subclone S40b [85]. They demonstrated that gemcitabine is an effective radiosensitiser in these BCa cell lines, with greater sensitisation in the radioresistant parental line [85]. Interestingly, MGHU1 cells did not show S-phase accumulation, which is the suggested radiosensitisation mechanism, but its subclone S40b did, despite both being radiosensitised by gemcitabine, indicating that S-phase accumulation is unlikely to be a major mechanism of radiosensitisation by gemcitabine [85]. However, also the MGHU1 cell line has been reported as contaminated by T24 cell line (Table S3) [86].

Further, Choudhury et al. used gemcitabine in combination with another targeted agent (imatinib). Imatinib (inhibitor of c-ABL, c-KIT, and platelet-derived growth factor receptor (PDGFR) tyrosine kinases) was found to be a sensitising agent to RT and gemcitabine-based CT treatment using RT112 luminal cell line in vitro (alongside a prostate cancer cell line in vitro and in vivo), concluding that imatinib can sensitise tumour cells to DNA damaging agents and induce mitotic catastrophe [87]. Two studies from Anne Kiltie team at the University of Oxford have investigated gemcitabine in vivo (Table 6) [55,88]. In the first one, Kerr et al. demonstrated that gemcitabine-resistant Calgem heterotopic xenografts were responsive to the combination of gemcitabine and irradiation [88]. In the second study, Groselj et al. showed that gemcitabine + RT resulted in more acute and late intestinal toxicity than HDAC inhibitor panobinostat + RT [55].

Table 6. Overview of studies using Gemcitabine in BCa preclinical studies in combination with RT.

	Groselj et al., 2018 [55]	Kerr et al., 2014 [88]
Cell lines	-	CALgem
Source/ Dose rate (Gy/min)	X-rays (220 kV)/ n/a	gamma (^{137}Cs)/ 1.7 Gy/min
IR regimen in vivo	1× 10, 12, or 14 Gy (acute toxicity) 5 × 5 Gy (late toxicity)	5 × 2 Gy
Gemcitabine dose	single 100 mg/kg injection	single 100 mg/kg injection
Normal tissue toxicity	yes (intestinal)	-

6.3. In Vivo Study Reporting/Design

We noted that the in vivo study design of studies using cisplatin and/or gemcitabine have been very few and heterogeneous. Different radiation schedules have been employed (single larger radiation dose vs. fractionated, more clinically relevant schedule). The source of irradiation was different (X-rays and gamma rays) and different dose rates were reported. It has been shown that a variation of X-ray energy and the dose rate can impact the relative biological effectiveness (RBE) both in cells in vitro and in vivo [89,90]. From the studies using gemcitabine and cisplatin, three studies had not reported the dose rate. Different choice of the radiation delivery schedule will prevent a direct comparison of the studies.

7. Targeted Agents to Improve RT Response in BCa

Following the studies of classical CT drugs, multiple agents have been tested as potential radiosensitisers since 2000. Nineteen studies using BCa xenografts (Table 7) and six studies of syngeneic mouse tumour models (Table 8) have reported a significant radiosensitisation. These studies are very heterogeneous, testing diverse agents ranging from RTK inhibitors, epigenetic modifiers, hypoxia- or angiogenesis targeting molecules, among others. Below, we comment on the data of few selected studies.

7.1. Epidermal Growth Factor Receptor

Receptor tyrosine kinases (RTKs) are frequently differentially expressed between normal and cancerous tissue. They mediate pro-proliferative, pro-survival pathways as well as DNA repair pathways, the activation of which could ultimately protect cancer cells from radiation-induced cell death. Furthermore, radiation-induced activation of several RTKs has been reported and belongs to the earliest events in response to DNA damage [91]; reviewed by [92]).

Epidermal growth factor receptor (EGFR) overexpression has been found in up to 70% of BCa tumours [93]. The team of François Radvanyi has been studying the role of several RTKs in BCa and has identified basal subtype BCa cell line dependency to EGFR when not mutated for a RAS family members [41,70]. EGFR is the most-studied RTK in the field of radiation oncology as it was the first RTK to be shown to be activated with RT [91,92]. A clinical trial combining EGFR blocking monoclonal antibody with RT versus RT alone significantly improved overall survival in head and neck squamous cell carcinoma patients [94].

In RT studies of BCa, Domniguez-Escrig et al. observed radiosensitising effect of the use of the EGFR inhibitor Gefitinib in vitro in two BCa cell lines. However, Gefitinib alone did not cause growth delay in the luminal RT112 xenograft in vivo, but was validated using the basal 253J B-V xenograft [95]. Colquhoun et al. demonstrated that radiation induced activation of EGFR and MAPK and Akt downstream effectors in two BCa cell lines. Further, Gefitinib + RT induced significant growth delay in non-luminal, non-basal J82 cell line xenografts compared to single treatment [70].

7.2. Chromatin Modifiers/Epigenetic Regulators

High levels of histone deacetylases (HDACs) have been detected in BCa tumours [96]. HDAC inhibitors have been tested alone in clinical trials in advanced solid tumours, including BCa; however, high reported toxicities to normal tissue have impeded their progress to clinics for example of the HDAC inhibitor mocetinostat [97]. The team of Anne Kiltie at Oxford University has been studying other HDAC inhibitors as potential radiosensitisers in BCa experimental and preclinical models and have found promising results using pan-HDAC inhibitor panobinostat [55] and more selective HDAC inhibitor romidepsin [57]. In addition, no increase in acute or late toxicity following mouse pelvic irradiation has been reported in [55].

Table 7. Targeted agents used as radiosensitisers in preclinical studies using BCa xenografts.

Class	Name	Target	Cell Line Subtype (According to [46])	Year	Ref.
TKI	Gefitinib (ZD1839)	EGFR	J82, non-luminal, non-basal	2007	[70]
	Afatinib, Erlotinib	EGFR/HER2, EGFR	NTUB1, class n/a	2015	[98]
PI3K	LY294002	PI3 kinase	T24, non-luminal, non-basal	2003	[66]
Epigenetic modifiers	Panobinostat	HDAC (histone deacetylase)	RT112, luminal	2018	[55]
	Romidepsin	HDAC (histone deacetylase)	RT112, luminal	2020	[57]
Heat shock protein inhibitors	17-AAG or 17-DMAG	Hsp90	UMUC3 non-luminal, non-basal	2011	[69]
Farnesyltransferase inhibitors	FTI-276 and L744832	Farnesyltransferase	T24, non-luminal, non-basal	2000	[67]
Hypoxia	AQ4N	Hypoxia	RT112, luminal	2009	[56]
Angiogenesis	SQAP	Angiogenesis	5637, basal	2016	[62]
Other	Chloroquine	Autophagy	T24, non-luminal, non-basal	2018	[64]
	HSA-MnO2-CQ nanoparticles	Autophagy	T24, non-luminal, non-basal	2020	[65]
	Ad-RSV-CD+5-FC	-	KK47, non-luminal, non-basal	2003	[71]
	shRNF8	DNA Damage Response	T24, non-luminal, non-basal	2016	[63]
	Caffeine	DNA Damage Response	RT4, luminal	2015	[60]
	siTUG1	HMGB1	SW780, luminal	2017	[61]
	shHMGB1	HMGB1	UMUC3 non-luminal, non-basal	2016	[68]
	Flutamide/shAR	AR	UMUC3 non-luminal, non-basal	2018	[53]
	Photofrin II	Angiogenesis	RT4, luminal	2001	[59]
	Low-fibre, soluble high-fibre, insoluble high-fibre, and mixed soluble/insoluble high-fibre diets	Metabolism	RT112, luminal	2020	[58]

Abbreviations: 17-AAG: 7-N-allylamino-17-demethoxygeldanamycin; 17-DMAG: 17-Dimethylaminoethylamino-17-demethoxygeldanamycin; AQ4N: banoxantrone dihydrochloride, topoisomerase II inhibitor; AR: androgen receptor; HER2: human erbB-2 receptor; HMGB1: high mobility group box 1; HSA-MnO2-CQ: MnO2 and chloroquine in human serum albumin (HSA)-based nanoplatform; RNF8: ring finger protein 8; SQAP: sulfoquinovosylacylpropanediol; TKI: tyrosine kinase inhibitor; TUG1: taurine-upregulated gene 1.

7.3. Radio-Immunotherapy

In BCa preclinical models, immune checkpoint inhibitors have been used with RT only in one study using an anti-PD-L1 antibody [72] (Table 8). Wu et al. demonstrated that RT upregulated PD-L1 expression in BCa tumour cells, correlating with radiation dose. Using heterotopic MB49 syngeneic mouse models, PD-L1 blockade induced a longer tumour growth delay following irradiation [72]. Bacillus Calmette-Guérin (BCG) bladder instillation is commonly used after local tumour resection for patients with superficial bladder cancer. In addition, BCG bacteria-induced immune response has been studied in

BCa to improve response to RT [75]. Invasive murine BCa cell line MB49-I was cultured in monolayers in 2D, in spheroids in vitro in 3D and inoculated in vivo in the syngeneic mice. BCG pre-treatment radio-sensitised spheroids, while no effect was shown in monolayers. In vivo, BCG improved the local response to RT and decreased the presence of lung metastasis. The combined BCG+RT treatment also resulted in abscopal effect, where second tumour development in the opposite flank was completely rejected, compared to the untreated or RT only arms [75].

Table 8. Targeted agents used as radiosensitisers in preclinical studies using syngeneic mice.

Class	Name	Target	Cell line (Subtype According to [46])	Ref.
TKI	Afatinib	EGFR/HER2	MBT-2, mice cell line (basal)	[77]
	Lapatinib	PDGF-R	MBT-2, mice cell line (basal)	[76]
HMGB1 inhibitor	Glycyrrhizin	HMGB1	MB49, mice cell line (basal)	[73]
Flavonoid	Silybin		MB49, MB49-I, mice cell lines (basal)	[74]
Immune checkpoint inhibitor	Anti-PD-L1 antibody	PD-L1	MB49, mice cell line (basal)	[72]
Non-specific immune stimulator	Bacillus Calmette-Guérin (BCG)	Immune system	MB49-I, mice cell line (basal)	[75]

Abbreviations: EGFR, epidermal growth factor receptor, HER2, human erbB-2 receptor; HMGB1, high mobility group box 1; PDGF-R, platelet-derived growth factor receptor; PD-L1, programmed death ligand 1; TKI: tyrosine kinase inhibitor.

8. Suggestions to Improve the Design of Future Experimental BCa Studies: New Agents and Relevant Models

In order to improve RT, its therapeutic index (i.e., the ratio: antitumour efficiency/toxic effects on surrounding healthy bladder tissues) must be increased. Modern radiotherapy has advanced spatial targeting of clinical tumour volume, including whole bladder and pelvic nodes while reducing side effects to pelvic organs other than the bladder. However, since the whole bladder is included in the clinical target volume (CTV), only strategies aiming at radiosensitising tumour tissue selectively and not or to a lesser extent of the normal bladder wall, will improve therapeutic index of RT.

8.1. Molecular Subtype Consideration

Gemcitabine and cisplatin are effective radiosensitisers, but other agents have shown superior effect in the few comparative studies in vitro and in vivo (Tables 5 and 6). Unfortunately, only one luminal cell line (RT112) and no basal cell line have been used in vivo to evaluate cisplatin- or gemcitabine-based CRT, thus not allowing the study of the differential effect observed in clinics between tumour response to CT of basal and luminal subtypes (where basal tumours are shown to be better responders). It would be important to compare CCRT/RT response in models of different subtypes and study the underlying mechanisms. For that, patient-derived xenografts (PDXs) are relevant models as they can conserve in mice the subtype of the original human tumour [99]. Briefly, PDX establishment consists of engrafting a fragment of a patient tumour directly into an immunocompromised animal, and then maintaining it through passaging from animal to animal, avoiding the in vitro selection and allowing one to conserve the initial histology. In future studies, using a PDX model would eliminate the non-luminal non-basal subtypes that most of the BCa cell lines used in RT studies are representing, but which do not clearly represent human tumours. Despite the possible tumour loss of distinct molecular features over time, PDX models in BCa warrant future efforts to be used in RT studies.

8.2. New Targeted Agents

In BCa, several experimental studies have shown the therapeutic efficacy of RTK inhibition other than EGFR. Fibroblast Growth Factor Receptor 3 (FGFR3) has been extensively studied in BCa due to its frequent mutations/translocations in the BCa, driving oncogenic dependency (more than 65% of NMIBCs and 15% of MIBCs) [100–102]. Pan-FGFR inhibitor has been recently shown to radiosensitise tumours in head and neck squamous cell carcinoma xenografts and PDX [103]. In the context of BCa subtypes, 50% of luminal BCa tumours harbour an FGFR3 alteration and therefore could potentially benefit the most from anti-FGFR3+RT treatment.

Recently, TYRO3, a member of the TAM family of RTKs (comprising TYRO3, AXL and MERTK) has been identified as a potential target in the BCa [104]. TYRO3 overexpression has been reported in 50% of MIBC and results in TYRO3-dependency for growth of BCa cancer cell lines [104]. However, FGFR3 and TYRO3 have never been explored as targets for radiosensitisation in BCa.

Using syngeneic models allows investigating the therapeutic index of different radiosensitising agents including the effect on tumour cells but also on immune response and other interactions restricted by species barrier. For example, inhibition of the TAM receptors could improve RT efficacy by increasing directly radiation-induced tumour cell killing and also by promoting innate immunity [105,106]. Currently, in the field of radiation oncology, there are many studies in syngeneic mouse models exploring different hypofractionated RT regimens in combination with immune checkpoint inhibitors, such as anti-PD-L1 and anti-T cell immunoreceptor with Ig and ITIM domains (TIGIT) in colorectal cancer [107]. The underlying interest is to combat PD-L1 expression which has been shown to be upregulated upon RT [108]. Further, also abscopal effect is being investigated and it has been shown that RT can promote a response of lung cancer to cytotoxic T-lymphocyte 4 (CTLA-4) blockade [109]. There is clinical evidence that basal subtype benefits from early aggressive management with CT agents and would benefit from T cell modulators (i.e., targeting CTLA-4) and EGFR, NFκB, Hif-1α/VEGF, and/or Stat-3-targeted agents will also be active within this subtype [110]. Such preclinical studies are needed in BCa to improve the RT-immunotherapy modalities.

8.3. Pelvic Toxisity Assessment

Experimental devices dedicated to accurate mice irradiation are currently available, which includes high resolution computerised tomography scanner for imaging, warmed beds to keep the model at a physiological temperature during longer irradiating sessions and possibility to target CTV more accurately. This allows studying the side effects of the RT on the bladder, giving the opportunity to compare acute radiation-induced toxicity versus long-term radiation-induced side effects such as fibrosis on the bladder. Although toxicity of normal tissue is one of the main limitations of use of RT in BCa, only two studies have considered pelvic toxicity. First, the study assessed pelvic irradiation-induced intestinal toxicity [56]. Second was an in vitro study using normal human urothelium (NHU) cells [69]. NHU cultures are relevant models to study bladder toxicity in vitro [111]. NHU cultures allow studying the impact on the normal urothelial cell proliferation from the RT or combined treatments as a first step before investigations in vivo. In addition, NHU can be differentiated into the non-proliferative phenotype, which is the physiological state of NHU cells in the human bladder [112,113]. Differentiated NHU cell monolayers would be a more relevant model to study radiation-induced toxicity in vitro.

8.4. Use of Orthotopic Mice Models

Currently, there is no preclinical study published using syngeneic or xenogeneic orthotopic graft in the field of BCa RT. However, Jäger et al. have developed a high-precision approach consisting of ultrasound-guided tumour cell inoculation within the bladder wall [114]. Another interesting model is the UPII-SV40T transgenic mouse model which expresses the SV40 large T antigen specifically in the urothelium and reliably develops BCa [115]. In addition, there is a simple chemically induced BCa model, developed by daily exposure to BBN (N-butyl-N-(4-hydroxybutyl)-nitrosamine) in drinking water (0.05%). Around 12 weeks of exposure to BBN result in development macroscopic lesions [116]. At the present time, all of these orthotopic models have never been used in preclinical RT studies.

8.5. Humanised Mouse Models

In recent years, humanised mouse models have been more widely used in the field of immunology. For example NOD scid gamma (NSG) highly immunodeficient mouse model have no B-, T-, and natural killer (NK) cells, and therefore allow the engraftment of tumour and immune cells of human origin [117,118]. Indeed, a recent study using this model established BCa xenografts in this humanised system and observed significant tumour growth delay using a pan-PI3K inhibitor in tumours bearing a PIK3CA mutation. Furthermore, pan-PI3K-treated PIK3CA-mutated BCa tumours were sensitive to PD-1 blockade. These results showed potential of combination of PI3K inhibitors with immune checkpoint inhibitors to overcome resistance to immune checkpoint inhibitors [119]. It would be interesting to include additional treatment arms combining such strategies with RT in the future.

9. Methods

Following the Preferred Reporting Items for Systematic Reviews and Meta-Analyses (PRISMA) guidelines for systematic reviews [120], we conducted a literature search limited to the database of PubMed. We used the following search string (((radiotherapy) OR (radiation therapy) OR (irradiation) OR (radiation) OR (electromagnetic radiation) OR (electromagnetic irradiation) OR (radiosensitisation) OR (radiosensitivity) OR (radioresistance) OR (radiosensitization) OR (radiation toxicity)) AND ((bladder cancer) OR (urinary bladder neoplasms) OR (Bladder Tumour) OR (urinary bladder) OR (urothelial carcinoma) OR (urothelium)) AND ((cell line) OR (xenograft) OR (syngeneic) OR (preclinical model) OR (pre-clinical model) OR (orthotopic) OR (cells) OR (NHU) OR (normal human urothelium) OR (urothelial cells) OR (mouse) OR (rodent)) NOT ((review) OR (case report) OR (systematic review) OR (meta-analysis))). All of the search was conducted between September and October 2020 without time restrictions. Relevant studies were identified between 1979 and 2020 (until October 31).

Initial study identification was carried out independently by L.S. and F.M. using the search tool in the PubMed with the following inclusion criteria: studies in English, studies with available abstracts, peer-reviewed journal articles only, not reviews.

The titles and abstracts of the obtained studies were further screened independently by L.S. and F.M. and excluded on the basis of following criteria: (1) no experimental models of bladder cancer used, (2) radiotherapy treatment not used or used as a single treatment. The lists were compared and the publications for which the two reviewers had a disagreement were reviewed together and, when needed, discussed with the third reviewer (F.M.C.). Then, the full text was obtained and assessed for eligibility, excluding only clinical studies and studies using photodynamic therapy. After careful reviewing of each full text article, additional studies were excluded where the experimental model used was only non-human BCa cell lines in vitro. The flow chart showing the numbers of the initial studies identified in PubMed and the steps leading to the final inclusion of 85 studies are depicted in Figure 2.

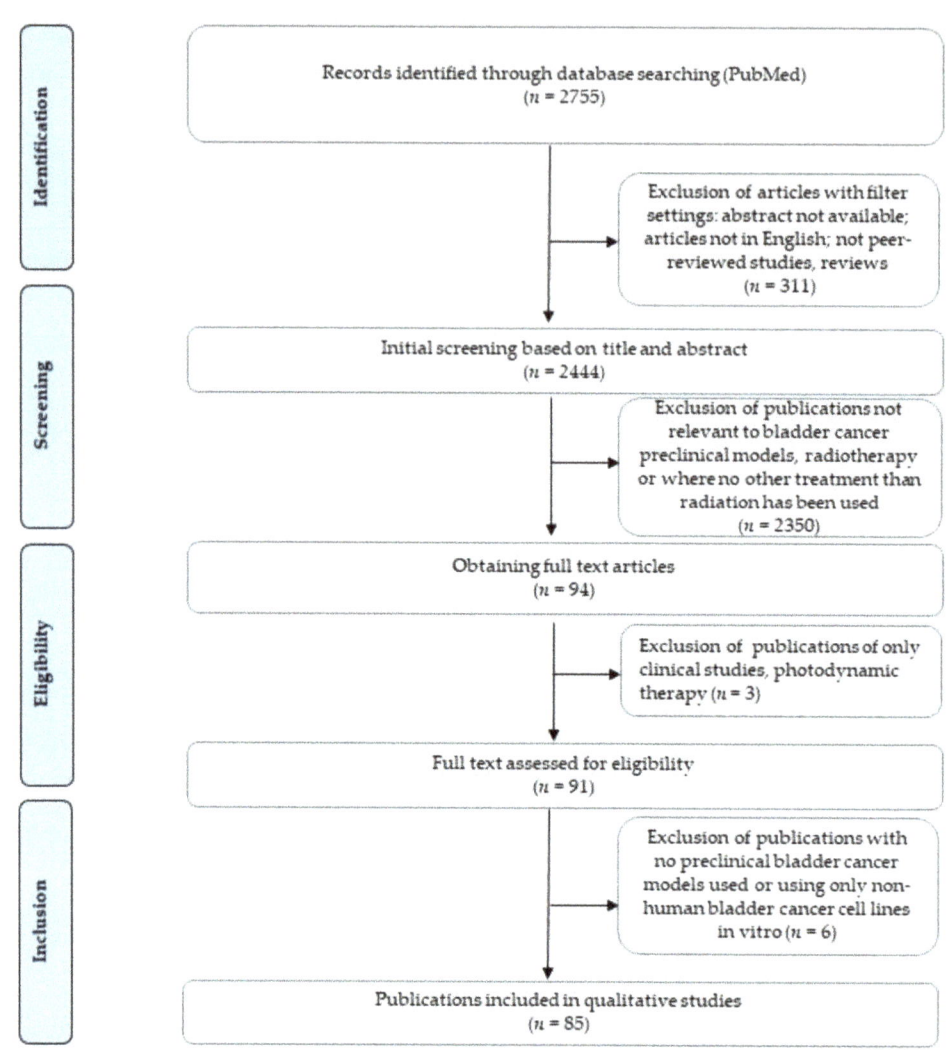

Figure 2. Flow chart showing the identification, screening, evaluation of eligibility and inclusion criteria of publications.

10. Conclusions

Increasing the use of bladder preserving radiation-based treatment needs an improved therapeutic index leading to reduced side effects. Experimental studies are needed to address this issue. This review first identified and analysed all experimental investigations on concurrent combination of radiation with different agents in bladder cancer and then provided suggestions aiding the selection of appropriate cell lines, mouse models, radiosensitising agents and radiotherapy regimen to improve the design of future experimental studies.

Supplementary Materials: The following are available online at https://www.mdpi.com/2072-6694/13/1/87/s1. Table S1: Clinical trials of concurrent chemoradiotherapy in bladder cancer, Table S2: Characteristics of BCa cell lines used in RT studies, Table S3: Characteristics of problematic human BCa cell lines used in experimental RT studies, Table S4: Preclinical studies of problematic cell lines used in vivo.

Author Contributions: Conceptualization, L.S., F.M. and P.V.; methodology, L.S. and P.V.; resources, L.S. and F.M.; writing—original draft preparation, L.S., F.M. and P.V.; writing—review and editing, L.S., F.M., G.C., I.B.-P., F.R., F.M.-C. and P.V.; visualization, L.S. and F.M.; supervision, P.V. All authors have read and agreed to the published version of the manuscript.

Funding: This study has been supported by a Fondation pour la Recherche Médicale (FRM) grant (agreement FDT201904008237) and a grant from Comité scientique/Comité de radioprotection from Électricité de France (EDF) (D4008/10.11.16/532) and by SIRIC-Curie INCa-DGOS-4654. This project received funding from the European Union's Horizon 2020 research and innovation program, under Marie Skłodowska-Curie grant agreement No. 666003.

Data Availability Statement: No new data were created or analysed in this study. Data sharing is not applicable to this article.

Acknowledgments: We thank Ayham Alnabulsi for valuable feedback on the manuscript.

Conflicts of Interest: The authors declare no conflict of interest.

References

1. Shipley, W.U.; Prout, G.R., Jr.; Einstein, A.B.; Coombs, L.J.; Wajsman, Z.; Soloway, M.S.; Englander, L.; Barton, B.A.; Hafermann, M.D. Treatment of Invasive Bladder Cancer by Cisplatin and Radiation in Patients Unsuited for Surgery. *JAMA* **1987**, *258*, 931–935. [CrossRef]
2. Housset, M.; Maulard, C.; Chretien, Y.; Dufour, B.; Delanian, S.; Huart, J.; Colardelle, F.; Brunel, P.; Baillet, F. Combined radiation and chemotherapy for invasive transitional-cell carcinoma of the bladder: A prospective study. *J. Clin. Oncol.* **1993**, *11*, 2150–2157. [CrossRef] [PubMed]
3. Coppin, C.M.; Gospodarowicz, M.K.; James, K.; Tannock, I.F.; Zee, B.; Carson, J.; Pater, J.; Sullivan, L.D. Improved local control of invasive bladder cancer by concurrent cisplatin and preoperative or definitive radiation. *J. Clin. Oncol.* **1996**, *14*, 2901–2907. [CrossRef]
4. James, N.D.; Hussain, S.A.; Hall, E.; Jenkins, P.; Tremlett, J.; Rawlings, C.; Crundwell, M.; Sizer, B.; Sreenivasan, T.; Hendron, C.; et al. Radiotherapy with or without Chemotherapy in Muscle-Invasive Bladder Cancer. *N. Engl. J. Med.* **2012**, *366*, 1477–1488. [CrossRef] [PubMed]
5. Jiang, D.M.; Chung, P.; Kulkarni, G.S.; Sridhar, S.S. Trimodality Therapy for Muscle-Invasive Bladder Cancer: Recent Advances and Unanswered Questions. *Curr. Oncol. Rep.* **2020**, *22*, 1–11. [CrossRef] [PubMed]
6. Thompson, C.; Joseph, N.; Sanderson, B.; Logue, J.; Wylie, J.; Elliott, T.; Lyons, J.; Anandadas, C.; Choudhury, A. Tolerability of Concurrent Chemoradiation Therapy With Gemcitabine (GemX), With and Without Prior Neoadjuvant Chemotherapy, in Muscle Invasive Bladder Cancer. *Int. J. Radiat. Oncol. Biol. Phys.* **2017**, *97*, 732–739. [CrossRef] [PubMed]
7. Hoskin, P.J.; Rojas, A.M.; Bentzen, S.M.; Saunders, M.I. Radiotherapy with concurrent carbogen and nicotinamide in bladder carcinoma. *J. Clin. Oncol.* **2010**, *28*, 4912–4918. [CrossRef] [PubMed]
8. Ajani, J.A.; Winter, K.A.; Gunderson, L.L.; Pedersen, J.; Benson, A.B.; Thomas, C.R.; Mayer, R.J.; Haddock, M.G.; Rich, T.A.; Willett, C. Fluorouracil, mitomycin, and radiotherapy vs fluorouracil, cisplatin, and radiotherapy for carcinoma of the anal canal: A randomized controlled trial. *JAMA J. Am. Med. Assoc.* **2008**, *299*, 1914–1921. [CrossRef]
9. Forastiere, A.A.; Zhang, Q.; Weber, R.S.; Maor, M.H.; Goepfert, H.; Pajak, T.F.; Morrison, W.; Glisson, B.; Trotti, A.; Ridge, J.A.; et al. Long-term results of RTOG 91-11: A comparison of three nonsurgical treatment strategies to preserve the larynx in patients with locally advanced larynx cancer. *J. Clin. Oncol.* **2013**, *31*, 845–852. [CrossRef] [PubMed]
10. Budach, V.; Stuschke, M.; Budach, W.; Baumann, M.; Geismar, D.; Grabenbauer, G.; Lammert, I.; Jahnke, K.; Stueben, G.; Herrmann, T.; et al. Hyperfractionated accelerated chemoradiation with concurrent fluorouracil-mitomycin is more effective than dose-escalated hyperfractionated accelerated radiation therapy alone in locally advanced head and neck cancer: Final results of the Radiotherapy Cooperative Clinical Trials Group of the German Cancer Society 95-06 prospective randomized trial. *J. Clin. Oncol.* **2005**, *23*, 1125–1135. [CrossRef]
11. El-Taji, O.M.S.; Alam, S.; Hussain, S.A. Bladder Sparing Approaches for Muscle-Invasive Bladder Cancers. *Curr. Treat. Options Oncol.* **2016**, *17*, 15. [CrossRef] [PubMed]
12. Flaig, T.W. NCCN Guidelines Updates: Management of Muscle-Invasive Bladder Cancer. *J. Natl. Compr. Canc. Netw.* **2019**, *17*, 591–593. [PubMed]
13. Witjes, J.A.; Bruins, H.M.; Cathomas, R.; Compérat, E.M.; Cowan, N.C.; Gakis, G.; Hernández, V.; Linares Espinós, E.; Lorch, A.; Neuzillet, Y.; et al. European Association of Urology Guidelines on Muscle-invasive and Metastatic Bladder Cancer: Summary of the 2020 Guidelines. *Eur. Urol.* **2020**, *79*, 82–104. [CrossRef] [PubMed]
14. Galsky, M.D.; Pal, S.K.; Chowdhury, S.; Harshman, L.C.; Crabb, S.J.; Wong, Y.N.; Yu, E.Y.; Powles, T.; Moshier, E.L.; Ladoire, S.; et al. Comparative effectiveness of gemcitabine plus cisplatin versus methotrexate, vinblastine, doxorubicin, plus cisplatin as neoadjuvant therapy for muscle-invasive bladder cancer. *Cancer* **2015**, *121*, 2586–2593. [CrossRef] [PubMed]

15. Donat, S.M.; Shabsigh, A.; Savage, C.; Cronin, A.M.; Bochner, B.H.; Dalbagni, G.; Herr, H.W.; Milowsky, M.I. Potential Impact of Postoperative Early Complications on the Timing of Adjuvant Chemotherapy in Patients Undergoing Radical Cystectomy: A High-Volume Tertiary Cancer Center Experience. *Eur. Urol.* **2009**, *55*, 177–186. [CrossRef] [PubMed]
16. Aziz, A.; May, M.; Burger, M.; Palisaar, R.-J.; Trinh, Q.-D.; Fritsche, H.-M.; Rink, M.; Chun, F.; Martini, T.; Bolenz, C.; et al. Prediction of 90-day Mortality After Radical Cystectomy for Bladder Cancer in a Prospective European Multicenter Cohort. *Eur. Urol.* **2014**, *66*, 156–163. [CrossRef] [PubMed]
17. Monteiro, L.L.; Kassouf, W. Radical cystectomy is the best choice for most patients with muscle-invasive bladder cancer? Opinion: Yes. *Int. Braz J. Urol.* **2017**, *43*, 184–187. [CrossRef]
18. Weldon, T.E.; Kursh, E.; Novak, L.J.; Persky, L. Combination radiotherapy and chemotherapy in murine bladder cancer. *Urology* **1979**, *14*, 47–52. [CrossRef]
19. Ploussard, G.; Daneshmand, S.; Efstathiou, J.A.; Herr, H.W.; James, N.D.; Rödel, C.M.; Shariat, S.F.; Shipley, W.U.; Sternberg, C.N.; Thalmann, G.N.; et al. Critical analysis of bladder sparing with trimodal therapy in muscle-invasive bladder cancer: A systematic review. *Eur. Urol.* **2014**, *66*, 120–137. [CrossRef]
20. Mak, R.H.; Hunt, D.; Shipley, W.U.; Efstathiou, J.A.; Tester, W.J.; Hagan, M.P.; Kaufman, D.S.; Heney, N.M.; Zietman, A.L. Long-term outcomes in patients with muscle-invasive bladder cancer after selective bladder-preserving combined-modality therapy: A pooled analysis of radiation therapy oncology group protocols 8802, 8903, 9506, 9706, 9906, and 0233. *J. Clin. Oncol.* **2014**, *32*, 3801–3809. [CrossRef]
21. Huddart, R.A.; Hall, E.; Hussain, S.A.; Jenkins, P.; Rawlings, C.; Tremlett, J.; Crundwell, M.; Adab, F.A.; Sheehan, D.; Syndikus, I.; et al. Randomized noninferiority trial of reduced high-dose volume versus standard volume radiation therapy for muscle-invasive bladder cancer: Results of the BC2001 Trial (CRUK/01/004). *Int. J. Radiat. Oncol. Biol. Phys.* **2013**, *87*, 261–269. [CrossRef] [PubMed]
22. Murthy, V.; Masodkar, R.; Kalyani, N.; Mahantshetty, U.; Bakshi, G.; Prakash, G.; Joshi, A.; Prabhash, K.; Ghonge, S.; Shrivastava, S. Clinical outcomes with dose-escalated adaptive radiation therapy for urinary bladder cancer: A prospective study. *Int. J. Radiat. Oncol. Biol. Phys.* **2016**, *94*, 60–66. [CrossRef]
23. Zhang, S.; Yu, Y.-H.; Zhang, Y.; Qu, W.; Li, J. Radiotherapy in muscle-invasive bladder cancer: The latest research progress and clinical application. *Am. J. Cancer Res.* **2015**, *5*, 854–868. [PubMed]
24. Daro-Faye, M.; Kassouf, W.; Souhami, L.; Marcq, G.; Cury, F.; Niazi, T.; Sargos, P. Combined radiotherapy and immunotherapy in urothelial bladder cancer: Harnessing the full potential of the anti-tumor immune response. *World J. Urol.* **2020**, *1*. [CrossRef] [PubMed]
25. Marks, L.B.; Carroll, P.R.; Dugan, T.C.; Anscher, M.S. The response of the urinary bladder, urethra, and ureter to radiation and chemotherapy. *Int. J. Radiat. Oncol. Biol. Phys.* **1995**, *31*, 1257–1280. [CrossRef]
26. Viswanathan, A.N.; Yorke, E.D.; Marks, L.B.; Eifel, P.J.; Shipley, W.U. Radiation Dose-Volume Effects of the Urinary Bladder. *Int. J. Radiat. Oncol. Biol. Phys.* **2010**, *76*, S116. [CrossRef]
27. Forker, L.J.; Choudhury, A.; Kiltie, A.E. Biomarkers of Tumour Radiosensitivity and Predicting Benefit from Radiotherapy. *Clin. Oncol.* **2015**, *27*, 561–569. [CrossRef]
28. Koga, F.; Takemura, K.; Fukushima, H. Biomarkers for predicting clinical outcomes of chemoradiation-based bladder preservation therapy for muscle-invasive bladder cancer. *Int. J. Mol. Sci.* **2018**, *19*, 2777. [CrossRef]
29. Desai, N.B.; Bagrodia, A. The challenge of matching assays to biology in DNA damage response biomarkers for response to radiotherapy in bladder cancer. *Transl. Androl. Urol.* **2019**, *8*, S514–S516. [CrossRef]
30. Choudhury, A.; Nelson, L.D.; Teo, M.T.W.; Chilka, S.; Bhattarai, S.; Johnston, C.F.; Elliott, F.; Lowery, J.; Taylor, C.F.; Churchman, M.; et al. MRE11 expression is predictive of cause-specific survival following radical radiotherapy for muscle-invasive bladder cancer. *Cancer Res.* **2010**, *70*, 7017–7026. [CrossRef]
31. Kawashima, A.; Nakayama, M.; Kakuta, Y.; Abe, T.; Hatano, K.; Mukai, M.; Nagahara, A.; Nakai, Y.; Oka, D.; Takayama, H.; et al. Excision repair cross-complementing group 1 may predict the efficacy of chemoradiation therapy for muscle-invasive bladder cancer. *Clin. Cancer Res.* **2011**, *17*, 2561–2569. [CrossRef]
32. Laurberg, J.R.; Brems-Eskildsen, A.S.; Nordentoft, I.; Fristrup, N.; Schepeler, T.; Ulhøi, B.P.; Agerbæk, M.; Hartmann, A.; Bertz, S.; Wittlinger, M.; et al. Expression of TIP60 (tat-interactive protein) and MRE11 (meiotic recombination 11 homolog) predict treatment-specific outcome of localised invasive bladder cancer. *BJU Int.* **2012**, *110*, E1228–E1236. [CrossRef] [PubMed]
33. Walker, A.K.; Karaszi, K.; Valentine, H.; Strauss, V.Y.; Choudhury, A.; McGill, S.; Wen, K.; Brown, M.D.; Ramani, V.; Bhattarai, S.; et al. MRE11 as a Predictive Biomarker of Outcome After Radiation Therapy in Bladder Cancer. *Int. J. Radiat. Oncol. Biol. Phys.* **2019**, *104*, 809–818. [CrossRef] [PubMed]
34. Efstathiou, J.A.; Spiegel, D.Y.; Shipley, W.U.; Heney, N.M.; Kaufman, D.S.; Niemierko, A.; Coen, J.J.; Skowronski, R.Y.; Paly, J.J.; McGovern, F.J.; et al. Long-term outcomes of selective bladder preservation by combined-modality therapy for invasive bladder cancer: The MGH experience. *Eur. Urol.* **2012**, *61*, 705–711. [CrossRef] [PubMed]
35. Mitin, T.; George, A.; Zietman, A.L.; Heney, N.M.; Kaufman, D.S.; Uzzo, R.G.; Dreicer, R.; Wallace, H.J.; Souhami, L.; Dobelbower, M.C.; et al. Long-term outcomes among patients who achieve complete or near-complete responses after the induction phase of bladder-preserving combined-modality therapy for muscle-invasive bladder cancer: A pooled analysis of NRG Oncology/RTOG 9906 and 0233. *Int. J. Radiat. Oncol. Biol. Phys.* **2016**, *94*, 67–74. [CrossRef]

36. Kamoun, A.; de Reyniès, A.; Allory, Y.; Sjödahl, G.; Robertson, A.G.; Seiler, R.; Hoadley, K.A.; Groeneveld, C.S.; Al-Ahmadie, H.; Choi, W.; et al. A Consensus Molecular Classification of Muscle-invasive Bladder Cancer[Formula presented]. *Eur. Urol.* **2020**, *77*, 420–433. [CrossRef]
37. Choi, W.; Porten, S.; Kim, S.; Willis, D.; Plimack, E.R.; Hoffman-Censits, J.; Roth, B.; Cheng, T.; Tran, M.; Lee, I.L.; et al. Identification of Distinct Basal and Luminal Subtypes of Muscle-Invasive Bladder Cancer with Different Sensitivities to Frontline Chemotherapy. *Cancer Cell* **2014**, *25*, 152–165. [CrossRef]
38. Rosenberg, J.E.; Hoffman-Censits, J.; Powles, T.; Van Der Heijden, M.S.; Balar, A.V.; Necchi, A.; Dawson, N.; O'Donnell, P.H.; Balmanoukian, A.; Loriot, Y.; et al. Atezolizumab in patients with locally advanced and metastatic urothelial carcinoma who have progressed following treatment with platinum-based chemotherapy: A single-arm, multicentre, phase 2 trial. *Lancet* **2016**, *387*, 1909–1920. [CrossRef]
39. Seiler, R.; Ashab, H.A.D.; Erho, N.; van Rhijn, B.W.G.; Winters, B.; Douglas, J.; Van Kessel, K.E.; Fransen van de Putte, E.E.; Sommerlad, M.; Wang, N.Q.; et al. Impact of Molecular Subtypes in Muscle-invasive Bladder Cancer on Predicting Response and Survival after Neoadjuvant Chemotherapy [Figure presented]. *Eur. Urol.* **2017**, *72*, 544–554. [CrossRef]
40. Mariathasan, S.; Turley, S.J.; Nickles, D.; Castiglioni, A.; Yuen, K.; Wang, Y.; Kadel, E.E.; Koeppen, H.; Astarita, J.L.; Cubas, R.; et al. TGFβ attenuates tumour response to PD-L1 blockade by contributing to exclusion of T cells. *Nature* **2018**, *554*, 544–548. [CrossRef]
41. Rebouissou, S.; Bernard-Pierrot, I.; De Reyniès, A.; Lepage, M.L.; Krucker, C.; Chapeaublanc, E.; Hérault, A.; Kamoun, A.; Caillault, A.; Letouzé, E.; et al. EGFR as a potential therapeutic target for a subset of muscle-invasive bladder cancers presenting a basal-like phenotype. *Sci. Transl. Med.* **2014**, *6*, 244ra91. [CrossRef] [PubMed]
42. Pawlik, T.M.; Keyomarsi, K. Role of cell cycle in mediating sensitivity to radiotherapy. *Int. J. Radiat. Oncol. Biol. Phys.* **2004**, *59*, 928–942. [CrossRef] [PubMed]
43. Horsman, M.R.; Overgaard, J. The impact of hypoxia and its modification of the outcome of radiotherapy. *J. Radiat. Res.* **2016**, *57* Suppl 1, i90–i98. [CrossRef]
44. Efstathiou, J.A.; Gibb, E.; Miyamoto, D.T.; Wu, C.L.; Drumm, M.R.; Lehrer, J.; Ashab, H.A.D.M.; Erho, N.G.; du Plessis, M.; Ong, K.; et al. Subtyping Muscle-Invasive Bladder Cancer to Assess Clinical Response to Trimodality Therapy. *Int. J. Radiat. Oncol.* **2017**, *99*, S118. [CrossRef]
45. Choudhury, A.; Yang, L.; Irlam, J.J.; Williamson, A.; Denley, H.; Hoskin, P.; West, C. A hypoxia transcriptomic signature to predict benefit from hypoxia-modifying treatment for muscle-invasive bladder cancer patients. *J. Clin. Oncol.* **2017**, *35*, 301. [CrossRef]
46. Shi, M.J.; Meng, X.Y.; Fontugne, J.; Chen, C.L.; Radvanyi, F.; Bernard-Pierrot, I. Identification of new driver and passenger mutations within APOBEC-induced hotspot mutations in bladder cancer. *Genome Med.* **2020**, *12*. [CrossRef]
47. Earl, J.; Rico, D.; Carrillo-de-Santa-Pau, E.; Rodríguez-Santiago, B.; Méndez-Pertuz, M.; Auer, H.; Gómez, G.; Grossman, H.B.; Pisano, D.G.; Schulz, W.A.; et al. The UBC-40 Urothelial Bladder Cancer cell line index: A genomic resource for functional studies. *BMC Genomics* **2015**, *16*, 403. [CrossRef]
48. Zuiverloon, T.C.M.; De Jong, F.C.; Costello, J.C.; Theodorescu, D. Systematic Review: Characteristics and Preclinical Uses of Bladder Cancer Cell Lines. *Bladder Cancer* **2018**, *4*, 169–183. [CrossRef]
49. Ruan, J.L.; Hsu, J.W.; Browning, R.J.; Stride, E.; Yildiz, Y.O.; Vojnovic, B.; Kiltie, A.E. Mouse Models of Muscle-invasive Bladder Cancer: Key Considerations for Clinical Translation Based on Molecular Subtypes. *Eur. Urol. Oncol.* **2019**, *2*, 239–247. [CrossRef]
50. Cellosaurus—SIB Swiss Institute of Bioinformatics | Expasy. Available online: https://www.expasy.org/resources/cellosaurus (accessed on 19 November 2020).
51. Zhang, Y. Understanding the gender disparity in bladder cancer risk: The impact of sex hormones and liver on bladder susceptibility to carcinogens. *J. Environ. Sci. Heal. Part C Environ. Carcinog. Ecotoxicol. Rev.* **2013**, *31*, 287–304. [CrossRef]
52. Dobruch, J.; Daneshmand, S.; Fisch, M.; Lotan, Y.; Noon, A.P.; Resnick, M.J.; Shariat, S.F.; Zlotta, A.R.; Boorjian, S.A. Gender and Bladder Cancer: A Collaborative Review of Etiology, Biology, and Outcomes. *Eur. Urol.* **2016**, *69*, 300–310. [CrossRef] [PubMed]
53. Ide, H.; Inoue, S.; Mizushima, T.; Jiang, G.; Chuang, K.H.; Oya, M.; Miyamoto, H. Androgen receptor signaling reduces radiosensitivity in bladder cancer. *Mol. Cancer Ther.* **2018**, *17*, 1566–1574. [CrossRef] [PubMed]
54. Yard, B.D.; Adams, D.J.; Chie, E.K.; Tamayo, P.; Battaglia, J.S.; Gopal, P.; Rogacki, K.; Pearson, B.E.; Phillips, J.; Raymond, D.P.; et al. A genetic basis for the variation in the vulnerability of cancer to DNA damage. *Nat. Commun.* **2016**, *7*, 1–14. [CrossRef] [PubMed]
55. Groselj, B.; Ruan, J.L.; Scott, H.; Gorrill, J.; Nicholson, J.; Kelly, J.; Anbalagan, S.; Thompson, J.; Stratford, M.R.L.; Jevons, S.J.; et al. Radiosensitization in vivo by histone deacetylase inhibition with no increase in early normal tissue radiation toxicity. *Mol. Cancer Ther.* **2018**, *17*, 381–392. [CrossRef]
56. Williams, K.J.; Albertella, M.R.; Fitzpatrick, B.; Loadman, P.M.; Shnyder, S.D.; Chinje, E.C.; Telfer, B.A.; Dunk, C.R.; Harris, P.A.; Stratford, I.J. In vivo activation of the hypoxia-targeted cytotoxin AQ4N in human tumor xenografts. *Mol. Cancer Ther.* **2009**, *8*, 3266–3275. [CrossRef]
57. Paillas, S.; Then, C.K.; Kilgas, S.; Ruan, J.L.; Thompson, J.; Elliott, A.; Smart, S.; Kiltie, A.E. The Histone Deacetylase Inhibitor Romidepsin Spares Normal Tissues While Acting as an Effective Radiosensitizer in Bladder Tumors in Vivo. *Int. J. Radiat. Oncol. Biol. Phys.* **2020**, *107*, 212–221. [CrossRef]
58. Then, C.K.; Paillas, S.; Wang, X.; Hampson, A.; Kiltie, A.E. Association of Bacteroides acidifaciens relative abundance with high-fibre diet-associated radiosensitisation. *BMC Biol.* **2020**, *18*. [CrossRef]
59. Schaffer, M.; Schaffer, P.M.; Corti, L.; Sotti, G.; Hofstetter, A.; Jori, G.; Dühmke, E. Photofrin II as an Efficient Radiosensitizing Agent in an Experimental Tumor. *Oncol. Res. Treat.* **2001**, *24*, 482–485. [CrossRef]

60. Zhang, Z.W.; Xiao, J.; Luo, W.; Wang, B.H.; Chen, J.M. Caffeine suppresses apoptosis of bladder cancer RT4 cells in response to ionizing radiation by inhibiting ataxia telangiectasia mutated-Chk2-p53 axis. *Chin. Med. J.* **2015**, *128*, 2938–2945. [CrossRef]
61. Jiang, H.; Hu, X.; Zhang, H.; Li, W. Down-regulation of LncRNA TUG1 enhances radiosensitivity in bladder cancer via suppressing HMGB1 expression. *Radiat. Oncol.* **2017**, *12*. [CrossRef]
62. Yagisawa, T.; Okumi, M.; Omoto, K.; Sawada, Y.; Morikawa, S.; Tanabe, K. Novel approach for bladder cancer treatment using sulfoquinovosylacylpropanediol as a radiosensitizer. *Int. J. Urol.* **2016**, *23*, 270–272. [CrossRef] [PubMed]
63. Zhao, M.J.; Song, Y.F.; Niu, H.T.; Tian, Y.X.; Yang, X.G.; Xie, K.; Jing, Y.H.; Wang, D.G. Adenovirus-mediated downregulation of the ubiquitin ligase RNF8 sensitizes bladder cancer to radiotherapy. *Oncotarget* **2016**, *7*, 8956–8967. [CrossRef]
64. Wang, F.; Tang, J.; Li, P.; Si, S.; Yu, H.; Yang, X.; Tao, J.; Lv, Q.; Gu, M.; Yang, H.; et al. Chloroquine Enhances the Radiosensitivity of Bladder Cancer Cells by Inhibiting Autophagy and Activating Apoptosis. *Cell. Physiol. Biochem.* **2018**, *45*, 54–66. [CrossRef] [PubMed]
65. Lin, T.; Zhang, Q.; Yuan, A.; Wang, B.; Zhang, F.; Ding, Y.; Cao, W.; Chen, W.; Guo, H. Synergy of tumor microenvironment remodeling and autophagy inhibition to sensitize radiation for bladder cancer treatment. *Theranostics* **2020**, *10*, 7683–7696. [CrossRef] [PubMed]
66. Gupta, A.K.; Cerniglia, G.J.; Mick, R.; Ahmed, M.S.; Bakanauskas, V.J.; Muschel, R.J.; McKenna, W.G. Radiation sensitization of human cancer cells in vivo by inhibiting the activity of PI3K using LY294002. *Int. J. Radiat. Oncol. Biol. Phys.* **2003**, *56*, 846–853. [CrossRef]
67. Cohen-Jonathan, E.; Muschel, R.J.; McKenna, W.G.; Evans, S.M.; Cerniglia, G.; Mick, R.; Kusewitt, D.; Sebti, S.M.; Hamilton, A.D.; Oliff, A.; et al. Farnesyltransferase inhibitors potentiate the antitumor effect of radiation on a human tumor xenograft expressing activated HRAS. *Radiat. Res.* **2000**, *154*, 125–132. [CrossRef]
68. Shrivastava, S.; Mansure, J.J.; Almajed, W.; Cury, F.; Ferbeyre, G.; Popovic, M.; Seuntjens, J.; Kassouf, W. The Role of HMGB1 in Radioresistance of Bladder Cancer. *Mol. Cancer Ther.* **2016**, *15*, 471–479. [CrossRef]
69. Yoshida, S.; Koga, F.; Tatokoro, M.; Kawakami, S.; Fujii, Y.; Kumagai, J.; Neckers, L.; Kihara, K. Low-dose Hsp90 inhibitors tumor-selectively sensitize bladder cancer cells to chemoradiotherapy. *Cell Cycle* **2011**, *10*, 4291–4299. [CrossRef]
70. Colquhoun, A.J.; Mchugh, L.A.; Tulchinsky, E.; Kriajevska, M.; Mellon, J.K. Combination Treatment with Ionising Radiation and Gefitinib ("Iressa", ZD1839), an Epidermal Growth Factor Receptor (EGFR) Inhibitor, Significantly Inhibits Bladder Cancer Cell Growth in vitro and in vivo. *J. Radiat. Res* **2007**, *48*, 351–360. [CrossRef]
71. Zhang, Z.; Shirakawa, T.; Hinata, N.; Matsumoto, A.; Fujisawa, M.; Okada, H.; Kamidono, S.; Matsuo, M.; Gotoh, A. Combination with CD/5-FC gene therapy enhances killing of human bladder-cancer cells by radiation. *J. Gene Med.* **2003**, *5*, 860–867. [CrossRef]
72. Wu, C.-T.; Chen, W.C.; Chang, Y.H.; Lin, W.Y.; Chen, M.F. The role of PD-L1 in the radiation response and clinical outcome for bladder cancer. *Sci. Rep.* **2016**, *6*. [CrossRef] [PubMed]
73. Ayoub, M.; Shinde-Jadhav, S.; Mansure, J.J.; Alvarez, F.; Connell, T.; Seuntjens, J.; Piccirillo, C.A.; Kassouf, W. The immune mediated role of extracellular HMGB1 in a heterotopic model of bladder cancer radioresistance. *Sci. Rep.* **2019**, *9*. [CrossRef] [PubMed]
74. Prack Mc Cormick, B.; Langle, Y.; Belgorosky, D.; Vanzulli, S.; Balarino, N.; Sandes, E.; Eiján, A.M. Flavonoid silybin improves the response to radiotherapy in invasive bladder cancer. *J. Cell. Biochem.* **2018**, *119*, 5402–5412. [CrossRef] [PubMed]
75. Prack Mc Cormick, B.; Belgorosky, D.; Langle, Y.; Balarino, N.; Sandes, E.; Eiján, A.M. Bacillus Calmette-Guerin improves local and systemic response to radiotherapy in invasive bladder cancer. *Nitric Oxide Biol. Chem.* **2017**, *64*, 22–30. [CrossRef]
76. Mu, Y.; Sun, D. Lapatinib, a dual inhibitor of epidermal growth factor receptor (EGFR) and HER-2, enhances radiosensitivity in mouse bladder tumor line-2 (MBT-2) cells in vitro and in vivo. *Med. Sci. Monit.* **2018**, *24*, 5811–5819. [CrossRef]
77. Tsai, Y.C.; Yeh, C.H.; Tzen, K.Y.; Ho, P.Y.; Tuan, T.F.; Pu, Y.S.; Cheng, A.L.; Cheng, J.C.H. Targeting epidermal growth factor receptor/human epidermal growth factor receptor 2 signalling pathway by a dual receptor tyrosine kinase inhibitor afatinib for radiosensitisation in murine bladder carcinoma. *Eur. J. Cancer* **2013**, *49*, 1458–1466. [CrossRef]
78. Flaig, T.W.; Spiess, P.E.; Agarwal, N.; Bangs, R.; Boorjian, S.A.; Buyyounouski, M.K.; Chang, S.; Downs, T.M.; Efstathiou, J.A.; Friedlander, M.; et al. Bladder Cancer, Version 3.2020, NCCN Clinical Practice Guidelines in Oncology. *J. Natl. Compr. Cancer Netw.* **2020**, *18*, 329–354. [CrossRef]
79. Kyriazis, A.P.; Kyriazis, A.A.; Yagoda, A. Time dependence of the radiation-modifying effect of cis-diamminedichloroplatinum II (cisplatin, DDP) on human urothelial cancer grown in nude mice. *Cancer Investig.* **1986**, *4*, 217–222. [CrossRef]
80. Bedford, P.; Shellard, S.A.; Walker, M.C.; Whelan, R.D.H.; Masters, J.R.W.; Hill, B.T. Differential expression of collateral sensitivity or resistance to cisplatin in human bladder carcinoma cell lines pre-exposedin vitro to either X-irradiation or cisplatin. *Int. J. Cancer* **1987**, *40*, 681–686. [CrossRef]
81. Hoskin, P.J.; Sibtain, A.; Daley, F.M.; Wilson, G.D. GLUT1 and CAIX as intrinsic markers of hypoxia in bladder cancer: Relationship with vascularity and proliferation as predictors of outcome of ARCON. *Br. J. Cancer* **2003**, *89*, 1290–1297. [CrossRef]
82. Caffo, O.; Thompson, C.; De Santis, M.; Kragelj, B.; Hamstra, D.A.; Azria, D.; Fellin, G.; Pappagallo, G.L.; Galligioni, E.; Choudhury, A. Concurrent gemcitabine and radiotherapy for the treatment of muscle-invasive bladder cancer: A pooled individual data analysis of eight phase I–II trials. *Radiother. Oncol.* **2016**, *121*, 193–198. [CrossRef] [PubMed]
83. Fechner, G.; Perabo, F.G.E.; Schmidt, D.H.; Haase, L.; Ludwig, E.; Schueller, H.; Blatter, J.; Müller, S.C.; Albers, P. Preclinical evaluation of a radiosensitizing effect of gemcitabine in p53 mutant and p53 wild type bladder cancer cells. *Urology* **2003**, *61*, 468–473. [CrossRef]

84. Pauwels, B.; Korst, A.E.C.; Pattyn, G.G.O.; Lambrechts, H.A.J.; Van Bockstaele, D.R.; Vermeulen, K.; Lenjou, M.; De Pooter, C.M.J.; Vermorken, J.B.; Lardon, F. Cell cycle effect of gemcitabine and its role in the radiosensitizing mechanism in vitro. *Int. J. Radiat. Oncol. Biol. Phys.* **2003**, *57*, 1075–1083. [CrossRef]
85. Sangar, V.K.; Cowan, R.; Margison, G.P.; Hendry, J.H.; Clarke, N.W. An evaluation of gemcitabines differential radiosensitising effect in related bladder cancer cell lines. *Br. J. Cancer* **2004**, *90*, 542–548. [CrossRef] [PubMed]
86. Capes-Davis, A.; Theodosopoulos, G.; Atkin, I.; Drexler, H.G.; Kohara, A.; MacLeod, R.A.F.; Masters, J.R.; Nakamura, Y.; Reid, Y.A.; Reddel, R.R.; et al. Check your cultures! A list of cross-contaminated or misidentified cell lines. *Int. J. Cancer* **2010**, *127*, 1–8. [CrossRef] [PubMed]
87. Choudhury, A.; Zhao, H.; Jalali, F.; Al Rashid, S.; Ran, J.; Supiot, S.; Kiltie, A.E.; Bristow, R.G. Targeting homologous recombination using imatinib results in enhanced tumor cell chemosensitivity and radiosensitivity. *Mol. Cancer Ther.* **2009**, *8*, 203–213. [CrossRef]
88. Kerr, M.; Scott, H.E.; Groselj, B.; Stratford, M.R.L.; Karaszi, K.; Sharma, N.L.; Kiltie, A.E. Deoxycytidine kinase expression underpins response to gemcitabine in bladder cancer. *Clin. Cancer Res.* **2014**, *20*, 5435–5445. [CrossRef]
89. Paget, V.; Ben Kacem, M.; Dos Santos, M.; Benadjaoud, M.A.; Soysouvanh, F.; Buard, V.; Georges, T.; Vaurijoux, A.; Gruel, G.; François, A.; et al. Multiparametric radiobiological assays show that variation of X-ray energy strongly impacts relative biological effectiveness: Comparison between 220 kV and 4 MV. *Sci. Rep.* **2019**, *9*. [CrossRef]
90. Ben Kacem, M.; Benadjaoud, M.A.; Dos Santos, M.; Soysouvanh, F.; Buard, V.; Tarlet, G.; Le Guen, B.; François, A.; Guipaud, O.; Milliat, F.; et al. Variation of 4 MV X-ray dose rate strongly impacts biological response both in vitro and in vivo. *Sci. Rep.* **2020**, *10*, 7021. [CrossRef]
91. Contessa, J.N.; Hampton, J.; Lammering, G.; Mikkelsen, R.B.; Dent, P.; Valerie, K.; Schmidt-Ullrich, R.K. Ionizing radiation activates Erb-B receptor dependent Akt and p70 S6 kinase signaling in carcinoma cells. *Oncogene* **2002**, *21*, 4032–4041. [CrossRef]
92. Schmidt-Ullrich, R.K.; Contessa, J.N.; Lammering, G.; Amorino, G.; Lin, P.S. ERBB receptor tyrosine kinases and cellular radiation responses. *Oncogene* **2003**, *22*, 5855–5865. [CrossRef] [PubMed]
93. Chaux, A.; Cohen, J.S.; Schultz, L.; Albadine, R.; Jadallah, S.; Murphy, K.M.; Sharma, R.; Schoenberg, M.P.; Netto, G.J. High epidermal growth factor receptor immunohistochemical expression in urothelial carcinoma of the bladder is not associated with EGFR mutations in exons 19 and 21: A study using formalin-fixed, paraffin-embedded archival tissues. *Hum. Pathol.* **2012**, *43*, 1590–1595. [CrossRef]
94. Bonner, J.A.; Harari, P.M.; Giralt, J.; Azarnia, N.; Shin, D.M.; Cohen, R.B.; Jones, C.U.; Sur, R.; Raben, D.; Jassem, J.; et al. Radiotherapy plus cetuximab for squamous-cell carcinoma of the head and neck. *N. Engl. J. Med.* **2006**, *354*, 567–578. [CrossRef] [PubMed]
95. Dominguez-Escrig, J.L.; Kelly, J.D.; Neal, D.E.; King, S.M.; Davies, B.R. Evaluation of the therapeutic potential of the epidermal growth factor receptor tyrosine kinase inhibitor gefitinib in preclinical models of bladder cancer. *Clin. Cancer Res.* **2004**, *10*, 4874–4884. [CrossRef] [PubMed]
96. Giannopoulou, A.F.; Velentzas, A.D.; Konstantakou, E.G.; Avgeris, M.; Katarachia, S.A.; Papandreou, N.C.; Kalavros, N.I.; Mpakou, V.E.; Iconomidou, V.; Anastasiadou, E.; et al. Revisiting histone deacetylases in human tumorigenesis: The paradigm of urothelial bladder cancer. *Int. J. Mol. Sci.* **2019**, *20*, 1291. [CrossRef] [PubMed]
97. Grivas, P.; Mortazavi, A.; Picus, J.; Hahn, N.M.; Milowsky, M.I.; Hart, L.L.; Alva, A.; Bellmunt, J.; Pal, S.K.; Bambury, R.M.; et al. Mocetinostat for patients with previously treated, locally advanced/metastatic urothelial carcinoma and inactivating alterations of acetyltransferase genes. *Cancer* **2019**, *125*, 533–540. [CrossRef] [PubMed]
98. Tsai, Y.-C.; Ho, P.-Y.; Tzen, K.-Y.; Tuan, T.-F.; Liu, W.-L.; Cheng, A.-L.; Pu, Y.-S.; Cheng, J.C.-H. Synergistic Blockade of EGFR and HER2 by New-Generation EGFR Tyrosine Kinase Inhibitor Enhances Radiation Effect in Bladder Cancer Cells. *Mol. Cancer Ther.* **2015**, *14*, 810–820. [CrossRef]
99. Inoue, T.; Terada, N.; Kobayashi, T.; Ogawa, O. Patient-derived xenografts as in vivo models for research in urological malignancies. *Nat. Rev. Urol.* **2017**, *14*, 267–283. [CrossRef]
100. Hedegaard, J.; Lamy, P.; Nordentoft, I.; Algaba, F.; Høyer, S.; Ulhøi, B.P.; Vang, S.; Reinert, T.; Hermann, G.G.; Mogensen, K.; et al. Comprehensive Transcriptional Analysis of Early-Stage Urothelial Carcinoma. *Cancer Cell* **2016**, *30*, 27–42. [CrossRef]
101. Robertson, A.G.; Kim, J.; Al-Ahmadie, H.; Bellmunt, J.; Guo, G.; Cherniack, A.D.; Hinoue, T.; Laird, P.W.; Hoadley, K.A.; Akbani, R.; et al. Comprehensive Molecular Characterization of Muscle-Invasive Bladder Cancer. *Cell* **2017**, *171*, 540–556.e25. [CrossRef]
102. Mahe, M.; Dufour, F.; Neyret-Kahn, H.; Moreno-Vega, A.; Beraud, C.; Shi, M.; Hamaidi, I.; Sanchez-Quiles, V.; Krucker, C.; Dorland-Galliot, M.; et al. An FGFR3/MYC positive feedback loop provides new opportunities for targeted therapies in bladder cancers. *EMBO Mol. Med.* **2018**, *10*. [CrossRef] [PubMed]
103. Fisher, M.M.; SenthilKumar, G.; Hu, R.; Goldstein, S.; Ong, I.M.; Miller, M.C.; Brennan, S.R.; Kaushik, S.; Abel, L.; Nickel, K.P.; et al. Fibroblast Growth Factor Receptors as Targets for Radiosensitization in Head and Neck Squamous Cell Carcinomas. *Int. J. Radiat. Oncol. Biol. Phys.* **2020**, *107*, 793–803. [CrossRef] [PubMed]
104. Dufour, F.; Silina, L.; Neyret-Kahn, H.; Moreno-Vega, A.; Krucker, C.; Karboul, N.; Dorland-Galliot, M.; Maillé, P.; Chapeaublanc, E.; Allory, Y.; et al. TYRO3 as a molecular target for growth inhibition and apoptosis induction in bladder cancer. *Br. J. Cancer* **2019**, *120*. [CrossRef] [PubMed]
105. Tormoen, G.W.; Crittenden, M.R.; Gough, M.J. The TAM family as a therapeutic target in combination with radiation therapy. *Emerg. Top. Life Sci.* **2017**, *1*, 493–500. [CrossRef] [PubMed]

106. Aguilera, T.A.; Rafat, M.; Castellini, L.; Shehade, H.; Kariolis, M.S.; Hui, A.B.Y.; Stehr, H.; Von Eyben, R.; Jiang, D.; Ellies, L.G.; et al. Reprogramming the immunological microenvironment through radiation and targeting Axl. *Nat. Commun.* **2016**, *7*. [CrossRef] [PubMed]
107. Grapin, M.; Richard, C.; Limagne, E.; Boidot, R.; Morgand, V.; Bertaut, A.; Derangere, V.; Laurent, P.A.; Thibaudin, M.; Fumet, J.D.; et al. Optimized fractionated radiotherapy with anti-PD-L1 and anti-TIGIT: A promising new combination. *J. Immunother. Cancer* **2019**, *7*. [CrossRef]
108. Dovedi, S.J.; Adlard, A.L.; Lipowska-Bhalla, G.; McKenna, C.; Jones, S.; Cheadle, E.J.; Stratford, I.J.; Poon, E.; Morrow, M.; Stewart, R.; et al. Acquired resistance to fractionated radiotherapy can be overcome by concurrent PD-L1 blockade. *Cancer Res.* **2014**, *74*, 5458–5468. [CrossRef]
109. Formenti, S.C.; Rudqvist, N.P.; Golden, E.; Cooper, B.; Wennerberg, E.; Lhuillier, C.; Vanpouille-Box, C.; Friedman, K.; Ferrari de Andrade, L.; Wucherpfennig, K.W.; et al. Radiotherapy induces responses of lung cancer to CTLA-4 blockade. *Nat. Med.* **2018**, *24*, 1845–1851. [CrossRef]
110. Choi, W.; Czerniak, B.; Ochoa, A.; Su, X.; Siefker-Radtke, A.; Dinney, C.; McConkey, D.J. Intrinsic basal and luminal subtypes of muscle-invasive bladder cancer. *Nat. Rev. Urol.* **2014**, *11*, 400–410. [CrossRef]
111. J Southgate; K A Hutton; D F Thomas; L K Trejdosiewicz Normal Human Urothelial Cells in Vitro: Proliferation and Induction of Stratification - PubMed. *Lab. Investig.* **1994**, *71*, 583–594.
112. Varley, C.L.; Stahlschmidt, J.; Lee, W.-C.; Holder, J.; Diggle, C.; Selby, P.J.; Trejdosiewicz, L.K.; Southgate, J. Role of PPARgamma and EGFR signalling in the urothelial terminal differentiation programme. *J. Cell Sci.* **2004**, *117*, 2029–2036. [CrossRef] [PubMed]
113. Cross, W.R.; Eardley, I.; Leese, H.J.; Southgate, J. A biomimetic tissue from cultured normal human urothelial cells: Analysis of physiological function. *Am. J. Physiol. Ren. Physiol.* **2005**, *289*. [CrossRef] [PubMed]
114. Jäger, W.; Moskalev, I.; Janssen, C.; Hayashi, T.; Awrey, S.; Gust, K.M.; So, A.I.; Zhang, K.; Fazli, L.; Li, E.; et al. Ultrasound-Guided Intramural Inoculation of Orthotopic Bladder Cancer Xenografts: A Novel High-Precision Approach. *PLoS ONE* **2013**, *8*. [CrossRef]
115. Zhang, Z.T.; Pak, J.; Shapiro, E.; Sun, T.T.; Wu, X.R. Urothelium-specific expression of an oncogene in transgenic mice induced the formation of carcinoma in situ and invasive transitional cell carcinoma. *Cancer Res.* **1999**, *59*, 3512–3517. [PubMed]
116. Vasoncelos-Nobrega, C.; Colaco, A.; Lopes, C.; Oliveira, P.A. BBN as an Urothelial Carcinogen. *In Vivo (Brooklyn).* **2012**, *26*, 727–739.
117. De La Rochere, P.; Guil-Luna, S.; Decaudin, D.; Azar, G.; Sidhu, S.S.; Piaggio, E. Humanized Mice for the Study of Immuno-Oncology. *Trends Immunol.* **2018**, *39*, 748–763. [CrossRef]
118. Shultz, L.D.; Ishikawa, F.; Greiner, D.L. Humanized mice in translational biomedical research. *Nat. Rev. Immunol.* **2007**, *7*, 118–130. [CrossRef]
119. Borcoman, E.; De La Rochere, P.; Richer, W.; Vacher, S.; Chemlali, W.; Krucker, C.; Sirab, N.; Radvanyi, F.; Allory, Y.; Pignot, G.; et al. Inhibition of PI3K pathway increases immune infiltrate in muscle-invasive bladder cancer. *Oncoimmunology* **2019**, *8*, e1581556. [CrossRef]
120. Moher, D.; Liberati, A.; Tetzlaff, J.; Altman, D.G. Preferred Reporting Items for Systematic Reviews and Meta-Analyses: The PRISMA Statement. *PLoS Med.* **2009**, *6*, e1000097. [CrossRef]

Article

Intravesical Bacillus Calmette–Guérin Treatment for T1 High-Grade Non-Muscle Invasive Bladder Cancer with Divergent Differentiation or Variant Morphologies

Makito Miyake [1,*], Nobutaka Nishimura [1], Kota Iida [1], Tomomi Fujii [2], Ryoma Nishikawa [3], Shogo Teraoka [3], Atsushi Takenaka [3], Hiroshi Kikuchi [4], Takashige Abe [4], Nobuo Shinohara [4], Eijiro Okajima [5], Takuto Shimizu [6], Shunta Hori [1,6], Norihiko Tsuchiya [7], Takuya Owari [1], Yasukiyo Murakami [8], Rikiya Taoka [9], Takashi Kobayashi [10], Takahiro Kojima [11], Naotaka Nishiyama [12], Hiroshi Kitamura [12], Hiroyuki Nishiyama [11] and Kiyohide Fujimoto [1,†]

1. Department of Urology, Nara Medical University, Kashihara, Nara 634-8522, Japan; ffxxxx.nqou@gmail.com (N.N.); kota1006iida@yahoo.co.jp (K.I.); horimaus@gmail.com (S.H.); tintherye@gmail.com (T.O.); kiyokun@naramed-u.ac.jp (K.F.)
2. Department of Diagnostic Pathology, Nara Medical University, Kashihara, Nara 634-8522, Japan; fujiit@naramed-u.ac.jp
3. Department of Urology, Faculty of Medicine, Tottori University, Yonago, Tottori 683-8503, Japan; ryoma_nishikawa@yahoo.co.jp (R.N.); teraoka-1119@tottori-u.ac.jp (S.T.); atake@med.tottori-u.ac.jp (A.T.)
4. Department of Urology, Graduate School of Medicine, Hokkaido University, Sapporo, Hokkaido 060-8638, Japan; hiroshikikuchi16@yahoo.co.jp (H.K.); takataka@rf6.so-net.ne.jp (T.A.); nozomis@mbj.nifty.com (N.S.)
5. Department of Urology, Nara City Hospital, Nara 630-8305, Japan; jiro_eddie@yahoo.co.jp
6. Department of Urology, Saiseikai Chuwa Hospital, Nara 633-0054, Japan; takutea19@gmail.com
7. Department of Urology, Faculty of Medicine, Yamagata University, Yamagata 990-9585, Japan; ntsuchiya@med.id.yamagata-u.ac.jp
8. Department of Urology, School of Medicine, Kitasato University, Sagamihara, Kanagawa 252-0374, Japan; yaskiyomura@yahoo.co.jp
9. Department of Urology, Faculty of Medicine, Kagawa University, Kagawa 761-0793, Japan; rikiya@med.kagawa-u.ac.jp
10. Department of Urology, Graduate School of Medicine, Kyoto University, Kyoto 606-8507, Japan; selecao@kuhp.kyoto-u.ac.jp
11. Department of Urology, Faculty of Medicine, University of Tsukuba, Tsukuba, Ibaraki 305-8576, Japan; tkojima@md.tsukuba.ac.jp (T.K.); nishiuro@md.tsukuba.ac.jp (H.N.)
12. Department of Urology, Faculty of Medicine, University of Toyama, Toyama 633-0054, Japan; nishiyan@med.u-toyama.ac.jp (N.N.); hkitamur@med.u-toyama.ac.jp (H.K.)
* Correspondence: makitomiyake@yahoo.co.jp; Tel.: +81-744-22-3051; Fax: +81-744-22-9282
† On behalf of the Japanese Urological Oncology Group.

Simple Summary: The 2016 World Health Organization classification system distinguishes between urothelial carcinomas (UCs) with divergent differentiation (DD) and those with variant morphologies (VMs), which until now had been considered to indicate highest-risk cases. Intravesical Bacillus Calmette–Guérin (BCG) treatment is an alternative therapeutic and adjuvant option after transurethral resection of a bladder tumor. However, data comparing oncological outcomes after intravesical BCG treatment among pure UC, UC with DD, and UC with VMs are sparse. This is a retrospective study to investigate the outcomes of bladder-preservation therapy using intravesical BCG treatment on cases of bladder UCs with DD or VMs. We followed the outcomes of 1490 patients with pure UCs, UC with DD, or UC with VM. We found that concomitant VMs, not DD, was more likely to result in cancer-specific death. The VM-associated risk was significant for cancer-specific mortality only, not for recurrence-free or progression-free survival rates.

Abstract: The 2016 World Health Organization classification newly described infiltrating urothelial carcinoma (UC) with divergent differentiation (DD) or variant morphologies (VMs). Data comparing oncological outcomes after bladder-preservation therapy using intravesical Bacillus Calmette–Guérin (BCG) treatment among T1 bladder pure UC (pUC), UC with DD (UC-DD), and UC with VMs (UC-VM) are limited. We evaluated 1490 patients with T1 high-grade bladder UC who received

intravesical BCG during 2000–2019. They were classified into three groups: 93.6% with pUC, 4.4% with UC-DD, and 2.0% with UC-VM. Recurrence-free, progression-free, and cancer-specific survival following intravesical BCG were compared among the groups using multivariate Cox regression analysis, also used to estimate inverse probability of treatment weighting-adjusted hazard ratio and 95% confidence interval for the outcomes. Glandular differentiation and micropapillary variant were the most common forms in the UC-DD and UC-VM groups, respectively. Of 1490 patients, 31% and 13% experienced recurrence and progression, respectively, and 5.0% died of bladder cancer. Survival analyses revealed the impact of concomitant VMs was significant for cancer-specific survival, but not recurrence-free and progression-free survival compared with that of pUC. Our analysis clearly demonstrated that concomitant VMs were associated with aggressive behavior in contrast to concomitant DD in patients treated with intravesical BCG.

Keywords: urinary bladder neoplasms; Bacillus Calmette–Guérin (BCG); immunotherapy; divergent differentiation; variant morphology; survival

1. Introduction

Bladder cancer is the 10th most common cancer, with roughly 549,000 initially diagnosed cases and 200,000 deaths reported annually according to 2018 worldwide cancer statistics [1]. Non-muscle-invasive bladder cancer (NMIBC; Ta, T1, and Tis) is a heterogeneous disease accounting for approximately 70% of initially diagnosed bladder cancers [2]. Although most of NMIBC can be treated with the combination of transurethral resection of a bladder tumor (TURBT) and intravesical treatment of chemotherapy and Bacillus Calmette–Guérin (BCG), some patients have repeated intravesical recurrence and disease progression.

The European Association of Urology guidelines 2019 specify characteristics of the highest-risk subset from high-risk non-muscle invasive bladder cancers (NMIBCs) [3]. Patients are stratified into highest-risk NMIBC when the tumor has at least one of the following five aggressive factors: i) T1/high-grade (HG) urothelial carcinoma (UC) with concomitant bladder carcinoma in situ (CIS); ii) multiple and/or large and/or recurrent T1/HG UC; iii) T1/HG UC with prostate-involving CIS; iv) lymphovascular invasion (LVI); v) histological variants of UC; or vi) BCG-unresponsive NMIBC. Although immediate radical cystectomy (RC) should be considered for highest-risk NMIBC, this highly invasive intervention may be associated with a significant risk of overtreatment. Clinical practice guidelines suggest that intravesical BCG treatment is still an alternative therapeutic and adjuvant option after transurethral resection of a bladder tumor (TURBT) for high-/highest-risk NMIBCs [4–6]. Clinicians may judge that not all the patients with highest-risk NMIBC require RC because of the highly heterogenic nature of highest-risk NMIBC [7].

The 2016 World Health Organization (WHO) classification of tumors of the urothelial tract newly described or better defined 'infiltrating UC with divergent differentiation (DD)' and 'infiltrating UC with variant morphologies (VMs),' implying that the two subsets should be labeled separately to optimize the therapeutic strategy [8]. The former includes squamous, glandular, and trophoblastic differentiation, whereas the latter includes nested, microcystic, micropapillary, lymphoepithelioma-like, plasmacytoid/signet ring cell/diffuse, sarcomatoid, giant cell, poorly differentiated, lipid-rich, and clear cell (glycogen-rich) variants. These unusual types of UC have been called 'histological variants' and clubbed together, which is one of the inclusion characteristics of highest-risk NMIBCs. Recent studies evaluated the frequency of DD and VMs in invasive UC, reporting that squamous differentiation was the most common in DD and micropapillary variant was the most common in VMs, respectively [9,10]. NMIBCs, MIBCs, and metastatic diseases with DDs and VMs respond poorly to intensive surgery with chemotherapy or radiotherapy, resulting in worse survival outcomes [11–15]. As some VMs are extremely rare, small sample size hinders appropriate evaluation of their clinical relevance, especially in response to intravesical BCG treatment [16–18].

Squamous and glandular differentiation is seen at a relatively high frequency among histological variants. However, there are only a few small studies or case reports regarding the efficacy of intravesical BCG against NMIBC with VMs [16,19–21]. There are only sparse data comparing oncological outcomes after intravesical BCG treatment among pure UC (without any histological variant), UC with DD, and UC with VMs considering the newly adopted WHO 2016 classification system. Given the little information regarding the efficacy of BCG in UC variants, this study aimed to shed more clarity on the mortality and survival outcomes by retrospectively examining the outcomes of 1490 patients with UC who were treated using intravesical BCG.

2. Materials and Methods

2.1. Data Collection

This retrospective multicenter study was approved by the institutional review board of each participating institute (reference protocol ID: NMU-2217) of the Japan Urological Oncology Group framework [2]. Informed consent was obtained from participants through posters and/or website using the opt-out method [22]. We reviewed 3226 patients who received intravesical BCG for pathologically diagnosed NMIBC and treatment during 2000–2019 at 31 collaborative hospitals. The clinicopathological data included oncological outcomes, treatments, age, sex, performance status, history of NMIBC, tumor multiplicity, tumor size, T category, tumor grade (per 2004 WHO classification), and presence of bladder CIS, prostate-involving CIS, LVI, DD, and VMs.

2.2. Intravesical BCG Treatment after TURBT

The intravesical BCG schedule included weekly instillations of Immunobladder (Tokyo-172 strain) or ImmuCyst (Connaught strain, currently unavailable) for 6–8 consecutive weeks for induction BCG (iBCG) with or without subsequent maintenance BCG (mBCG). Planned maintenance protocol was BCG doses once a week for 3 weeks at 3, 6, 12, 18, 24, 30, and 36 months after iBCG initiation [23]. Adequate BCG' therapy is when a patient has received at least five of six doses in induction phase followed by at least one maintenance (two of three dose) for clinical trials [24]. Based on the definition of adequate BCG, at least two of three doses in the first mBCG round at 3 months was considered mBCG implementation in this study.

2.3. Patient Selection

Figure 1 shows a flowchart of the patient selection process. Of 3226 patients, the cohort was first restricted to 1490 (46.1%) including only T1 HG tumors. Next, 76 patients (2.3%) with critical missing data were excluded, leaving 1490 patients (46.1%) eligible for the analysis. Among 1490 patients, 1395 (93.6%) had pure UC (pUC group), 65 (4.4%) had UC with DD (UC-DD group), and only 30 (2.0%) had UC with VMs (UC-VM group).

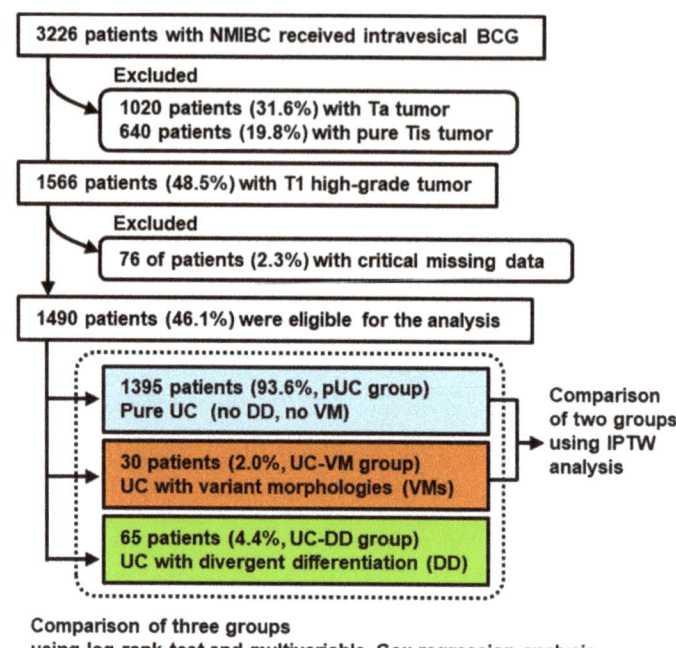

Figure 1. Flow chart for creation of the patient cohort dataset. From the original dataset, the cohort was restricted to T1 high-grade urothelial carcinoma (UC). Abbreviations: NMIBC, non-muscle invasive bladder cancer; BCG, Bacillus Calmette–Guérin; pUC, pure urothelial carcinoma; DD, divergent differentiation; VM, variant morphology; IPTW, inverse probability of treatment weighting.

2.4. Surveillance after Intravesical BCG and During mBCG

Although the surveillance protocol varied across institutions, patients were generally followed up using white-light cystoscopy and urinary cytology every 3 months for 2 years, then every 6 months in the third and fourth years, and annually thereafter [6]. Recurrence was defined as the presence of recurrent tumors of pathologically proven UC in the bladder and prostatic urethra. Neither recurrence of upper urinary tract nor positive result of urinary cytology without pathologically proven UC was considered recurrence. Progression was defined as recurrent disease with invasion into the muscularis propria ($\geq$T2), positive regional lymph nodes, and/or distant metastases.

2.5. Statistical Analysis

Statistical analyses and were performed and plots were generated using GraphPad Prism version 7.00 (GraphPad Software, San Diego, CA, USA). All reported p values are two-sided, and statistical significance was set at $p < 0.05$. Clinicopathological characteristics were compared using Kruskal–Wallis tests or chi-square tests, as appropriate. Intravesical recurrence-free survival (RFS), progression-free survival (PFS), and cancer-specific survival (CSS) were calculated from the date of administration of the initial iBCG dose. Survival rates were analyzed using the Kaplan–Meier method and compared using the log-rank test among the pUC, UC-DD, and UC-VM groups. Variables that potentially affected prognosis ($p < 0.1$) in univariate analysis were included in a stepwise Cox proportional hazards regression model. Hazard ratio (HR) with 95% confidence interval (CI) was calculated to identify independent prognostic variables.

To minimize selection bias, inverse probability of treatment weighting (IPTW) analysis was performed using R version 4.0.0 (R Development Core Team, Vienna, Austria). IPTW, which is a form of propensity score analysis, uses weighting by the inverse of the propensity score to reduce imbalance in possible confounders between the pUC and UC-VM groups [25]. The baseline characteristics were matched between the pUC and UC-VM groups by calculating the propensity score for each patient using a multivariable logistic regression model based on covariates such as age, sex, multiplicity, tumor size, presence of concomitant bladder CIS, prostate-involving CIS, LVI on TUR specimen, implementation of second TUR, and implementation of mBCG. Differences in means and proportions between the two groups was quantified using standardized mean difference. Standardized mean difference more than 0.20 was considered to indicate potentially relevant imbalances between the two groups [25]. Multivariable Cox regression analysis was used to estimate the IPTW-adjusted HRs and 95% CIs as outcomes of the two groups.

3. Results

3.1. DD and VMs Detected in ur Cohort

Table 1 lists the distribution of unusual histology findings in the UC-DD and UC-VM groups. In the UC-DD group, 38 (69%) and 27 (41%) patients had glandular differentiation and squamous differentiation, respectively, whereas trophoblastic differentiation was not seen in our cohort. The most common VM was micropapillary variant (13/30; 43% in the UC-VM group). The second most common VM was nested variant (9/30, 30%). The third, sarcomatoid variant, was observed in 4 of 30 patients (13%), followed by clear cell variant (2/30, 6.7%), microcystic variant (1/30, 3.3%), and giant cell variant (1/30, 3.3%). Other VMs such as lymphoepithelioma-like, plasmacytoid, poorly differentiated, and lipid-rich variants were not present in our cohort. Representative hematoxylin & eosin-stained images of DD and VMs are shown in Figure 2.

Table 1. Distribution of divergent differentiation (DD) and variant morphologies (VMs) in this cohort.

Histology	UC with DD (UC-DD Group)	UC with VMs (UC-VM Group)
N	65 (100%)	30 (100%)
Glandular differentiation	38 (69%)	-
Squamous differentiation	27 (41%)	-
Micropapillary variant	-	13 (43%)
Nested variant	-	9 (30%)
Sarcomatoid variant	-	4 (13%)
Clear cell variant	-	2 (6.7%)
Microcystic variant	-	1 (3.3%)
Giant cell variant	-	1 (3.3%)

Abbreviations: UC, urothelial carcinoma.

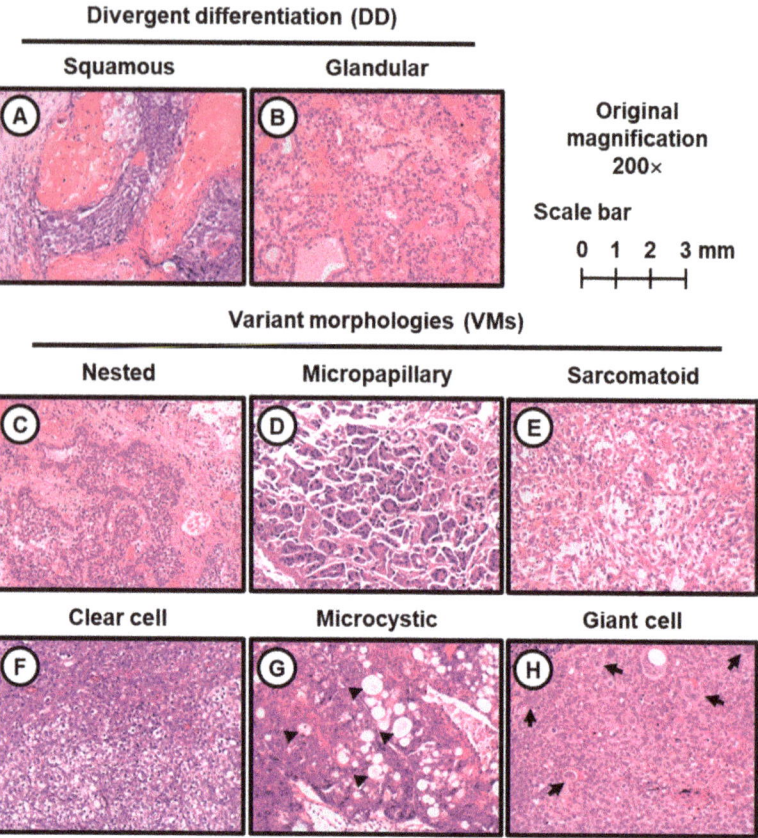

Figure 2. Divergent differentiation and variant morphologies in T1 high-grade urothelial carcinoma of cases from this study. All the illustrative images of hematoxylin-eosin staining were captured at original magnification 200×. (**A**) Squamous differentiation; (**B**) Glandular differentiation. This tumor has enteric features and mucin production; (**C**) Nested variant; (**D**) Micropapillary variant; (**E**) Sarcomatoid variant; (**F**) Clear cell variant; (**G**) Microcystic variant. This tumor forms neoplastic cystic structure. The lumina tend to be empty, but some of them contain necrotic cells, granular eosinophilic debris, or mucin (arrowheads); (**H**) Giant cell variant. This tumor has pleomorphic giant cells (black arrows) with cytoplasmic vacuoles.

3.2. Comparison of Patient Characteristics and Outcomes among the Groups

Table S1 summarizes the patient characteristics of the three groups. Statistical comparisons showed significant differences in age, LVI, and implementation of second TUR. The positivity rate of LVI in TUR specimens and rate of second TUR implementation was higher in patients in the UC-VM group than in the pUC and UC-DD groups. Both the European Organization for Research and Treatment of Cancer (EORTC) [26] and the Spanish Urological Club for Oncological Treatment (CUETO) [27] risk tables for recurrence and progression incorporated six parameters to categorize patients into risk groups according to the summed scores. These score models were applied to our cohort to calculate each patient's risk after intravesical BCG treatment, demonstrating that no significant difference was observed among each group (Table S1). In this cohort, 241 (16%) of 1490 patients received maintenance BCG and the median number of BCG doses in maintenance phase was 6 times (interquartile range, 3 to 9) and the mean number was 7.9 times. Of the 1490 patients, 466 (31%) and 199 (13%) experienced bladder cancer recurrence and progression, respectively, and 74 (5.0%) died of bladder cancer, with a median follow-up of 50 months (interquartile range, 27–79) after BCG initiation. We compared the clinical outcomes after induction of iBCG between Tokyo-172 strain (1141 patients) and Connaught strain (349 patients), demonstrating that no significant difference was observed for any of three endpoints. To investigate the impact of DD and VMs on oncological outcomes, RFS, PFS, and CSS were compared among the three groups using the Kaplan–Meier method and log-rank test (Figure 3 and Table S2). No significant impact of concomitant DD and VMs was seen for bladder recurrence. Compared with the pUC group, the UC-DD and UC-VM groups were associated with favorable prognosis and poor prognosis for progression ($p = 0.08$) and cancer-specific death ($p < 0.01$), respectively. Multivariate analysis revealed that concomitant VMs, not DD, was a strong independent factor for cancer-specific death (HR, 3.89, 95% CI, 1.55–9.77).

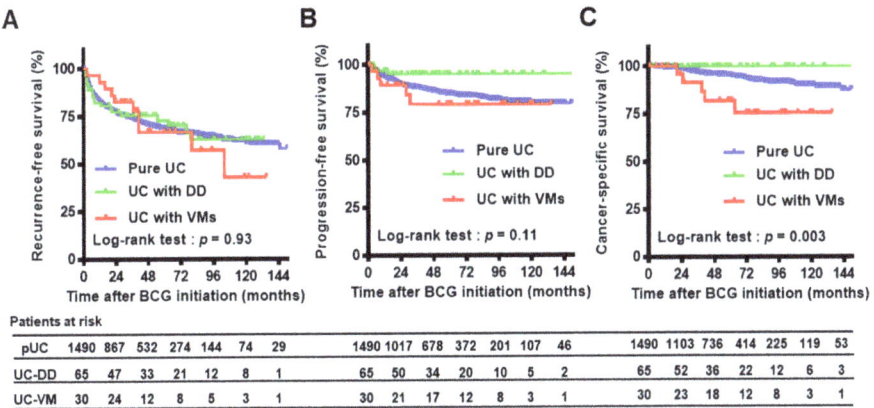

Figure 3. Survival curves of outcomes after initiation of intravesical BCG treatment among pure UC, UC with DD, and UC-VMs. Bladder recurrence-free survival (**A**), progression-free survival (**B**), and cancer-specific survival (**C**) were plotted and compared among three groups. Abbreviations: BCG, Bacillus Calmette–Guérin; UC, urothelial carcinoma; DD, divergent differentiation; VM, variant morphology.

The UC-DD group displayed better survival outcomes than the pUC group (Figure 3). Based on this finding, we decided to further explore the clinical impact of VMs in terms of response to intravesical BCG. The background and outcomes of 30 patients in the UC-VM group are listed in Table S3. Of them, eight (27%) and five (17%) patients experienced bladder recurrence and progression, respectively, and five of the latter five patients experiencing progression died of bladder cancer. Notably, unresectable metastatic lesions occurred suddenly in three (no. 8, 15, and 24) of these five patients.

3.3. IPTW-Adjusted Comparison of Outcomes between the pUC and UC-VM Groups

IPTW analysis was applied to adjust for patient characteristics between the pUC and UC-VM groups and decrease the influence of possible confounding factors (Table 2). All weighted baseline characteristics included in the propensity score model were closely balanced between the two groups. Univariate and multivariate Cox regression analyses for RFS, PFS, and CSS with unadjusted cohort and the IPTW-adjusted model are shown in Tables S4–S6. Cox regression analysis using IPTW adjustment demonstrated that the impact of concomitant VMs was significant for CSS (Table 3; multivariate analysis; HR, 3.38; 95% CI, 1.92–5.93; $p < 0.01$) but not RFS (univariate analysis; HR, 0.86; 95% CI, 0.25–2.95; $p = 0.81$) and PFS (univariate analysis; HR, 1.88; 95% CI, 0.49–7.21; $p = 0.36$).

Table 2. Baseline characteristics of patients with T1 high-grade tumor who received intravesical Bacillus Calmette–Guérin (BCG) treatment—before and after adjustment.

Valiables		Unweighted Population (n)				IPTW Population		
		Pure UC (pUC)	UC with VMs (UC-VM)	p Value	SMD	Pure UC (pUC)	UC with VMs (UC-VM)	SMD
N		1395	30			1425	1442	
Age, mean ± SD		70.7 ± 9.5	67.4 ± 9.8	0.06 #	0.35	70.7 ± 9.4	69.8 ± 1.4	0.13
Sex				1.00 ##	0.014			0.16
	Male	1155 (83%)	25 (83%)			83%	88%	
	Female	240 (17%)	5 (17%)			17%	12%	
Multiplicity				0.13 ##	0.28			0.19
	Single	503 (36%)	15 (50%)			36%	27%	
	Multiple	892 (64%)	15 (50%)			64%	73%	
Tumor size				0.83 ##	0.021			0.047
	<3 cm	1082 (78%)	23 (77%)			77%	79%	
	≥3 cm	313 (22%)	7 (23%)			23%	21%	
Bladder CIS				0.85 ##	0.044			0.132
	No	854 (61%)	19 (63%)			61%	45%	
	Yes	541 (39%)	11 (37%)			39%	55%	
Prostate-involving CIS				1.00 ##	0.199			0.197
	No	1368 (98%)	30 (100%)			98%	100%	
	Yes	27 (1.9%)	0 (0%)			1.9%	0%	
LVI				<0.01 ##	0.64			0.030
	No	1275 (91%)	20 (67%)			91%	92%	
	Yes	120 (9.1%)	10 (33%)			9.1%	8.3%	
Second TUR				0.14 ##	0.32			0.026
	No	629 (45%)	9 (30%)			45%	46%	
	Yes	766 (55%)	21 (70%)			55%	54%	
Maintenance BCG				0.81 ##	0.085			0.20
	No	1167 (84%)	26 (87%)			84%	76%	
	Yes	228 (16%)	4 (13%)			16%	24%	

NMIBC, non-muscle invasive bladder cancer; IPTW, inverse probability of treatment weighting; UC, urothelial carcinoma; VMs, variant morphologies; SMD, standardized mean difference; SD, standard deviation; CIS, carcinoma in situ; LVI, lymphovascular invasion; TUR, transurethral resection; # Kruskal-Wallis test; ## Chi-square test.

Table 3. Inverse probability of treatment weighting (IPTW)-adjusted multivariate Cox proportional hazard regression analysis for oncological outcomes.

Oncological Outcomes	Variables		HR	95% CI	p Value
Bladder Recurrence-Free Survival					
	Age	≥70 yo/<70 yo	1.31	1.07–1.60	<0.01
	Multiplicity	multiple/solitary	1.46	1.18–1.80	<0.01
	Prostate-involving CIS	yes/no	2.96	1.82–4.79	<0.01
	LVI	yes/no	1.45	1.06–1.97	0.02
	Second TUR	yes/no	0.73	0.60–0.89	<0.01
	Maintenance BCG	yes/no	0.49	0.36–0.68	<0.01
Progression-free survival					
	Age	≥ 70 yo/<70 yo	1.25	0.93–1.67	0.14
	Prostate-involving CIS	yes/no	3.38	1.92–5.93	<0.01
	LVI	yes/no	1.56	1.02–2.37	0.04
	Second TUR	yes/no	0.75	0.56–1.00	0.05
	Maintenance BCG	yes/no	0.59	0.37–0.94	0.03

Table 3. Cont.

Oncological Outcomes	Variables		HR	95% CI	p Value
Cancer-specific survival					
	Age	≥ 70 yo/<70 yo	1.90	1.15–3.13	0.01
	Histology type	UC-VM/pUC	3.38	1.92–5.93	<0.01
	Second TUR	yes/no	0.58	0.34–0.97	0.04

HR, hazard ratio; CI, confidence interval; CIS, carcinoma in situ; LVI, lymphovascular invasion; TUR, transurethral resection; BCG, Bacillus Calmette-Guérin; UC-VM, urothelial carcinoma with variant morphologies; pUC, pure urothelial carcinoma.

4. Discussion

We investigated the impact of concomitant DD and VMs in TUR specimens in T1/HG NMIBC patients who received intravesical BCG treatment. In our cohort, the frequency of some forms of DD or VMs was 6.4% (95/1490), which seemed to be much lower than that in previous studies [6,7,9,10]. This would be strongly attributed to biased patient selection. As our cohort included only T1/HG NMIBC patients treated with intravesical BCG treatment, we counted neither patients with NMIBC undergoing immediate RC nor MIBC patients. The European Association of Urology states that micropapillary, plasmacytoid, and sarcomatoid variants are associated with poorer prognosis [3]. In addition, the National Comprehensive Cancer Network Clinical guidelines recommend immediate RC for T1 NMIBC with micropapillary, plasmacytoid, and sarcomatoid variants [4]. However, this real-world data displayed treatment outcomes of a few cases with rare VMs, such as sarcomatoid, clear cell, microcytic, and giant cell variants. Although there were no patients with the plasmacytoid variant here, our findings highlight the favorable response to intravesical BCG against T1 disease with rare VMs.

Metastatic lesion occurs usually after local progression to MIBC in patients with T1 bladder UC. In our cohort, unresectable metastatic lesions occurred suddenly in three (60%) of five UC-VMs patients who experienced disease progression. Concomitant VMs can be characterized by a highly metastatic potential. As to micropapillary variant, a large multi-institutional cohort study demonstrated that micropapillary variant was associated with higher pathologic stage and LVI in MIBC patients who underwent radical cystectomy [28]. This finding strongly implies UC with micropapillary variant presents higher capability of invasion and metastasis. Regarding the prognosis of T1 UC with micropapillary variant, previous study by Willis et al. [19] concluded improved survival was seen in those patients who underwent immediate RC, while some patients may respond to intravesical BCG. However, future precision medicine such as detailed molecular analysis would be required to identify subsets who can be managed by bladder-preservation therapy. To date, various types of analyses have been reported to evaluate the efficacy of intravesical BCG for NMIBC with DD or VMs. Yorozuya et al. [29] compared the outcomes between BCG-treated and non-BCG-treated patients with UC with DD, concluding that intravesical BCG may provide clinical benefit. Suh et al. conducted comparative analysis of immediate RC, intravesical BCG, and no treatment groups for UC with DD, demonstrating that intravesical BCG could be an appropriate treatment option [30]. Furthermore, Shapur and Gofrit et al. reported unfavorable outcomes of 22 NMIBC patients with concomitant DD or VMs treated with intravesical BCG [21]. The same group later compared the outcomes after intravesical BCG between 41 patients with NMIBC with DD or VMs and 140 control patients with pure UC, interpreting that concomitant DD or VMs was associated with significantly worse prognosis [20]. However, these retrospective studies did not separately evaluate the UC-DD and UC-VM groups, and thus, do not provide insight into their differences. Our study adopted a two-step prognostic analysis: conventional multivariate Cox regression analysis for all three groups (pUC, UC-DD, and UC-VM) and subsequent IPTW-adjusted analysis for the selected two groups (pUC and UC-VM). Our first-step analysis revealed favorable survival outcomes of UC-DD groups that were consistent with the results from previous studies [29,30]. Notably, Gofrit et al. [20] reported that none of nine patients with glandular differentiation experienced bladder recurrence, progression, or cancer-specific death. Similarly, no patient with glandular or squamous differentiation in our cohort died of bladder cancer progression (Figure 3C).

Another issue to be discussed is underdiagnosis and misdiagnosis of rare histological UC variants in pathological reports for TUR specimens. A previous study demonstrated that the involvement of experienced uropathologists in the process of pathological diagnosis is of great importance, increasing the detection rate of UC variants, especially for sarcomatoid variants [31]. Regarding the concordance between histological diagnoses for TUR and RC specimens, the detection rate of UC variants was 6.4% and 14.1% of TUR and RC specimens, respectively [32]. Despite significant lack of concordance in UC variants between TUR and RC specimens, concomitant DDs or VMs were still associated with unfavorable clinical outcomes. Unfortunately, the involvement of experienced uropathologists did not increase the concordance rate between histological diagnoses of UC variants for TUR and RC specimens [31], suggesting that underdiagnosis of UC variants at TUR specimens might be unavoidable even with the involvement of experienced uropathologists in the diagnostic process.

This study had several limitations. First, its retrospective nature had an inherent potential for selection bias; for example, the criteria, dose, and schedule of BCG treatment depended on the institutional protocol and physician's discretion. The cohort was derived from multiple institutions, which could introduce inconsistencies in surgical skills, clinical interpretations, and pathological diagnoses. Second, because all the collaborative institutes are academic hospitals, it was assumed that the pathologists were well-experienced with urogenital cancer diagnosis. However, the pathological skill and insight could vary among institutes. Third, we did not include patients who underwent immediate RC and did not evaluate how beneficial intravesical BCG was when compared with the RC-treated or the untreated group. The rate of selection of immediate RC and other alternative treatment and the outcome of those patients are not available. Finally, we did not evaluate the possible impact of the amount of concomitant DD or VMs on the diagnosis. The 2016 WHO blue book recommends that pathologists report the percentage of variants in the pathology report.

5. Conclusions

Except for some rare variants, DD or VMs are often seen in bladder UC. Our two-step prognostic analysis clearly demonstrated that UC-VM was associated with poor outcomes in comparison to UC-DD. We conducted IPTW-adjusted analysis to investigate the real clinical impact of concomitant VMs. This aggressive subset should be clearly separated from UC-DD in terms of response to intravesical BCG. Our study should help both urologists and pathologists better understand the clinical and biological behavior of UC-VM and the potential of bladder-preservation therapy using intravesical BCG. We believe that our findings will help establish optimized treatment strategies such as precision medicine for this aggressive UC subset.

Supplementary Materials: The following are available online at https://www.mdpi.com/article/10.3390/cancers13112615/s1, Table S1: Clinicopathological variables of patients with T1 high-grade tumor treated with intravesical Bacillus Calmette–Guérin (BCG) and comparison according to divergent differentiation (DD) and variant morphologies (VM)., Table S2: Prognostic variables in patients with T1 high-grade tumor treated using intravesical Bacillus Calmette–Guérin (BCG)., Table S3: Background and outcomes of 30 patients with T1 high-grade tumor with variant morphologies (VMs). Table S4: Cox proportional hazard regression models for bladder recurrence-free survival., Table S5: Cox proportional hazard regression models for progression-free survival., Table S6: Cox proportional hazard regression models for cancer-specific survival.

Author Contributions: Conceptualization, M.M., R.T., and T.K. (Takashi Kobayashi); methodology, N.N. (Nobutaka Nishimura); validation, T.O., S.H., and T.S.; formal analysis, N.N. (Nobutaka Nishimura); investigation, R.N., S.T., A.T., H.K. (Hiroshi Kikuchi), T.A., N.S., E.O. and Y.M.; data curation, K.I.; writing—original draft preparation, M.M.; writing—review and editing, K.F. and T.K. (Takahiro Kojima); visualization, T.F.; supervision, N.T., N.N. (Naotaka Nishiyama) and H.K. (Hiroshi Kitamura); project administration, H.N. All authors have read and agreed to the published version of the manuscript.

Funding: This research received no external funding.

Institutional Review Board Statement: The study was conducted according to the guidelines of the Declaration of Helsinki, and approved on 23 May 2019 by the Institutional Review Board of each participating institute (reference protocol ID: NMU-2217) of the Japan Urological Oncology Group framework. Official names of institutional review board of each collaborating institute are listed here: the Medical ethics committee at Osaka University; the ethics committee of Kagawa University Faculty of Medicine; the Ethics Committee on Clinical Research, Kagoshima University; the Medical Ethics Committee for Kitasato University; the Kyushu University Hospital Ethics Committee; the Ethics Committee Graduate School and Faculty of Medicine Kyoto University; the Ethics committee of The Jikei University School of Medicine for Biomedical Research; the Ethics Committee of Shimane University; the Ethics Committee of Chiba University; the Ethics in Human Subject Research at the University of Tsukuba; the Ethics Committee of University of Toyama; the Ethical committee of Harasanshin Hospital; the Committee of Medical Ethics of Hirosaki University Graduate School of Medicine; the Independent Ethics Committee of Hokkaido University Graduate School of Medicine; the Ethical Review Committee of Yamagata University Faculty of Medicine; the Research Ethics Committee of University of Miyazaki; the Medical Ethics Review board of the Kyoto Prefectural University of Medicine; the Ethical committee of National Cancer Center Research Institute; the Ethical committee National Cancer Center Research Institute East; the Shikoku Cancer Center Ethics Committee; the Akita University School of Medicine Ethics Committee; Tottori University Faculty of Medicine Ethics Committee; the Tohoku University Graduate School of Medicine Ethics Committee; the Ethics Committee of Hamamatsu University School of Medicine; the Ethics Committee of Nara Prefecture General Medical Center; the Ethics Committee of Kokuho Chuo Hospital; the Ethics Committee of Nara City Hospital; the Ethics Committee of Yamato Koriyama Hospital; the Ethics Committee of Hirao Hospital; the Ethics Committee of Saiseikai Chuwa Hospital.

Informed Consent Statement: An opt-out approach was used to obtain informed consent from the patients and personal information was protected during data collection.

Data Availability Statement: The data presented in this study are available in present article and supplementary.

Acknowledgments: The clinicopathological statistics are based on the results of contributions from several institutions in Japan. We thank the contributions of many urologists who are not listed as co-authors.

Conflicts of Interest: The authors declare no conflict of interest.

References

1. Bray, F.; Ferlay, J.; Soerjomataram, I.; Siegel, R.L.; Torre, L.A.; Jemal, A. Global cancer statistics 2018: GLOBOCAN estimates of incidence and mortality worldwide for 36 cancers in 185 countries. *CA Cancer J. Clin.* **2018**, *68*, 394–424. [CrossRef]
2. Miyake, M.; Iida, K.; Nishimura, N.; Miyamoto, T.; Fujimoto, K.; Tomida, R.; Matsumoto, K.; Numakura, K.; Inokuchi, J.; Morizane, S.; et al. Non-maintenance intravesical Bacillus Calmette–Guérin induction therapy with eight doses in patients with high- or highest-risk non-muscle invasive bladder cancer: A retrospective non-randomized comparative study. *BMC Cancer* **2021**, *21*, 266. [CrossRef] [PubMed]
3. Babjuk, M.; Burger, M.; Compérat, E.M.; Gontero, P.; Mostafid, A.H.; Palou, J.; van Rhijn, B.W.G.; Rouprêt, M.; Shariat, S.F.; Sylvester, R.; et al. European Association of Urology Guidelines on Non-muscle-invasive Bladder Cancer (TaT1 and Carcinoma In Situ)—2019 Update. *Eur. Urol.* **2019**, *76*, 639–657. [CrossRef] [PubMed]
4. Flaig, T.; Spiess, P.; Agarwal, N.; Bangs, R.; Boorjian, S.A.; Buyyounouski, M.K.; Chang, S.; Downs, T.M.; Efstathiou, J.A.; Friedlander, T.; et al. Bladder Cancer, Version 3.2020, NCCN Clinical Practice Guidelines in Oncology. *J. Natl. Compr. Cancer Netw.* **2020**, *18*, 329–354. [CrossRef] [PubMed]

5. Chang, S.S.; Bochner, B.H.; Chou, R.; Dreicer, R.; Kamat, A.M.; Lerner, S.P.; Lotan, Y.; Meeks, J.J.; Michalski, J.M.; Morgan, T.M.; et al. Treatment of Non-Metastatic Muscle-Invasive Bladder Cancer: AUA/ASCO/ASTRO/SUO Guideline. *J. Urol.* **2017**, *198*, 552–559. [CrossRef] [PubMed]
6. Matsumoto, H.; Shiraishi, K.; Azuma, H.; Inoue, K.; Uemura, H.; Eto, M.; Ohyama, C.; Ogawa, O.; Kikuchi, E.; Kitamura, H.; et al. Clinical Practice Guidelines for Bladder Cancer 2019 update by the Japanese Urological Association: Summary of the revision. *Int. J. Urol.* **2020**, *27*, 702–709. [CrossRef] [PubMed]
7. Miyamoto, T.; Miyake, M.; Toyoshima, Y.; Fujii, T.; Shimada, K.; Nishimura, N.; Iida, K.; Nakahama, T.; Hori, S.; Gotoh, D.; et al. Clinical outcomes after intravesical bacillus Calmette–Guérin for the highest-risk non-muscle invasive bladder cancer newly defined in the Japanese Urological Association Guidelines 2019. *Int. J. Urol.* **2021**. [CrossRef] [PubMed]
8. Humphrey, P.A.; Moch, H.; Cubilla, A.L.; Ulbright, T.M.; Reuter, V.E. The 2016 WHO Classification of Tumours of the Urinary System and Male Genital Organs-Part B: Prostate and Bladder Tumours. *Eur. Urol.* **2016**, *70*, 106. [CrossRef]
9. Santana, S.C.; de Souza, M.F.; Amaral, M.E.P.; Athanazio, D.A. Divergent differentiation and variant morphology in invasive urothelial carcinomas–association with muscle-invasive disease. *Surg. Exp. Pathol.* **2020**, *3*, 14. [CrossRef]
10. Shah, R.B.; Montgomery, J.S.; Montie, J.E.; Kunju, L.P. Variant (divergent) histologic differentiation in urothelial carcinoma is under-recognized in community practice: Impact of mandatory central pathology review at a large referral hospital. *Urol. Oncol.* **2013**, *31*, 1650–1655. [CrossRef]
11. Amin, M.B. Histological variants of urothelial carcinoma: Diagnostic, therapeutic and prognostic implications. *Mod. Pathol.* **2009**, *22*, S96–S118. [CrossRef]
12. Black, P.C.; Brown, G.A.; Dinney, C.P. The impact of variant histology on the outcome of bladder cancer treated with curative intent. *Urol. Oncol.* **2009**, *27*, 3–7. [CrossRef] [PubMed]
13. Meeks, J.J.; Taylor, J.M.; Matsushita, K.; Herr, H.W.; Donat, S.M.; Bochner, B.H.; Dalbagni, G. Pathological response to neoadjuvant chemotherapy for muscle-invasive micropapillary bladder cancer. *BJU Int.* **2013**, *111*, E325–E330. [CrossRef] [PubMed]
14. Siefker-Radtke, A.O.; Dinney, C.P.; Shen, Y.; Williams, D.L.; Kamat, A.M.; Grossman, H.B.; Millikan, R.E. A phase 2 clinical trial of sequential neoadjuvant chemotherapy with ifosfamide, doxorubicin, and gemcitabine followed by cisplatin, gemcitabine, and ifosfamide in locally advanced urothelial cancer: Final results. *Cancer* **2013**, *119*, 540–547. [CrossRef]
15. Dayyani, F.; Czerniak, B.A.; Sircar, K.; Munsell, M.F.; Millikan, R.E.; Dinney, C.P.; Siefker-Radtke, A.O. Plasmacytoid urothelial carcinoma, a chemosensitive cancer with poor prognosis, and peritoneal carcinomatosis. *J. Urol.* **2013**, *189*, 1656–1661. [CrossRef] [PubMed]
16. Burger, M.; Kamat, A.M.; McConkey, D. Does Variant Histology Change Management of Non-muscle-invasive Bladder Cancer? *Eur. Urol. Oncol.* **2019**. [CrossRef]
17. Pang, K.H.; Noon, A.P. Selection of patients and benefit of immediate radical cystectomy for non-muscle invasive bladder cancer. *Transl. Androl. Urol.* **2019**, *8*, 101–107. [CrossRef] [PubMed]
18. Baumeister, P.; Zamboni, S.; Mattei, A.; Antonelli, A.; Simeone, C.; Mordasini, L.; DiBona, C.; Moschini, M. Histological variants in non-muscle invasive bladder cancer. *Transl. Androl. Urol.* **2019**, *8*, 34–38. [CrossRef] [PubMed]
19. Willis, D.L.; Fernandez, M.I.; Dickstein, R.J.; Parikh, S.; Shah, J.B.; Pisters, L.L.; Guo, C.C.; Henderson, S.; Czerniak, B.A.; Grossman, H.B.; et al. Clinical outcomes of cT1 micropapillary bladder cancer. *J. Urol.* **2015**, *193*, 1129–1134. Available online: https://www.ncbi.nlm.nih.gov/pubmed/25254936 (accessed on 22 April 2021). [CrossRef]
20. Gofrit, O.N.; Yutkin, V.; Shapiro, A.; Pizov, G.; Zorn, K.C.; Hidas, G.; Gielchinsky, I.; Duvdevani, M.; Landau, E.H.; Pode, D.; et al. The Response of Variant Histology Bladder Cancer to Intravesical Immunotherapy Compared to Conventional Cancer. *Front Oncol.* **2016**, *6*, 43. [CrossRef]
21. Shapur, N.K.; Katz, R.; Pode, D.; Shapiro, A.; Yutkin, V.; Pizov, G.; Appelbaum, L.; Zorn, K.C.; Duvdevan, I.M.; Landau, E.H.; et al. Is radical cystectomy mandatory in every patient with variant histology of bladder cancer. *Rare Tumors* **2011**, *3*, e22. [CrossRef] [PubMed]
22. Vellinga, A.; Cormican, M.; Hanahoe, B.; Bennett, K.; Murphy, A.W. Opt-out as an acceptable method of obtaining consent in medical research: A short report. *BMC Med. Res. Methodol.* **2011**, *11*, 40. [CrossRef] [PubMed]
23. Lamm, D.L.; Blumenstein, B.A.; Crissman, J.D.; Montie, J.E.; Gottesman, J.E.; Lowe, B.A.; Sarosdy, M.F.; Bohl, R.D.; Grossman, H.B.; Beck, T.M.; et al. Maintenance bacillus Calmette-Guerin immunotherapy for recurrent TA, T1 and carcinoma in situ transitional cell carcinoma of the bladder: A randomized Southwest Oncology Group Study. *J. Urol.* **2000**, *163*, 1124–1129. [CrossRef]
24. Kamat, A.M.; Sylvester, R.J.; Böhle, A.; Palou, J.; Lamm, D.L.; Brausi, M.; Soloway, M.; Persad, R.; Buckley, R.; Colombel, M.; et al. Definitions, End Points, and Clinical Trial Designs for Non-Muscle-Invasive Bladder Cancer: Recommendations from the International Bladder Cancer Group. *J. Clin. Oncol.* **2016**, *34*, 1935–1944. [CrossRef]
25. Austin, P.C.; Stuart, E.A. Moving towards best practice when using inverse probability of treatment weighting (IPTW) using the propensity score to estimate causal treatment effects in observational studies. *Stat. Med.* **2015**, *34*, 3661–3679. [CrossRef] [PubMed]
26. Sylvester, R.J.; van der Meijden, A.P.M.; Oosterlinck, W.; Witjes, J.A.; Bouffioux, C.; Denis, L.; Newling, D.W.; Kurth, K. Predicting recurrence and progression in individual patients with stage Ta T1 bladder cancer using EORTC risk tables: A combined analysis of 2596 patients from seven EORTC trials. *Eur. Urol.* **2006**, *49*, 466–475. [CrossRef] [PubMed]

27. Fernandez-Gomez, J.; Madero, R.; Solsona, E.; Unda, M.; Martinez-Piñeiro, L.; Gonzalez, M.; Portillo, J.; Ojea, A.; Pertusa, C.; Rodriguez-Molina, J.; et al. Predicting nonmuscle invasive bladder cancer recurrence and progression in patients treated with Bacillus Calmette–Guerin: The Cueto scoring model. *J. Urol.* **2009**, *182*, 2195–2203. [CrossRef]
28. Mitra, A.P.; Fairey, A.S.; Skinner, E.C.; Boorjian, S.A.; Frank, I.; Schoenberg, M.P.; Bivalacqua, T.J.; Hyndman, M.E.; Reese, A.C.; Steinberg, G.D.; et al. Implications of micropapillary urothelial carcinoma variant on prognosis following radical cystectomy: A multi-institutional investigation. *Urol. Oncol.* **2019**, *37*, 48–56. [CrossRef]
29. Yorozuya, W.; Nishiyama, N.; Shindo, T.; Kyoda, Y.; Itoh, N.; Sugita, S.; Hasegawa, T.; Masumori, N. Bacillus Calmette-Guérin may have clinical benefit for glandular or squamous differentiation in non-muscle invasive bladder cancer patients: Retrospective multicenter study. *Jpn. J. Clin. Oncol.* **2018**, *48*, 661–666. [CrossRef]
30. Suh, J.; Moon, K.C.; Jung, J.H.; Lee, J.; Song, W.H.; Kang, Y.J.; Jeong, C.W.; Kwak, C.; Kim, H.H.; Ku, J.H. BCG instillation versus radical cystectomy for high-risk NMIBC with squamous/glandular histologic variants. *Sci. Rep.* **2019**, *9*, 15268. [CrossRef] [PubMed]
31. Mantica, G.; Simonato, A.; Du Plessis, D.E.; Maffezzini, M.; De Rose, A.F.; van der Merwe, A.; Terrone, C. The pathologist's role in the detection of rare variants of bladder cancer and analysis of the impact on incidence and type detection. *Minerva Urol. Nefrol.* **2018**, *70*, 594–597. [CrossRef] [PubMed]
32. Cai, T.; Tiscione, D.; Verze, P.; Pomara, G.; Racioppi, M.; Nesi, G.; Barbareschi, M.; Brausi, M.; Gacci, M.; Luciani, L.G.; et al. Concordance and clinical significance of uncommon variants of bladder urothelial carcinoma in transurethral resection and radical cystectomy specimens. *Urology* **2014**, *84*, 1141–1146. [CrossRef] [PubMed]

Article

Prognostic Impact of AHNAK2 Expression in Patients Treated with Radical Cystectomy

Dai Koguchi, Kazumasa Matsumoto *, Yuriko Shimizu, Momoko Kobayashi, Shuhei Hirano, Masaomi Ikeda, Yuichi Sato and Masatsugu Iwamura

Department of Urology, Kitasato University School of Medicine, 1-15-1 Kitasato Minami-ku Sagamihara, Kanagawa 252-0374, Japan; dai.k@med.kitasato-u.ac.jp (D.K.); yulico@med.kitasato-u.ac.jp (Y.S.); momoko_dus@yahoo.co.jp (M.K.); s.hirano@med.kitasato-u.ac.jp (S.H.); ikeda.masaomi@grape.plala.or.jp (M.I.); sato.yuichi@kobal.co.jp (Y.S.); miwamura@med.kitasato-u.ac.jp (M.I.)
* Correspondence: kazumasa@cd5.so-net.ne.jp; Tel.: +81-42-778-9091; Fax: +81-42-778-9374

Simple Summary: Unfavorable results following radical cystectomy for bladder cancer (BCa) highlights a critical need for a novel prognostic molecular biomarker with potential therapeutic benefits. In the present study, the expression levels of AHNAK2 in specimens obtained by radical cystectomy were classified as "low expression" or "high expression" by immunohistochemical staining. Then, we retrospectively evaluated associations between the two AHNAK2 expression patterns and the prognoses in terms of recurrence-free survival (RFS) and cancer-specific survival (CSS). Our multivariate analysis, adjusting for the effects of clinicopathological features, showed that the high expression level of AHNAK2 was an independent risk factor for RFS and CSS. The present study showed that AHNAK2 acts as a novel prognostic biomarker in patients with radical cystectomy for BCa.

Abstract: Data regarding expression levels of AHNAK2 in bladder cancer (BCa) have been very scarce. We retrospectively reviewed clinical data including clinicopathological features in 120 patients who underwent radical cystectomy (RC) for BCa. The expression levels of AHNAK2 in the specimens obtained by RC were classified as low expression (LE) or high expression (HE) by immunohistochemical staining. Statistical analyses were performed to compare associations between the two AHNAK2 expression patterns and the prognoses in terms of recurrence-free survival (RFS) and cancer-specific survival (CSS). A Kaplan–Meier analysis showed that patients with HE had a significantly worse RFS and CSS than those with LE (hazard ratio [HR]: 1.78, 95% confidence interval [CI]: 1.02–2.98, $p = 0.027$ and HR: 1.91, 95% CI: 1.08–3.38, $p = 0.023$, respectively). In a multivariate analysis, independent risk factors for worse RFS and CSS were shown as HE (HR: 1.96, 95% CI: 1.08–3.53, $p = 0.026$ and HR: 2.22, 95% CI: 1.14–4.31, $p = 0.019$, respectively) and lymph node metastasis (HR: 2.04, 95% CI: 1.09–3.84, $p = 0.026$ and HR: 1.19, 95% CI: 1.25–4.97, $p = 0.009$, respectively). The present study showed that AHNAK2 acts as a novel prognostic biomarker in patients with RC for BCa.

Keywords: bladder cancer; radical cystectomy; AHNAK2; prognosis

Citation: Koguchi, D.; Matsumoto, K.; Shimizu, Y.; Kobayashi, M.; Hirano, S.; Ikeda, M.; Sato, Y.; Iwamura, M. Prognostic Impact of AHNAK2 Expression in Patients Treated with Radical Cystectomy. *Cancers* **2021**, *13*, 1748. https://doi.org/10.3390/cancers13081748

Academic Editor: José I. López

Received: 7 March 2021
Accepted: 6 April 2021
Published: 9 April 2021

Publisher's Note: MDPI stays neutral with regard to jurisdictional claims in published maps and institutional affiliations.

Copyright: © 2021 by the authors. Licensee MDPI, Basel, Switzerland. This article is an open access article distributed under the terms and conditions of the Creative Commons Attribution (CC BY) license (https://creativecommons.org/licenses/by/4.0/).

1. Introduction

Bladder cancer (BCa) is the most common malignancy of the urinary tract and the fourth most common cancer in men [1]. For the last three decades, radical cystectomy (RC) has been the gold-standard treatment in patients with muscle-invasive BCa (MIBC) and the non-muscle-invasive BCa (NMIBC) that is refractory to intravesical therapy. Despite advances in surgical techniques and an improved understanding of the role of pelvic lymphadenectomy, recurrences after RC usually occur within the first 2–3 years, giving a 5-year survival for only about 50% of patients [2,3]. Moreover, once MIBC metastasizes, a five-year survival rate was dismal at less than 10% even with salvage treatments [3].

To improve such unfavorable results following RC [4], great efforts have been made for the investigation of prognostic factors related to the surgery [5]. Currently, management of BCa still relies on histopathological parameters such as tumor stage, lymph node status, and lymphovascular invasion (LVI) [5]. Although these prognostic variables have been helpful in estimating the recurrence risk and survival outcomes of BCa, they do not largely play predictive roles in the individual strategy. For example, some studies have shown that the effect of neoadjuvant chemotherapy did not correlate with T stage [6,7]. Meanwhile, in several other cancers, the use of molecular biomarkers as a guide to personalized treatment has become standard, improving patients' survival, especially in breast cancer [8]. Therefore, there is an urgent need to identify potential molecular markers of BCa, which can serve as not only prognostic values but act as potential therapeutic targets.

AHNAK2, also known as C14orf78, is a member of the AHNAK family and was originally found in mouse heart tissue extract, encoding a giant protein of more than 600 kilodaltons (kDa) [9]. Over the last seven years, overexpression of AHNAK2 has been reported to be associated with poor prognosis in clear cell renal cell carcinoma, pancreatic ductal adenocarcinoma, uveal melanoma, papillary thyroid carcinoma, and lung adenocarcinoma [10–14]. Furthermore, previous studies indicated AHNKA2 as a possible new therapeutic target in some cancers because it would play an important role in regulating multiple tumor progression pathways, including *mitogen-activated protein kinase (MAPK)/extracellular signal-regulated kinase (ERK)*, phosphatidylinositol 3-kinase (PI3K)/protein *kinase* B (AKT), hypoxia inducible factor-1α (HIF-1α), and transforming growth factor-β (TGF-β)/Smad3 [10,12,15,16]. However, there are no data from investigations of the expression levels of AHNAK2 in patients following RC that consider clinicopathological features.

We previously examined AHNAK in BCa tissues and successfully identified AHNAK2 in patients with RC for BCa [17]. The present study evaluates AHNAK2 expression levels in patients and retrospectively investigates an association between AHNAK2 expression levels and the prognosis adjusted by pathological variables obtained by RC.

2. Results

2.1. Tissues Immunostained for AHNAK2

Figure 1 shows representative tissue sections immunostained for AHNAK2 in normal urothelial and tumor tissues (200×). In non-neoplastic tissues, AHNAK2 was observed in the cytoplasm of smooth muscle cells in the muscular layer, peripheral nerve cells, endothelial cells, macrophages, and tumor stromal fibroblasts. No, or only a weak, expression was observed in the cytoplasm of normal urothelial cells (Figure 1A). In tumor tissues, AHNAK2 was variously observed in the cytoplasm and/or plasma and nuclear membrane of tumor cells (Figure 1B–D).

(A)

(B)

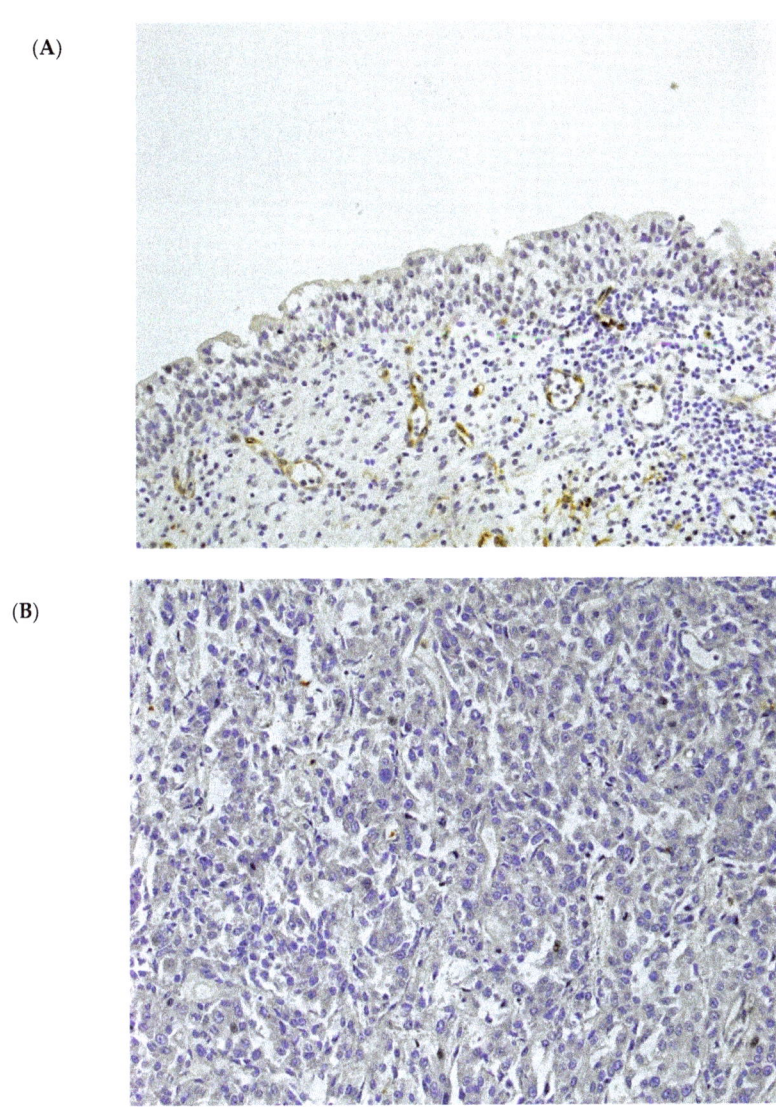

Figure 1. *Cont.*

(C)

(D)

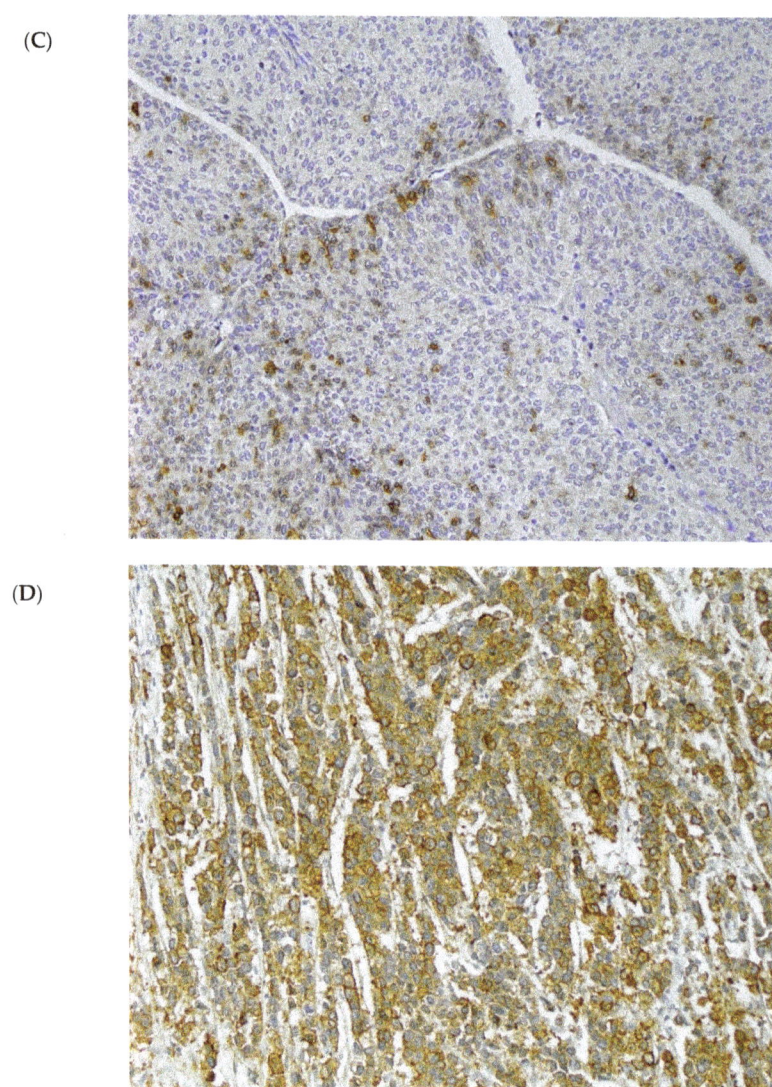

Figure 1. Immunohistochemical analysis of AHNAK2 expression in normal urothelial and bladder carcinoma tissues. Microscopic images are representative normal urothelial and bladder cancer (BCa) tissues for AHNAK2 staining (200× magnification). (**A**) AHNAK2-negative normal urothelial tissues. (**B**) AHNAK2-negative BCa tissues: score 0. (**C**) AHNAK2-negative BCa tissues: Score 2. (**D**) AHNAK2-positive BCa tissues: Score 12.

2.2. Patient Characteristics

This study included 97 (80.8%) men and 23 (19.2%) women. The median time to a follow-up appointment was 38.8 months (range: 0.7–283.3 months; mean: 69.3 months). The patients' characteristics are listed in Table 1. Patients with low expression (LE) and high expression (HE) of AHNAK2 accounted for 46.7% (n = 56) and 53.3% (n = 64), respectively. Of all patients, 50.8% (n = 61) had tumor recurrence and 42.5% (n = 51) experienced cancer death, with a significantly higher proportion in the HE group than in the LE group in terms of both parameters (recurrence: 57.8% vs. 42.9%, respectively, $p = 0.027$; cancer death: 50.0% vs. 33.9%, respectively, $p = 0.023$).

Table 1. Comparison of clinical and pathological characteristics of patients with either low or high expressions of AHNAK2.

	LE ($n = 56$)	HE ($n = 64$)	p-Value
Age			
≤65	29 (51.8)	33 (51.6)	0.98
>65	27 (48.2)	31 (48.4)	
Sex			
Male	51 (91.1)	46 (71.2)	0.01
Female	5 (8.9)	18 (28.8)	
T stage			
pTa	2 (3.6)	0	
pTis	2 (3.6)	1 (1.6)	
pT1	11 (19.6)	7 (10.9)	0.047
pT2	18 (32.1)	11 (17.2)	
pT3	16 (28.6)	30 (46.9)	
pT4	7 (12.5)	15 (23.4)	
N stage			
pN0	45 (80.3)	43 (67.2)	
≥pN1	10 (17.8)	16 (25.0)	0.027
Unknown	1 (1.8)	5 (7.8)	
Grade			
G1/2	23 (41.1)	23 (35.9)	
G3	32 (57.1)	41 (64.1)	0.57
Unknown	1 (1.8)	0	
LVI			
Negative	24 (42.9)	15 (23.4)	
Positive	28 (50.0)	43 (67.2)	0.003
Unknown	4 (7.1)	6 (9.4)	
CIS			
Negative	49 (87.5)	55 (85.9)	
Positive	7 (12.5)	8 (12.5)	0.97
Unknown	0	1 (1.6)	
Adjuvant chemotherapy			
Yes	8 (14.3)	12 (18.8)	0.51
No	48 (85.7)	52 (81.2)	
Salvage chemotherapy			
Response	2 (22.2)	3 (27.3)	0.80
No Response	7 (77.8)	8 (72.7)	
Recurrence			
Yes	24 (42.9)	37 (57.8)	0.027
No	32 (57.1)	27 (42.2)	
Cancer death			
Yes	19 (33.9)	32 (50.0)	0.023
No	37 (66.1)	32 (50.0)	
Follow-up, months (IQR)	51.0 (21–133)	20.0 (11–98.5)	0.075

Unless otherwise stated, values are medians with ranges in parentheses or numbers of patients with percentages in parentheses. LE, low expression; HE, high expression; LVI, lymphovascular invasion; CIS, carcinoma in situ; IQR, interquartile range.

Numbers of each clinical T stage before RC in the LE and the HE group were as follows; Ta: 1 (1.8%) and 1 (1.6%), Tis: 3 (5.4%) and 1 (1.6%), T1: 15 (26.8%) and 10 (15.6%), T2: 14 (25.0%) and 9 (14.1%), T3: 15 (26.8%) and 28 (43.7%), and T4: 8 (14.2%) and 15 (23.4%), respectively. In terms of clinical N stage, patients with node involvement were 5 in the LE and 11 in the HE group.

With the exclusion of a small number of unknown cases due to unavailable data, patients in the HE group had a significantly higher proportion of MIBC, lymph node metastasis, and LVI ($p = 0.047$, $p = 0.027$, and $p = 0.003$, respectively) than patients in the LE group.

Not all patients received adjuvant chemotherapy (AC), but all those who did a platinum-based chemotherapy. The recurrence rate in the HE group was not significantly different between patients with AC and those who without AC (n = 7: 58.3% and n = 30: 57.7%, respectively, $p = 0.97$); the results were similar in the LE group (n = 4: 50% and n = 20: 41.7%, respectively, $p = 0.68$). In terms of patients with AC (n = 20), there was no significant difference in the recurrence rate between the two groups ($p = 0.71$). Of all the patients who received salvage chemotherapy (SC) (n = 20), 80% (n = 16; n = 7: 35% in the LE group and n = 9: 45% in the HE group) received platinum-based chemotherapy for the disease progression after RC. The response rates of SC between the HE and LE groups were not significantly different (27.3% and 22.2%, respectively, $p = 0.80$), and 80% (4/5) of all patients with the response experienced cancer death thereafter.

2.3. Survival Analysis Using Kaplan-Meier Methods for RFS and CSS in Terms of Two Types of AHNAK2 Expression

A Kaplan–Meier analysis showed that patients in the HE group had a significantly worse recurrence-free survival (RFS) than those in the LE group (hazard ratio [HR]: 1.78, 95% confidence interval [CI]: 1.02–2.98, $p = 0.027$; Figure 2). The cumulative RFS rates for patients in the HE and LE groups were 62.1% and 85.2% at one year, 46.7% and 73.6% at two years, and 41.2% and 56.3% at five years, respectively. The mean times to recurrence after RC for patients in the HE and LE groups were 14.9 months and 25.2 months ($p = 0.003$), respectively. In terms of cancer death, a Kaplan–Meier analysis showed that patients in the HE group had a significantly worse cancer-specific survival (CSS) than those in the LE group (HR: 1.91, 95% CI: 1.08–3.38, $p = 0.023$; Figure 3). The cumulative CSS rates for patients in the HE and LE groups were 74.9% and 90.7% at one year, 52.3% and 80.0% at two years, and 46.7% and 67.7% at five years, respectively. The mean times to cancer death from RC in patients in the HE and LE groups were 15.4 months and 29.2 months ($p = 0.006$), respectively.

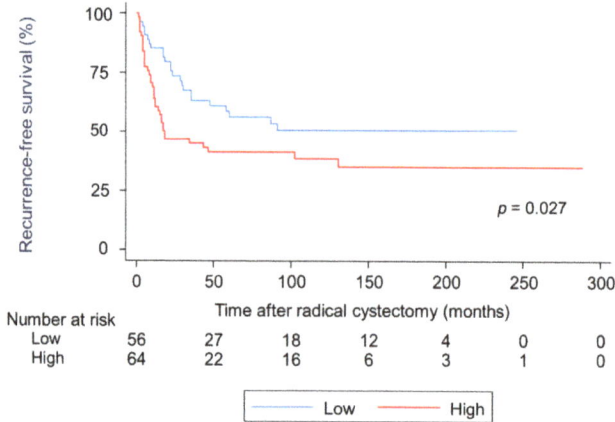

Figure 2. Survival analysis using Kaplan–Meier methods for recurrence-free survival in terms of two types of AHNAK2 expression. Patients in the high expression (HE) group showed a significantly worse recurrence-free survival than those in the low expression (LE) group.

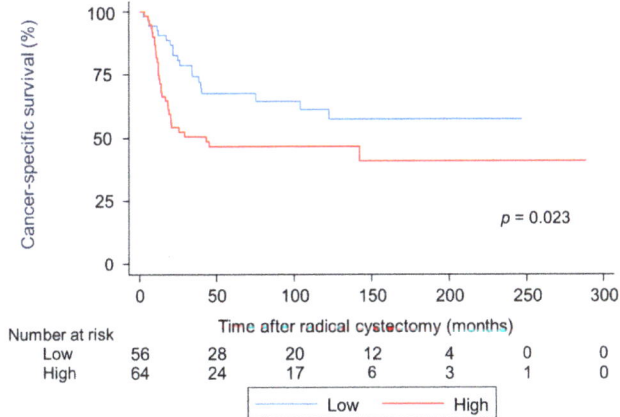

Figure 3. Survival analysis using Kaplan–Meier methods for cancer-specific survival in terms of two types of AHNAK2 expression. Patients in the high expression (HE) group showed a significantly worse cancer-specific survival than those in the low expression (LE) group.

2.4. Univariate and Multivariate Analyses of Prognostic Factors for RFS and CSS

A univariate analysis showed that the recurrence was associated with HE, MIBC, lymph node metastasis, and LVI; a multivariate analysis adjusted for the effects of clinicopathological features showed that HE and lymph node metastasis were independent risk factors for the recurrence (Table 2). In terms of cancer death, a univariate analysis showed that cancer death was associated with HE, lymph node metastasis, and LVI; a multivariate analysis showed that HE and lymph node metastasis were independent risk factors for cancer death (Table 3).

Table 2. Univariate and multivariate analyses for worse recurrence-free survival. AHNAK2, T stage, N stage, grade, lymphovascular invasion (LVI), and carcinoma in situ (CIS) were evaluated, and statistically significant values are highlighted in bold.

Variable	Category	Univariate Analysis			Multivariate Analysis		
		HR	95% CI	p-Value	HR	95% CI	p-Value
AHNAK2	HE	1.78	1.02–2.98	0.027	1.96	1.08–3.53	0.026
	LE	1.0			1.0		
T stage	MIBC	2.24	1.14–4.41	0.019	1.66	0.64–4.28	0.30
	NMIBC	1.0			1.0		
N stage	≥pN1	2.86	1.67–4.89	<0.001	2.04	1.09–3.84	0.026
	pN0	1.0			1.0		
Grade	G3	1.47	0.83–2.60	0.18	1.12	0.55–2.27	0.75
	G1/2	1.0			1.0		
LVI	Positive	2.52	1.35–4.71	0.004	1.12	0.55–2.27	0.75
	Negative	1.0			1.0		
CIS	Positive	0.64	0.29–1.41	0.27	1.16	0.48–2.84	0.73
	Negative	1.0			1.0		

HR, hazard ratio; CI, confidence interval; HE, high expression; LE, low expression; MIBC, muscle-invasive bladder cancer; NMIBC, nonmuscle-invasive bladder cancer; LVI, lymphovascular invasion; CIS, carcinoma in situ.

Table 3. Univariate and multivariate analyses for worse cancer-specific survival. AHNAK2, T stage, N stage, grade, LVI, and CIS were evaluated, and statistically significant values are highlighted in bold.

Variable	Category	Univariate Analysis			Multivariate Analysis		
		HR	95% CI	p-Value	HR	95% CI	p-Value
AHNAK2	HE	1.91	1.08–3.38	**0.023**	2.22	1.14–4.31	**0.019**
	LE	1.0			1.0		
T stage	MIBC	2.01	0.98–4.13	0.057	1.44	0.50–4.11	0.68
	NMIBC	1.0			1.0		
N stage	≥pN1	3.03	1.69–5.45	**<0.001**	1.19	1.25–4.97	**0.009**
	pN0	1.0			1.0		
Grade	G3	1.65	0.87–3.15	0.13	1.17	0.54–2.56	0.69
	G1/2	1.0			1.0		
LVI	Positive	2.43	1.23–4.80	**0.01**	1.19	0.53–2.64	0.68
	Negative	1.0			1.0		
CIS	Positive	0.42	0.15–1.18	0.101	0.57	0.17–1.91	0.37
	Negative	1.0			1.0		

HR, hazard ratio; CI, confidence interval; HE, high expression; LE, low expression; MIBC, muscle-invasive bladder cancer; NMIBC, nonmuscle-invasive bladder cancer; LVI, lymphovascular invasion; CIS, carcinoma in situ.

2.5. Subgroup Analysis of Associations between AHNAK2 and Some Proteins

Table 4 shows associations in patients between the two types of AHNAK2 expressions and other proteins including S100A2, S100A4, S100A8, S100A9, and nestin. Patients in the HE group had a significantly higher proportion of S100A4, S100A8, S100A9, and nestin than those in the LE group.

Table 4. Associations in patients between the two types of AHNAK2 expressions and S100A2, S100A4, S100A8, S100A9, and nestin. Statistically significant values are highlighted in bold.

Protein	LE (n = 56)	HE (n = 64)	p-Value
S100A2			
Normal	16 (28.6)	12 (18.8)	0.15
Abnormal	19 (33.9)	31 (48.4)	
S100A4			
Normal	22 (39.3)	15 (23.4)	**0.022**
Abnormal	13 (23.2)	28 (43.8)	
S100A8			
Normal	24 (42.9)	20 (31.1)	**0.015**
Abnormal	7 (12.5)	22 (34.4)	
S100A9			
Normal	28 (50)	20 (31.1)	**<0.001**
Abnormal	3 (5.4)	22 (34.4)	
Nestin			
Negative	36 (64.3)	35 (54.6)	**0.005**
Positive	1 (1.8)	12 (18.8)	

Values are numbers of patients with percentages in parentheses. The numbers (percentages) of patients in the LE and HE groups for whom protein information was not available are as follows: S100A2/4: 21 (37.5) and 21 (32.8), respectively; S100A8/9: 25 (44.6) and 22 (34.4), respectively; Nestin, 19 (33.9) and 17 (26.6), respectively. LE, low expression; HE, high expression.

3. Discussion

Currently, an evaluation of AHNAK2 in BCa has been very limited. As a diagnostic marker, AHNAK2 could immunohistochemically differentiate between inflammatory changes and carcinoma in situ (CIS) [18]. In terms of the prognostic value, only a few studies have been performed and reported that a higher expression of AHNAK2 was associated with shorter overall survival in BCa [13,19]. However, these previous studies were based on gene enrichment analyses and focused on an association between gene types and survival duration without adjusting clinical information. Hence, the present study retrospectively investigated the clinicopathological features and prognosis in 120 patients treated with RC in terms of the protein expression levels of AHNAK2. Consequently, we found that in patients with HE, the level of expression was related to biological aggressiveness such as for MIBC, lymph node metastasis, and LVI, with worse RFS and CSS and with about a year less time to recurrence and cancer death in comparison with those with LE. Furthermore, the multivariate analysis using the Cox proportional hazards regression model showed that HE and lymph node metastasis were independent predictors of worse RFS and CSS. Taken together, these results highlight the significant prognostic value of AHNAK2 in patients with BCa.

Pan-cancer analyses revealed the functional roles of AHNAK2 in the epithelial-mesenchymal transition (EMT), which plays a key role as a tumor promoter by allowing epithelial cells to gain a range of mesenchymal characteristics [13,14]. Analyses based on clinical specimens showed that AHNAK2 regulated the EMT via a TGF β/Smad3 pathway in lung adenocarcinoma and a hypoxia inducible factor-1α/zinc finger E-box-binding homeobox 1 (HIF-1/ZEB1) pathway in clear cell renal cell carcinoma [10,16]. In terms of BCa, basic experiments with lung metastasis of BCa showed that such EMT activation via TGF-β/Smad3 and HIF-1α/ZEB1 was indeed shown to strongly contribute to the invasion and metastasis of BCa [20,21]. Because most of the S100 proteins have been considered as EMT facilitators in certain carcinoma cell lines [21], the present study showed that HE of AHNAK2 had a significantly higher proportion of S100A4, S100A8, and S100A9 expressions. Therefore, the EMT pathway potentially driven by TGF-β/Smad3 and HIF-1α/ZEB1 possibly explains how the high level of AHNAK2 may be an independent prognostic factor in the poor survival rates of patients treated with RC.

The poor prognostic value of AHNAK2 in BCa may also result from fibroblast growth factor-1 (FGF1) signaling. The Cancer Genome Atlas (TGCA) project recently indicated that the MAPK/ERK and PI3K/AKT pathways were potential BCa driver genes and that these downstream cascades were mainly activated by FGF receptors (FGFRs), especially FGFR1 and FGFR3 in BCa [22]. In particular, AHNAK2 is required for non-classical secretion of FGF1, which can universally activate all FGFRs [23,24]. Because the HE of AHNAK2 correlated with a significantly higher proportion of S100A8/9 and nestin in the current study, several in vitro studies indicated that S100A8/9 activated MAPK pathway in breast, colon, and prostate cancer, and nestin activated PI3K phosphorylation in glioblastoma and embryogenesis [25,26]. In fact, these potential relations between AHNAK2 and the MAPK/ERK and PI3K/AKT pathways were reported in lung adenocarcinoma and uveal melanoma [12,15]. Therefore, given the accelerated approval of FGFR inhibitors for patients with advanced BCa by the U.S. Food and Drug Administration, the antitumor effects of a combination of therapies targeting AHNAK2 and FGFRs might be worth analyzing in the future [22].

Additionally, we evaluated differences in clinical outcomes following AC and SC between the two AHNAK2 groups. In terms of AC, although the recurrence rate in the HE group was higher than that in the LE group, the difference did not reach statistical significance. Moreover, the response rate of SC in both groups was only about 25%, and almost all the responders experienced cancer death thereafter. The expression level of AHNAK2 in the RC specimen thus did not show predictive value in such cisplatin-based chemotherapy. However, emerging evidence suggests that immune evasion induced by EMT may largely contribute to cisplatin resistance, and the small sample size of patients with the postoperative chemotherapy in the present study may have insufficient power to draw any conclusion [27]. Notably, EMT may also be related to the resistance of immune checkpoint inhibitor (ICI) in some malignant cells, and a dataset of BCa from TCGA recently showed an association between a higher EMT-related gene expression and a lower response rate of nivolumab (the programmed death-1 inhibitor) [28,29]. Considering the possible oncogenic role of AHNAK2 via EMT described in the present study, further studies should be warranted to verify the predictive value of AHNAK2 for cisplatin and ICI in a large population in the future.

This study had some limitations. First, the study was based on the retrospective design with the small number of patients, which may have limited proper assessment of the correlation between expression level of AHNAK2 in the RC specimen and prognosis in patients with RC. Second, RC was performed by multiple surgeons, and the management of the postoperative chemotherapy such as the treatment intensity was decided by each doctor in charge, and these differences may have influenced our results. Third, in terms of the subgroup analysis, the times during which we investigated the expression levels of the five proteins in patients treated with RC varied. Subsequently, the total number of proteins investigated in each patient chronologically increased. Fourth, although we did not conduct an experiment to investigate the actual mechanism of AHNAK2, the results of our subgroup analysis may be helpful to infer the possible oncogenic role of AHNAK2 in BCa. Fifth, we did not include some patient characteristics—including smoking status—which potentially affect prognosis in BCa. However, we believe that a focus on pathological findings, when added to the AHNAK2 expression patterns, may simply allow us to explain the differences in the prognosis.

4. Materials and Methods

4.1. Patient Population

We retrospectively reviewed the clinical data of 161 consecutive patients with BCa who underwent RC with pelvic and iliac lymphadenectomy from 1990 to 2015 at Kitasato University Hospital, Japan. RC was performed for patients with non-MIBC that had been refractory to intravesical therapy and MIBC without distant metastasis. We excluded 10 patients who had histological variants of BCa, including squamous cell carcinoma, adenocarcinoma, and small cell carcinoma; 15 who had been previously treated with neoadjuvant chemotherapy; and 16 who were lost to follow-up. None of the remaining patients were treated preoperatively with either systemic chemotherapy or radiation therapy. The Ethics Committee of Kitasato University School of Medicine and Hospital approved the study (B17-010, B18-149). All participants were approached on the basis of approved ethical guidelines. The patients could refuse entry and discontinue participation at any time.

4.2. Patient Characteristics

The following data on patient characteristics were collected from the patients' medical charts: Age at the RC; sex; pathological status including pT stage, pN stage, grade, LVI; CIS; history of AC; history of SC; recurrence; and mortality. BCa with $\geq$pT2 and $\leq$pT1 were also classified as MIBC and NMIBC, respectively. Tumor grade was assessed according to the 1973 World Health Organization grading system. The tumor stage was assessed according to the 2002 TNM classification of malignant tumors. LVI indicated the presence of cancer cells within the endothelial space. Cancer cells that merely invaded a vascular lumen were considered negative [30]. The chemotherapeutic response was evaluated by the Response Evaluation Criteria in Solid Tumors (RECIST) version 1.1 [31]. We categorized the patients as either responsive (complete response or partial response) or non-responsive (stable disease or disease progression).

4.3. Immunohistochemistry and Scoring

Formalin-fixed, paraffin-embedded tissue blocks representing the most invasive areas of each tumor were collected for further investigation. Normal urothelium was harvested from cystectomized specimens.

Three-micron-thick sections were immunostained using the BOND-MAX automated immunohistochemistry system and Bond Polymer Refine Detection kit DS 9800 (Leica Biosystems, Newcastle, UK), following our previous studies with minor modifications [32]; tissues obtained by RC were deparaffinized and pretreated with Bond Epitope Retrieval Solution 2 (Leica Biosystems) at 100 °C for 20 min. After washing and peroxidase blocking for 10 min, tissues were re-washed, and immunohistochemistry was performed for AHNAK2 on the specimens by rabbit polyclonal anti-AHNAK2 antibody (HPA004145; Sigma-Aldrich,

St. Louis, MO, USA, diluted 1:1000). Sections were incubated with EnVision FLEX+ Rabbit Linker (Dako, Glostrup, Denmark) for 15 min. Finally, the sections were incubated with BOND Polymer (Leica Biosystems) for 10 min, developed with 3,3′-diaminobenzidine (DAB) chromogen for 10 min, and counterstained with hematoxylin for 5 min. Sections treated with BOND Primary Antibody Diluent (Leica Biosystems) replacing the primary antibody were used as negative controls.

AHNAK2 was located in the cytoplasm and/or plasma and nuclear membrane of tumor cells, and the staining of tumor cells in these locations was considered positive. We scored the expression of AHNAK2 in the tumor cell using the following scheme: The intensity of staining was scored as 0 (negative), 1 (weak), 2 (moderate), or 3 (strong). The extent of staining was scored according to the percentage of positive tumor cells: 0 (none), 1 (1–25%), 2 (26–50%), 3 (51–75%), and 4 (76–100%). The two scores were multiplied, and the products (score) ranged from 0 to 12. A score between 0 and was considered as LE, whereas a score of 3 or more was considered as HE based on the median score of 3 [14]. All the immunostained sections were reviewed by two investigators (D.K. and Y.S.) without any knowledge of the clinical data. Discordant cases were reviewed and discussed until a consensus was reached.

4.4. Subgroup Analysis of Associations between AHNAK2 and Some Proteins

We additionally examined associations between AHNAK2 and some proteins, which we had previously reported in BCa, such as S100A2, S100A4, S100A8, S100A9, and nestin, for a better understanding of the mechanism of AHNAK2. The expressions of these proteins were calculated by a sum index score and categorized as previously described; S100 families were categorized as normal or abnormal and nestin as negative or positive [33–35].

4.5. Statistical Analysis

Comparisons of clinicopathological futures between the LE and HE of AHNAK2 were performed using the chi-square test (or Fisher's exact test, if appropriate) for categorical variables, and using the Mann–Whitney U test for continuous variables. RFS and CSS were estimated by the Kaplan–Meier method with the log-rank test. Multivariate analyses were performed with the Cox proportional hazards regression model, controlling for the effects of clinicopathological parameters. All statistical analyses were performed with Stata ver. 14 for Windows (Stata, Chicago, IL, USA). All p values were two-sided, and $p < 0.05$ was considered statistically significant.

5. Conclusions

The present study found that high expression levels of AHNAK2 were associated with aggressive pathological findings obtained by RC and were an independent predictor of worse RFS and CSS in patients with RC. Hence, we believe that AHNAK2 may act as a novel prognostic biomarker in the patients. Furthermore, associations between HE and S100A4, S100A8, S100A9, and nestin may highlight AHNAK2 as a novel therapeutic target of BCa. Further studies are warranted to elucidate the reported complex mechanisms of AHNAK2 in BCa.

Author Contributions: Conceptualization, D.K., K.M., and Y.S. (Yuichi Sato); methodology, K.M. and Y.S. (Yuichi Sato); formal analysis, D.K., K.M., and Y.S. (Yuichi Sato); investigation, K.M., Y.S. (Yuriko Shimizu), M.K., S.H., and M.I. (Masaomi Ikeda); writing—original draft preparation, D.K. and K.M.; supervision, M.I. (Masatsugu Iwamura); All authors have read and agreed to the published version of the manuscript.

Funding: This study was supported in part by the Japan Society for the Promotion of Science Grants-in-Aid for Scientific Research (KAKENHI), grant number JP18K09206.

Institutional Review Board Statement: The study was conducted according to the guidelines of the Declaration of Helsinki and approved by the Institutional Review Board of Kitasato University School of Medicine and Hospital (B17-010, B18-149).

Informed Consent Statement: Informed consent was obtained from some subjects involved with the study, or the information on the opportunity to opt out was provided to participants via our website and posters.

Data Availability Statement: The datasets used and/or analyzed during the current study are available from the corresponding author on reasonable request.

Conflicts of Interest: The authors declare no conflict of interest.

References

1. Siegel, R.; Ma, J.; Zou, Z.; Jemal, A. Cancer Statistics. *CA Cancer J. Clin.* **2014**, *64*, 9–29. [CrossRef]
2. Zehnder, P.; Studer, U.E.; Skinner, E.C.; Thalmann, G.N.; Miranda, G.; Roth, B.; Cai, J.; Birkhäuser, F.D.; Mitra, A.P.; Burkhard, F.C.; et al. Oncological Outcomes of Radical Cystectomy over Three Decades. *Bju Int.* **2013**, *112*, 51–58. [CrossRef]
3. Stein, J.P.; Skinner, D.G. Radical Cystectomy for Invasive Bladder Cancer: Long-Term Results of a Standard Procedure. *World J. Urol.* **2006**, *24*, 296–304. [CrossRef]
4. Mitra, A.P.; Quinn, D.I.; Dorff, T.B.; Skinner, E.C.; Schuckman, A.K.; Miranda, G.; Gill, I.S.; Daneshmand, S. Factors Influencing Post-Recurrence Survival in Bladder Cancer Following Radical Cystectomy. *BJU Int.* **2012**, *109*, 846–854. [CrossRef]
5. Witjes, J.A.; Lebret, T.; Compérat, E.M.; Cowan, N.C.; Santis, M.D.; Bruins, H.M.; Hernández, V.; Espinós, E.L.; Dunn, J.; Rouanne, M.; et al. Updated 2016 EAU Guidelines on Muscle-Invasive and Metastatic Bladder Cancer. *Eur. Urol.* **2017**, *71*, 462–475. [CrossRef]
6. Seiler, R.; Ashab, H.A.D.; Erho, N.; van Rhijn, B.W.G.; Winters, B.; Douglas, J.; Kessel, K.E.V.; van de Putte, E.E.F.; Sommerlad, M.; Wang, N.Q.; et al. Impact of Molecular Subtypes in Muscle-Invasive Bladder Cancer on Predicting Response and Survival after Neoadjuvant Chemotherapy. *Eur. Urol.* **2017**, *72*, 544–554. [CrossRef] [PubMed]
7. Zargar, H.; Espiritu, P.N.; Fairey, A.S.; Mertens, L.S.; Dinney, C.P.; Mir, M.C.; Krabbe, L.-M.; Cookson, M.S.; Jacobsen, N.-E.; Gandhi, N.M.; et al. Multicenter Assessment of Neoadjuvant Chemotherapy for Muscle-Invasive Bladder Cancer. *Eur. Urol.* **2015**, *67*, 241–249. [CrossRef] [PubMed]
8. Rosa, M. Advances in the Molecular Analysis of Breast Cancer: Pathway toward Personalized Medicine. *Cancer Control* **2015**, *22*, 211–219. [CrossRef] [PubMed]
9. Komuro, A.; Masuda, Y.; Kobayashi, K.; Babbitt, R.; Gunel, M.; Flavell, R.A.; Vincent, T.M. The AHNAKs are a Class of Giant Propeller-like Proteins that Associate with Calcium Channel Proteins of Cardiomyocytes and Other Cells. *Proc. Natl. Acad. Sci. USA* **2004**, *102*, 4053–4058. [CrossRef] [PubMed]
10. Wang, M.; Li, X.; Zhang, J.; Yang, Q.; Chen, W.; Jin, W.; Huang, Y.-R.; Yang, R.; Gao, W.-Q. AHNAK2 is a Novel Prognostic Marker and Oncogenic Protein for Clear Cell Renal Cell Carcinoma. *Theranostics* **2017**, *7*, 1100–1113. [CrossRef]
11. Lu, D.; Wang, J.; Shi, X.; Yue, B.; Hao, J. AHNAK2 is a Potential Prognostic Biomarker in Patients with PDAC. *Oncotarget* **2014**, *5*, 31775–31784. [CrossRef] [PubMed]
12. Li, M.; Liu, Y.; Meng, Y.; Zhu, Y. AHNAK Nucleoprotein 2 Performs a Promoting Role in the Proliferation and Migration of Uveal Melanoma Cells. *Cancer Biother. Radiopharm.* **2019**, *34*, 626–633. [CrossRef] [PubMed]
13. Xie, Z.; Lun, Y.; Li, X.; He, Y.; Wu, S.; Wang, S.; Sun, J.; He, Y.; Zhang, J. Bioinformatics Analysis of the Clinical Value and Potential Mechanisms of AHNAK2 in Papillary Thyroid Carcinoma. *Aging* **2020**, *12*, 18163–18180. [CrossRef] [PubMed]
14. Zhang, S.; Lu, Y.; Qi, L.; Wang, H.; Wang, Z.; Cai, Z. AHNAK2 is Associated with Poor Prognosis and Cell Migration in Lung Adenocarcinoma. *BioMed Res. Int.* **2020**, *2020*, 8571932. [CrossRef]
15. Wang, D.-W.; Zheng, H.-Z.; Cha, N.; Zhang, X.-J.; Zheng, M.; Chen, M.-M.; Tian, L.-X. Down-Regulation of AHNAK2 Inhibits Cell Proliferation, Migration and Invasion Through Inactivating the MAPK Pathway in Lung Adenocarcinoma. *Technol. Cancer Res. Treatment* **2020**, *19*, 153303382095700. [CrossRef] [PubMed]
16. Liu, G.; Guo, Z.; Zhang, Q.; Liu, Z.; Zhu, D. AHNAK2 Promotes Migration, Invasion, and Epithelial-Mesenchymal Transition in Lung Adenocarcinoma Cells via the TGF-β/Smad3 Pathway. *Oncotargets Ther.* **2020**, *13*, 12893–12903. [CrossRef]
17. Okusa, H.; Kodera, Y.; Oh-Ishi, M.; Minamida, Y.; Tsuchida, M.; Kavoussi, N.; Matsumoto, K.; Sato, T.; Iwamura, M.; Maeda, T.; et al. Searching for New Biomarkers of Bladder Cancer Based on Proteomic Analysis. *J. Electrophor.* **2008**, *52*, 19–24. [CrossRef]
18. Witzke, K.E.; Großerueschkamp, F.; Jütte, H.; Horn, M.; Roghmann, F.; von Landenberg, N.; Bracht, T.; Kallenbach-Thieltges, A.; Käfferlein, H.; Brüning, T.; et al. Integrated Fourier Transform Infrared Imaging and Proteomics for Identification of a Candidate Histochemical Biomarker in Bladder Cancer. *Am. J. Pathol.* **2019**, *189*, 619–631. [CrossRef]
19. Luo, Y.; Zeng, G.; Wu, S. Identification of Microenvironment-Related Prognostic Genes in Bladder Cancer Based on Gene Expression Profile. *Front. Genet.* **2019**, *10*, 1187. [CrossRef]
20. Chen, Z.; He, S.; Zhan, Y.; He, A.; Fang, D.; Gong, Y.; Li, X.; Zhou, L. TGF-β-Induced Transgelin Promotes Bladder Cancer Metastasis by Regulating Epithelial-Mesenchymal Transition and Invadopodia Formation. *EBioMedicine* **2019**, *47*, 208–220. [CrossRef]
21. Al-Ismaeel, Q.; Neal, C.P.; Al-Mahmoodi, H.; Almutairi, Z.; Al-Shamarti, I.; Straatman, K.; Jaunbocus, N.; Irvine, A.; Issa, E.; Moreman, C.; et al. ZEB1 and IL-6/11-STAT3 Signalling Cooperate to Define Invasive Potential of Pancreatic Cancer Cells via Differential Regulation of the Expression of S100 Proteins. *Br. J. Cancer* **2019**, *121*, 65–75. [CrossRef]

22. Scholtes, M.; Akbarzadeh, M.; Zwarthoff, E.; Boormans, J.; Mahmoudi, T.; Zuiverloon, T. Targeted Therapy in Metastatic Bladder Cancer: Present Status and Future Directions. *Appl. Sci.* **2020**, *10*, 7102. [CrossRef]
23. Sluzalska, K.D.; Slawski, J.; Sochacka, M.; Lampart, A.; Otlewski, J.; Zakrzewska, M. Intracellular Partners of Fibroblast Growth Factors 1 and 2—Implications for Functions. *Cytokine Growth Facor Rev.* **2020**, *57*, 93–111. [CrossRef] [PubMed]
24. Kwon, C.H.; Moon, H.J.; Park, H.J.; Choi, J.H.; Park, D.Y. S100A8 and S100A9 Promotes Invasion and Migration through P38 Mitogen-Activated Protein Kinase-Dependent NF-KB Activation in Gastric Cancer Cells. *Mol. Cell* **2013**, *35*, 226–234. [CrossRef]
25. Kirov, A.; Kacer, D.; Conley, B.A.; Vary, C.P.H.; Prudovsky, I. AHNAK2 Participates in the Stress-Induced Nonclassical FGF1 Secretion Pathway. *J. Cell Biochem.* **2015**, *116*, 1522–1531. [CrossRef]
26. Eckerdt, F.D.; Bell, J.B.; Gonzalez, C.; Oh, M.S.; Perez, R.E.; Mazewski, C.; Fischietti, M.; Goldman, S.; Nakano, I.; Platanias, L.C. Combined PI3Kα-MTOR Targeting of Glioma Stem Cells. *Sci. Rep.* **2020**, *10*, 21873. [CrossRef]
27. Ashrafizadeh, M.; Zarrabi, A.; Hushmandi, K.; Kalantari, M.; Mohammadinejad, R.; Javaheri, T.; Sethi, G. Association of the Epithelial–Mesenchymal Transition (EMT) with Cisplatin Resistance. *Int. J. Mol. Sci.* **2020**, *21*, 4002. [CrossRef] [PubMed]
28. Jiang, Y.; Zhan, H. Communication between EMT and PD-L1 Signaling: New Insights into Tumor Immune Evasion. *Cancer Lett.* **2019**, *468*, 72–81. [CrossRef]
29. Wang, L.; Saci, A.; Szabo, P.M.; Chasalow, S.D.; Castillo-Martin, M.; Domingo-Domenech, J.; Siefker-Radtke, A.; Sharma, P.; Sfakianos, J.P.; Gong, Y.; et al. EMT- and Stroma-Related Gene Expression and Resistance to PD-1 Blockade in Urothelial Cancer. *Nat. Commun.* **2018**, *9*, 3503. [CrossRef]
30. Matsumoto, K.; Ikeda, M.; Sato, Y.; Kuruma, H.; Kamata, Y.; Nishimori, T.; Tomonaga, T.; Nomura, F.; Egawa, S.; Iwamura, M. Loss of Periplakin Expression is Associated with Pathological Stage and Cancer-Specific Survival in Patients with Urothelial Carcinoma of the Urinary Bladder. *BioMed Res.* **2014**, *35*, 201–206. [CrossRef]
31. Eisenhauer, E.A.; Therasse, P.; Bogaerts, J.; Schwartz, L.H.; Sargent, D.; Ford, R.; Dancey, J.; Arbuck, S.; Gwyther, S.; Mooney, M.; et al. New Response Evaluation Criteria in Solid Tumours: Revised RECIST Guideline (Version 1.1). *Eur. J. Cancer* **2009**, *45*, 228–247. [CrossRef] [PubMed]
32. Amano, N.; Matsumoto, K.; Shimizu, Y.; Nakamura, M.; Tsumura, H.; Ishii, D.; Sato, Y.; Iwamura, M. High HNRNPA3 Expression is Associated with Lymph Node Metastasis and Poor Prognosis in Patients Treated with Radical Cystectomy. *Urol. Oncol. Semin. Orig. Investig.* **2020**, *39*, 196. [CrossRef]
33. Matsumoto, K.; Irie, A.; Satoh, T.; Ishii, J.; Iwabuchi, K.; Iwamura, M.; Egawa, S.; Baba, S. Expression of S100A2 and S100A4 Predicts for Disease Progression and Patient Survival in Bladder Cancer. *Urology* **2007**, *70*, 602–607. [CrossRef] [PubMed]
34. Minami, S.; Sato, Y.; Matsumoto, T.; Kageyama, T.; Kawashima, Y.; Yoshio, Y.; Ishii, J.-I.; Matsumoto, K.; Nagashio, R.; Okayasu, I. Proteomic Study of Sera from Patients with Bladder Cancer: Usefulness of S100A8 and S100A9 Proteins. *Cancer Genom. Proteom.* **2010**, *7*, 181–189.
35. Tabata, K.; Matsumoto, K.; Minami, S.; Ishii, D.; Nishi, M.; Fujita, T.; Saegusa, M.; Sato, Y.; Iwamura, M. Nestin is an Independent Predictor of Cancer-Specific Survival after Radical Cystectomy in Patients with Urothelial Carcinoma of the Bladder. *PLoS ONE* **2014**, *9*, e91548. [CrossRef]

Article

Targeting *S1PR1* May Result in Enhanced Migration of Cancer Cells in Bladder Carcinoma

Chin-Li Chen [1], En Meng [1], Sheng-Tang Wu [1], Hsing-Fan Lai [2], Yi-Shan Lu [2], Ming-Hsin Yang [1], Chih-Wei Tsao [1], Chien-Chang Kao [1] and Yi-Lin Chiu [2,*]

1. Division of Urology, Department of Surgery, Tri-Service General Hospital, National Defense Medical Center, Taipei 11490, Taiwan; iloveyou@mail.ndmctsgh.edu.tw (C.-L.C.); mengen@mail.ndmctsgh.edu.tw (E.M.); doc20283@mail.ndmctsgh.edu.tw (S.-T.W.); yangming@mail.ndmctsgh.edu.tw (M.-H.Y.); wei@mail.ndmctsgh.edu.tw (C.-W.T.); everyyoung@mail.ndmctsgh.edu.tw (C.-C.K.)
2. Department of Biochemistry, National Defense Medical Center, Taipei 11490, Taiwan; 809302001@mail.ndmctsgh.edu.tw (H.-F.L.); 608040006@mail.ndmctsgh.edu.tw (Y.-S.L.)
* Correspondence: yilin1107@mail.ndmctsgh.edu.tw; Tel.: +886-2-8792-3100 (ext. 18828)

Simple Summary: Metastasis is critical to the prognosis of patients with bladder cancer, and it is important to understand the mechanism of its occurrence. *S1PR1* expression is thought to be associated with poor prognosis, but it is unknown whether it is associated with tumor metastasis. Analysis of clinical gene expression data suggests that endothelial or immune cells in tumor tissue may be the source of *S1PR1* expression. Comparative analysis of clinical tumor tissues with bladder cancer cells suggests that *S1PR1* expression is associated with cellular adhesion. In vitro experiments demonstrated that *S1PR1* expression was negatively correlated with cancer cell motility, and that *S1PR1* inhibition by FTY-720 may cause an increase in cancer cell motility, suggesting that the use of *S1PR1* inhibition as a synergistic therapy requires additional observations and considerations.

Abstract: Clinical bladder tumor histological analysis shows that high expression of *S1PR1* is associated with poor patient prognosis. However, there are no studies that describe the underlying mechanism. To investigate the relative distribution and actual function of *S1PR1* in bladder tumors, we analyzed multiple clinical databases in combination with tumor purity and immune cell infiltration simulations, as well as databases of well-defined histological phenotypes of bladder cancer, and single-cell sequencing of adjacent normal tissues and bladder tumors, and further compared them with bladder cancer cell lines. The results showed that S1PR1 expression was generally higher in normal tissues than in bladder cancer tissues, and its distribution was mainly in endothelial cells or immune cells. The association between high S1PR1 expression and poor prognosis may be due to tumor invasion of adjacent normal tissues, where highly expressed S1PR1 may affect prognostic interpretation. The effect of *S1PR1* itself on cancer cells was associated with cell adhesion, and in bladder cancer cells, *S1PR1* expression was negatively correlated with cell motility. Moreover, the use of FTY-720 will cause an increased metastatic ability of bladder cancer cells. In conclusion, we suggest that the use of *S1PR1*-specific inhibition as a synergistic treatment requires more observation and consideration.

Keywords: sphingosine 1-phosphate receptor 1; bladder carcinoma; cell migration; epithelial–mesenchymal transition; FTY-720

1. Introduction

Urothelial carcinoma of the bladder (UCB) is among the top 10 most common cancers in the world, with an estimated 80,000 new cases and 17,000 deaths in the United States each year [1–3]. Significant advances have been made in the management of bladder cancer since the 1990s. More accurate staging has been achieved with refined tissue imaging, and advances in surgical techniques have been combined with improved chemotherapy regimens. Even more, the 5-year survival rate for patients with non-muscle invasive UCB is over 90% [1], and radical cystectomy is the treatment of choice for patients with surgically resectable disease without evidence of metastatic disease. However, patients with muscle-invasive bladder cancer or disseminated disease have a much lower survival rate [4,5], suggesting that the occurrence of metastasis has a significant impact on the prognosis of patients with bladder cancer. Considering the impact of metastatic disease on treatment options and patient prognosis, the importance of timely detection and prevention of metastasis in UCB cannot be overemphasized.

Sphingosine 1-phosphate receptor 1 (*S1PR1*) is a biologically active sphingolipid metabolite receptor, whose ligand Sphingosine 1-phosphate (S1P) is known to modulate cell survival, migration, immune cell transport, angiogenesis, and vascular barrier function [6]. Physiologically, *S1PR1*, which is abundantly expressed in vascular endothelial cells (EC), is important for embryonic vascular development and maturation [6]. In addition, *S1PR1* expression in immune cells is thought to be associated with the regulation of traffic between tissues, including B cells, T cells, natural killer cells, macrophage, monocyte, and neutrophil [7–12]. In cancer progression, *S1PR1* is thought to be highly expressed in bladder cancer cells and is associated with poor patient prognosis [13]. S1P can promote cancer cell viability, survival, growth, and transformation by activating *S1PR1* [14]. In addition, S1PR1 overexpression is associated with the convening of regulatory T cells (T_{reg}) [15], suggesting S1PR1 as a potential prognostic biomarker and therapeutic target for UCB patients. Clinically, FTY-720 (Fingolimod) is used as an *S1PR1* inhibitor and is widely used in multiple sclerosis or as an immunomodulator [16,17]. FTY-720 is reported to promote apoptosis of bladder cancer cells [18]. As *S1PR1* is important for the regulation of immune cell movement, we wondered whether it might have a similar role in the metastasis of bladder cancer cells. However, given the complexity of tumor tissue composition, such as endothelial cells and immune cells, which are rich in *S1PR1* expression, it is still unknown whether *S1PR1* is actually overexpressed in bladder cancer cells or whether *S1PR1* expression has a substantial effect on bladder cancer cells themselves.

In this study, we analyze the association between S1PR1 expression and patient prognosis by exploring several bladder cancer clinical databases. In addition, the association of S1PR1 with tumor purity and immune cell infiltration was comprehensively analyzed. The main distribution of S1PR1 in bladder tumors was analyzed in microarray and single cell sequencing databases with well-defined histological patterns. Using comparative analysis of bladder cancer cell lines and clinical tissues, we confirmed the association of *S1PR1* with cell adhesion ability in bladder cancer cells. A negative correlation between *S1PR1* expression and cell motility was also confirmed in bladder cancer cell lines. Lastly, the effect of FTY-720 on promoting bladder cancer cell motility was confirmed in cell lines and patient-derived tumor cultures. This article first revealed that loss of S1PR1 expression in bladder cancer cells is associated with increased cell motility and that inhibition of *S1PR1* activity with FTY-720 may cause a similar phenomenon. We suggest that the use of FTY-720 as a synergistic strategy for other treatments requires more evaluation and observation.

2. Materials and Methods

2.1. Patient Specimens and Clinical Data

The bladder tissue obtained for the study was obtained in accordance with the relevant guidelines and regulations and was approved by the Institutional Review Board of Triservice General Hospital, National Defense Medical Center (IRB approval ID:1-108-05-130). Tumors were obtained from patients who underwent transurethral resection of bladder tumor (TURBT) and signed an informed consent form. Patient information and related clinical information were de-identified.

2.2. Gene Expression Database Collection and Analysis

We searched the NCBI-GEO database (http://www.ncbi.nlm.nih.gov/geo/, accessed on 18 June 2018) for gene expression studies related to urothelial carcinoma of the bladder using the keyword "urothelial carcinoma". The following criteria were used for selection: (1) datasets included S1PR1 gene probes; (2) had overall survival status and survival time; and (3) for each bladder urothelial cancer tissue dataset, the total number of available samples was greater than 40. To investigate the prognostic relevance of S1PR1 grouping without a predetermined stance, we used Evaluate Cutpoints [19] to define the best cutpoints in terms of S1PR1 mRNA expression, survival status, and survival time. Meta-analyses were performed on TCGA BLCA (The Cancer Genome Atlas-Bladder Urothelial Carcinoma) (424 patients), GSE5287 (49 patients), GSE13507 (164 patients), GSE31684 (93 patients), GSE32894 (307 patients), and GSE48277 (159 patients).

For the expression of S1PR1 in different cancer types in TCGA, we searched for "S1PR1" in "Gene DE" in the "Exploration" tab of TIMER2.0 [20]. For the single cell sequencing database of bladder cancer published by Chen et al., we analyzed and plotted the distribution of S1PR1 expression in R using the scripts provided by the authors. The distribution of tumor grade and cell type was directly quoted from the authors' publication [21]. Cancer cell line encyclopedia (CCLE) and TCGA BLCA whole gene expression matrix (FPKM) with corresponding clinical parameters for patients or methylation expression profiles and copy number for cell lines were downloaded from UCSC XENA (https://xenabrowser.net/, accessed on 26 March 2021) [22]. The hierarchical cluster function provided in Morpheus (Broad Institute, https://software.broadinstitute.org/morpheus/, accessed on 4 March 2021) was used to cluster S1PR1 expression, methylation, and copy number data obtained from the Cancer Cell Line Encyclopedia (CCLE) database [23].

2.3. Evaluation of Tumor Purity and Immune Cell Simulated Infiltration

For the obtained database, we used the ESTIMATE R software package to calculate the ESTIMATE scores and used the formula of Yoshihara et al. to calculate the purity [24]. The ESTIMATE scores of TCGA BLCA were downloaded directly from the authors' website (https://bioinformatics.mdanderson.org/estimate/disease.html, accessed on 26 March 2021). For the in silico simulation of immune cell infiltration analysis, we used the quanTIseq function provided in immunedeconv and set the parameters according to the authors' instructions to adapt to different types of databases [25,26]. For S1PR1 gene expression and purity and multiple immune cell infiltration scores, we calculated Spearman's correlation and p-values, and graphed the correlation matrix with OriginPro 2021b (Origin Lab, Northampton, MA, USA).

2.4. Enrichment Map Visualization

The rationale and operation were performed as previously described [27,28]. Briefly, the TCGA BLCA and GSE13507 databases were chosen based on (1) the largest number of patient samples and (2) the divergence of S1PR1 expression and prognosis. We used the BP:GO bioprocess (7530 gene sets) downloaded from Molecular Signatures Database (MsigDB) for gene-set enrichment analysis (GSEA). Samples were categorized as "S1PR1 high vs. S1PR1 low" according to the annotation elsewhere in this study. All nodes presented have passed the screening of $|NES| > 1.5$, $FDR < 0.01$ (NES: normalized enrichment score, FDR: false discovery rate).

2.5. Messenger RNA Expression Analysis

Total RNA from tumor and adjacent tissues or cultured cells was isolated using a Qiagen RNeasy Mini Kit (Qiagen, Hilden, Germany). RNA concentration and quality were assessed using SpectraMax iD3 (Molecular Devices, San Jose, CA, USA). The cDNA was generated from 2 µg of RNA by reverse transcription using MMLV high performance reverse transcriptase (Epicentre Technologies, Madison, WI, USA) according to the manufacturer's instructions. Real-time PCR was performed using an CFX96 Touch Real-Time PCR Detection System (Bio-Rad, Hercules, CA, USA). The cycling condition was 95 °C for 12 min, followed by 40 cycles of 95 °C for 15 s, 60 °C for 20 s, and 72 °C for 20 s. The housekeeping gene Glyceraldehyde 3-phosphate dehydrogenase (GAPDH) was measured as an internal control. The expression level of target genes was analyzed by the relative quantity (RQ) value calculated using the ΔΔCt method [Δ(Ct$_{TARGET}$ − Ct$_{GAPDH}$)$_{sample}$ − Δ(Ct$_{TARGET}$ − Ct$_{GAPDH}$)$_{calibrator}$] in triplicate. All primer sequences are listed in the following Table 1.

Table 1. Primer sequences used in the qPCR assay.

Gene	Forward Primer	Reverse Primer
SLUG	AAGCATTTCAACGCCTCCAAA	GGATCTCTGGTTGTGGTATGACA
CDH1	GCCTCCTGAAAAGAGAGTGGAAG	TGGCAGTGTCTCTCCAAATCCG
GAPDH	CATCACTGCCACCCAGAAGACTG	ATGCCAGTGAGCTTCCCGTTCAG
CDH2	TGCGGTACAGTGTAACTGGG	GAAACCGGGCTATCTGCTCG
SNAI1	TCGGAAGCCTAACTACAGCGA	AGATGAGCATTGGCAGCGAG
FN1	CGGTGGCTGTCAGTCAAAG	AAACCTCGGCTTCCTCCATAA
S1PR1	ATCATGGGCTGGAACTGCATCA	CGAGTCCTGACCAAGGAGTAGAT

2.6. Immunoblotting

The bladder cancer cell pellets or pretreated tumor specimens were washed twice with phosphate-buffered saline (PBS), lysed in radioimmunoprecipitation assay (RIPA) buffer, and quantified by Bradford protein assay (Bio-Rad, Hercules, CA, USA; #500-0006). Firstly, 30–50 μg of quantified total protein lysate was loaded into each well of the gel, analyzed in 8–10% SDS-PAGE under reducing conditions, and then transferred to a nitrocellulose blotting membrane (PALL Corpo., Pansacola, FL, USA) followed by blocking in 5% skim milk. The membrane was stained with primary antibody as follows: S1PR1(Abclonal Inc., Woburn, MA, USA; A3997); E-cadherin (Cell signaling, Danvers, MA, USA; #3195); Vimentin (Abclonal Inc., Woburn, MA, USA; A11952); N-cadherin (Cell signaling #4061); Fibronectin (Finetest, Wuhan, China; fnab03122); SNAI1 (Cell signaling, #3879); Slug (Cell signaling, #9585); and internal control GAPDH (Cell signaling, #5174), prepared in 1% bovine serum albumin (BSA) in Tris-buffered saline and 0.1% Tween®20 (TBST) at 4 °C overnight. Then, the membrane was washed and incubated with secondary antibody at room temperature for 1 h. Signals were detected for 1–10 min using an enhanced chemiluminescence solution (Advansta, Menlo Park, CA, USA) and iBright FL1500 Imaging System (Thermo Fisher Scientific., Waltham, MA, USA). All experiments were performed in duplicate.

2.7. Cell Culture and Establishment of Stably Expressed shRNA Cell Lines

The J82 human bladder cancer cell line was purchased from Bioresource Collection and Research Center (BCRC, Hsinchu, Taiwan). Cells were cultured in Dulbecco's modified Eagle's medium and supplemented with 10% fetal bovine serum (FBS, Thermo Fisher Scientific) under a humidified atmosphere of 5% CO_2 at 37 °C. For the subcultures, cells were trypsinized with 0.05% Trypsin-EDTA (Thermo Fisher Scientific).

All the short hairpin RNA (shRNA) clones were obtained from National RNAi Core Facility (Genomics Research Center, Academia Sinica, Taipei, Taiwan). The shRNA against S1PR1 target sequence was 5′-GACAACCCAGAGACCATTATG-3′ (clone ID: TRCN0000356960, shS1PR1#1), 5′-CCCATGTGAAAGCGTCTCTTT-3′ (clone ID: TRCN0000221119, shS1PR1#2), and 5′- TCCTAAGGTTAAGTCGCCCTCG-3′ (clone ID: TRCN0000072249, shLuc) for firefly luciferase as the negative control. The shRNA plasmids were transfected into J82 cell lines with Lipofectamine 3000 (Thermo Fisher Scientific.) according to the manufacturer's instructions. The stably expressed shRNA cell lines were established with the screening of puromycin 2 μg/mL for 1 week. Knockdown efficiency of S1PR1 was confirmed by quantitative reverse transcriptase-PCR (data not shown) and Western blot analysis at 24 to 48 h post-transfection (Uncropped Western blot images were provided in Supplementary File S1).

2.8. Wound Healing Assay and LIVE Cell Imaging

Migration was evaluated in duplicate by seeding cells on both sides of an Ibidi culture insert (Ibidi, Munich, Germany) with a 500 μm separation gap. J82 shLuc bladder cancer cell line transfected with S1PR1/pcDNA 3.1(+) (1 or 3 μg) or pcDNA 3.1(+) (3 μg) ws grown for 24 h, then the growth medium was changed for complete RPMI1640 supplied with 0.5% FBS for 24 h before wound healing assay to diminish the potential interfere of cell proliferation. The gaps of J82 cells were time-lapse photographed every 30 min for 24 h and 48 h using Lumascope 620 with a 10× objective (Etaluma, San Diego, CA, USA), and all experiments were performed in duplicate. For multi-dose FTY-720 (MedChemexpress, Monmouth Junction, NJ, USA; HY-12005) treatment gap healing analysis, we used the ImageXpress Pico system (Molecular Devices) to detect hourly changes in the total number of cells in the gap over a 48 h period. Each dose was duplicated, and the curve results were presented as mean only to minimize interpretation interference. Migration ability of cancer cells was evaluated by Chemotaxis and Migration Tool 2.0 (Ibidi) and Manual Tracking plug-in in ImageJ 1.53h (National Institutes of Health, Stapleton, NY, USA).

2.9. Transwell Migration Assay

For the Transwell migration assay, cells were collected and transfected at a density of 10^5/well for 24 h. Cells were placed in 200 μL of serum-free medium and inoculated in the upper compartment of the chamber. Then, 1 mL of complete medium containing 10% FBS was added to the lower compartment. After 24 h of incubation, the chambers were removed and the cells on the upper surface of the membrane were wiped off with a cotton swab. Then, the cells invading the microporous membrane were washed three times with PBS, fixed with 4% paraformaldehyde solution for 30 min, and stained with 0.1% crystal violet (Sigma-Aldrich, St. Louis, MO, USA) for 20 min. Finally, the cells were observed with a microscope (CKX53, Olympus, Tokyo, Japan) and images were taken for further imageJ analysis. All experiments were performed in duplicate.

2.10. Patient-Derived Tumor Primary Culture (PDC)

The experimental procedure is based on the publication of van de Merbel et al. with minor modifications [29]. Briefly, freshly collected bladder tissues were divided into multiple 1 mm^3 slices, and the sliced samples were evenly divided into four aliquots and collected for analysis after 48 h of incubation at different concentrations of FTY-720.

2.11. Statistical Analysis

GraphPad Prism 9.1.2 (GraphPad Software, San Diego, CA, USA) or OriginPro 2021b was used for data analysis and graph production. Student's t-test was used for analysis of measurement data between two groups, and one-way analysis of variance (ANOVA) was used for multiple group comparisons. Survival curves were plotted using Kaplan–Meier analysis and logarithmic tests. The univariate Cox proportional hazards regression model was used to estimate the odds ratio (OR). Data are expressed as mean ± standard deviation (SD) and all experiments were repeated in duplicate independently at least.

3. Results

3.1. Retrospective Evaluation of the Association between S1PR1 Expression and Bladder Cancer Prognosis Shows Divergent Results in Different Databases

Several publications have demonstrated the association of S1PR1 overexpression with worse prognosis in bladder cancer patients. To comprehensively evaluate the association between S1PR1 expression and prognosis of bladder cancer patients, we collected six available databases including GSE5287, GSE13507, GSE31684, GSE32894, GSE48075, and TCGA BLCA. In order to unbiasedly group the S1PR expression, the "Evaluate Cutpoints" application in R was used to find the best cut-off point for the lowest p-value in survival of S1PR1 expression [19]. The results showed that high S1PR1 expression in TCGA BLCA (OR: 1.922, p-value: 0.0002), GSE 32894 (OR: 1.955, p-value: 0.0757), and GSE31684 (OR: 2.94, p-value: 0.0223) was associated with worse prognosis. On the contrary, high *S1PR1* expression was associated with better prognosis in GSE5287 (OR: 0.2913, *p*-value: 0.02), GSE48075 (OR: 0.5491, *p*-value: 0.0824), and GSE13507 (OR: 0.705, *p*-value: 0.0709) (Figure 1). The meta-analysis showed that the overall odds ratio still reached 1.681, indicating that, in general, high *S1PR1* expression was associated with a worse prognosis in bladder cancer.

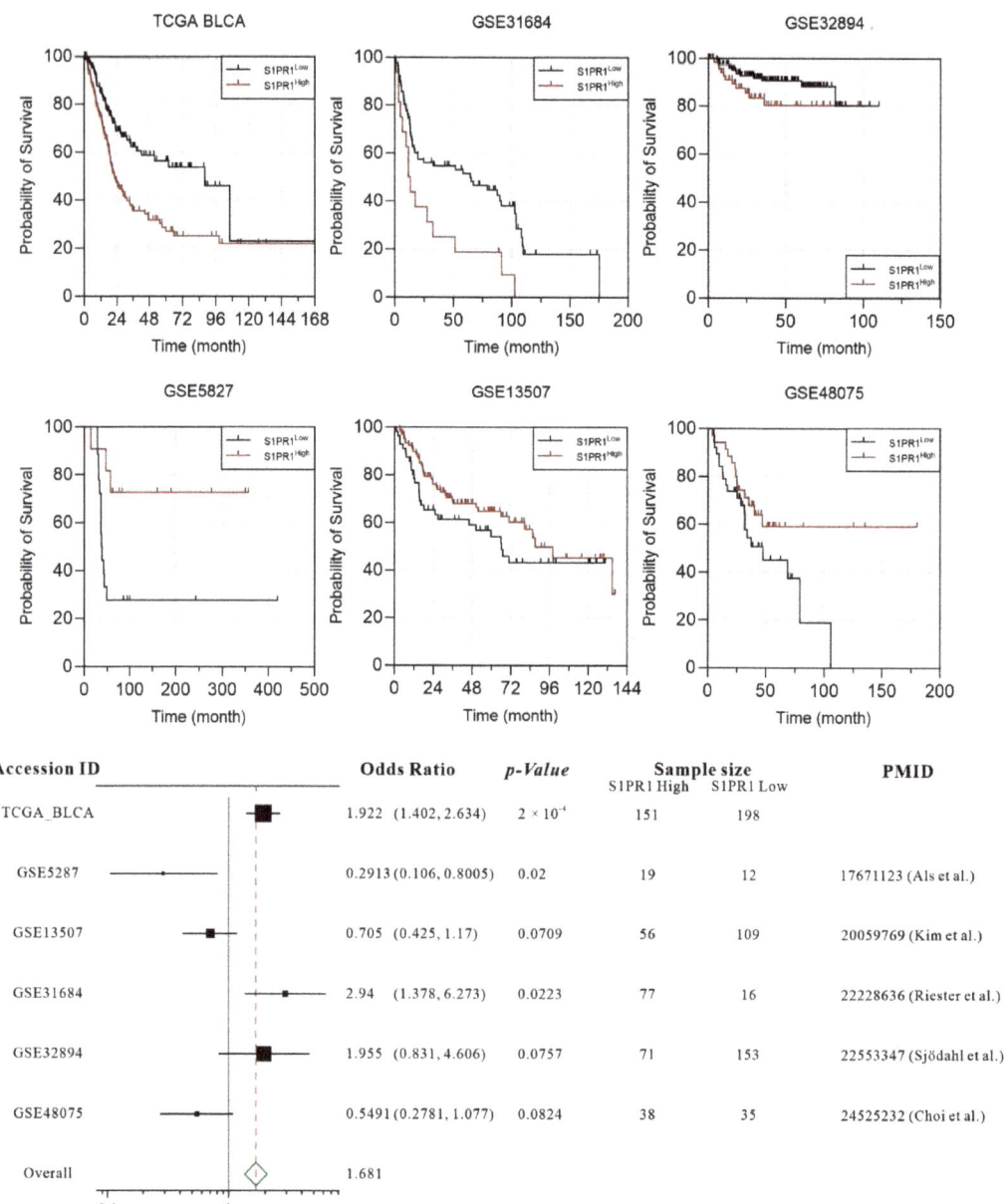

Figure 1. Individual and integrated analysis of the association between *S1PR1* mRNA expression and patient prognosis in published bladder cancer clinical databases [30–34].

3.2. Differences in S1PR1 Expression and Prognosis of Patients with Bladder Cancer May Be Related to the Degree of Neutrophil Infiltration

Considering that *S1PR1* is an important receptor related to the regulation of migration by various immune cells [12], clinical specimens may have different levels of immune cell infiltration affecting the *S1PR1* mRNA expression in bulky tumors. Evaluation of the association between tumor purity and *S1PR1* expression using the estimate score strategy showed that *S1PR1* expression was negatively correlated with tumor purity in all databases except GSE13507 (Figure 2), suggesting that high *S1PR1* expression in bulk tissue may be associated with enriched immune or stromal cell infiltration.

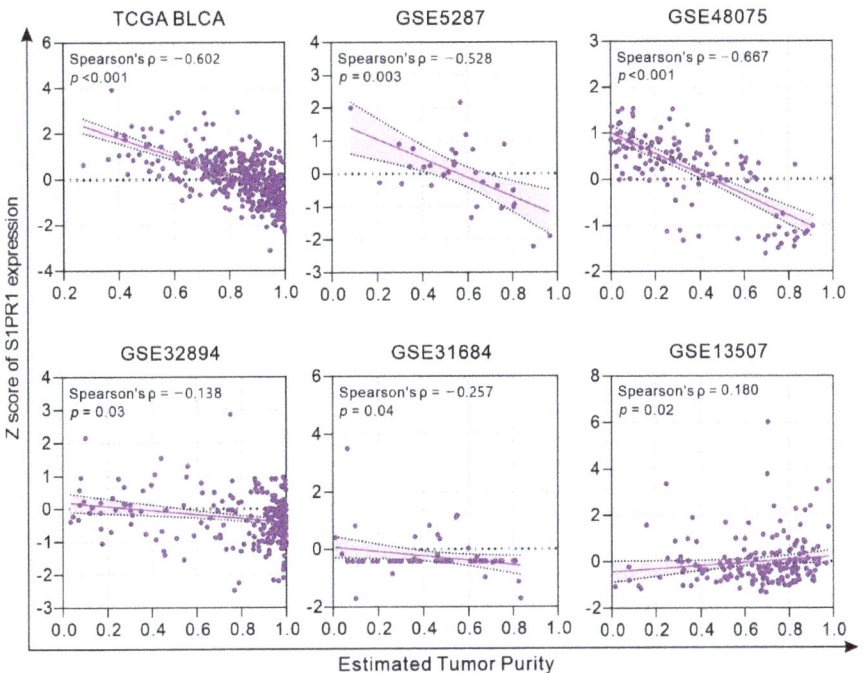

Figure 2. Association of *S1PR1* expression with tumor purity.

Further evaluation of the correlation between *S1PR1* and immune cell populations by QUANTISEQ showed that *S1PR1* expression was positively correlated with B cells, macrophage (M1 and M2), and regulatory T cells in all databases (Figure 3). Interestingly, although not all correlations were significant, *S1PR1* was positively correlated with neutrophil infiltration in all three databases with better prognosis (GSE5287: $\rho = 0.47$, GSE13507: $\rho = 0.16$, GSE48075: $\rho = 0.082$). In contrast, *S1PR1* was negatively correlated with neutrophil (TCGA BLCA: $\rho = -0.094$, GSE32894: $\rho = -0.047$, GSE31684: $\rho = -0.21$) in the three databases where *S1PR1* was associated with poorer prognosis. This suggests that the enriched neutrophil infiltration may directly affect the prognosis prediction of bladder cancer patients using *S1PR1* expression.

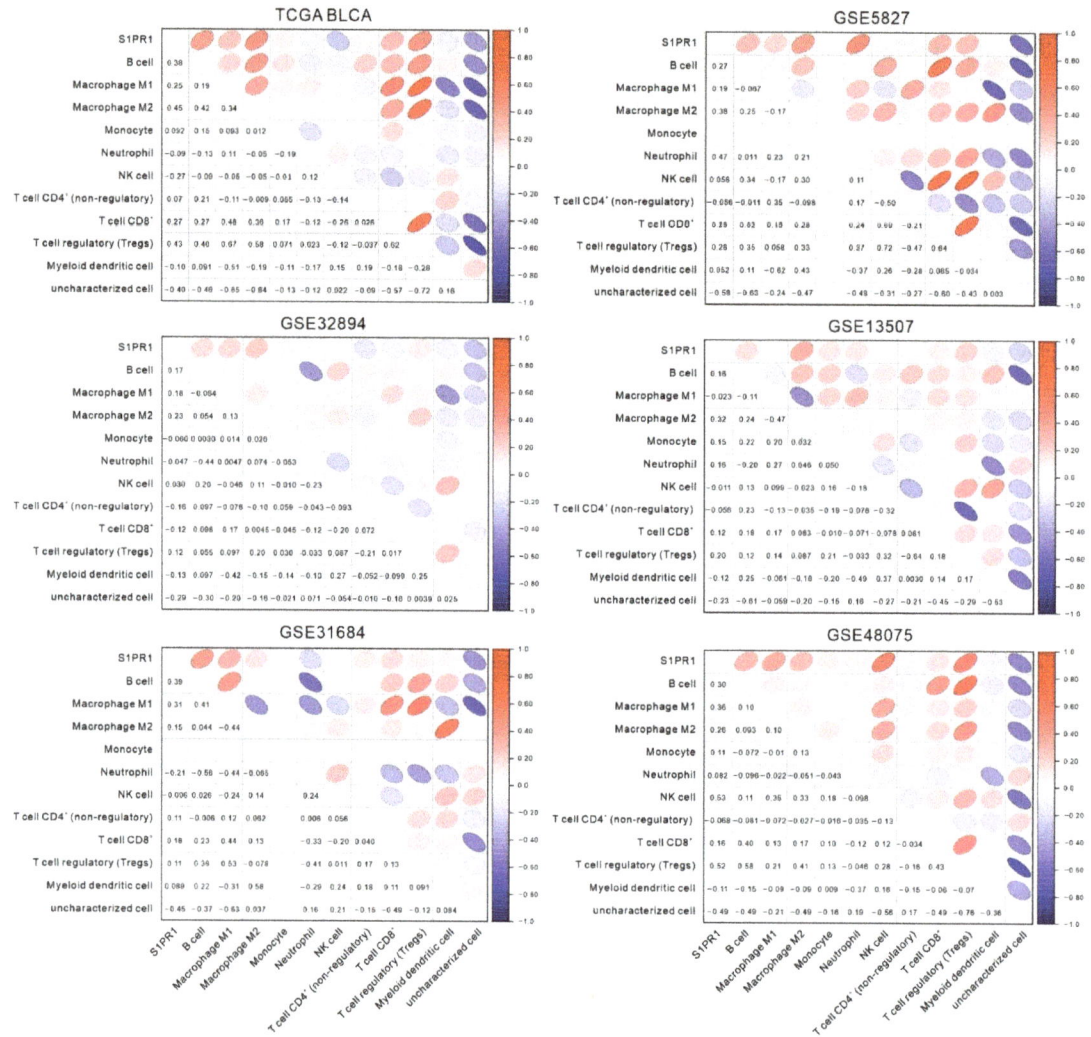

Figure 3. Association of *S1PR1* expression with various immune cell infiltrations. The association of *S1PR1* expression in bladder cancer tissues with QUANTISEQ simulated multiple immune cell infiltration was assessed using Spearman's correlation.

3.3. Comprehensive Assessment of S1PR1 Expression Differences between Bladder Cancer Tumors and Normal Tissue

Given the negative correlation between *S1PR1* expression and estimated tumor purity, we suggest that the proportion of normal tissue adulterated in tumor samples may also affect *S1PR1* expression. To clarify this issue, we attempted to identify the differential expression of *S1PR1* in normal tissue and bladder cancer tumor tissue. Gene expression analysis of the TCGA database provided by Timer 2.0 [20] showed that *S1PR1* expression was significantly higher in normal tissue than in tumor tissue in most cancers, including BLCA (Figure 4A). Further, the association of *S1PR1* expression differences with tissue types was evaluated in the clinical database of bladder cancer provided by Sanchez-Carbayo et al. and Lee et al. (Figure 4B) [35,36], showing that *S1PR1* expression was higher in normal bladder tissue and decreased as the tumor histology became more defined. For example,

S1PR1 expression was significantly lower in superficial bladder cancer or primary bladder cancer than in infiltrating bladder urothelial carcinoma or bladder mucosae surrounding cancer, suggesting that the actual expression of *S1PR1* may be affected when the tumor sample contains normal tissue. To further characterize the expression distribution of *S1PR1* in clinical bladder carcinoma samples, we analyzed *S1PR1* expression using the bladder urothelial carcinoma single-cell RNA sequencing database (Figure 4C) published by Chen et al. [21]. The results showed that *S1PR1* was mainly expressed in endothelial cells and to a lesser extent in various immune cells. In addition, endothelial cells were mainly found in normal and high-grade bladder urothelial carcinoma rather than low-grade (refer to Figure 1b in the publication of Chen et al., 2020), suggesting that high-grade tumors contain a high proportion of endothelial cells, which may be related to the high expression of endothelial and *S1PR1* due to tumor invasion of normal tissues.

We evaluated *S1PR1* expression in normal or tumor tissues from four clinical cases and analyzed the epithelial–mesenchymal transition (EMT) marker together, which is thought to be regulated by *S1PR1* expression, but with different effects in different tissues or spatial and temporal contexts [37,38]. The results showed that *S1PR1* mRNA and protein expression were mostly amplified in normal tissues, similar to E-cadherin expression, while N-cadherin, FN, SLUG, and SNAI1, which are mesenchymal markers, were higher in tumor tissues, suggesting a preliminary association between *S1PR1* expression and EMT in bladder cancer (Figure 4D).

3.4. Comparison of S1PR1 Expression in Bladder Cancer Cell Lines with Clinical Databases Reveals its Potential Function in Cell Adhesion

There are many factors in clinical bulk tissue that may affect *S1PR1* expression, such as the infiltration of endothelial or immune cells, which may lead to misinterpretation of the biological response of *S1PR1* expression in bladder cancer cells. To understand the direct effect of *S1PR1* expression on bladder cancer cells, 21 bladder cancer-related cell lines were screened from the CCLE database for *S1PR1* expression analysis, showing that the high *S1PR1* expression group generally had lower methylation and the low *S1PR1* expression group had higher methylation in addition to lower copy number variation (Figure 5A). Further, GSEA was performed after clustering cell lines with high or low *S1PR1* expression. On the other hand, GSEA was performed with *S1PR1* expression-related genes in TCGA BLCA (high *S1PR1* associated with poor prognosis) and GSE13507 (low *S1PR1* associated with poor prognosis) (illustrated as Figure 5B), and the results of the three GSEAs were visualized using the Enrichment map in Cytoscape (Figure 5C). The results showed that the node cluster associated with immune cell activation in the clinical database was not enriched in the cell lines, suggesting that the biological response associated with *S1PR1* expression in clinical tissues is indeed influenced by the immune microenvironment. On the contrary, we observed that angiogenesis and cell adhesion gene clusters were positively associated with *S1PR1* expression in all three, suggesting that the real effect of *S1PR1* on bladder cancer is related to these gene clusters.

3.5. S1PR1 Expression Shows an Opposite Association with the Promoting of Epithelial–Mesenchymal Transition

Among the 21 uroepithelial cancer cell lines, only four cell lines, T24, J82, JMSU1, and SCABER, expressed high amounts of S1PR1, and its mRNA expression was not even detected in most cell lines. The expression of S1PR1 was positively correlated with the variation of copy number, especially in JMSU1 and SCABER, which had a high copy number of the S1PR1 gene (Figure 4A). Only the J82 bladder cancer cell line has a near normal copy number and moderate mRNA expression of the S1PR1 gene. Therefore, to avoid potential interference with genetic abnormalities, the J82 bladder cancer cell line was used to establish a stable expression of S1PR1 targeting shRNA clone (Figure 6A–C). In addition, the control cells (J82 shLuc) were transfected with pcDNA3.1-*S1PR1* plasmids for overexpression (Figure 6D–F).

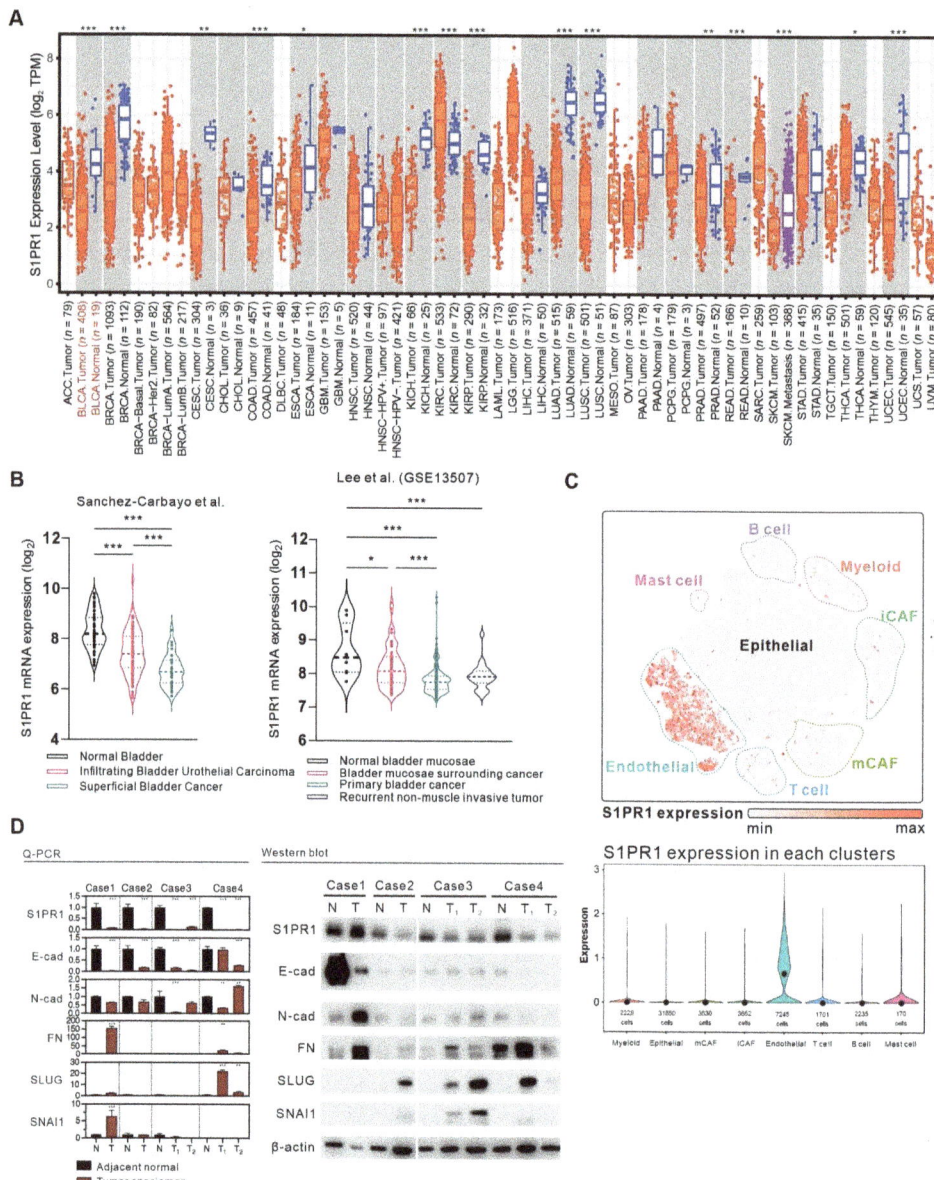

Figure 4. A comprehensive analysis of the differences in *S1PR1* expression in bladder cancer tumors versus normal tissues. (**A**) Differences in *S1PR1* expression in tumors and normal tissues among the 34 cancer types of TCGA adopted from TIMER 2.0 searching "*S1PR1*". (**B**) Differences in *S1PR1* expression in normal versus clinically defined histopathological bladder cancer tissues from the bladder cancer database published by Sanchez-Carbayo et al. and Lee et al [31,32]. (GSE13507). (**C**) Bladder urothelial carcinoma single-cell RNA sequencing database adopted from Chen et al. showed that *S1PR1* was mostly expressed in endothelial cells, followed by immune cells. (**D**) mRNA and protein expression of *S1PR1* and EMT marker in four bladder cancer tumors and adjacent normal tissues. Student's t-test or one-way ANOVA was utilized to analyze the statistical significance of the differences in *S1PR1* expression between groups. * $p < 0.05$, ** $p < 0.01$, *** $p < 0.001$. (N: adjacent normal tissue, T: tumor tissue, T1/2: samples from two separated tumors tissues, E-cad: E-cadherin, N-cad: N-cadherin, FN: Fibronectin).

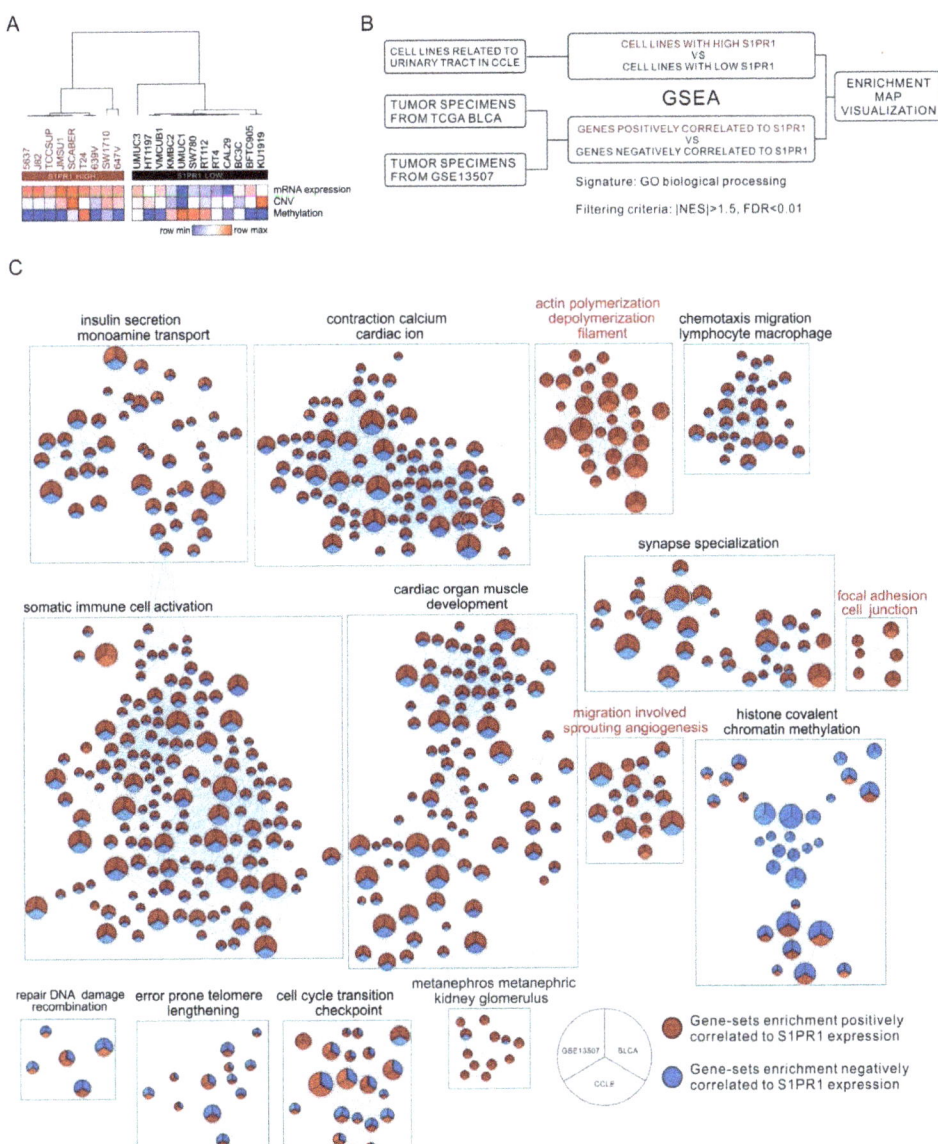

Figure 5. Investigation of *S1PR1* expression in bladder cancer cells for authentic biological response enrichment associations. (**A**) Heat map demonstrating *S1PR1* mRNA expression (fragments per kilobase per million, FPKM), copy number variation (log value), and methylation (β value) of 21 uroepithelial carcinomas in the CCLE database with unsupervised hierarchical clustering. (**B**) Flow chart presenting comparison of *S1PR1* expression and biological response association in CCLE bladder cancer cell lines or clinical bladder cancer tumors (TCGA BLCA, GSE13507). (**C**) Enrichment map visualization showing the GSEA scores of gene-sets significantly enriched in (**B**). The color represents the degree of normalized enrichment score (NES); red means the gene-sets are enriched in high *S1PR1* samples (NES > 1.5) and blue means the gene-sets are enriched in low *S1PR1* samples (NES < −1.5). All presented enriched gene sets have passed the screening criteria of p-value < 0.05, FDR < 0.01. Gene-sets without edge linkage were excluded to increase the ease of visualization of the results.

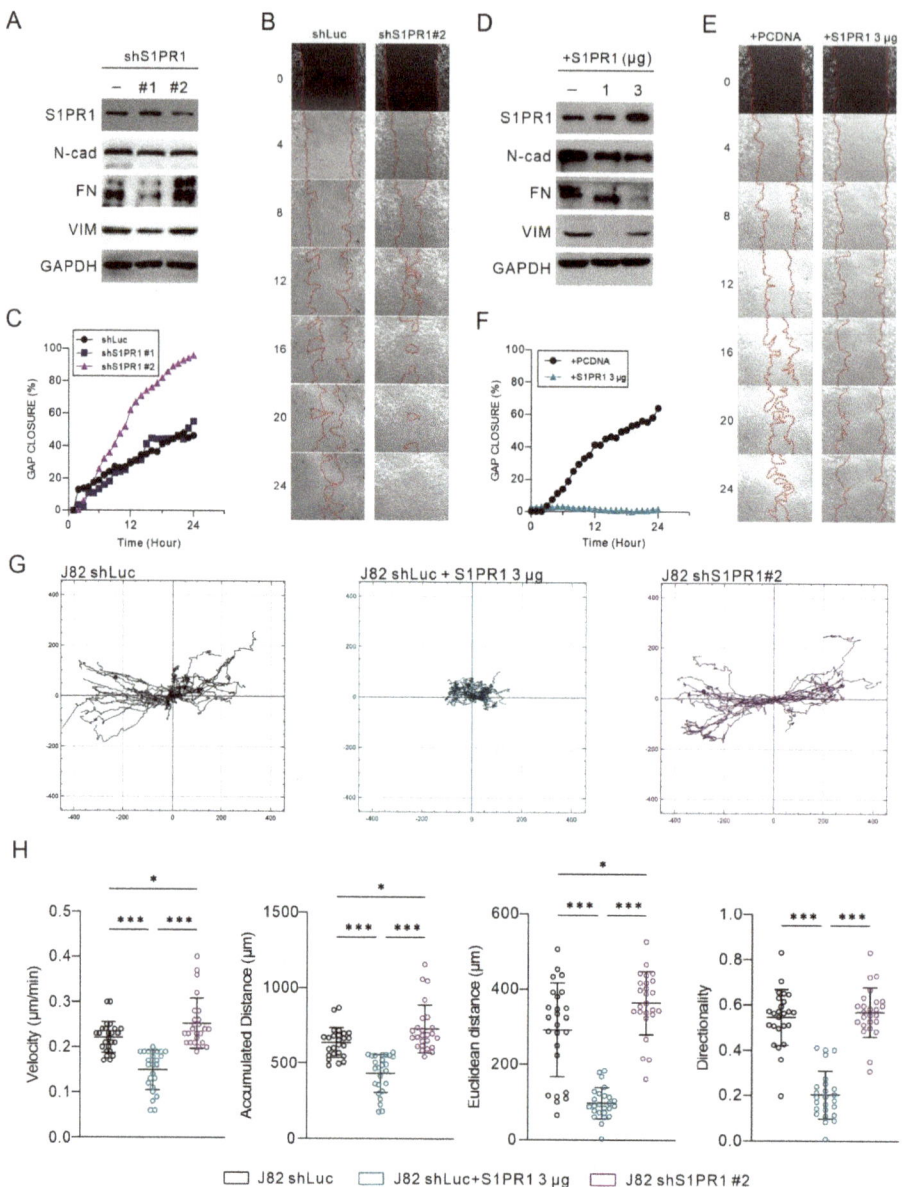

Figure 6. Negative correlation between *S1PR1* expression and bladder cancer cell motility. (**A,D**) Western blot demonstrating the efficacy of *S1PR1* expression manipulation and matching EMT marker expression. (**B,E**) Twenty-four hour live cell image tracking showed the progressions in gap area at multiple time points. (**C,F**) Gap closure (%) presents the effect of manipulating *S1PR1* expression on wound healing. (**G**) Migration tracking plot for J82 shLuc, shLuc + *S1PR1* (3 µg), and J82 sh*S1PR1*#2. Cell migration was tracked for 48 h after the wound healing assay started, with cell positions determined every 30 min. In each panel, the center indicates the starting point. (**H**) Statistical analysis of the J82 tracking cell migration rate, distance, and directionality, with lines showing mean and standard deviation. J82 shLuc (n = 29); J82 + *S1PR1* 3 µg (n = 30); J82 sh*S1PR1*#2 (n = 30). * $p < 0.05$, *** $p < 0.001$. (VIM: vimentin, GAPDH: glyceraldehyde 3 phosphate dehydrogenase) (for uncropped Western Blot images, please refer to Supplementary File S1).

Confirming the successful manipulation of *S1PR1* expression in J82 cells, it was found that *S1PR1* inhibition was associated with the enhancement of EMT (Figure 6A), while *S1PR1* over-expression inhibited EMT (Figure 6D). A further 24 h live cell imaging showed that the rate of gap healing was significantly increased when *S1PR1* expression was inhibited (Figure 6B). In particular, there was a significant upregulation in the clone that significantly inhibited *S1PR1* expression (sh*S1PR1*#2) (Figure 6C). In contrast, the rate of gap healing was significantly reduced upon overexpression of *S1PR1* (3 µg) (Figure 6E,F). As assessed by the cell migration trajectory (Figure 6G), overexpression of *S1PR1* significantly inhibited the migration distance of J82. Further evaluation of velocity, accumulated distance, Euclidean distance, and directionality showed that the over-expression of *S1PR1* significantly inhibited cell mobility, suggesting that *S1PR1* may affect cell movement by modulating cell adhesion (Figure 6H).

3.6. The Administration of FTY-720 Promotes EMT in Bladder Carcinoma

FTY-720 is identified to inhibit cell proliferation and promote apoptosis by regulating *S1PR1*; however, the effect of FTY-720 on EMT in bladder cancer remains unknown. Transwell migration assay showed that the metastatic capacity of J82 increased with increasing FTY-720 treatment dose (Figure 7A), and the total cell coverage area was significantly increased with 2 and 5 µM FTY-720 treatment (Figure 7B). Similarly, FTY720 treatment inhibited E-cadherin expression and promoted mesenchymal marker expression, a phenomenon that was disturbed by the addition of S1P, but not in RT4 cells that did not express *S1PR1* (Figure 7C). In the wound healing assay, treatment with FTY720 promoted gap closure, especially at 48 h, with a significant difference. In contrast, FTY720 treatment in the presence of S1P had no significant effect on gap closure (Figure 7D), suggesting that inhibition of *S1PR1* by FTY720 promoted EMT in bladder cancer cells. Finally, to understand the overall effect of FTY720 treatment on human bladder cancer tumors, we established a patient-derived tumor culture model [29] (Figure 7E). A decrease in E-cadherin and increase in mesenchymal marker due to FTY720 treatment was observed in all four cases (Figure 7F), suggesting that bladder cancer tumors and cell lines respond similarly to FTY-720 treatment. In particular, in the case of PDC#4, the response to FTY720 in adjacent normal tissues and bladder cancer tumors showed an opposite trend, implying that FTY-720 may have divergent responses in different types of tissues or cancers.

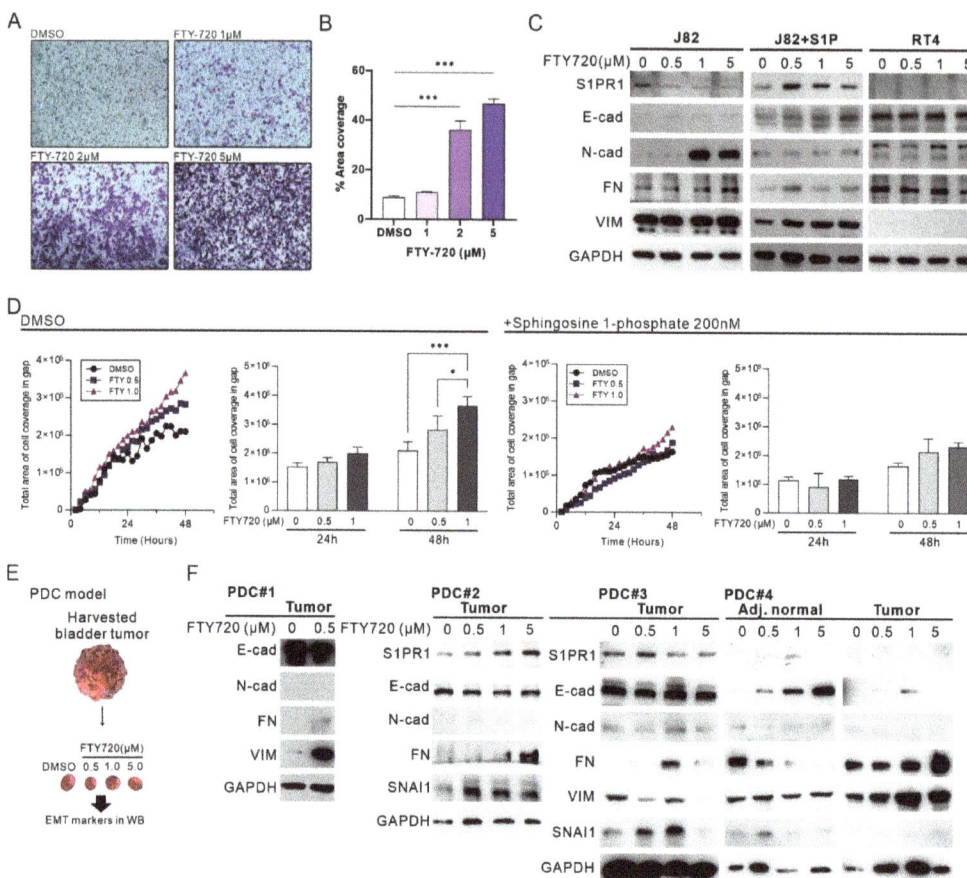

Figure 7. The administration of FTY-720 promotes EMT in bladder carcinoma. (**A**) Transwell migration assay showing the effect of FTY-720 treatment for 24 h (×100). (**B**) Area coverage (%) was calculated by quantifying the ratio of cell coverage per field using ImageJ. (**C**) Western blot showed the expression of *S1PR1*- and EMT-related markers after different dosage treatment of FTY-720 in J82 and RT4. Sphingosine 1-phosphate (200 nM) was added at the same time of FTY-720 treatment. (**D**) J82 cell gap coverage curves treated with FTY-720 in the presence and absence of S1P and statistical analysis of the number of cells in the gap at 24 h and 48 h. The curve data are presented as mean values only to reduce reading interference. (**E**) Illustration of patient-derived tumor culture (PDC) model. (**F**) Western blot showed four sets of clinical bladder cancer tumors with *S1PR1*- and EMT-related marker expression measured 48 h after FTY-720 addition. Bar charts are presented as the mean ± SD based on three independent experiments. * $p < 0.05$, *** $p < 0.001$ (for uncropped Western Blot images, please refer to Supplementary File S1).

4. Discussion

The physiological function and importance of *S1PR1* as a G protein-coupled receptor in vascular endothelial cells have been well established [39–42]. In the immune system, the expression of *S1PR1* is associated with selective in vivo recruitment, egress, and activation of various immune cells [10,43–50]. However, the role of *S1PR1* in cancer remains controversial. Correlation of *S1PR1* with pathological grade in tumors suggests its potential as a prognostic tool for patients with bladder cancer as well as liver and gallbladder cancers [37,51–53]. On the other hand, low *S1PR1* expression is suggested to be linked to poor prognosis in breast and lung cancer [54,55]. Given the complexity of immune cell infiltration and tumor purity, the exact expression and inhibitory effect of *S1PR1* in bulky bladder cancer tumor still needs to be validated.

In this study, we collected six databases of urothelial carcinoma with accompanying survival status and follow-up time for the association between *S1PR1* expression and patient prognosis (Figure 1). Overall, high *S1PR1* expression was associated with poorer prognosis, but the prognosis of patient survival showed a divergent trend in individual databases. Analysis by tumor purity assessment showed that *S1PR1* expression was negatively correlated with tumor purity in most databases (Figure 2A), as observed by Zhong et al. in breast and lung cancer [55]. Further, in silico simulation of immune cell infiltration showed that S1PR1 expression was generally positively correlated with B cells, macrophage, and regulatory T cells, suggesting that assessment of immune cell infiltration may help to clarify the source or function of S1PR1 expression in tumor tissues (Figure 2B) [56,57]. For instance, Liu et al. reported the correlation between *S1PR1* expression in bladder cancer cells convening regulatory T cells and poor prognosis [15]. In addition, S1PR1 expression was positively correlated with neutrophil infiltration in three databases, where *S1PR1* was associated with better prognosis, suggesting that better prognosis may due to higher neutrophil infiltration [58,59], implying that *S1PR1* expression may be susceptible to the degree of immune cell infiltration in the tumor microenvironment.

The TCGA database shows that *S1PR1* expression is significantly higher in normal tissues than in tumor tissues for most cancers (Figure 3A). The work of Sanchez-Carbayo et al. and Lee et al. showed that *S1PR1* expression was reduced at low grade tumor (Figure 3B). Further, the single-cell RNA sequencing results reported by Chen et al. showed that most *S1PR1* expression was from endothelial cells and a few from multiple immune cells, both of which were underrepresented in low grade tumor (Figure 3C). These results suggest that the expression of *S1PR1* might be deeply affected by the purity of tumor cell composition in the lesioned tissue. Samples collected from high-grade tumor cells are more likely to be adulterated with normal tissue, which may explain the association of high *S1PR1* expression with poor prognosis observed in some databases. Nevertheless, whether *S1PR1* expression in tumor cell may affect the generation of tumor-associated endothelial cells requires further investigation. As our analysis shows that *S1PR1* expression is associated with angiogenesis (Figure 3C), it is possible that bladder cancer cells overexpressing *S1PR1* may affect tumor progression by altering microenvironmental angiogenesis [50,56,57]. When *S1PR1* expression was manipulated in J82, it was shown that overexpression of *S1PR1* had an inhibitory effect on bladder cancer cell migration, possibly associated with enhanced cellular apposition, echoing the analysis in Figure 3C. Conversely, shRNA interference with *S1PR1* expression or inhibition of *S1PR1* by FTY-720 accelerated bladder cancer cell migration, and the addition of S1P antagonized the effect of FTY-720. Moreover, a similar phenomenon could be observed in the patient-derived tumor primary culture model, suggesting that clinical inhibition of *S1PR1* may cause accelerated metastasis of bladder cancer cells.

High expression of *S1PR1* was significantly associated with poor prognosis in multiple cancer databases, raising the possibility of its potential role in promoting tumorigenesis. Based on this inference, FTY-720 has been reported and demonstrated to induce apoptosis in a variety of cancer cells, including bladder cancer [18,60]. Moreover, the mechanism of EMT inhibition by FTY-720 in cholangiocarcinoma and glioblastoma has also been proposed [61,62]. However, our study clearly indicates that reduction of *S1PR1* expression by human manipulation may cause a promotion in EMT in bladder cancer cells and patient tumor tissue, a phenomenon consistent with FTY-720 treatment. Different responses to FTY-720 in normal or tumor tissues also indicate divergences in response between various types of tissues or cancers; this may be due to the fact that normal tissue is usually rich in endothelium [63,64]. Although a majority of the literature has confirmed the induction of apoptosis in cancer cells by FTY-720, we still need to pay attention to its potential role in EMT induction. Furthermore, whether the apoptosis induction caused by FTY-720 is related to the EMT-induced anoikis needs to be further clarified. The inhibition of *S1PR1* contributes to the reduced interaction with the ECM, thus allowing a higher migration ability. Cell proliferation may be reduced as a result and apoptosis may occur owing to separation from the matrix [65]. Furthermore, cancer cells may thus develop an anti-anoikis mechanism and become more resistant to chemotherapy or accelerate the progression of metastasis [66–68].

5. Conclusions

In conclusion, in general, high *S1PR1* expression is associated with poor prognosis, but this observation may be interfered with by endothelial or immune cell infiltration, so the accuracy of *S1PR1* expression for clinical diagnosis needs to be further evaluated. *S1PR1* expression promotes cancer cell adhesion and, conversely, inhibition of *S1PR1* by genetic manipulation or FTY-720 may increase bladder cancer cell migration ability. Although it is known that inhibition of *S1PR1* has multiple mechanisms to counteract tumor growth, the resulting risk of metastasis should not be overlooked. Therefore, the use of FTY-720 as a concurrent treatment strategy for bladder cancer requires further evaluation and observation.

Supplementary Materials: The following are available online at https://www.mdpi.com/article/10.3390/cancers13174474/s1, File S1: Uncropped Western blot figures.

Author Contributions: Conceptualization, Y.-L.C. and C.-L.C.; methodology, H.-F.L. and Y.-S.L.; software, H.-F.L. and Y.-S.L.; validation, H.-F.L. and Y.-S.L.; formal analysis, H.-F.L. and Y.-S.L.; investigation, C.-L.C.; resources, C.-L.C., M.-H.Y., C.-W.T. and C.-C.K.; data curation, Y.-L.C.; writing—original draft preparation, Y.-L.C. and C.-L.C.; writing—review and editing, E.M. and S.-T.W.; visualization, Y.-L.C.; supervision, E.M. and S.-T.W.; project administration, Y.-L.C.; funding acquisition, Y.-L.C. and C.-L.C. All authors have read and agreed to the published version of the manuscript.

Funding: This study was funded by the Ministry of National Defense-Medical Affairs Bureau (MND-MAB-110-122 to C.-L.C. and MND-MAB-110-058 to Y.-L.C.) and Ministry of Science and Technology, Taiwan (MOST109-2320-B-016-004 to Y.-L.C.).

Institutional Review Board Statement: The study was conducted according to the guidelines of the Declaration of Helsinki and approved by the Institutional Review Board of Tri-Service General Hospital, National Defense Medical Center (TSGHIRB No.:1-108-05-130, Date of approval: 23 July 2019).

Informed Consent Statement: Informed consent was obtained from all subjects involved in the study.

Data Availability Statement: GSE5287, GSE13507, GSE19915, GSE31684, and GSE32894 whole gene expression and clinical data used in this study were obtained from NCBI Gene Expression Omnibus. TCGA BLCA and CCLE whole gene expression and corresponding data were obtained from UCSC Xena (https://xenabrowser.net/datapages/, accessed on 26 March 2021). The whole gene expression databases of Sanchez-Carbayo et al. and Chen et al. are available from the supplemental data of the authors' publications.

Conflicts of Interest: The authors declare no conflict of interest.

References

1. Lenis, A.T.; Lec, P.M.; Chamie, K.; Mshs, M.D. Bladder Cancer: A Review. *JAMA* **2020**, *324*, 1980–1991. [CrossRef]
2. Ferlay, J.; Colombet, M.; Soerjomataram, I.; Mathers, C.; Parkin, D.M.; Piñeros, M.; Znaor, A.; Bray, F. Estimating the global cancer incidence and mortality in 2018: GLOBOCAN sources and methods. *Int. J. Cancer* **2019**, *144*, 1941–1953. [CrossRef]
3. Richters, A.; Aben, K.K.H.; Kiemeney, L. The global burden of urinary bladder cancer: An update. *World J. Urol.* **2020**, *38*, 1895–1904. [CrossRef] [PubMed]
4. Witjes, J.A.; Comperat, E.; Cowan, N.C.; De Santis, M.; Gakis, G.; Lebret, T.; Ribal, M.J.; Van der Heijden, A.G.; Sherif, A. EAU guidelines on muscle-invasive and metastatic bladder cancer: Summary of the 2013 guidelines. *Eur. Urol.* **2014**, *65*, 778–792. [CrossRef]
5. Brausi, M.; Witjes, J.A.; Lamm, D.; Persad, R.; Palou, J.; Colombel, M.; Buckley, R.; Soloway, M.; Akaza, H.; Bohle, A. A review of current guidelines and best practice recommendations for the management of nonmuscle invasive bladder cancer by the International Bladder Cancer Group. *J. Urol.* **2011**, *186*, 2158–2167. [CrossRef]
6. Cartier, A.; Hla, T. Sphingosine 1-phosphate: Lipid signaling in pathology and therapy. *Science* **2019**, *366*. [CrossRef]

7. Riese, J.; Gromann, A.; Luhrs, F.; Kleinwort, A.; Schulze, T. Sphingosine-1-Phosphate Receptor Type 4 (S1P4) Is Differentially Regulated in Peritoneal B1 B Cells upon TLR4 Stimulation and Facilitates the Egress of Peritoneal B1a B Cells and Subsequent Accumulation of Splenic IRA B Cells under Inflammatory Conditions. *Int. J. Mol. Sci.* **2021**, *22*, 3465. [CrossRef] [PubMed]
8. Abarca-Zabalia, J.; Garcia, M.I.; Lozano Ros, A.; Marin-Jimenez, I.; Martinez-Gines, M.L.; Lopez-Cauce, B.; Martin-Barbero, M.L.; Salvador-Martin, S.; Sanjurjo-Saez, M.; Garcia-Dominguez, J.M.; et al. Differential Expression of SMAD Genes and *S1PR1* on Circulating CD4+ T Cells in Multiple Sclerosis and Crohn's Disease. *Int. J. Mol. Sci.* **2020**, *21*, 676. [CrossRef] [PubMed]
9. Allende, M.L.; Tuymetova, G.; Lee, B.G.; Bonifacino, E.; Wu, Y.P.; Proia, R.L. S1P1 receptor directs the release of immature B cells from bone marrow into blood. *J. Exp. Med.* **2010**, *207*, 1113–1124. [CrossRef] [PubMed]
10. Benechet, A.P.; Menon, M.; Xu, D.; Samji, T.; Maher, L.; Murooka, T.T.; Mempel, T.R.; Sheridan, B.S.; Lemoine, F.M.; Khanna, K.M. T cell-intrinsic *S1PR1* regulates endogenous effector T-cell egress dynamics from lymph nodes during infection. *Proc. Natl. Acad. Sci. USA* **2016**, *113*, 2182–2187. [CrossRef]
11. Gonzalez, L.; Qian, A.S.; Tahir, U.; Yu, P.; Trigatti, B.L. Sphingosine-1-Phosphate Receptor 1, Expressed in Myeloid Cells, Slows Diet-Induced Atherosclerosis and Protects against Macrophage Apoptosis in Ldlr KO Mice. *Int. J. Mol. Sci.* **2017**, *18*, 2721. [CrossRef] [PubMed]
12. Bryan, A.M.; Del Poeta, M. Sphingosine-1-phosphate receptors and innate immunity. *Cell Microbiol.* **2018**, *20*, e12836. [CrossRef] [PubMed]
13. Go, H.; Kim, P.J.; Jeon, Y.K.; Cho, Y.M.; Kim, K.; Park, B.H.; Ku, J.Y. Sphingosine-1-phosphate receptor 1 (*S1PR1*) expression in non-muscle invasive urothelial carcinoma: Association with poor clinical outcome and potential therapeutic target. *Eur. J. Cancer* **2015**, *51*, 1937–1945. [CrossRef]
14. Patmanathan, S.N.; Wang, W.; Yap, L.F.; Herr, D.R.; Paterson, I.C. Mechanisms of sphingosine 1-phosphate receptor signalling in cancer. *Cell Signal.* **2017**, *34*, 66–75. [CrossRef] [PubMed]
15. Liu, Y.N.; Zhang, H.; Zhang, L.; Cai, T.T.; Huang, D.J.; He, J.; Ni, H.H.; Zhou, F.J.; Zhang, X.S.; Li, J. Sphingosine 1 phosphate receptor-1 (S1P1) promotes tumor-associated regulatory T cell expansion: Leading to poor survival in bladder cancer. *Cell Death Dis.* **2019**, *10*, 50. [CrossRef] [PubMed]
16. Chun, J.; Hartung, H.P. Mechanism of action of oral fingolimod (FTY720) in multiple sclerosis. *Clin. Neuropharmacol.* **2010**, *33*, 91–101. [CrossRef] [PubMed]
17. Chiba, K. FTY720, a new class of immunomodulator, inhibits lymphocyte egress from secondary lymphoid tissues and thymus by agonistic activity at sphingosine 1-phosphate receptors. *Pharmacol. Ther.* **2005**, *108*, 308–319. [CrossRef]
18. Azuma, H.; Takahara, S.; Horie, S.; Muto, S.; Otsuki, Y.; Katsuoka, Y. Induction of apoptosis in human bladder cancer cells in vitro and in vivo caused by FTY720 treatment. *J. Urol.* **2003**, *169*, 2372–2377. [CrossRef]
19. Ogłuszka, M.; Orzechowska, M.; Jędroszka, D.; Witas, P.; Bednarek, A.K. Evaluate Cutpoints: Adaptable continuous data distribution system for determining survival in Kaplan-Meier estimator. *Comput. Methods Programs Biomed.* **2019**, *177*, 133–139. [CrossRef]
20. Li, T.; Fu, J.; Zeng, Z.; Cohen, D.; Li, J.; Chen, Q.; Li, B.; Liu, X.S. TIMER2.0 for analysis of tumor-infiltrating immune cells. *Nucleic Acids Res.* **2020**, *48*, W509–W514. [CrossRef]
21. Chen, Z.; Zhou, L.; Liu, L.; Hou, Y.; Xiong, M.; Yang, Y.; Hu, J.; Chen, K. Single-cell RNA sequencing highlights the role of inflammatory cancer-associated fibroblasts in bladder urothelial carcinoma. *Nat. Commun.* **2020**, *11*, 5077. [CrossRef] [PubMed]
22. Goldman, M.J.; Craft, B.; Hastie, M.; Repecka, K.; McDade, F.; Kamath, A.; Banerjee, A.; Luo, Y.; Rogers, D.; Brooks, A.N.; et al. Visualizing and interpreting cancer genomics data via the Xena platform. *Nat. Biotechnol.* **2020**, *38*, 675–678. [CrossRef] [PubMed]
23. Broadinstitute. Morpheus. Available online: https://software.broadinstitute.org/morpheus (accessed on 20 August 2021).
24. Yoshihara, K.; Shahmoradgoli, M.; Martínez, E.; Vegesna, R.; Kim, H.; Torres-Garcia, W.; Treviño, V.; Shen, H.; Laird, P.W.; Levine, D.A.; et al. Inferring tumour purity and stromal and immune cell admixture from expression data. *Nat. Commun.* **2013**, *4*, 2612. [CrossRef] [PubMed]
25. Finotello, F.; Mayer, C.; Plattner, C.; Laschober, G.; Rieder, D.; Hackl, H.; Krogsdam, A.; Loncova, Z.; Posch, W.; Wilflingseder, D.; et al. Molecular and pharmacological modulators of the tumor immune contexture revealed by deconvolution of RNA-seq data. *Genome Med.* **2019**, *11*, 34. [CrossRef]
26. Sturm, G.; Finotello, F.; List, M. Immunedeconv: An R Package for Unified Access to Computational Methods for Estimating Immune Cell Fractions from Bulk RNA-Sequencing Data. *Methods Mol. Biol.* **2020**, *2120*, 223–232. [CrossRef]
27. Wu, Y.Y.; Wang, S.H.; Wu, C.H.; Yen, L.C.; Lai, H.F.; Ho, C.L.; Chiu, Y.L. In silico immune infiltration profiling combined with functional enrichment analysis reveals a potential role for naive B cells as a trigger for severe immune responses in the lungs of COVID-19 patients. *PLoS ONE* **2020**, *15*, e0242900. [CrossRef]
28. Su, H.C.; Wu, S.C.; Yen, L.C.; Chiao, L.K.; Wang, J.K.; Chiu, Y.L.; Ho, C.L.; Huang, S.M. Gene expression profiling identifies the role of Zac1 in cervical cancer metastasis. *Sci. Rep.* **2020**, *10*, 11837. [CrossRef]
29. van de Merbel, A.F.; van der Horst, G.; van der Mark, M.H.; van Uhm, J.I.M.; van Gennep, E.J.; Kloen, P.; Beimers, L.; Pelger, R.C.M.; van der Pluijm, G. An ex vivo Tissue Culture Model for the Assessment of Individualized Drug Responses in Prostate and Bladder Cancer. *Front. Oncol.* **2018**, *8*, 400. [CrossRef]
30. Als, A.B.; Dyrskjot, L.; von der Maase, H.; Koed, K.; Mansilla, F.; Toldbod, H.E.; Jensen, J.L.; Ulhoi, B.P.; Sengelov, L.; Jensen, K.M.; et al. Emmprin and survivin predict response and survival following cisplatin-containing chemotherapy in patients with advanced bladder cancer. *Clin. Cancer Res.* **2007**, *13*, 4407–4414. [CrossRef]

31. Choi, W.; Porten, S.; Kim, S.; Willis, D.; Plimack, E.R.; Hoffman-Censits, J.; Roth, B.; Cheng, T.; Tran, M.; Lee, I.L.; et al. Identification of distinct basal and luminal subtypes of muscle-invasive bladder cancer with different sensitivities to frontline chemotherapy. *Cancer Cell* **2014**, *25*, 152–165. [CrossRef] [PubMed]
32. Kim, W.J.; Kim, E.J.; Kim, S.K.; Kim, Y.J.; Ha, Y.S.; Jeong, P.; Kim, M.J.; Yun, S.J.; Lee, K.M.; Moon, S.K.; et al. Predictive value of progression-related gene classifier in primary non-muscle invasive bladder cancer. *Mol. Cancer* **2010**, *9*, 3. [CrossRef] [PubMed]
33. Riester, M.; Taylor, J.M.; Feifer, A.; Koppie, T.; Rosenberg, J.E.; Downey, R.J.; Bochner, B.H.; Michor, F. Combination of a novel gene expression signature with a clinical nomogram improves the prediction of survival in high-risk bladder cancer. *Clin. Cancer Res.* **2012**, *18*, 1323–1333. [CrossRef]
34. Sjodahl, G.; Lauss, M.; Lovgren, K.; Chebil, G.; Gudjonsson, S.; Veerla, S.; Patschan, O.; Aine, M.; Ferno, M.; Ringner, M.; et al. A molecular taxonomy for urothelial carcinoma. *Clin. Cancer Res.* **2012**, *18*, 3377–3386. [CrossRef]
35. Sanchez-Carbayo, M.; Socci, N.D.; Lozano, J.; Saint, F.; Cordon-Cardo, C. Defining molecular profiles of poor outcome in patients with invasive bladder cancer using oligonucleotide microarrays. *J. Clin. Oncol.* **2006**, *24*, 778–789. [CrossRef]
36. Lee, J.S.; Leem, S.H.; Lee, S.Y.; Kim, S.C.; Park, E.S.; Kim, S.B.; Kim, S.K.; Kim, Y.J.; Kim, W.J.; Chu, I.S. Expression signature of E2F1 and its associated genes predict superficial to invasive progression of bladder tumors. *J. Clin. Oncol.* **2010**, *28*, 2660–2667. [CrossRef] [PubMed]
37. Yokota, T.; Nojima, H.; Kuboki, S.; Yoshitomi, H.; Furukawa, K.; Takayashiki, T.; Takano, S.; Ohtsuka, M. Sphingosine-1-phosphate Receptor-1 Promotes Vascular Invasion and EMT in Hepatocellular Carcinoma. *J. Surg. Res.* **2021**, *259*, 200–210. [CrossRef] [PubMed]
38. Wang, W.; Wang, A.; Luo, G.; Ma, F.; Wei, X.; Bi, Y. S1P1 receptor inhibits kidney epithelial mesenchymal transition triggered by ischemia/reperfusion injury via the PI3K/Akt pathway. *Acta Biochim. Biophys. Sin.* **2018**, *50*, 651–657. [CrossRef] [PubMed]
39. Masana, M.I.; Brown, R.C.; Pu, H.; Gurney, M.E.; Dubocovich, M.L. Cloning and characterization of a new member of the G-protein coupled receptor EDG family. *Recept. Channels* **1995**, *3*, 255–262.
40. Zondag, G.C.; Postma, F.R.; Etten, I.V.; Verlaan, I.; Moolenaar, W.H. Sphingosine 1-phosphate signalling through the G-protein-coupled receptor Edg-1. *Biochem. J.* **1998**, *330 Pt 2*, 605–609. [CrossRef]
41. Lee, M.J.; Thangada, S.; Claffey, K.P.; Ancellin, N.; Liu, C.H.; Kluk, M.; Volpi, M.; Sha'afi, R.I.; Hla, T. Vascular endothelial cell adherens junction assembly and morphogenesis induced by sphingosine-1-phosphate. *Cell* **1999**, *99*, 301–312. [CrossRef]
42. Liu, Y.; Wada, R.; Yamashita, T.; Mi, Y.; Deng, C.X.; Hobson, J.P.; Rosenfeldt, H.M.; Nava, V.E.; Chae, S.S.; Lee, M.J.; et al. Edg-1, the G protein-coupled receptor for sphingosine-1-phosphate, is essential for vascular maturation. *J. Clin. Investig.* **2000**, *106*, 951–961. [CrossRef] [PubMed]
43. Aoki, M.; Aoki, H.; Ramanathan, R.; Hait, N.C.; Takabe, K. Sphingosine-1-Phosphate Signaling in Immune Cells and Inflammation: Roles and Therapeutic Potential. *Mediat. Inflamm.* **2016**, *2016*, 8606878. [CrossRef]
44. Chi, H.; Flavell, R.A. Cutting edge: Regulation of T cell trafficking and primary immune responses by sphingosine 1-phosphate receptor 1. *J. Immunol.* **2005**, *174*, 2485–2488. [CrossRef]
45. Roviezzo, F.; Del Galdo, F.; Abbate, G.; Bucci, M.; D'Agostino, B.; Antunes, E.; De Dominicis, G.; Parente, L.; Rossi, F.; Cirino, G.; et al. Human eosinophil chemotaxis and selective in vivo recruitment by sphingosine 1-phosphate. *Proc. Natl. Acad. Sci. USA* **2004**, *101*, 11170–11175. [CrossRef]
46. Chu, P.; Pardo, J.; Zhao, H.; Li, C.C.; Pali, E.; Shen, M.M.; Qu, K.; Yu, S.X.; Huang, B.C.; Yu, P.; et al. Systematic identification of regulatory proteins critical for T-cell activation. *J. Biol.* **2003**, *2*, 21. [CrossRef]
47. Dorsam, G.; Graeler, M.H.; Seroogy, C.; Kong, Y.; Voice, J.K.; Goetzl, E.J. Transduction of multiple effects of sphingosine 1-phosphate (S1P) on T cell functions by the S1P1 G protein-coupled receptor. *J. Immunol.* **2003**, *171*, 3500–3507. [CrossRef] [PubMed]
48. Xiong, Y.; Piao, W.; Brinkman, C.C.; Li, L.; Kulinski, J.M.; Olivera, A.; Cartier, A.; Hla, T.; Hippen, K.L.; Blazar, B.R.; et al. CD4 T cell sphingosine 1-phosphate receptor (S1PR)1 and S1PR4 and endothelial S1PR2 regulate afferent lymphatic migration. *Sci. Immunol.* **2019**, *4*. [CrossRef] [PubMed]
49. Resop, R.S.; Douaisi, M.; Craft, J.; Jachimowski, L.C.; Blom, B.; Uittenbogaart, C.H. Sphingosine-1-phosphate/sphingosine-1-phosphate receptor 1 signaling is required for migration of naive human T cells from the thymus to the periphery. *J. Allergy Clin. Immunol.* **2016**, *138*, 551–557. [CrossRef] [PubMed]
50. Sassoli, C.; Pierucci, F.; Tani, A.; Frati, A.; Chellini, F.; Matteini, F.; Vestri, A.; Anderloni, G.; Nosi, D.; Zecchi-Orlandini, S.; et al. Sphingosine 1-Phosphate Receptor 1 Is Required for MMP-2 Function in Bone Marrow Mesenchymal Stromal Cells: Implications for Cytoskeleton Assembly and Proliferation. *Stem Cells Int.* **2018**, *2018*, 5034679. [CrossRef]
51. Palangi, A.; Shakhssalim, N.; Parvin, M.; Bayat, S.; Allameh, A. Differential expression of S1P receptor subtypes in human bladder transitional cell carcinoma. *Clin. Transl. Oncol.* **2019**, *21*, 1240–1249. [CrossRef] [PubMed]
52. Lin, Q.; Wei, Y.; Zhong, Y.; Zhu, D.; Ren, L.; Xu, P.; Zheng, P.; Feng, Q.; Ji, M.; Lv, M.; et al. Aberrant expression of sphingosine-1-phosphate receptor 1 correlates with metachronous liver metastasis and poor prognosis in colorectal cancer. *Tumour Biol.* **2014**, *35*, 9743–9750. [CrossRef]
53. Yuan, L.W.; Liu, D.C.; Yang, Z.L. Correlation of S1P1 and ERp29 expression to progression, metastasis, and poor prognosis of gallbladder adenocarcinoma. *Hepatobiliary Pancreat. Dis. Int.* **2013**, *12*, 189–195. [CrossRef]
54. Liu, S.; Ni, C.; Zhang, D.; Sun, H.; Dong, X.; Che, N.; Liang, X.; Chen, C.; Liu, F.; Bai, J.; et al. S1PR1 regulates the switch of two angiogenic modes by VE-cadherin phosphorylation in breast cancer. *Cell Death Dis.* **2019**, *10*, 200. [CrossRef]

55. Zhong, L.; Xie, L.; Yang, Z.; Li, L.; Song, S.; Cao, D.; Liu, Y. Prognostic value of *S1PR1* and its correlation with immune infiltrates in breast and lung cancers. *BMC Cancer* **2020**, *20*, 766. [CrossRef] [PubMed]
56. Syed, S.N.; Raue, R.; Weigert, A.; von Knethen, A.; Brune, B. Macrophage *S1PR1* Signaling Alters Angiogenesis and Lymphangiogenesis During Skin Inflammation. *Cells* **2019**, *8*, 785. [CrossRef] [PubMed]
57. Weichand, B.; Popp, R.; Dziumbla, S.; Mora, J.; Strack, E.; Elwakeel, E.; Frank, A.C.; Scholich, K.; Pierre, S.; Syed, S.N.; et al. *S1PR1* on tumor-associated macrophages promotes lymphangiogenesis and metastasis via NLRP3/IL-1beta. *J. Exp. Med.* **2017**, *214*, 2695–2713. [CrossRef] [PubMed]
58. Zhou, L.; Xu, L.; Chen, L.; Fu, Q.; Liu, Z.; Chang, Y.; Lin, Z.; Xu, J. Tumor-infiltrating neutrophils predict benefit from adjuvant chemotherapy in patients with muscle invasive bladder cancer. *Oncoimmunology* **2017**, *6*, e1293211. [CrossRef]
59. Mandelli, G.F.; Missale, F.; Bresciani, D.; Gatta, L.B.; Scapini, P.; Caveggion, E.; Roca, E.; Bugatti, M.; Monti, M.; Cristinelli, L.; et al. Tumor Infiltrating Neutrophils Are Enriched in Basal-Type Urothelial Bladder Cancer. *Cells* **2020**, *9*, 291. [CrossRef] [PubMed]
60. Ubai, T.; Azuma, H.; Kotake, Y.; Inamoto, T.; Takahara, K.; Ito, Y.; Kiyama, S.; Sakamoto, T.; Horie, S.; Muto, S.; et al. FTY720 induced Bcl-associated and Fas-independent apoptosis in human renal cancer cells in vitro and significantly reduced in vivo tumor growth in mouse xenograft. *Anticancer Res.* **2007**, *27*, 75–88.
61. Zhang, L.; Wang, H.; Zhu, J.; Ding, K.; Xu, J. FTY720 reduces migration and invasion of human glioblastoma cell lines via inhibiting the PI3K/AKT/mTOR/p70S6K signaling pathway. *Tumour Biol.* **2014**, *35*, 10707–10714. [CrossRef] [PubMed]
62. Lu, Z.; Wang, J.; Zheng, T.; Liang, Y.; Yin, D.; Song, R.; Pei, T.; Pan, S.; Jiang, H.; Liu, L. FTY720 inhibits proliferation and epithelial-mesenchymal transition in cholangiocarcinoma by inactivating STAT3 signaling. *BMC Cancer* **2014**, *14*, 783. [CrossRef]
63. Nair, A.; Ingram, N.; Verghese, E.T.; Wijetunga, I.; Markham, A.F.; Wyatt, J.; Prasad, K.R.; Coletta, P.L. CD105 is a prognostic marker and valid endothelial target for microbubble platforms in cholangiocarcinoma. *Cell Oncol.* **2020**, *43*, 835–845. [CrossRef]
64. Mei, X.; Chen, Y.S.; Chen, F.R.; Xi, S.Y.; Chen, Z.P. Glioblastoma stem cell differentiation into endothelial cells evidenced through live-cell imaging. *Neuro. Oncol.* **2017**, *19*, 1109–1118. [CrossRef] [PubMed]
65. Liu, Y.; Zhi, Y.; Song, H.; Zong, M.; Yi, J.; Mao, G.; Chen, L.; Huang, G. *S1PR1* promotes proliferation and inhibits apoptosis of esophageal squamous cell carcinoma through activating STAT3 pathway. *J. Exp. Clin. Cancer Res.* **2019**, *38*, 369. [CrossRef] [PubMed]
66. Paoli, P.; Giannoni, E.; Chiarugi, P. Anoikis molecular pathways and its role in cancer progression. *Biochim. Biophys Acta* **2013**, *1833*, 3481–3498. [CrossRef] [PubMed]
67. Zhang, N.; Dai, L.; Qi, Y.; Di, W.; Xia, P. Combination of FTY720 with cisplatin exhibits antagonistic effects in ovarian cancer cells: Role of autophagy. *Int. J. Oncol.* **2013**, *42*, 2053–2059. [CrossRef]
68. Kim, Y.N.; Koo, K.H.; Sung, J.Y.; Yun, U.J.; Kim, H. Anoikis resistance: An essential prerequisite for tumor metastasis. *Int. J. Cell Biol.* **2012**, *2012*, 306879. [CrossRef] [PubMed]

Article

Serum Epiplakin Might Be a Potential Serodiagnostic Biomarker for Bladder Cancer

Soichiro Shimura [1], Kazumasa Matsumoto [1,*], Yuriko Shimizu [1], Kohei Mochizuki [2], Yutaka Shiono [1], Shuhei Hirano [1], Dai Koguchi [1], Masaomi Ikeda [1], Yuichi Sato [1] and Masatsugu Iwamura [1]

[1] Department of Urology, Kitasato University School of Medicine, Sagamihara Campus, 1-15-1 Kitasato, Sagamihara 252-0374, Kanagawa, Japan; sou-telb89@live.jp (S.S.); yulico@med.kitasato-u.ac.jp (Y.S.); shiono-y@med.kitasato-u.ac.jp (Y.S.); s.hirano@med.kitasato-u.ac.jp (S.H.); dai.k@med.kitasato-u.ac.jp (D.K.); ikeda.masaomi@grape.plala.or.jp (M.I.); yuichi@med.kitasato-u.ac.jp (Y.S.); iwamura@med.kitasato-u.ac.jp (M.I.)
[2] International University of Health and Welfare Atami Hospital, 13-1 Higashikaigan Town, Atami 413-0012, Shizuoka, Japan; kmochi14338@iuhw.ac.jp
* Correspondence: kazumasa@cd5.so-net.ne.jp; Tel.: +81-427-78-9091

Simple Summary: Improving early diagnosis and long-term postoperative monitoring of bladder cancer has become a focus of international research. In the present study, we evaluated the epiplakin expression levels in sera from patients with bladder cancer via a micro-dot blot array. Serum epiplakin levels were significantly higher in patients with bladder cancer than in those with stone disease and in healthy volunteers. Furthermore, serum epiplakin levels did not differ between patients with non-muscle-invasive and muscle-invasive bladder cancer. Immunohistochemistry revealed no association between staining scores, clinicopathological findings, and patients' outcomes. In summary, our findings showed that serum epiplakin might be a potential diagnostic biomarker for patients with bladder cancer.

Abstract: Tumor markers that can be detected at an early stage are needed. Here, we evaluated the epiplakin expression levels in sera from patients with bladder cancer (BC). Using a micro-dot blot array, we evaluated epiplakin expression levels in 60 patients with BC, 20 patients with stone disease, and 28 healthy volunteers. The area under the curve (AUC) and best cut-off point were calculated using receiver-operating characteristic (ROC) analysis. Serum epiplakin levels were significantly higher in patients with BC than in those with stone disease ($p = 0.0013$) and in healthy volunteers ($p < 0.0001$). The AUC-ROC level for BC was 0.78 (95% confidence interval (CI) = 0.69–0.87). Using a cut-off point of 873, epiplakin expression levels exhibited 68.3% sensitivity and 79.2% specificity for BC. However, the serum epiplakin levels did not significantly differ by sex, age, pathological stage and grade, or urine cytology. We performed immunohistochemical staining using the same antibody on another cohort of 127 patients who underwent radical cystectomy. Univariate and multivariate analysis results showed no significant differences between epiplakin expression, clinicopathological findings, and patient prognoses. Our results showed that serum epiplakin might be a potential serodiagnostic biomarker in patients with BC.

Keywords: bladder cancer; radical cystectomy; epiplakin; diagnosis

1. Introduction

Bladder cancer (BC) is one of the most common genitourinary tumors [1]. At initial diagnosis, most patients have non-muscle-invasive bladder cancer (NMIBC) and are treated with transurethral resection of the bladder tumor (TURBT) [2]. Up to 70% of NMIBC patients eventually relapse, and 10–20% experience disease progression to muscle-invasive bladder cancer (MIBC) [3]. Cystoscopy and urine cytology are typical modalities for diagnosing and surveilling BC. Cystoscopy helps detect tumor lesions but is painful and invasive even if flexible. Although urine cytology is less invasive, one of its limitations is low sensitivity [4]. Improving early diagnosis and long-term postoperative monitoring of BC have become a focus of international research. Some tumor markers (e.g., BTA and NMP22) may be useful for BC [5,6]. NMP22 is reported to have 32–92% sensitivity and 51–94% specificity; BTA is reported to have 51–94% sensitivity and 53–89% specificity. However, the clinical effectiveness of these markers is modest.

Specific genes have been investigated and evaluated in BC tissues, and we previously reported some BC-related proteins [7]. A comprehensive study of high molecular mass (HMM) protein expression in bladder cancer tissue was investigated by agarose two-dimensional gel electrophoresis followed by the analysis of liquid chromatography tandem mass spectroscopy. As a result of a literature search on the association between proteins' expression and bladder cancer outcomes, eight proteins, including epiplakin, were selected for new biomarkers. These proteins were not previously reported in terms of any relation of bladder cancer. One candidate protein was epiplakin. Epiplakin is a 552-kDa protein originally identified as an autoantigen in the serum of a patient with a subepidermal blistering disease [8]. Epiplakin is involved in wound healing and mechanical skin strengthening [9,10] and is reported in the UALCAN database to be expressed in normal bladder tissue [11]. However, there are few articles that have shown a link between epiplakin and cancer, including BC.

Here, we assessed whether the dynamics of serum epiplakin could be used to diagnose BC and predict the outcomes in patients with BC. We conducted this study to investigate the circulating levels of epiplakin in sera from patients with BC, patients with stone disease, and healthy volunteers. We also assessed whether the serum epiplakin and immunohistochemical staining of surgical specimens would be associated with clinicopathological findings and patient prognoses.

2. Materials and Methods

2.1. Patients

We retrospectively analyzed 60 patients with BC who were treated at Kitasato University Hospital from March 2010 to September 2014. The study group comprised 46 men (77%) and 14 women (23%), with a median age of 74.5 years (range, 29–88 years). Fifty-five patients were treated via transurethral resection (TUR). Of the patients initially treated with TUR, 20 subsequently underwent radical cystectomy, and 4 underwent partial cystectomy. The five patients who did not undergo TUR were initially treated with radical cystectomy and bilateral pelvic lymphadenectomy.

Serum epiplakin levels were measured preoperatively. No anticoagulant was used in the measurement of serum epiplakin. Surgery was the initial treatment in patients with BC. Blood tests, chest X-rays, and computed tomography or magnetic resonance imaging were routinely performed, and no patients had distant metastases. Tumor-node-metastasis (TNM) staging was based on the 2002 TNM Classification of the International Union for Cancer Control and American Joint Committee on Cancer Guidelines [12]. Tumor grading was assessed according to the 1998 World Health Organization/International Society of Urologic Pathology consensus [13]. The median follow-up time was 51.3 months (mean: 46.0 months; range: 3.3–92.8 months) for those patients still alive at the last follow-up. No patients had previous radiation or systemic chemotherapy before surgical treatment, and none had histories of other cancers, skin diseases or pulmonary diseases.

We also measured serum epiplakin levels in 20 patients with stone disease and 28 healthy volunteers. The ethics committee of Kitasato University School of Medicine and Hospital (B17-010, B18-149) approved the study. All participants were treated as per the approved ethical guidelines. Patients could refuse entry or discontinue participation at any time.

2.2. Measurement of Serum Epiplakin

Patient and control sera were kept at −80 °C until use. Monoclonal antibodies specific for epiplakin were gifted from Drs. Tsuchisaka and Hashimoto [14]. Serum epiplakin was measured using reverse-phase protein array (RPPA) analysis [15,16]. Serum epiplakin levels were detected using an automated micro-dot blot array spotBot3 (Arrayit Corp., Sunnyvale, CA, USA). Serum samples were diluted 1:800 with 0.01% Triton X-100/phosphate-buffered saline (PBS) without bivalent ions and spotted onto high-density amino-group-induced glass slides with dimethyl sulfoxide (SDM0011; Matsunami Glass Ind., Ltd., Osaka, Japan). The glass slides were then blocked with 0.5% casein sodium (Wako Pure Chemical Industries, Osaka, Japan) for 1 h at room temperature (RT), then reacted with 400 times diluted rabbit anti-epiplakin antibody with 0.5% casein sodium for 2 h at RT. After being washed with 0.01% Triton X-100/PBS, the slides were incubated with biotinylated anti-rabbit IgG diluted 1:100 (BA-1000, Vector Laboratories, Burlingame, CA, USA) for 1 h at RT and diluted 1:1000 with streptavidin-horseradish peroxidase conjugate (GE Healthcare Bio-Sciences, Pittsburgh, PA, USA) for 30 min at RT. Peroxidase activity was detected using the Tyramide Signal Amplification Cyanine 5 System (PerkinElmer Life Sciences, Boston, MA, USA) diluted 1:100 for 20 min at RT. The slides were counterstained with Alexa Fluor 546-labeled goat anti-human IgG diluted 1:2000 (Life Technologies, Carlsbad, CA, USA) for 5 min at RT. Finally, the stained slides were scanned on a microarray scanner (Genepix 4000B; Molecular Devices, Sunnyvale, CA, USA). The fluorescence intensity, defined as the median net value of quadruple samples, was determined using the Genepix pro 6.0 software package (Molecular Devices).

Data were analyzed using DotBlotChip System software, version 4.0 (Dynacom Co., Ltd., Chiba, Japan). Normalized signals are presented as the positive intensity minus background intensity around the spot.

2.3. Immunohistochemistry and Scoring

We performed immunohistochemistry for radical cystectomized tissues using the same antibody in 127 consecutive cases at Kitasato University Hospital from October 1995 to June 2015. Paraffin-embedded 3 μm-thick sections of the harvested samples were deparaffinized in xylene, rehydrated in a descending ethanol series, and treated with 3% hydrogen peroxide for 15 min. After blocking with Protein Block Serum-Free (Agilent Technologies, Santa Clara, CA, USA) for 10 min, the sections were reacted with anti-epiplakin monoclonal antibody diluted 1:200 at RT for 1 h. After rinsing three times in Tris-buffered saline for 5 min each, the sections were incubated with Histfine Simple Stain MAX Peroxidase (Nichirei, Tokyo, Japan) at RT for 30 min. The sections were subsequently stained with stable DAB solution (Agilent Technologies) and counterstained with Mayer's hematoxylin.

Immunohistochemistry was evaluated semiquantitatively by incorporating both the staining intensity and percentage of positive tumor cells (labeling frequency). The percentages of positive cells were scored as 0 for 0%, 1+ for 1–25%, 2+ for 26–50%, 3+ for 51–75%, or 4+ for 76–100%. The staining intensity was also scored as 1+ (weakly positive), 2+ (moderately positive), or 3+ (strongly positive). The multiply index was obtained by totaling the intensity and percentage scores. Epiplakin expression scores were stratified further as low ($\leq$6) or high (>6) for the prognostic analyses. Two investigators (S.S. and Yuichi Sato) who were blinded to the clinical and pathological data reviewed all slides. Discordant cases were reviewed and discussed until a consensus was reached.

2.4. Statistical Analyses

The serum epiplakin levels between patients with BC and controls, including those with stone disease and healthy volunteers, and clinicopathological findings were compared via the analysis of variance and Mann–Whitney U test. The area under the curve (AUC) and best cut-off point were calculated using receiver-operating characteristic (ROC) analysis.

Association of the clinicopathological findings with immunohistochemistry of epiplakin expression was assessed using the chi-square test (or Fisher's exact test, if appropriate) for categorical variables. Recurrence-free survival and cancer-specific survival were estimated with the log-rank test.

Univariate and multivariate analyses were performed using the Cox proportional hazards regression model, controlling for the effects of epiplakin and clinicopathological parameters. The statistical significance level was set at $p < 0.05$. Stata v. 16 for Windows (Stata, Chicago, IL, USA) was used for all analyses.

3. Results

3.1. Validation of Serum Epiplakin Levels

Figure 1 shows the serum epiplakin levels in patients with BC, those with stone disease, and healthy volunteers. Serum epiplakin levels were significantly increased in patients with BC compared with those with stone disease ($p = 0.0013$) and healthy volunteers ($p < 0.0001$). No significant differences were found between NMIBC and MIBC ($p = 0.63$). Patients with NMIBC also had significantly higher serum epiplakin levels than did those with stone disease ($p = 0.0016$) and healthy volunteers ($p < 0.0001$). No significant difference was found between patients with stone disease and healthy volunteers ($p = 0.28$).

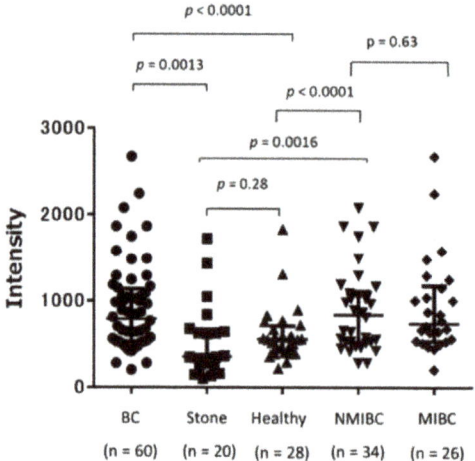

Figure 1. Serum epiplakin expression levels in patients with stone disease, healthy volunteers, and patients with bladder cancer. Serum epiplakin levels were significantly higher in patients with bladder cancer than in those with stone diseases and in healthy volunteers. Serum epiplakin levels were also significantly higher in patients with NMIBC than in those with stone diseases and in healthy volunteers. NMIBC: non-muscle-invasive bladder cancer, MIBC: muscle-invasive bladder cancer.

ROC analysis was used to compare the serum epiplakin levels in BC patients with those of stone disease and healthy volunteers. The AUC for all BC patients was 0.78 (Figure 2a). Using an optimal cut-off point of 873 (95% confidence interval (CI) = 0.69–0.87) revealed that serum epiplakin levels exhibited 68.3% sensitivity and 79.2% specificity for BC. If specificity was increased, serum epiplakin levels for BC showed as follows: 40% sensitivity and 90% specificity, 31.7% sensitivity and 95% specificity, and 10% sensitivity and 98% specificity, respectively. The AUC for NMIBC patients only was 0.70 (Figure 2b). Using an optimal cut-off point of 1051 (95% CI = 0.69–0.89), serum epiplakin levels exhibited 64.7% sensitivity and 81.3% specificity for NMIBC.

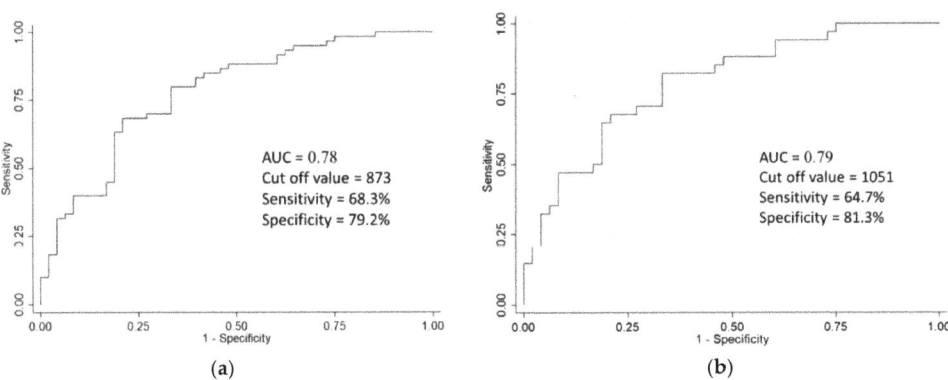

Figure 2. Receiver-operating characteristic (ROC) analysis of serum epiplakin expression levels (**a**) ROC analysis of serum epiplakin expression levels in patients with bladder cancer. The area under the ROC curve level was 0.78. The sensitivity and specificity were 68.3% and 79.2%, respectively, using a cut-off point of 873. (**b**) ROC analysis of serum epiplakin expression levels in patients with NMIBC. The area under the curve-ROC level for bladder cancer was 0.79. The sensitivity and specificity were 64.7% and 81.3%, respectively, using a cut-off point of 1051.

3.2. Association of Serum Epiplakin Levels with Clinicopathological Characteristics

Table 1 shows the relationships between the serum epiplakin levels and clinicopathological features. Serum epiplakin levels did not significantly differ by sex, age, pathological stage and grade, or urine cytology.

Table 1. The relationships between of serum epiplakin levels and clinicopathological features.

Characteristics	No. of Pts.	Serum Epiplakin Level			p-Value
		Median	Mean	Range	
Age(years)					0.11
≤65	11	761.8	1136.6	373.5–3209	
>65	49	1302.8	1844.2	263.8–4955	
Sex					0.28
Male	45	1192.8	1813.4	346–6485.8	
Female	14	1181	1384	263.8–3501	
T stage					0.50
≤pT1	34	1250.9	1806.1	346–6845.8	
≥pT2	26	1175.6	1575.2	263.8–4955	
Grade					0.40
G1/2	32	1430	1849.5	383.5–6845.8	
G3	28	1177	1542.1	263.8–4955	
Urine cytology					0.51
<classIIIb	16	1171	1544.7	373.5–3982	
≥classIIIb	43	1303	1800	263.8–6845.8	

No.: number. pts.: patients.

3.3. Association of Serum Epiplakin Level with BC Recurrence

BC recurred in 13 of 34 patients (38%) with pT1 ($n = 17$) or less ($n = 17$) (median time to recurrence, 19.5 months; range, 5–44 months). Univariate and multivariate analyses for predicting BC recurrence demonstrated that epiplakin was not a significant factor (Table S1).

3.4. Immunohistochemistry of Epiplakin

Epiplakin was expressed at various levels mainly in the tumor cell membrane and cytoplasm (Figure 3). Univariate and multivariate analyses revealed that epiplakin expression, clinicopathological findings, and patient outcomes did not significantly differ (Tables S2–S4; Figures S1 and Figure S2).

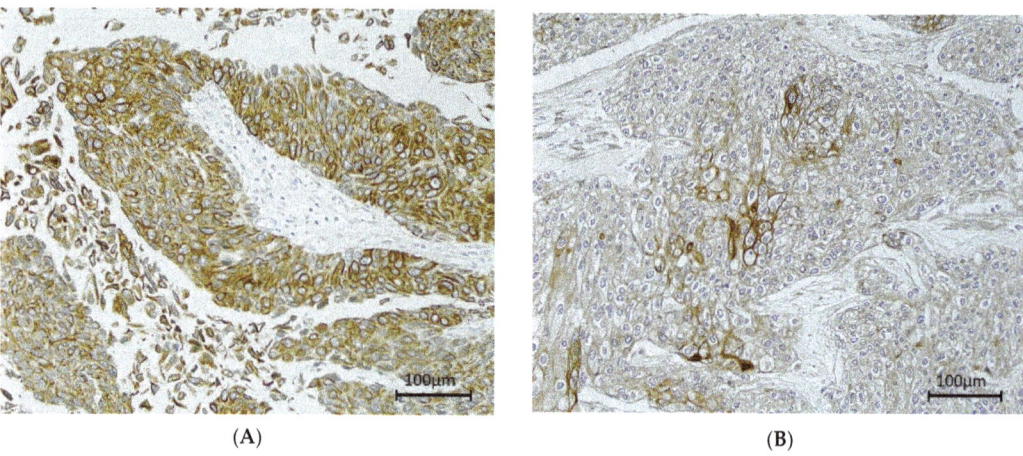

Figure 3. Immunohistochemical staining of epiplakin expression in bladder cancer. Microscopic images are representative tumor cells (200× magnification). (**A**) High expression (score = 12); (**B**) Low expression (score = 1).

4. Discussion

Several studies have been conducted to find molecular markers that might identify progression from normal urothelium to BC. Both protein and gene analyses have frequently been conducted on BC tissues, and reports have suggested a relationship with mutagenesis and protein gene mutation [17]. However, no circulating molecular biomarkers or genetic mutations have been approved for clinical use. Here, we used a micro-dot blot array to show that serum epiplakin levels were higher in patients with BC than in those with stone disease and in healthy volunteers. However, no association was found between staining scores for BC tissue, clinicopathological findings, and patient prognoses. Furthermore, serum epiplakin levels did not differ between patients with NMIBC and those with MIBC. Hence, serum epiplakin might be a potential diagnostic marker in patients with BC, perhaps even in the early stages.

Plakins are a family of gigantic proteins that make up the cell cytoskeleton in cell-to-cell junctions mediated by cadherin. Some plakins are specific for cell junctions in the epithelium that are linked to intermediate filaments [8]. Plakins play various roles in cancer [18–20] and are correlated with urothelial cancer. We previously showed that periplakin expression was significantly correlated with the aggressive pathology and cancer-specific survival in patients with BC [21]. Kudo et al. reported that, in maxillary sinus cancer, p53-mutant tumors exhibited increased expression of cell adhesion genes containing epiplakin [22]. These data suggest that cell adhesion proteins are involved in cancer development and progression.

However, few articles have reported an association between epiplakin and cancer. Yoshida et al. reported immunohistochemical findings for epiplakin in pancreatic ductal adenocarcinoma precursor lesions [23] and demonstrated that epiplakin was expressed in precancerous lesions but not in pancreatic cancer. Dong et al. reported KLF5-mediated upregulation of epiplakin in tissues can activate the p38 signaling pathway to promote the proliferation of cervical cells [24]. Furthermore, this result suggested that epiplakin may be a target factor for the treatment of cervical cancer. Another possible mechanism of action is that epiplakin connects intermediate filaments (IFs) [25]. IF regulates a variety of cellular processes from cell migration to apoptosis and proliferation [26,27]. Keratin belongs to a family of IF proteins expressed in all epithelial cells and provides important structural support for mechanical and non-mechanical stress. Keratin is immunologically altered because cancer cells are derived from epithelial cells. Studies in epiplakin-deficient mice have shown that epiplakin plays a role in keratin filament rearrangement in response to stress [28,29]. Taken together, it is suggested that the association between epiplakin and carcinogenesis may be mediated by keratin. Our immunohistochemical study revealed no correlation between epiplakin expression and clinicopathological findings. However, to our knowledge, no study has reported a relationship between serum epiplakin levels and cancers. While serum epiplakin expression levels were not associated with clinicopathological findings or outcomes, these levels were significantly increased in patients with BC compared with those in patients with stone disease and in healthy volunteers. Although the mechanism for elevated serum epiplakin in BC and the relationship between immunohistochemical expressions in BC tissues are unclear, this study is the first to show that serum epiplakin might be a potential biomarker for diagnosing BC.

Cystoscopy and urine cytology are effective tools for diagnosing BC. Although flexible cystoscopy has made examinations easier for patients, it remains an invasive procedure. This study was conducted to investigate the circulating levels of epiplakin in sera as a potential diagnostic marker for BC. We found that epiplakin presented adequate sensitivity (64.7%) and specificity (81.3%) for NMIBC. Tilki et al. reported that the sensitivity and specificity of BTA ranged from 57–83% and 60–92%, respectively, and the sensitivity and specificity of NMP22 ranged from 47–100% and 60–90%, respectively. However, BTA and NMP22 are reported to yield high false-positive rates for urinary stones [30]. Although epiplakin showed similar sensitivity and specificity to those of BTA and NMP22, these markers differed significantly between patients with BC and those with stone disease. Thus, we think that serum epiplakin estimates might aid and overcome the problems of established BC markers.

Recent studies have reported that liquid biopsy by analysis of cell-free DNA (cfDNA) using next-generation sequencing (NGS) is useful [31]. Genetic panel assays for BCs such as Uroseek are also in development [32,33]. Patrice et al. reported that the sensitivity and specificity for detection for BCs using the singleplex assay UroMuTERT, which detects mutations in the promoter of the telomerase reverse transcriptase gene (TERT), was very high [34]. However, although NGS is more reproducible and accurate than the enzyme-linked immuno solvent assay (ELISA), it is expensive and requires a high degree of expertise and a high level of experimental equipment [35]. We think that micro-dot arrays are one of the simple methods. Compared to other cancers, the field of biomarker in bladder cancer is still insufficient. Although epiplakin currently has many issues to be used in daily clinical practice, it could be a promising biomarker.

This study has several limitations. First, epiplakin is located not only in the urothelium but in the epithelium and other organs [11]. Second, the role of serum epiplakin expression must be validated in other diseases, including cancerous and inflammatory lesions. One study reported that plakin deletion in the lungs caused overexpression of anti-inflammatory cytokines [36]. In terms of urinary tract infection, epiplakin was not measured in this study. It invalidates biomarker results because it can cause false-positive findings. When screening with biomarkers, people with urinary tract infection have to be treated first for their infection before they can be re-examined with cancer biomarkers. However, differentiation from urinary tract infection is still one of the issues. Third, although we performed immunohistochemical staining of BC tissues, the patient cohorts differed between the serum and immunohistochemistry groups. The dynamics of epiplakin between serum and immunohistochemistry findings must be determined. Fourth, the serum levels of epiplakin were only measured using micro-dot blot analysis. It will be necessary to compare with other measurement methods such as Western blotting and ELISA, considering the stable measurement of epiplakin in clinical application. Finally, serum epiplakin was not associated with BC recurrence on multivariate analysis. Serum epiplakin levels at the time of diagnosis may reflect only existence of a tumor but not the biological aggressiveness of the tumor. At the moment, epiplakin is a possible candidate that could be tested in a study for their ability to detect bladder cancer.

5. Conclusions

In conclusion, patients with BC demonstrated significantly increased serum epiplakin expression compared with those in patients with stone disease and in healthy volunteers. Multi-institutional evaluations of serum epiplakin in a large patient population are warranted before serum epiplakin can be included as a biomarker for routine clinical use for early diagnosis of BC. Serum epiplakin levels might be a suitable non-invasive diagnostic method as an adjunct to urinary cytology and cystoscopy for diagnosis of bladder cancer once proven in a prospective cohort study.

6. Patents

Epiplakin has been patented in Japan as a diagnostic marker for bladder cancer (JP. PAT. 6060425). The Kitasato Institute, Diagnosis of Bladder Cancer, Japan, No. 6060425.

Supplementary Materials: The following are available online at https://www.mdpi.com/article/10.3390/cancers13205150/s1, Table S1: Univariate and multivariate Cox proportional hazards regression analyses using serum samples to predict NMIBC recurrence, Table S2: Clinical and pathological characteristics of patients with low and high epiplakin expressions according to immunohistochemistry, Table S3: Univariate and multivariate Cox proportional hazards regression analyses using immunohistochemical staining to predict recurrence, Table S4: Univariate and multivariate Cox proportional hazards regression analyses using immunohistochemical staining to predict cancer-specific survival, Figure S1: Survival analysis using a Kaplan–Meier curve to determine recurrence-free survival for high and low epiplakin expression, Figure S2: Survival analysis using a Kaplan–Meier curve to determine cancer-specific survival for high and low epiplakin expression.

Author Contributions: Conceptualization, S.S., K.M. (Kazumasa Matsumoto); Methodology, Y.S. (Yuriko Shimizu), Y.S. (Yuichi Sato); Validation, M.I. (Masaomi Ikeda); Formal analysis, S.S.; Investigation, S.S., K.M. (Kohei Mochizuki); Resources, S.S., K.M. (Kazumasa Matsumoto); Data collection, Y.S. (Yuriko Shimizu), K.M. (Kohei Mochizuki), Y.S. (Yutaka Shiono), D.K., S.S., S.H.; Writing—original draft preparation, S.S.; Writing—review and editing, K.M. (Kazumasa Matsumoto), Y.S. (Yuichi Sato); Supervision, M.I. (Masatsugu Iwamura). All authors have read and agreed to the published version of the manuscript.

Funding: This work was supported in part by a Parent's Association (Keyaki Kai) Grant of Kitasato University School of Medicine (to S.S.) and by a research grant for young doctors and healthcare professionals Grant of SRL Co., Ltd. (to S.S.) by JSPS KAKENHI Grant Number JP18K09206 and JP21K09355.

Institutional Review Board Statement: The study was conducted according to the guidelines of the Declaration of Helsinki and approved by the Institutional Review Board of Kitasato University School of Medicine and Hospital (protocol code B17-010, B18-149).

Informed Consent Statement: Informed consent was obtained from some subjects involved with the study, or the information on the opportunity to opt out was provided to participants via our website and posters.

Data Availability Statement: The datasets used and/or analyzed during the current study are available from the corresponding author on reasonable request.

Conflicts of Interest: The authors declare no conflict of interest.

References

1. Bray, F.; Colombet, M.; Mery, L.; Piñeros, M.; Znaor, A.; Zanetti, R.; Ferlay, J. *Cancer Incidence in Five Continents, Vol. XI (Electronic Version)*; IARC Press: Lyon, France, 2017; Available online: https://ci5.iarc.fr (accessed on 31 August 2021).
2. Babjuk, M.; Bohle, A.; Burger, M.; Capoun, O.; Cohen, D.; Comperat, E.M.; Hernandez, V.; Kaasinen, E.; Palou, J.; Rouprêt, M.; et al. EAU guidelines on non-muscle-invasive urothelial carcinoma of the bladder: Update 2016. *Eur. Urol.* **2017**, *71*, 447–461. [CrossRef]
3. Sylvester, R.J.; van der Meijden, A.P.; Oosterlinck, W.; Witjes, J.A.; Bouffioux, C.; Denis, L.; Newling, D.W.; Kurth, K. Predicting recurrence and progression in individual patients with stage Ta T1 bladder cancer using eortc risk tables: A combined analysis of 2596 patients from seven eortc trials. *Eur. Urol.* **2006**, *49*, 466–477. [CrossRef] [PubMed]
4. Muus Ubago, J.; Mehta, V.; Wojcik, E.M.; Barkan, G.A. Evaluation of atypical urine cytology progression to malignancy. *Cancer Cytopathol.* **2013**, *121*, 387–391. [CrossRef] [PubMed]
5. Babjuk, M.; Soukup, V.; Pesl, M.; Kostirova, M.; Drncova, E.; Smolova, H.; Szakacsova, M.; Getzenberg, R.; Pavlik, I.; Dvoracek, J. Urinary cytology and quantitative BTA and UBC tests in surveillance of patients with ptapt1 bladder urothelial carcinoma. *Urology* **2008**, *71*, 718–722. [CrossRef] [PubMed]
6. Nguyen, C.T.; Jones, J.S. Defining the Role of Nmp22 in Bladder Cancer Surveillance. *World J. Urol.* **2008**, *26*, 51–58. [CrossRef] [PubMed]
7. Okusa, H.; Kodera, Y.; Oh-ishi, M.; Minamida, S.; Tsuchida, M.; Kavoussi, N.; Matsumoto, K.; Sato, T.; Iwamura, M.; Maeda, T.; et al. Searching for new biomarkers of bladder cancer based on proteomics analysis. *J. Electrophor.* **2008**, *52*, 19–24. [CrossRef]
8. Jefferson, J.J.; Leung, C.L.; Liem, R.K. Plakins: Goliaths that link cell junctions and the cytoskeleton. *Nat. Rev. Mol. Cell Biol.* **2004**, *5*, 542–553. [CrossRef]
9. Ishikawa, K.; Sumiyoshi, H.; Matsuo, N.; Takeo, N.; Goto, M.; Okamoto, O.; Tatsukawa, S.; Kitamura, H.; Fujikura, Y.; Yoshioka, H.; et al. Epiplakin Accelerates the Lateral Organization of Keratin Filaments During Wound Healing. *J. Dermatol. Sci.* **2010**, *60*, 95–104. [CrossRef]
10. Andra, K.; Lassmann, H.; Bittner, R.; Shorny, S.; Fassler, R.; Propst, F.; Wiche, G. Targeted inactivation of plectin reveals essential function in maintaining the integrity of skin, muscle, and heart cytoarchitecture. *Genes Dev.* **1997**, *11*, 3143–3156. [CrossRef]
11. UALCAN. Available online: http://ualcan.path.uab.edu/index.html (accessed on 31 August 2021).
12. Greene, F.L.; Page, D.L.; Fleming, I.D.; Fritz, A.G.; Balch, C.M.; Haller, D.G.; Morrow, M. *AJCC Cancer Staging Manual*, 6th ed; Springer: Chicago, IL, USA, 2002.
13. Epstein, J.I.; Amin, M.B.; Reuter, V.R.; Mostofi, F.K. The World Health Organization/International Society of Urological Pathology consensus classification of urothelial(transitional cell)neoplasms of the urinary bladder. Bladder Consensus Conference Committee. *Am. J. Surg. Pathol.* **1998**, *22*, 1435–1448. [CrossRef]
14. Kobayashi, M.; Nagashio, R.; Jiang, S.X.; Saito, K.; Tsuchiya, B.; Ryuge, S.; Katono, K.; Nakashima, H.; Fukuda, E.; Goshima, N.; et al. Calnexin Is a Novel Sero-Diagnostic Marker for Lung Cancer. *Lung Cancer* **2015**, *90*, 342–345. [CrossRef] [PubMed]
15. Yanagita, K.; Nagashio, R.; Jiang, S.X.; Kuchitsu, Y.; Hachimura, K.; Ichinoe, M.; Igawa, S.; Nakashima, H.; Fukuda, E.; Goshima, N.; et al. Cytoskeleton-Associated Protein 4 is a Novel Serodiagnostic Marker for Lung Cancer. *Am. J. Pathol.* **2018**, *188*, 1328–1333. [CrossRef]
16. Tsuchisaka, A.; Numata, S.; Teye, K.; Natsuki, Y.; Kawakami, T.; Takeda, Y.; Wang, W.; Ishikawa, K.; Goto, M.; Koga, H.; et al. Epiplakin Is a Paraneoplastic Pemphigus Autoantigen and Related to Bronchiolitis Obliterans in Japanese Patients. *J. Investig. Dermatol.* **2016**, *136*, 399–408. [CrossRef] [PubMed]
17. Robertson, A.G.; Kim, J.; Al-Ahmadie, H.; Bellmunt, J.; Guo, G.; Cherniack, A.D.; Hinoue, T.; Laird, P.W.; Hoadley, K.A.; Akbani, R.; et al. Comprehensive Molecular Characterization of Muscle-Invasive Bladder Cancer. *Cell* **2017**, *171*, 540–556. [CrossRef] [PubMed]
18. Young, G.D.; Winokur, T.S.; Cerfolio, R.J.; Van Tine, B.A.; Chow, L.T.; Okoh, V.; Garver, R.I., Jr. Differential expression and biodistribution of cytokeratin 18 and desmoplakins in non-small cell lung carcinoma subtypes. *Lung Cancer* **2002**, *36*, 133–141. [CrossRef]
19. Narayana, N.; Gist, J.; Smith, T.; Tylka, D.; Trogdon, G.; Wahl, J.K. Desmosomal Component Expression in Normal, Dysplastic, and Oral Squamous Cell Carcinoma. *Dermatol. Res. Pract.* **2010**, *2010*, 649731. [CrossRef] [PubMed]
20. Li, X.; Zhang, G.; Wang, Y.; Elgehama, A.; Sun, Y.; Li, L.; Gu, Y.; Guo, W.; Xu, Q. Loss of Periplakin Expression Is Associated with the Tumorigenesis of Colorectal Carcinoma. *Biomed. Pharm.* **2017**, *87*, 366–374. [CrossRef]
21. Matsumoto, K.; Ikeda, M.; Sato, Y.; Kuruma, H.; Kamata, Y.; Nishimori, T.; Tomonaga, T.; Nomura, F.; Egawa, S.; Iwamura, M. Loss of periplakin expression is associated with pathological stage and cancerspecific survival in patients with urothelial carcinoma of the urinary bladder. *Biomed. Res.* **2014**, *35*, 201–206. [CrossRef]
22. Kudo, I.; Esumi, M.; Kusumi, Y.; Furusaka, T.; Oshima, T. Particular Gene Upregulation and P53 Heterogeneous Expression in Tp53-Mutated Maxillary Carcinoma. *Oncol. Lett.* **2017**, *14*, 4633–4640. [CrossRef]
23. Yoshida, T.; Shiraki, N.; Baba, H.; Goto, M.; Fujiwara, S.; Kume, K.; Kume, S. Expression Patterns of Epiplakin1 in Pancreas, Pancreatic Cancer and Regenerating Pancreas. *Genes Cells* **2008**, *13*, 667–678. [CrossRef] [PubMed]
24. Ma, D.; Pan, Z.; Chang, Q.; Zhang, J.J.; Liu, X.; Hua, N.; Li, G.H. Klf5-Mediated Eppk1 Expression Promotes Cell Proliferation in Cervical Cancer Via the P38 Signaling Pathway. *BMC Cancer* **2021**, *21*, 377. [CrossRef] [PubMed]

25. Spazierer, D.; Fuchs, P.; Reipert, S.; Fischer, I.; Schmuth, M.; Lassmann, H.; Wiche, G. Epiplakin is dispensable for skin barrier function and for integrity of keratin network cytoarchitecture in simple and stratified epithelia. *Mol. Cell. Biol.* **2006**, *26*, 559–568. [CrossRef] [PubMed]
26. Kim, S.; Coulombe, P.A. Intermediate filament scaffolds fulfill mechanical, organizational, and signaling functions in the cytoplasm. *Genes Dev.* **2007**, *21*, 1581–1597. [CrossRef] [PubMed]
27. Etienne-Manneville, S. Cytoplasmic Intermediate Filaments in Cell Biology. *Annu. Rev. Cell Dev. Biol.* **2018**, *34*, 1–28. [CrossRef] [PubMed]
28. Wang, W.; Sumiyoshi, H.; Yoshioka, H.; Fujiwara, S. Interactions between Epiplakin and Intermediate Filaments. *J. Dermatol.* **2006**, *33*, 518–527. [CrossRef] [PubMed]
29. Goto, M.; Sumiyoshi, H.; Sakai, T.; Fassler, R.; Ohashi, S.; Adachi, E.; Yoshioka, H.; Fujiwara, S. Elimination of Epiplakin by Gene Targeting Results in Acceleration of Keratinocyte Migration in Mice. *Mol. Cell. Biol.* **2006**, *26*, 548–558. [CrossRef] [PubMed]
30. Tilki, D.; Burger, M.; Dalbagni, G.; Grossman, H.B.; Hakenberg, O.W.; Palou, J.; Reich, O.; Roupret, M.; Shariat, S.F.; Zlotta, A.R. Urine Markers for Detection and Surveillance of Non-Muscle-Invasive Bladder Cancer. *Eur. Urol.* **2011**, *60*, 484–492. [CrossRef]
31. Ferro, M.; La Civita, E.; Liotti, A.; Cennamo, M.; Tortora, F.; Buonerba, C.; Crocetto, F.; Lucarelli, G.; Busetto, G.M.; Del Giudice, F.; et al. Liquid Biopsy Biomarkers in Urine: A Route Towards Molecular Diagnosis and Personalized Medicine of Bladder Cancer. *J. Pers. Med.* **2021**, *11*, 237. [CrossRef]
32. Wong, R.; Rosser, C.J. Uroseek Gene Panel for Bladder Cancer Surveillance. *Transl. Androl. Urol.* **2019**, *8*, 546–549. [CrossRef]
33. Eich, M.L.; Rodriguez Pena, M.D.C.; Springer, S.U.; Taheri, D.; Tregnago, A.C.; Salles, D.C.; Bezerra, S.M.; Cunha, I.W.; Fujita, K.; Ertoy, D.; et al. Incidence and Distribution of Uroseek Gene Panel in a Multi-Institutional Cohort of Bladder Urothelial Carcinoma. *Mod. Pathol.* **2019**, *32*, 1544–1550. [CrossRef]
34. Avogbe, P.H.; Manel, A.; Vian, E.; Durand, G.; Forey, N.; Voegele, C.; Zvereva, M.; Hosen, M.I.; Meziani, S.; De Tilly, B.; et al. Urinary Tert Promoter Mutations as Non-Invasive Biomarkers for the Comprehensive Detection of Urothelial Cancer. *EBioMedicine* **2019**, *44*, 431–438. [CrossRef] [PubMed]
35. Boonham, N.; Kreuze, J.; Winter, S.; van der Vlugt, R.; Bergervoet, J.; Tomlinson, J.; Mumford, R. Methods in Virus Diagnostics: From Elisa to Next Generation Sequencing. *Virus Res.* **2014**, *186*, 20–31. [CrossRef] [PubMed]
36. Besnard, V.; Dagher, R.; Madjer, T.; Joannes, A.; Jaillet, M.; Kolb, M.; Bonniaud, P.; Murray, L.A.; Sleeman, M.A.; Crestani, B. Identification of Periplakin as a Major Regulator of Lung Injury and Repair in Mice. *JCI Insight* **2018**, *3*, e90163. [CrossRef] [PubMed]

MDPI
St. Alban-Anlage 66
4052 Basel
Switzerland
Tel. +41 61 683 77 34
Fax +41 61 302 89 18
www.mdpi.com

Cancers Editorial Office
E-mail: cancers@mdpi.com
www.mdpi.com/journal/cancers

www.ingramcontent.com/pod-product-compliance
Lightning Source LLC
LaVergne TN
LVHW070052120526
838202LV00102B/2053